Thermodynamics and Chemistry

Howard DeVoe

Department of Chemistry and Biochemistry
University of Maryland

Prentice Hall, Upper Saddle River, NJ 07458

Library of Congress Cataloging-in-Publication Data

DeVoe, Howard.
  Thermodynamics and chemistry / Howard DeVoe
    p. cm.
  Includes bilbliographical references and index
  ISBN 0-02-328741-1
    1. Thermodynamics. I. Title.
  QD504 . D48 2001
  541.3'69--dc21

00-051489

Executive Editor: John Challice
Editorial Assistant: Gillian Buonanno
Production Editor: Kim Dellas
Manufacturing Manager: Trudy Pisciotti
Manufacturing Buyer: Michael Bell
Art Director: Jayne Conte
Cover Designer: Karen Salzbach

 ©2001 by Prentice-Hall, Inc.
Upper Saddle River, NJ 07458

All rights reserved. No part of this book may be reproduced, in any form or by any means, without permission in writing from the publisher.

Printed in the United States of America

10 9 8 7 6 5 4 3 2 1

ISBN 0-02-328741-1

Prentice-Hall International (UK) Limited, London
Prentice-Hall of Australia Pty. Limited, Sydney
Prentice-Hall Canada Inc., Toronto
Prentice-Hall Hispanoamericana, S.A., Mexico City
Prentice-Hall of India Private Limited, New Delhi
Prentice-Hall of Japan Inc., Tokyo
Pearson Education Asia Pte. Ltd.
Editora Prentice-Hall do Brasil, Ltda., Rio de Janeiro

Non-SI derived units

Physical quantity	Unit	Symbol	Definition of unit
volume	liter	L	$1\,\text{L} = 1\,\text{dm}^3 = 10^{-3}\,\text{m}^3$
pressure	bar	bar	$1\,\text{bar} = 10^5\,\text{Pa}$
pressure	atmosphere	atm	$1\,\text{atm} = 101{,}325\,\text{Pa} = 1.01325\,\text{bar}$
pressure	torr	Torr	$1\,\text{Torr} = (1/760)\,\text{atm} = (101{,}325/760)\,\text{Pa}$
energy	calorie	cal	$1\,\text{cal} = 4.184\,\text{J}$

Fundamental physical constants

Physical quantity	Symbol	Value in SI units
Avogadro constant	N_A	$6.022\,141\,99(47) \times 10^{23}\,\text{mol}^{-1}$
elementary charge	e	$1.602\,176\,462(63) \times 10^{-19}\,\text{C}$
Faraday constant	F	$9.648\,534\,15(39) \times 10^4\,\text{C}\,\text{mol}^{-1}$
gas constant	R	$8.314\,472(15)\,\text{J}\,\text{K}^{-1}\,\text{mol}^{-1}$
permeability of vacuum	μ_0	$4\pi \times 10^{-7}\,\text{N}\,\text{A}^{-2}$ (exact)
permittivity of vacuum	ϵ_0	$(10^7\,\text{C}^2\,\text{m}\,\text{J}^{-1}\,\text{s}^{-2})/4\pi c_0^2$ (exact) $= 8.854\,187\,816\ldots \times 10^{-12}\,\text{C}^2\,\text{J}^{-1}\,\text{m}^{-1}$
speed of light in vacuum	c_0	$2.997\,924\,58 \times 10^8\,\text{m}\,\text{s}^{-1}$ (exact)
standard acceleration of free fall	g_n	$9.806\,65\,\text{m}\,\text{s}^{-2}$ (exact)

To Stephanie, John, Trish, Donald, and Dorothy

CONTENTS

PREFACE xiv

1 INTRODUCTION 1
   1.1 Units 1
       1.1.1 Amount of substance and amount 3
   1.2 Quantity Calculus 4
   1.3 Dimensional Analysis 6

2 SOME BASIC PROPERTIES AND THEIR MEASUREMENT 8
   2.1 General Considerations 8
       2.1.1 The macroscopic viewpoint 8
       2.1.2 Extensive and intensive properties 9
       2.1.3 Phases 10
       2.1.4 Physical states of matter 10
       2.1.5 Phase coexistence and phase transitions 11
       2.1.6 Fluids 12
       2.1.7 The equation of state of a fluid 13
       2.1.8 Virial equations of state for pure gases 14
       2.1.9 Solids 16
   2.2 Mass 17
   2.3 Volume 17
   2.4 Density 18
   2.5 Pressure 19
   2.6 Temperature 20
       2.6.1 Temperature scales 20
       2.6.2 The International Temperature Scale of 1990 22

		2.6.3	Equilibrium systems for fixed temperatures	22
		2.6.4	Gas thermometry	22
		2.6.5	Practical thermometers	25

3 THE FIRST LAW — 27

- 3.1 Terminology — 27
  - 3.1.1 The system, surroundings, and boundary — 27
  - 3.1.2 The state of the system — 28
  - 3.1.3 Equilibrium states — 30
  - 3.1.4 Processes and paths — 32
- 3.2 Energy — 34
  - 3.2.1 The energy of the system — 34
  - 3.2.2 The internal energy of an ideal gas — 35
- 3.3 Heat, Work, and the First Law — 36
  - 3.3.1 The concept of thermodynamic work — 37
  - 3.3.2 Heat and work as path functions — 38
- 3.4 Spontaneous and Reversible Processes — 40
- 3.5 Heat Transfer — 41
- 3.6 Expansion and Compression — 43
  - 3.6.1 Conditions for spontaneity and equilibrium — 43
  - 3.6.2 Expansion work — 45
  - 3.6.3 Reversible isothermal expansion of an ideal gas — 47
  - 3.6.4 Reversible adiabatic expansion of an ideal gas — 47
  - 3.6.5 Indicator diagrams — 48
  - 3.6.6 Spontaneous adiabatic expansion — 49
  - 3.6.7 Expansion into a vacuum — 50
- 3.7 Work with Dissipated Energy — 50
- 3.8 Electrical Work — 53
  - 3.8.1 Electrical heating — 54
  - 3.8.2 Electrical work with a galvanic cell — 55
- 3.9 Gravitational Work — 56
- 3.10 Spontaneous and Reversible Processes: Generalities — 57
- 3.11 Enthalpy — 59
- 3.12 Heat Capacity — 60

4 THE SECOND AND THIRD LAWS — 64

- 4.1 Types of Processes — 64

	4.2		Statements of the Second Law	66
	4.3		Concepts Developed with Carnot Engines	69
		4.3.1	Carnot engines and Carnot cycles	69
		4.3.2	The equivalence of the Clausius and Kelvin–Planck statements	71
		4.3.3	The efficiency of a Carnot engine	72
		4.3.4	Thermodynamic temperature	75
	4.4		Derivation of the Mathematical Form of the Second Law	76
		4.4.1	The existence of the entropy function	76
		4.4.2	Using reversible processes to define the entropy	79
		4.4.3	Some properties of the entropy	81
		4.4.4	Irreversible processes	82
	4.5		Adiabatic Processes: Examples	84
		4.5.1	Adiabatic processes with work	84
		4.5.2	Isolated systems	84
		4.5.3	Internal heat flow	85
		4.5.4	Free expansion of an ideal gas	85
	4.6		In Retrospect: Processes and Entropy Changes	86
	4.7		What is the "Meaning" of Entropy?	87
		4.7.1	Entropy as a measure of unavailable energy	87
		4.7.2	Entropy as a measure of disorder	88
	4.8		The Helmholtz Energy and Gibbs Energy	89
	4.9		Combining the First and Second Laws	90
	4.10		The Third Law	91
	4.11		Molar Entropies	92
		4.11.1	Molar entropy in a pure phase	92
		4.11.2	The molar entropy of an ideal gas from statistical mechanics	96
	4.12		Cryogenics	97
		4.12.1	Joule–Thomson expansion	97
		4.12.2	Adiabatic demagnetization	99
5	**PURE SUBSTANCES IN SINGLE PHASES**			**103**
	5.1		System Variables	103
	5.2		Volume Properties	104
	5.3		Energy Functions	106
	5.4		Internal Pressure	108
	5.5		Thermal Properties	110

		5.5.1	The relation between constant-volume and constant-pressure heat capacities	110
		5.5.2	The measurement of heat capacities	111
		5.5.3	Typical values	117
	5.6	Heating at Constant Volume or Pressure		117
	5.7	Partial Derivatives with Respect to Temperature, Pressure, and Volume		120
		5.7.1	Tables of partial derivatives	120
		5.7.2	The Joule–Thomson coefficient	122
	5.8	Isothermal Pressure and Volume Changes		125
		5.8.1	Ideal gases	125
		5.8.2	Condensed phases	126
	5.9	Standard States of Pure Substances		127
	5.10	Chemical Potential and Fugacity		127
		5.10.1	Gases	128
		5.10.2	Liquids and solids	130
	5.11	Standard Molar Functions of a Gas		131
	5.12	Additional Work Coordinates		133
		5.12.1	Surface tension	133
		5.12.2	Magnetization	135
	5.13	Open Systems		136
6	**PHASE TRANSITIONS AND EQUILIBRIA OF PURE SUBSTANCES**			**142**
	6.1	Phase Equilibria		142
		6.1.1	Conditions for spontaneity and equilibrium	142
		6.1.2	Equilibrium in a two-phase system	143
		6.1.3	A simple derivation of equilibrium conditions	144
		6.1.4	Gas pressure in a gravitational field	144
		6.1.5	The pressure in a liquid droplet	147
		6.1.6	The number of independent variables	147
		6.1.7	The Gibbs phase rule for a pure substance	148
	6.2	Phase Diagrams of Pure Substances		149
		6.2.1	Features of phase diagrams	149
		6.2.2	Two-phase equilibrium	153
		6.2.3	The critical point	154
		6.2.4	The lever rule	155
		6.2.5	Volume properties	158
	6.3	Phase Transitions		159

		6.3.1	Extensive transition quantities	159
		6.3.2	Molar transition quantities	160
		6.3.3	Calorimetric measurement of transition enthalpies	161
		6.3.4	Standard molar transition quantities	161
	6.4	Coexistence Curves		162
		6.4.1	Chemical potential surfaces	162
		6.4.2	The Clapeyron equation	164
		6.4.3	The Clausius–Clapeyron equation	166

7 MIXTURES — 170

	7.1	Composition Variables		170
		7.1.1	Species and substances	170
		7.1.2	Mixtures in general	170
		7.1.3	Solutions	171
		7.1.4	Binary solutions	171
		7.1.5	The composition of a mixture	173
	7.2	Partial Molar Quantities		173
		7.2.1	Partial molar volume	173
		7.2.2	The total differential of the volume in an open system	175
		7.2.3	Measuring partial molar volumes	177
		7.2.4	General relations	179
	7.3	The Chemical Potential of a Species in a Mixture		180
		7.3.1	Equilibrium conditions	181
		7.3.2	Relations involving partial molar quantities	182
	7.4	Gas Mixtures		183
		7.4.1	Partial pressure	184
		7.4.2	The ideal gas mixture	184
		7.4.3	Partial molar quantities in ideal gas mixtures	184
		7.4.4	Real gas mixtures	187
	7.5	Liquid and Solid Mixtures of Nonelectrolytes		190
		7.5.1	Raoult's law	190
		7.5.2	Ideal mixtures	192
		7.5.3	Partial molar quantities in ideal mixtures	193
		7.5.4	Henry's law	194
		7.5.5	The ideal-dilute solution	196
		7.5.6	Solvent behavior in the ideal-dilute solution	198
		7.5.7	Partial molar quantities in an ideal-dilute solution	199

7.6	Activity Coefficients		201
	7.6.1	Ideal mixtures	201
	7.6.2	Real mixtures	202
	7.6.3	Nonideal dilute solutions	204
	7.6.4	Activity coefficients from gas fugacities	205
	7.6.5	Activity coefficients from the Gibbs–Duhem equation	208
	7.6.6	Activity coefficients from osmotic coefficients	209
	7.6.7	Evaluation of the osmotic coefficient	212
7.7	Activities		213
	7.7.1	Standard states of mixture constituents	213
	7.7.2	The pressure factor Γ_i	216
7.8	Electrolyte Solutions		218
	7.8.1	Single-ion quantities	218
	7.8.2	Symmetrical electrolytes	220
	7.8.3	Electrolytes in general	222
	7.8.4	The Debye–Hückel theory	224
	7.8.5	Derivation of the Debye–Hückel equation	227
	7.8.6	Mean ionic activity coefficients from osmotic coefficients	229
7.9	Gravitational and Centrifugal Fields		230
	7.9.1	Ideal gas mixture in a gravitational field	231
	7.9.2	Liquid solution in a centrifugal field	231
8	**REACTIONS AND OTHER CHEMICAL PROCESSES**		**239**
8.1	The Advancement and Molar Reaction Quantities		239
	8.1.1	An example	240
	8.1.2	Molar reaction quantities in general	242
	8.1.3	Standard molar reaction quantities	244
8.2	Mixing Processes		244
	8.2.1	Mixtures in general	244
	8.2.2	Ideal mixtures	246
	8.2.3	Excess quantities	247
	8.2.4	The entropy change to form an ideal gas mixture	249
	8.2.5	Molecular model of a liquid mixture	249
	8.2.6	Phase separation of a liquid mixture	251
8.3	Reaction Enthalpies		254
	8.3.1	Standard molar enthalpies of formation	255
	8.3.2	Effect of temperature on molar reaction enthalpies	257

8.4	Enthalpies of Solution and Dilution		258
	8.4.1	Enthalpy of solution	259
	8.4.2	Enthalpy of dilution	261
	8.4.3	Molar enthalpies of solute formation	262
	8.4.4	Evaluation of relative partial molar enthalpies	262
	8.4.5	Electrolyte solutions at infinite dilution	266
8.5	Reaction Calorimetry		269
	8.5.1	The constant-pressure reaction calorimeter	270
	8.5.2	The bomb calorimeter	272
	8.5.3	Other calorimeters	277
8.6	Adiabatic Flame Temperature		278
8.7	Gibbs Energy and Reaction Equilibrium		279
	8.7.1	The molar reaction Gibbs energy	279
	8.7.2	Spontaneity and equilibrium	280
	8.7.3	Pure phases	282
	8.7.4	Reaction in an ideal gas mixture	282
	8.7.5	The standard molar reaction Gibbs energy	285
8.8	The Thermodynamic Equilibrium Constant		287
	8.8.1	Definition	287
	8.8.2	Reaction in a gas phase	289
	8.8.3	Reaction in solution	290
	8.8.4	Pressure-factor quotient of a reaction in solution	291
	8.8.5	Evaluation of K	292
	8.8.6	Free-energy functions	294
8.9	Effects of Temperature and Pressure on Equilibrium Position		296

9 EQUILIBRIUM CONDITIONS IN MULTICOMPONENT SYSTEMS — 306

9.1	Effects of Temperature		306
	9.1.1	Variation of μ_i/T with temperature	306
	9.1.2	Variation of μ_i°/T with temperature	307
	9.1.3	Variation of $\ln K$ with temperature	308
9.2	Solvent Chemical Potentials from Phase Equilibria		310
	9.2.1	Freezing-point measurements	310
	9.2.2	Osmotic-pressure measurements	312
9.3	Binary Mixture in Equilibrium with a Pure Phase		314
9.4	Colligative Properties of a Dilute Solution		315
	9.4.1	Freezing-point depression	317

		9.4.2	Boiling-point elevation	319
		9.4.3	Vapor-pressure lowering	320
		9.4.4	Osmotic pressure	321
	9.5	Solid–Liquid Equilibria		322
		9.5.1	Freezing points of ideal binary liquid mixtures	322
		9.5.2	Solubility of a solid nonelectrolyte	323
		9.5.3	Ideal solubility of a solid	325
		9.5.4	Solid compounds	326
		9.5.5	Solubility of a solid electrolyte	328
	9.6	Liquid–Liquid Equilibria		330
		9.6.1	Pure-liquid reference states	331
		9.6.2	Solubility of a liquid	331
		9.6.3	Distribution of a solute between two solvents	333
		9.6.4	Membrane equilibria	334
	9.7	Liquid–Gas Equilibria		338
		9.7.1	Effect of liquid pressure on gas fugacity	338
		9.7.2	Effect of liquid composition on gas fugacities	339
		9.7.3	The Duhem–Margules equation	342
		9.7.4	Gas solubility	343
		9.7.5	Effect of pressure on Henry's law constants	345
	9.8	Reaction Equilibria		346
	9.9	Evaluation of Standard Molar Quantities		347

10 THE PHASE RULE AND PHASE DIAGRAMS FOR MULTICOMPONENT SYSTEMS — 355

	10.1	The Gibbs Phase Rule for Multicomponent Systems		355
		10.1.1	Degrees of freedom	356
		10.1.2	Species approach to the phase rule	357
		10.1.3	Components approach to the phase rule	358
		10.1.4	Examples	359
	10.2	Phase Diagrams: Binary Systems		363
		10.2.1	Generalities	363
		10.2.2	Solid–liquid systems	364
		10.2.3	Partially miscible liquids	368
		10.2.4	Liquid–gas systems	369
		10.2.5	Solid–gas systems	374
		10.2.6	Systems at high pressure	376

		10.3 Phase Diagrams: Ternary Systems	378
		10.3.1 Three liquids	380
		10.3.2 Two solids and a solvent	381

11 GALVANIC CELLS — 387

- 11.1 Cell Reactions and Cell Diagrams — 387
- 11.2 The Cell Potential — 389
- 11.3 Molar Reaction Quantities — 391
- 11.4 The Nernst Equation — 394
- 11.5 Evaluation of the Standard Potential — 396
- 11.6 Standard Electrode Potentials — 397

A DEFINITIONS OF THE SI BASE UNITS — 402

B PHYSICAL CONSTANTS — 403

C SYMBOLS FOR PHYSICAL QUANTITIES — 404

D MISCELLANEOUS ABBREVIATIONS AND SYMBOLS — 408

- D.1 Physical States — 408
- D.2 Subscripts for Chemical Processes — 408
- D.3 Superscripts — 409

E CALCULUS REVIEW — 410

- E.1 Derivatives — 410
- E.2 Integrals — 410

F MATHEMATICAL PROPERTIES OF STATE FUNCTIONS — 412

G FORCES, ENERGY, AND WORK — 414

H STANDARD MOLAR THERMODYNAMIC PROPERTIES — 422

I GENERAL REFERENCES — 425

J ANSWERS TO SELECTED PROBLEMS — 427

INDEX — 431

PREFACE

Classical thermodynamics, the subject of this book, is concerned with macroscopic aspects of the interaction of matter with energy in its various forms. This book is designed as a text for a one-semester course for senior undergraduate or graduate students who have already been introduced to thermodynamics in an undergraduate physical chemistry course.

Anyone who studies and uses thermodynamics knows that a deep understanding of this subject does not come easily. There are subtleties and interconnections that are difficult to grasp at first. The more times one goes through a thermodynamics course (as a student or a teacher), the more insight one gains. Thus, this text will reinforce and extend the knowledge gained from an earlier exposure to thermodynamics. To this end, there is fairly intense discussion of some basic topics, such as the nature of spontaneous and reversible processes, and inclusion of a number of advanced topics, such as the reduction of bomb calorimetry measurements to standard-state conditions.

This book makes no claim to be an exhaustive treatment of thermodynamics. It concentrates on derivations of fundamental relations starting with the thermodynamic laws and on applications of these relations in various areas of interest to chemists. Although classical thermodynamics treats matter from a purely macroscopic viewpoint, the book discusses connections with molecular properties when appropriate.

In deriving equations, I have strived for rigor, clarity, and a minimum of mathematical complexity. I have attempted to clearly state the conditions under which each theoretical relation is valid because only by understanding the assumptions and limitations of a derivation can one know when to use the relation and how to adapt it for special purposes. I have taken care to be consistent in the use of symbols for physical properties. The choice of symbols follows the current recommendations of the International Union of Pure and Applied Chemistry (IUPAC) with a few exceptions made to avoid ambiguity.

To reduce the number of topics covered, or to adapt the text to a first-time introduction to thermodynamics, the instructor might consider skipping material in some of the following sections: 4.12, 5.12, 7.6.6, 7.6.7, 7.8.5, 7.8.6, 7.9, 8.2.5, 8.4.4, 8.4.5, 9.5.4, 9.6.4, 9.7, 10.2.6, and 10.3.

A Solutions Manual (ISBN 0-02-328742-X) is available to faculty who have adopted the text. Please speak to your Prentice Hall representative if you would like a copy.

I owe much to J. Arthur Campbell, Luke E. Steiner, and William Moffitt, gifted teachers who introduced me to the elegant logic and practical utility of thermodynamics. I am immensely grateful to my wife Stephanie for her continued encouragement and patience during the period this book went from concept to reality.

I would also like to acknowledge the help of the following reviewers: James L. Copeland, Kansas State University; Lee Hansen, Brigham Young University; Reed Howald, Montana State University–Bozeman; David W. Larsen, University of Missouri–St. Louis; Mark Ondrias, University of New Mexico; Philip H. Rieger, Brown University; Leslie Schwartz, St. John Fisher College; Allan L. Smith, Drexel University; and Paul E. Smith, Kansas State University.

I welcome comments and suggestions for improving this book. My e-mail address appears below.

Howard DeVoe
College Park, Maryland
e-mail: hd5@umail.umd.edu

Chapter 1

INTRODUCTION

Thermodynamics is a quantitative subject. It allows us to derive relations between the values of numerous physical quantities. Some physical quantities, such as a mole fraction, are dimensionless; the value of one of these quantities is a pure number. Most quantities, however, are not dimensionless and we must express their values with one or more *units*. This chapter reviews the SI system of units, which are the preferred units in science applications. The chapter then discusses some useful mathematical manipulations of physical quantities using quantity calculus and certain general aspects of dimensional analysis.

1.1 UNITS

There is international agreement that the units used for physical quantities in science and technology should be those of the International System of Units, or SI (standing for the French **Système International d'Unités**). The Physical Chemistry Division of the International Union of Pure and Applied Chemistry, or IUPAC, produces a manual of recommended symbols and terminology for physical quantities and units based on the SI. The manual has become known as the Green Book (from the color of its cover) and is referred to here as the IUPAC Green Book. We shall, with a few exceptions, use symbols recommended in the 1993 edition of the IUPAC Green Book[1]; these symbols are listed for convenient reference in Appendices C and D.

The SI is built on the seven **base units** listed in Table 1.1. These base units are independent physical quantities that are sufficient to describe all other physical quantities. One of the seven quantities, luminous intensity, is not used in this book and is usually not needed in thermodynamics. The official definitions of the base units are given in Appendix A.

Table 1.2 lists derived units for some additional physical quantities used in thermodynamics. The derived units have exact definitions in terms of SI base units, as given in the last column of the table.

The units listed in Table 1.3 are sometimes used in thermodynamics but are not part of the SI. They do, however, have exact definitions in terms of SI units and so offer no problems of numerical conversion to or from SI units.

Any of the symbols for units listed in Tables 1.1–1.3, except kg and °C, may be preceded

---
[1] See reference in Appendix I.

Table 1.1 SI base units

Physical quantity	SI unit	Symbol
length	meter[a]	m
mass	kilogram	kg
time	second	s
thermodynamic temperature	kelvin	K
amount of substance	mole	mol
electric current	ampere	A
luminous intensity	candela	cd

[a] or metre

Table 1.2 SI derived units

Physical quantity	Unit	Symbol	Definition of unit
force	newton	N	$1\,\text{N} = 1\,\text{m}\,\text{kg}\,\text{s}^{-2}$
pressure	pascal	Pa	$1\,\text{Pa} = 1\,\text{N}\,\text{m}^{-2} = 1\,\text{kg}\,\text{m}^{-1}\,\text{s}^{-2}$
Celsius temperature	degree Celsius	°C	$t/°\text{C} = T/\text{K} - 273.15$
energy	joule	J	$1\,\text{J} = 1\,\text{N}\,\text{m} = 1\,\text{m}^2\,\text{kg}\,\text{s}^{-2}$
power	watt	W	$1\,\text{W} = 1\,\text{J}\,\text{s}^{-1} = 1\,\text{m}^2\,\text{kg}\,\text{s}^{-3}$
frequency	hertz	Hz	$1\,\text{Hz} = 1\,\text{s}^{-1}$
electric charge	coulomb	C	$1\,\text{C} = 1\,\text{A}\,\text{s}$
electric potential	volt	V	$1\,\text{V} = 1\,\text{J}\,\text{C}^{-1} = 1\,\text{m}^2\,\text{kg}\,\text{s}^{-3}\,\text{A}^{-1}$
electric resistance	ohm	Ω	$1\,\Omega = 1\,\text{V}\,\text{A}^{-1} = 1\,\text{m}^2\,\text{kg}\,\text{s}^{-3}\,\text{A}^{-2}$

Table 1.3 Non-SI derived units

Physical quantity	Unit	Symbol	Definition of unit
volume	liter[a]	L[b]	$1\,\text{L} = 1\,\text{dm}^3 = 10^{-3}\,\text{m}^3$
pressure	bar	bar	$1\,\text{bar} = 10^5\,\text{Pa}$
pressure	atmosphere	atm	$1\,\text{atm} = 101{,}325\,\text{Pa} = 1.01325\,\text{bar}$
pressure	torr	Torr	$1\,\text{Torr} = (1/760)\,\text{atm} = (101{,}325/760)\,\text{Pa}$
energy	calorie	cal	$1\,\text{cal} = 4.184\,\text{J}$

[a] or litre
[b] or l

Table 1.4 SI prefixes

Fraction	Prefix	Symbol	Multiple	Prefix	Symbol
10^{-1}	deci	d	10	deka	da
10^{-2}	centi	c	10^2	hecto	h
10^{-3}	milli	m	10^3	kilo	k
10^{-6}	micro	μ	10^6	mega	M
10^{-9}	nano	n	10^9	giga	G
10^{-12}	pico	p	10^{12}	tera	T
10^{-15}	femto	f	10^{15}	peta	P
10^{-18}	atto	a	10^{18}	exa	E
			10^{21}	zetta	Z
			10^{24}	yotta	Y

by one of the prefix symbols of Table 1.4 to construct a decimal fraction or multiple of the unit. (The symbol g may be preceded by a prefix symbol to construct a fraction or multiple of the gram.) The combination of prefix symbol and unit symbol is taken as a new symbol that can be raised to a power. Examples are mg, cm, and cm^3.

1.1.1 Amount of substance and amount

The physical quantity formally called **amount of substance** is a counting quantity for particles, such as atoms or molecules, or for other chemical entities. We could use any counting unit for this purpose, but in practice it is invariably the **mole**. The mole is defined as the amount of substance containing as many particles as the number of atoms in exactly 12 grams of pure carbon-12 nuclide, ^{12}C. See Appendix A for the wording of the official IUPAC definition. This definition is such that one mole of H_2O molecules, for example, has a mass of 18.0153 grams (where 18.0153 is the relative molecular mass of H_2O) and contains 6.022137×10^{23} molecules (where $6.022137 \times 10^{23}\,mol^{-1}$ is the *Avogadro constant* to seven significant digits). We can make the same statement for any other substance if we replace 18.0153 by the appropriate atomic mass or molecular mass value.

The symbol for amount of substance is n. It is admittedly awkward to refer to $n(H_2O)$ as "the amount of substance of water." This book simply shortens "amount of substance" to **amount**, a common usage that is condoned by the IUPAC.[2] Thus, "the amount of water in the system" refers not to the mass or volume of water, but to the *number* of H_2O molecules in the system expressed in a counting unit such as the mole.

---

[2] Mills, I. M. *J. Chem. Educ.* **1989**, *66*, 887–889. An alternative name suggested for n is "chemical amount."

1.2 QUANTITY CALCULUS

This section gives examples of how we may manipulate physical quantities by the rules of algebra. The method is called *quantity calculus*, although a better term might be "quantity algebra."

Quantity calculus is based on the concept that a physical quantity, unless it is dimensionless, has a value equal to the product of a *numerical value* (a pure number) and one or more *units*:

$$\text{physical quantity} = \text{numerical value} \times \text{units}$$

(If the quantity is dimensionless, it is equal to a pure number without units.) We may denote the physical property by a symbol, but the symbol does *not* imply a particular choice of units. For instance, this book uses the symbol ρ for density, but ρ can be expressed in *any* units having the dimensions of mass divided by volume.

A simple example illustrates the use of quantity calculus. We may express the density of water at 25 °C to four significant digits in SI base units by the equation

$$\rho = 9.970 \times 10^2 \, \text{kg m}^{-3}$$

and in different density units by the equation

$$\rho = 0.9970 \, \text{g cm}^{-3}$$

We may divide both sides of the last equation by $1 \, \text{g cm}^{-3}$ to obtain a new equation

$$\rho / \text{g cm}^{-3} = 0.9970$$

Now the pure number 0.9970 appearing in this equation is the number of grams in one cubic centimeter of water, so we may call the ratio $\rho/\text{g cm}^{-3}$ "the number of grams per cubic centimeter." By the same reasoning, $\rho/\text{kg m}^{-3}$ is the number of kilograms per cubic meter. In general, a physical quantity divided by particular units for the physical quantity is a pure number representing the number of those units.

> Just as it would be incorrect to call ρ "the number of grams per cubic centimeter," because that would refer to a particular choice of units for ρ, the common practice of calling n "the number of moles" is also strictly speaking not correct. It is actually the ratio n/mol that is the number of moles.

In a table, the ratio $\rho/\text{g cm}^{-3}$ makes a convenient heading for a column of density values because the column can then show pure numbers. Likewise, it is convenient to use $\rho/\text{g cm}^{-3}$ as the label of a graph axis and to show pure numbers at the grid marks of the axis. You may find many examples of this usage in the tables and figures of this book.

A major advantage of using SI base units and SI derived units is that they are *coherent*. That is, values of a physical quantity expressed in different combinations of these units have the same numerical value.

Section 1.2 Quantity Calculus

For example, suppose we wish to evaluate the pressure of a gas according to the ideal gas equation[3]

$$p = \frac{nRT}{V} \tag{1.2.1}$$
(ideal gas)

In this equation, p, n, T, and V are the symbols for the physical quantities pressure, amount (amount of substance), thermodynamic temperature, and volume, respectively, and R is the gas constant.

The calculation of p for 5.000 moles of the gas at a temperature of 298.15 kelvins, in a volume of 4.000 cubic meters, is

$$p = \frac{(5.000\,\text{mol})(8.3145\,\text{J K}^{-1}\,\text{mol}^{-1})(298.15\,\text{K})}{4.000\,\text{m}^3} = 3.099 \times 10^3\,\text{J m}^{-3}$$

The mole and kelvin units cancel, and we are left with units of J m^{-3}, a combination of an SI derived unit (the joule) and an SI base unit (the meter). The units J m^{-3} must have dimensions of pressure, but are not commonly used to express pressure.

To convert J m^{-3} to the SI derived unit of pressure, the pascal (Pa), we can use the following relations in Table 1.2:

$$1\,\text{J} = 1\,\text{N m} \qquad 1\,\text{Pa} = 1\,\text{N m}^{-2}$$

When we divide both sides of the first relation by 1 J and divide both sides of the second relation by $1\,\text{N m}^{-2}$, we obtain the two new relations

$$1 = (1\,\text{N m/J}) \qquad (1\,\text{Pa/N m}^{-2}) = 1$$

The ratios in parentheses are *conversion factors*. When a physical quantity is multiplied by a conversion factor that, like these, is equal to the pure number 1, the physical quantity changes its units but not its value. When we multiply the equation for p by both of these conversion factors, all units cancel except Pa:

$$p = (3.099 \times 10^3\,\text{J m}^{-3}) \times (1\,\text{N m/J}) \times (1\,\text{Pa/N m}^{-2})$$
$$= 3.099 \times 10^3\,\text{Pa}$$

This example illustrates the fact that to calculate a physical quantity we can simply enter into a calculator numerical values expressed in SI units and the result is the numerical value of the calculated quantity expressed in SI units. In other words, as long as we use only SI base units and SI derived units (without prefixes), all conversion factors have a numerical value of unity.

---

[3]This is the first numbered equation in this book. Like many of the numbered equations, it shows conditions of validity in parentheses immediately below the equation number at the right. Thus, Eq. 1.2.1 is valid for an ideal gas.

Of course we do not have to limit the calculation to SI units. Suppose we wish to express the calculated pressure in torrs, a non-SI unit. In this case, using a conversion factor obtained from the definition of the torr in Table 1.3, the calculation becomes

$$p = (3.099 \times 10^3 \, \text{Pa}) \times (760 \, \text{Torr}/101,325 \, \text{Pa})$$
$$= 23.24 \, \text{Torr}$$

1.3 DIMENSIONAL ANALYSIS

Sometimes we can catch an error in the form of an equation or expression, or in values used for a calculation, by checking for dimensional consistency. Among the things to check are that both sides of an equation have the same dimensions, that all terms of a sum have the same dimensions, that the argument of a logarithm is dimensionless, and that any quantity used as a power or logarithm is dimensionless.

In this book the *differential* of a function, such as df, refers to an *infinitesimal* quantity. If one side of an equation is an infinitesimal quantity, the other side must also be. Thus, the equation $df = a\,dx + b\,dy$ (where ax and by have the same dimensions as f) makes mathematical sense, but $df = ax + b\,dy$ does not.

Derivatives, partial derivatives, and integrals have dimensions that we must take into account when determining the overall dimensions of an expression that includes them. For instance, the derivative dp/dT and the partial derivative $(\partial p/\partial T)_V$ both have the same dimensions as p/T. The partial second derivative $(\partial^2 p/\partial T^2)_V$ has the same dimensions as p/T^2. The integral $\int T\,dT$ has the same dimensions as T^2.

Some examples of applying these principles are given here using symbols described in the preceding section.

Example 1. Since the gas constant R may be expressed in units of $J\,K^{-1}\,mol^{-1}$, it has dimensions of energy divided by thermodynamic temperature and amount. Thus, RT has dimensions of energy divided by amount, and nRT has dimensions of energy. These quantities appear frequently in thermodynamic expressions.

Example 2. What are the dimensions of the quantity $nRT\ln(p/p°)$ and of $p°$ in this expression? The quantity has the same dimensions as nRT (or energy) because the logarithm is dimensionless. Furthermore $p°$ in this expression has dimensions of pressure to make the argument of the logarithm, $p/p°$, dimensionless.

Example 3. Find the dimensions of the constants a and b in the van der Waals equation

$$p = \frac{nRT}{V - nb} - \frac{n^2 a}{V^2}$$

Dimensional analysis tells us that, because $-nb$ is added to V, nb has dimensions of volume and therefore b has dimensions of volume/amount. Furthermore, since the right side of the equation is a sum of two terms, these terms have the same dimensions as the left side, which is pressure. Therefore, the second term $n^2 a/V^2$ has dimensions of pressure and a has dimensions of pressure $\cdot$ volume^2 $\cdot$ amount^{-2}.

Example 4. Consider an equation of the form

$$Pd\ln xTp = \frac{y}{R}$$

where x is a dimensionless quantity. What are the SI units of y? The left side of the equation has the dimensions of $1/T$, and therefore SI units of K^{-1}. The SI units of the right side are therefore also K^{-1}. Since R has the units $J\,K^{-1}\,mol^{-1}$, the SI units of y are $J\,K^{-2}\,mol^{-1}$.

PROBLEMS

1.1 Consider the following equations for the pressure of a real gas. For each equation, find the dimensions of the constants a and b and express these dimensions in SI units.

(a) The Dieterici equation:

$$p = \frac{RTe^{-(an/VRT)}}{(V/n) - b}$$

(b) The Redlich–Kwong equation:

$$p = \frac{RT}{(V/n) - b} - \frac{an^2}{T^{1/2}V(V + nb)}$$

Chapter 2

SOME BASIC PROPERTIES AND THEIR MEASUREMENT

This chapter begins by discussing macroscopic properties of matter in general and properties distinguishing different physical states of matter in particular. Virial equations of state of a pure gas are introduced. The chapter then goes on to discuss some fundamental macroscopic properties including mass, volume, density, pressure, and temperature, with examples of how these properties can be measured. As we see in later chapters, thermodynamic theory allows us to use these measurements for the evaluation of many other useful quantities.

2.1 GENERAL CONSIDERATIONS

A thermodynamic **system** is any three-dimensional region of physical space on which we wish to focus our attention. The term **body** usually implies a system, or part of a system, whose mass and chemical composition are constant over time.

2.1.1 The macroscopic viewpoint

Chemists are interested in systems containing matter—that which has mass and occupies physical space. As mentioned in the Preface, classical thermodynamics looks at *macroscopic* aspects of matter. It deals with the properties of aggregates of vast numbers of microscopic particles (molecules,[1] atoms, and ions). The macroscopic viewpoint, in fact, treats matter as a *continuous* material medium rather than as the collection of discrete microscopic particles we know are actually present. Although this book is an exposition of classical thermodynamics, it at times points out connections between macroscopic properties and molecular structure and behavior.

---

[1] Some molecules, actually, are macroscopic in size—for example, a network covalent crystal such as diamond or quartz (SiO_2) is a single macroscopic molecule. The microscopic particles in these molecules are atoms.

2.1.2 Extensive and intensive properties

A quantitative property of a system describes some macroscopic feature that, although it may vary with time, has a particular value at any given instant of time. Most of the properties studied by thermodynamics may be classified as either extensive or intensive. We can distinguish these two types of properties by the following considerations.

If we imagine the system to be divided by an imaginary surface into two parts, any property of the system that is the sum of the property for the two parts is an **extensive property**. That is, an additive property is extensive. Examples are mass, volume, amount, energy, and the surface area of a solid.

> Sometimes a more restricted definition of an extensive property is used: The property must be not only additive, but also proportional to the mass or the amount when intensive properties remain constant. According to this definition, mass, volume, amount, and energy are extensive, but surface area is not.

If we imagine a homogeneous region of space to be divided into two or more parts of arbitrary size, any property that has the same value in each part and the whole is an **intensive property**; for example density, concentration, pressure (in a fluid), and temperature. The value of an intensive property is the same everywhere in a homogeneous region but may vary from point to point in a heterogeneous region.

Since classical thermodynamics treats matter as a continuous medium, whereas matter actually contains discrete microscopic particles, the value of an intensive property at a point is a statistical average of the behavior of many particles. For instance, the density of a gas at one point in space is the average mass of a small volume element at that point, large enough to contain many molecules, divided by the volume of that element.

Some properties are defined as the ratio of two extensive quantities. If both extensive quantities refer to a homogeneous region of the system or to a small volume element, the ratio is an *intensive* property; for example, concentration = amount/volume. A mathematical derivative of one such extensive quantity with respect to another is also intensive. A special case is an extensive quantity divided by the mass, giving an intensive **specific quantity**; for example

$$\text{Specific volume} = \frac{\text{volume}}{\text{mass}}$$

Another special case that we encounter frequently in this book is an extensive quantity for a pure, homogeneous substance divided by the amount n; the resulting intensive quantity is called, in general, a **molar quantity**. To symbolize a molar quantity, this book follows the recommendation of the IUPAC: The symbol of the extensive quantity is followed by subscript m, and optionally the identity of the substance is indicated either by a subscript or a formula in parentheses. Examples are

$$\text{Molar volume} = \frac{V}{n} = V_\text{m} \qquad (2.1.1)$$

$$\text{Molar volume of substance } i = \frac{V}{n_i} = V_{\text{m},i}$$

$$\text{Molar volume of } H_2O = V_m(H_2O)$$

In the past, especially in the United States, molar quantities were commonly denoted with an overbar (e.g., $\overline{V}_i$).

2.1.3 Phases

A **phase** is a region of space in which each intensive property (such as temperature and pressure) has at each instant either the same value throughout (a *uniform* or *homogeneous* phase) or else a value that varies continuously from one point to another. Whenever this book mentions a phase, it is a *uniform* phase unless otherwise stated. Two different phases meet at an **interface** surface, where intensive properties have a discontinuity or change over a small distance.

Some intensive properties (e.g., refractive index and polarizability) can have directional characteristics. A uniform phase may be either *isotropic*, exhibiting the same values of these properties in all directions, or *anisotropic*, as in the case of some solids and liquid crystals. A vacuum is a uniform phase of zero density.

Suppose we have to deal with a *nonuniform* region in which intensive properties vary continuously in space along one or more directions—for example, a tall column of gas in a gravitational field whose density decreases with increasing altitude. There are two ways we may treat such a nonuniform, continuous region: either as a single nonuniform phase or else as an infinite number of uniform phases, each of infinitesimal size in one or more dimensions.

2.1.4 Physical states of matter

We are used to labeling phases by physical state, or state of aggregation. It is common to say that a phase is a *solid* if it is relatively rigid, a *liquid* if it is easily deformed and relatively incompressible, and a *gas* if it is easily deformed and easily compressed. Since these descriptions of responses to external forces differ only in degree, they are inadequate to classify intermediate cases.

We can take a more rigorous approach by making a primary distinction between a *solid* and a *fluid*, based on the phase's response to an applied shear stress, and then use other criteria to classify a fluid as a *liquid, gas,* or *supercritical fluid*. **Shear stress** is a tangential force per unit area exerted by matter on one side of an interior plane on matter on the other side. We can produce shear stress in a phase by applying tangential forces to parallel surfaces of the phase as shown in Fig. 2.1 on the next page.

A **solid** responds to shear stress by undergoing momentary relative motion of its parts, resulting in *deformation*—a change of shape. If the applied shear stress is constant and small (not large enough to cause creep or fracture), the solid quickly reaches a certain degree of deformation that depends on the magnitude of the stress and maintains this deformation without further change as long as the shear stress continues to be applied. On the microscopic level, deformation requires relative movement of adjacent layers of particles (atoms, molecules, or ions). The shape of an unstressed

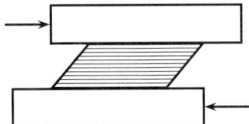

Figure 2.1 Experimental procedure for producing shear stress in a phase (shaded). Blocks at the upper and lower surfaces of the phase are pushed in opposite directions, dragging the adjacent portions of the phase with them.

solid is determined by the attractive and repulsive forces between the particles; these forces make it difficult for adjacent layers to slide past one another, so that the solid resists deformation.

A **fluid** responds to shear stress differently by undergoing continuous relative motion (flow) of its parts. The flow continues as long as there is any shear stress, no matter how small, and stops only when the shear stress is removed. A fluid at rest cannot support shear stress.

Thus, a constant applied shear stress causes a fixed deformation in a solid and continuous flow in a fluid. We say that a phase under constant shear stress is a solid if, after the initial deformation, we are unable to detect a further change in shape during the period we observe the phase.

Usually this criterion allows us to unambiguously classify a phase as a solid or a fluid. However, over a sufficiently long time period, detectable flow occurs in *any* material under shear stress of *any* magnitude. Thus, the distinction between solid and fluid actually depends on the time scale of observation. This fact is obvious when we observe the behavior of certain materials (such as Silly Putty or a paste of water and cornstarch) that exhibit solid-like behavior over a short time period and fluidlike behavior over a longer period. Such materials, which resist deformation by a suddenly applied shear stress but undergo flow over an ordinary time period of observation, say several minutes, are called *viscoelastic solids*.

2.1.5 Phase coexistence and phase transitions

Before going on to classify fluids, we need to consider some general characteristics of systems containing more than one phase.

Suppose we bring two uniform phases of the same constituents into physical contact at an interface surface. If we find that the phases remain unchanged for an indefinite period while both have the same temperature and the same pressure, but differ in other intensive properties such as density and composition, we say that they **coexist** in equilibrium with one another. The conditions for such phase coexistence are the subject of later sections in this book, but they tend to be quite restricted. For instance, the liquid and gas phases of pure H_2O at a pressure of 1 bar can coexist at only one temperature, 99.61 °C.

A **phase transition** of a pure substance is a change in which there is a continuous transfer of the substance from one phase to another. Eventually one phase can completely disappear, and the substance has been completely transferred to the other phase. If both phases coexist

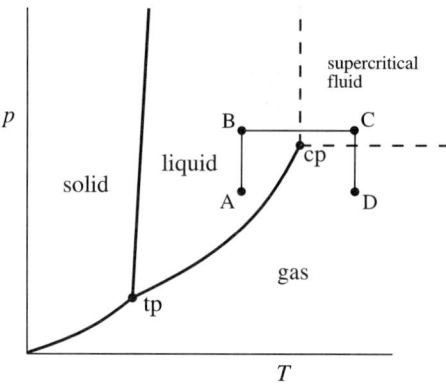

Figure 2.2 Pressure–temperature phase diagram of a pure substance (schematic). Point cp is the critical point, and point tp is the triple point. Each area is labeled with the physical state that is stable under the pressure-temperature conditions that fall within the area. A solid curve (coexistence curve) separating two areas is the locus of pressure-temperature conditions that allow the phases of these areas to coexist at equilibrium. Path ABCD illustrates *continuity of states*.

in equilibrium with one another, and the temperature and pressure of both phases remain equal and constant during the phase transition, the change is an *equilibrium* or *reversible* phase transition. For example, H_2O at 99.61 °C and 1 bar can undergo an equilibrium phase transition from liquid to gas (vaporization) or from gas to liquid (condensation). During an equilibrium phase transition, there is a transfer of energy in the form of heat or work between the substance and its surroundings.

2.1.6 Fluids

We are used to classifying a fluid as either a *liquid* or a *gas*. The distinction is important for a pure substance because the choice determines how we treat the phase's standard state (see Sec. 5.9). To complicate matters, a fluid at high pressure may be a *supercritical fluid*. Sometimes a *plasma* (a highly ionized, electrically conducting medium) is considered a separate kind of fluid state; it is the state found in the earth's ionosphere and in stars.

In general, and provided the pressure is not high enough for supercritical phenomena to exist (this is usually true of pressures below 25 bar except in the case of He or H_2), we can make the distinction between liquid and gas simply on the basis of density. A **liquid** has a relatively high density, which is relatively insensitive to changes in temperature and pressure. A **gas**, on the other hand, has a relatively low density, which is sensitive to temperature and pressure and which approaches zero as pressure is reduced at constant temperature.

This simple distinction between liquids and gases fails at high pressures, where liquid and gas phases may have similar densities at the same temperature. Figure 2.2 shows how we can classify stable fluid states of a pure substance in relation to a liquid–gas coexistence curve and a critical point. If raising the temperature of a fluid at constant pressure causes a phase transition to a second fluid phase, the original fluid was a liquid and the transition occurs at the liquid–gas *coexistence curve*. This curve ends at a **critical point**, at which all intensive properties of the coexisting liquid and gas phases become identical.[2] The fluid state of a pure substance at a temperature greater than the critical temperature and a pressure greater than the critical pressure is called a **supercritical fluid**.

The designation of a supercritical fluid state of a substance is used more for convenience than because of any unique properties compared to a liquid or gas. If we vary the temperature or pressure in such a way that the substance changes from what we call a liquid to what we call a supercritical fluid, we observe only a continuous change of a single phase, and no phase transition with two coexisting phases. The same is true for a change from a supercritical fluid to a gas. Thus, by making the changes described by the path ABCD shown in Fig. 2.2, we can transform a pure substance from a liquid to a gas at the same pressure without ever observing an interface between two coexisting phases! This curious phenomenon is called *continuity of states*.

Chapter 6 takes up the discussion of further aspects of the physical states of pure substances.

If we are dealing with a fluid *mixture* (instead of a pure substance) at a high pressure, it may be difficult to classify the phase as either liquid or gas. The complexity of classification at high pressure is illustrated by the *barotropic effect*, observed in some mixtures, in which a small change of temperature or pressure causes what was the more dense of two coexisting fluid phases to become the less dense phase. In a gravitational field, the two phases switch positions.

2.1.7 The equation of state of a fluid

When we prepare a uniform fluid phase containing a known amount n_i of each constituent substance i, with the temperature T and pressure p adjusted to definite known values, we expect the phase to have a definite, fixed volume V. If we change any one of the properties T, p, or n_i, there is usually a change in V. The value of V is dependent on the other properties and cannot be varied independently of them. Thus, for a given substance or mixture of substances in a uniform fluid phase, V is a unique function of T, p, and n_i. We may be able to express this relation in an explicit equation: $V = f(T, p, n_i)$. This equation (or a rearranged form) that gives a relation among V, T, p, and n_i, is the **equation of state** of the fluid.

We may solve the equation of state, implicitly or explicitly, for any one of the quantities V, T, p, and n_i in terms of the other quantities. Thus, of the $3 + s$ quantities (where s is the number of substances), only $2 + s$ are independent.

The *ideal gas equation*, $p = nRT/V$ (Eq. 1.2.1 on page 5), is an equation of state. It is found experimentally that the behavior of any gas in the limit of low pressure, as temperature is held constant, approaches this equation of state. This limiting behavior is also predicted by the kinetic-molecular theory (Sec. 3.2.2).

If the fluid has only one constituent (i.e., is a pure substance rather than a mixture), then at a fixed T and p the volume is proportional to the amount. In this case, the equation of

---

[2] Apparently no critical points exist involving a fluid phase and a solid phase.

state may be expressed as a relation among T, p, and the molar volume $V_m = V/n$. The equation of state for a pure ideal gas may be written $p = RT/V_m$.

The *Redlich–Kwong equation* is a two-parameter equation of state frequently used to describe, to good accuracy, the behavior of a pure gas at a pressure where the ideal gas equation fails:

$$p = \frac{RT}{V_m - b} - \frac{a}{V_m(V_m + b)T^{1/2}} \qquad (2.1.2)$$

In this equation, a and b are constants that are independent of temperature and depend on the substance.

The next section describes features of *virial* equations, an important class of equations of state for real (nonideal) gases.

2.1.8 Virial equations of state for pure gases

We shall have occasion to apply thermodynamic derivations to virial equations of state of a pure gas or gas mixture. These formulas accurately describe the gas at low and moderate pressures using empirically determined, temperature-dependent parameters. The equations may be derived from statistical mechanics, and so have a theoretical as well as empirical foundation. There are two forms of virial equations for a pure gas: one a series in powers of $1/V_m$:

$$pV_m = RT\left(1 + \frac{B}{V_m} + \frac{C}{V_m^2} + \cdots\right) \qquad (2.1.3)$$

and the other a series in powers of p:

$$pV_m = RT\left(1 + B'p + C'p^2 + \cdots\right) \qquad (2.1.4)$$

The parameters B, C, ... in Eq. 2.1.3 are called the *second, third, ... virial coefficients*; their values depend on the substance and are functions of temperature. (The *first* virial coefficient is 1 because pV_m must approach RT as $1/V_m$ approaches zero.) Coefficients beyond the third virial coefficient are small and rarely evaluated. To find the relation between these virial coefficients and the parameters B', C', ... in Eq. 2.1.4, we solve Eq. 2.1.3 for p in terms of V_m

$$p = RT\left(\frac{1}{V_m} + \frac{B}{V_m^2} + \cdots\right)$$

and substitute in the right side of Eq. 2.1.4:

$$pV_m = RT\left[1 + B'RT\left(\frac{1}{V_m} + \frac{B}{V_m^2} + \cdots\right)\right.$$
$$\left. + C'(RT)^2\left(\frac{1}{V_m} + \frac{B}{V_m^2} + \cdots\right)^2 + \cdots\right] \qquad (2.1.5)$$

Then we equate coefficients of equal powers of $1/V_m$ in Eqs. 2.1.3 and 2.1.5 (since both equations must yield the same value of pV_m for any value of $1/V_m$):

$$B = RTB' \tag{2.1.6}$$

$$C = B'RTB + C'(RT)^2 = (RT)^2(B'^2 + C')$$

In the last equation, we have substituted for B from Eq. 2.1.6.

At pressures up to at least one bar, the terms beyond $B'p$ in the power series of Eq. 2.1.4 are negligible; then pV_m may be approximated by $RT(1 + B'p)$, giving, with the help of Eq. 2.1.6, the simple approximate equation of state[3]

$$V_m \approx \frac{RT}{p} + B \tag{2.1.7}$$
(pure gas, $p \leq 1$ bar)

The **compression factor** (or compressibility factor) Z of a gas is defined by

$$Z \equiv \frac{pV}{nRT} = \frac{pV_m}{RT} \tag{2.1.8}$$
(gas)

When a gas at a particular temperature and pressure satisfies the ideal gas equation, the value of Z is 1. The virial equations rewritten using Z are

$$Z = 1 + \frac{B}{V_m} + \frac{C}{V_m^2} + \cdots$$

$$Z = 1 + B'p + C'p^2 + \cdots$$

These equations show that the second virial coefficient B is the initial slope of the curve of a plot of Z versus $1/V_m$ at constant T, and B' is the initial slope of Z versus p at constant T.

The way in which Z varies with p at different temperatures is shown for the case of carbon dioxide in Fig. 2.3 on the next page. A temperature at which the initial slope is zero is called the **Boyle temperature**, which for CO_2 is 710 K. Both B and B' must be zero at the Boyle temperature. At lower temperatures B and B' are negative, and at higher temperatures they are positive (see Fig. 2.4 on the next page). This kind of temperature dependence is typical. Experimentally, and also according to statistical mechanical theory, B and B' for a gas can be zero only at a single Boyle temperature.

> The fact that at any temperature other than the Boyle temperature B is nonzero is significant since it means that in the limit as p approaches zero at constant T and the gas approaches ideal-gas behavior, the *difference* between the actual molar volume V_m and the ideal-gas molar volume RT/p does not approach zero. Instead, $V_m - RT/p$ approaches the nonzero value B (see Eq. 2.1.7). However, the *ratio* of the actual and ideal molar volumes, $V_m/(RT/p)$, approaches unity in this limit.

Virial equations of gas *mixtures* are discussed in Sec. 7.4.4.

---

[3] Guggenheim (reference in Appendix I) calls a gas with this equation of state a *slightly imperfect gas*.

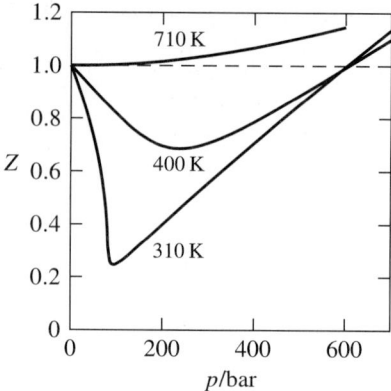

Figure 2.3 The compression factor for CO_2 as a function of pressure at three temperatures. At 710 °C, the Boyle temperature, the initial slope is zero.

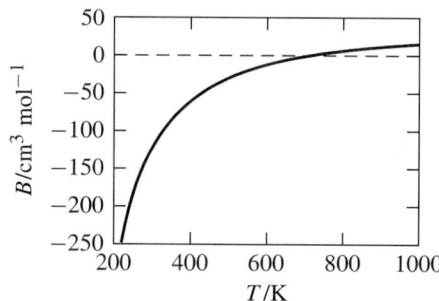

Figure 2.4 The second virial coefficient of CO_2 as a function of temperature.

2.1.9 Solids

A solid phase responds to a small applied stress by undergoing a small *elastic* deformation. When the stress is removed, the solid returns to its initial shape and the properties return to those of the unstressed solid. Under these conditions of small stress, the solid has an equation of state just as a fluid does, in which p is the pressure of a fluid surrounding the solid (the hydrostatic pressure) as explained in Sec. 2.5. A stress is an additional independent variable. For example, the length of a metal spring that is elastically deformed is a unique function of the temperature, the pressure of the surrounding air, and the stretching force.

However, if the stress applied to the solid exceeds its elastic limit, the response is a *plastic* deformation. This deformation persists when the stress is removed, and the unstressed solid no longer has its original properties. Plastic deformation is a kind of hysteresis and is caused by such microscopic behavior as the slipping of crystal planes past one another in a crystal subjected to shear stress and conformational rearrangements about single bonds in a stretched macromolecular fiber. Properties of the solid then depend on its past history and are not unique functions of a set of independent variables. There is no equation of state.

2.2 MASS

We may measure the mass of an object with a balance utilizing the downward force exerted on the object by the earth's gravitational field. The classic balance has a beam and knife-edge arrangement to compare the gravitational force on the body of interest with the gravitational force on a weight of known mass. A modern balance (strictly speaking a *scale*) incorporates a strain gauge or comparable device to directly measure the gravitational force on the unknown mass; this type must be calibrated with known masses. For the most accurate measurements, we must take into account the effect of the buoyancy of the body and the calibration masses in air. The accuracy of the calibration masses should be traceable to a national standard kilogram (which in the United States is maintained at the National Institute of Standards and Technology, formerly the National Bureau of Standards, in Gaithersburg, Maryland) and ultimately to the international prototype of the kilogram located at the International Bureau of Weights and Measures in Sèvres, France.

From the measured mass of a sample of a pure substance, we can calculate the amount of substance (called simply the *amount* in this book). The SI base unit for amount is the mole (Sec. 1.1.1). Chemists are familiar with the fact that, although the mole is a counting unit, an amount in moles is measured not by counting but by weighing. This is because one mole is defined as the amount of atoms in exactly 12 grams of carbon-12, the most abundant isotope of carbon (Appendix A). One mole of a substance has a mass of M_r grams, where M_r is the relative molecular mass (or molecular weight) of the substance, a dimensionless quantity.

A quantity related to molecular weight is the **molar mass** of a substance, defined as the mass divided by the amount:

$$\text{Molar mass} = M = \frac{m}{n} \tag{2.2.1}$$

(The symbol M for molar mass is an exception to the rule given on page 9 that a subscript m is used to indicate a molar quantity.) The numerical value of the molar mass expressed in units of g mol^{-1} is equal to the relative molecular mass:

$$M/\text{g mol}^{-1} = M_r$$

2.3 VOLUME

We commonly measure liquid volumes with precision volumetric glassware such as burets, pipets, and volumetric flasks. The National Institute of Standards and Technology in the United States has established specifications for "Class A" glassware; two examples are listed in Table 2.1 on the next page. We may accurately determine the volume of a vessel at one temperature from the mass of a liquid of known density, such as water, that fills the vessel at this temperature.

The SI unit of volume is the cubic meter, but chemists commonly express volumes in units of liters and milliliters. The liter is defined as one cubic decimeter (Table 1.3). One cubic meter is the same as 10^3 liters and 10^6 milliliters. The milliliter is identical to the cubic centimeter.

Table 2.1 Representative measurement methods

Physical quantity	Method	Typical value	Approximate uncertainty
Mass	analytical balance	100 g	0.1 mg
	microbalance	20 mg	0.1 μg
Volume	pipet, Class A	10 mL	0.02 mL
	volumetric flask, Class A	1 L	0.3 mL
Density	pycnometer, 25-mL capacity	1 g mL^{-1}	2 mg mL^{-1}
	magnetic float densimeter	1 g mL^{-1}	0.1 mg mL^{-1}
Pressure	mercury manometer or barometer	760 Torr	0.001 Torr
	diaphragm gauge	100 Torr	1 Torr
Temperature	constant-volume gas thermometer	10 K	0.001 K
	mercury-in-glass thermometer	300 K	0.01 K
	platinum resistance thermometer	300 K	0.0001 K
	monochromatic optical pyrometer	1300 K	0.03 K

Before 1964, the liter had a different definition: it was the volume of 1 kilogram of water at 3.98 °C, the temperature of maximum density. This definition made one liter equal to 1.000028 dm^3. Thus, a numerical value of volume (or density) reported before 1964 and based on the liter as then defined may need a small correction to be consistent with the present definition of the liter.

2.4 DENSITY

Density, an intensive property, is defined as the ratio of the two extensive properties mass and volume:

$$\rho = \frac{m}{V} \tag{2.4.1}$$

The molar volume V_m of a homogeneous pure substance is inversely proportional to its density. From Eqs. 2.1.1, 2.2.1, and 2.4.1, we obtain the relation

$$V_m = \frac{M}{\rho} \tag{2.4.2}$$

Various methods are available for determining the density of a phase, many of them based on the measurement of the mass of a known volume or on a buoyancy technique. Examples of apparatus for liquid density measurements using these methods are shown in Figs. 2.5 and 2.6.

Similar apparatus may be used for gases. The density of a solid may be determined from the volume of a nonreacting liquid (e.g., mercury) displaced by a known mass of the solid or from the loss of weight due to buoyancy when the solid is suspended by a thread in a liquid of known density.

Section 2.5 Pressure 19

Figure 2.5 A glass pycnometer for measuring liquid density. The vessel is filled with the liquid at a temperature below the temperature required for the measurement, and the capillary stopper is inserted. The pycnometer is then thermostated in a bath at the desired temperature. The liquid overflows from the tip of the capillary tube and is wiped off. The pycnometer is cooled, the outside is dried, and the assembly is weighed. The capacity and empty weight are determined separately.

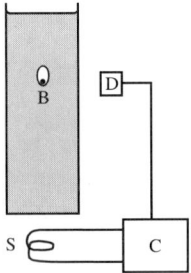

Figure 2.6 A magnetic float densimeter for measuring liquid density. A buoy B containing a magnet is suspended in the liquid. The buoy is kept in position with a solenoid S by means of a position detector D and a servo control system C. (The buoy here weighs less than the liquid it displaces and is pulled down by the solenoid.) The solenoid current required to keep the buoy in place is a measure of the liquid density (Greer, S. G.; Moldover, M. R.; Hocken, R. *Rev. Sci. Instrum.* **1974**, *45*, 1462).

2.5 PRESSURE

Pressure is a force per unit area. Specifically, it is the normal component of stress exerted by an isotropic fluid on a surface element.[4] The surface can be an interface surface between the fluid and another phase, or an imaginary dividing plane within the fluid.

Pressure is usually a positive quantity. Because cohesive forces exist in a liquid, it may be possible to place the liquid under tension and create a *negative* pressure. For instance, the pressure is negative at the top of a column of liquid mercury suspended below the closed end of a capillary tube that has no vapor bubble. Negative pressure in a liquid is an unstable condition that can result in spontaneous vaporization.

The SI unit of pressure is the **pascal**. Its symbol is Pa. One pascal is a force of one

---
[4] A liquid crystal and a polar liquid in a electric field are examples of fluids that are not isotropic—that is, that have different macroscopic properties in different directions.

newton per square meter (Table 1.2).

Chemists are accustomed to using the non-SI units of millimeters of mercury, torr, and atmosphere. One millimeter of mercury (symbol mmHg) is the pressure exerted by a column exactly 1 mm high of a fluid of density equal to exactly $13.5951 \text{ g cm}^{-3}$ (the density of mercury at $0\,°\text{C}$) in a place where the acceleration of free fall has its standard value g_n (see Appendix B). One atmosphere is defined as exactly 1.01325×10^5 Pa (Table 1.3). The torr is defined by letting one atmosphere equal exactly 760 Torr. One atmosphere is approximately 760 mmHg. In other words, the millimeter of mercury and the torr are practically identical; they differ from one another by less than 2×10^{-7} Torr.

Another non-SI pressure unit is the **bar**, equal to exactly 10^5 Pa. A pressure of one bar is approximately one percent smaller than one atmosphere. This book often refers to a **standard pressure**, $p°$. In the past, the value of $p°$ was usually taken to be 1 atm, but since 1982 the IUPAC has recommended the value $p° = 1$ bar.

A variety of manometers and other devices is available to measure the pressure of a fluid, each type useful in a particular pressure range. Some measure the pressure of the fluid directly. Others measure the differential pressure between the fluid and the atmosphere; the fluid pressure is obtained by combining this measurement with the atmospheric pressure measured with a barometer.

Within a *solid*, we cannot define pressure simply as a force per unit area. Macroscopic forces at a point within a solid are described by the nine components of a stress tensor. When we say that a solid *has* or *is at* a certain pressure, we mean that this is the hydrostatic pressure exerted on the solid's exterior surface. Thus, a solid immersed in a uniform isotropic fluid of pressure p is at pressure p; if the fluid pressure is constant, the solid is at constant pressure.

2.6   TEMPERATURE

2.6.1   Temperature scales

Temperature and thermometry are of fundamental importance in thermodynamics. Unlike the other physical quantities discussed in this chapter, temperature does not have a single definition. The definition we choose, whatever it may be, requires a *temperature scale* described by an operational method of measuring temperature values. For the scale to be useful, the values should increase monotonically with the increase of what we experience physiologically as the degree of "hotness." We can define a satisfactory scale with any measuring method that satisfies this requirement. The values on a particular temperature scale correspond to a particular physical quantity and a particular temperature unit.

For example, suppose you construct a simple liquid-in-glass thermometer with equally spaced marks along the stem and number the marks consecutively. To define a temperature scale and a temperature unit, you could place the thermometer in thermal contact with a body whose temperature is to be measured, wait until the indicating liquid reaches a stable position, and read the meniscus position by linear interpolation between two marks.[5]

Thermometry is based on the principle that the temperatures of different bodies may be

---

[5] Of course, placing the thermometer and body in thermal contact may affect the body's temperature. The measured temperature is that of the body *after* thermal equilibrium is achieved.

compared with a thermometer. For example, if you find by separate measurements with your thermometer that two bodies give the same reading, you know that within experimental error both have the same temperature. The significance of two bodies having the same temperature (on any scale) is that if they are placed in thermal contact with one another, they will prove to be in thermal equilibrium with one another as evidenced by the absence of any changes in their properties. This principle is sometimes called the **zeroth law of thermodynamics**, and was first stated as follows by J. C. Maxwell (1872): "Bodies whose temperatures are equal to that of the same body have themselves equal temperatures."[6]

We shall use two particular temperature scales extensively. The **ideal-gas temperature scale** is defined by gas thermometry measurements, as described in Sec. 2.6.4. The **thermodynamic temperature scale** is defined by the behavior of a theoretical Carnot engine, as explained in Sec. 4.3.4. These temperature scales correspond to the physical quantities called ideal-gas temperature and thermodynamic temperature, respectively. Although the two scales are defined differently, the two temperatures turn out (Sec. 4.3.4) to be proportional to one another. Their values become identical when we use the same unit of temperature for both. Thus, we define the *kelvin* by specifying that a system containing the solid, liquid, and gaseous phases of H_2O coexisting at equilibrium with one another (the triple point of water) has a thermodynamic temperature of exactly 273.16 kelvins. If we let the ideal-gas temperature of this system also be 273.16 kelvins, the temperatures measured on the two scales are identical.

Formally, the symbol T refers to thermodynamic temperature. Strictly speaking, we should use a different symbol for ideal-gas temperature. Since the two kinds of temperatures have identical values, we use the symbol T for both and refer to both physical quantities simply as "temperature" except when it is necessary to make a distinction.

Why is the temperature of the triple point of water taken to be 273.16 kelvins? This value is chosen arbitrarily to make the steam point approximately one hundred kelvins greater than the ice point.

The **ice point** is the temperature at which ice and air-saturated water coexist in equilibrium at a pressure of one atmosphere. The **steam point** is the temperature at which liquid and gaseous H_2O coexist in equilibrium at one atmosphere. Neither of these temperatures has sufficient reproducibility for high-precision work. The temperature of the ice-water-air system used to define the ice point is affected by air bubbles in the ice and by varying concentrations of air in the water around each piece of ice. The steam point is uncertain because the temperature of coexisting liquid and gas is a sensitive function of the experimental pressure.

The obsolete **centigrade scale** was defined to give a value of exactly 0 degrees centigrade at the ice point and a value of exactly 100 degrees centigrade at the steam point, and to be a linear function of an ideal-gas temperature scale.

The centigrade scale has been replaced by the **Celsius scale**, which is based on the triple point of water rather than on the less reproducible ice point and steam point. The Celsius scale is the thermodynamic (or ideal-gas) temperature scale shifted by exactly 273.15

---

[6]L. A. Turner (*Am. J. Phys.* **1961**, *29*, 71–76) argues that the "zeroth law" is a consequence of the first and second laws and therefore is not a separate assumption in the axiomatic framework of thermodynamics. The term "law" for this principle is also questioned by O. Redlich (*J. Chem. Educ.* **1970**, *47*, 740–741).

kelvins. The temperature unit is the degree Celsius (°C), identical in size to the kelvin. Thus, Celsius temperature t is related to thermodynamic temperature T by

$$t/°C = T/K - 273.15 \qquad (2.6.1)$$

On the Celsius scale, the triple point of water is exactly 0.01 °C. The ice point is 0 °C to within 0.0001 °C, and the steam point is 99.97 °C.

2.6.2 The International Temperature Scale of 1990

The International Temperature Scale of 1990 (abbreviated ITS-90) is the most recent scale devised for practical high-precision temperature measurements.[7] This scale defines the physical quantity called international temperature, with symbol T_{90}. Each value of T_{90} is intended to be very close to the corresponding thermodynamic temperature T.

The ITS-90 is defined over a very wide temperature range, from 0.65 K up to at least 1358 K. There is a specified procedure for each measurement of T_{90}. Depending on the range in which T falls, the measurement is made by either vapor-pressure thermometry (0.65–5.0 K), gas thermometry[8] (3.0–24.5561 K), platinum-resistance thermometry (13.8033–1234.93 K), or optical pyrometry (above 1234.93 K). For vapor-pressure thermometry, the ITS-90 provides formulas for T_{90} in terms of the vapor pressure of the helium isotopes ^{3}He and ^{4}He. For the other methods, it assigns values of several fixed calibration temperatures. The fixed temperatures are achieved with the reproducible equilibrium systems listed in Table 2.2 on the next page.

2.6.3 Equilibrium systems for fixed temperatures

If two different phases of a pure substance coexist at a controlled, fixed pressure, the temperature has a fixed value. Eight of the fixed temperatures on the ITS-90 are obtained experimentally with coexisting solid and liquid phases at 1 bar (Table 2.2). These temperatures are the melting points at 1 bar.

Triple-point systems provide the most reproducible temperatures. The ITS-90 uses six of these systems (Table 2.2). A triple-point cell contains coexisting solid, liquid, and gas phases of a pure substance. Depending on the substance, both the temperature and pressure have definite fixed values. Fig. 2.7 on page 24 illustrates a triple-point cell for water capable of a reproducibility within 10^{-4} K. When ice, liquid water, and water vapor are in equilibrium in this cell, the cell is at the triple point of water (273.16 K).

2.6.4 Gas thermometry

Only the triple point of water has a defined value on the thermodynamic temperature scale. How are the values of other fixed temperatures of a scale such as the ITS-90 determined?

The fundamental method is gas thermometry using a **constant-volume gas thermometer**. This device consists of a bulb containing a gas and a means of adjusting the pressure

---

[7]McGlashan, M. L. *J. Chem. Thermodynamics* **1990**, *22*, 653–663.
[8]Gas thermometry is discussed in Sec. 2.6.4.

Section 2.6 Temperature

Table 2.2 Fixed temperatures of the International Temperature Scale of 1990

T_{90}/K	Equilibrium system for obtaining fixed temperature
13.8033	H_2 triple point
24.5561	Ne triple point
54.3584	O_2 triple point
83.8058	Ar triple point
234.3156	Hg triple point
273.16	H_2O triple point
302.9146	Ga melting point at 1 bar
429.7485	In melting point at 1 bar
505.078	Sn melting point at 1 bar
692.677	Zn melting point at 1 bar
933.473	Al melting point at 1 bar
1234.93	Ag melting point at 1 bar
1337.33	Au melting point at 1 bar
1357.77	Cu melting point at 1 bar

so as to bring the gas volume to a reproducible reference value (see Fig. 2.8 on the next page). The gas is usually helium, which has minimal deviations from ideal-gas behavior. A series of paired measurements is made, each pair as follows. First the gas is brought into thermal equilibrium with a system of known temperature T_1 (such as one of the systems listed in Table 2.2), the gas volume is adjusted to the reference value, and the gas pressure p_1 is obtained from the height of mercury Δh and the atmospheric pressure. Then the gas is brought into thermal equilibrium with the system whose temperature T_2 is to be measured, the volume is readjusted to the reference value, and the pressure p_2 is obtained. In both measurements of the pair, the amount of gas is the same and the volume is the same (except for a small change due to thermal expansion of the bulb); the temperatures and pressures are different.

If the gas exactly obeyed the ideal gas equation in both measurements, we would have $nR = p_1V_1/T_1 = p_2V_2/T_2$ or $T_2 = T_1(p_2V_2/p_1V_1)$. Since, however, the gas approaches ideal behavior only in the limit of low pressure, it is necessary to make a series of the paired measurements, pumping some of the gas from the bulb before each new pair so as to reduce the measured pressures. Thus, the operational equation for evaluating the unknown temperature is

$$T_2 = T_1 \lim_{p_1 \to 0} \frac{p_2 V_2}{p_1 V_1} \quad (2.6.2)$$
$$\text{(gas)}$$

(The ratio V_2/V_1 differs from unity only because of any change in the bulb volume when T changes.) The limiting value of p_2V_2/p_1V_1 may be obtained by plotting this quantity

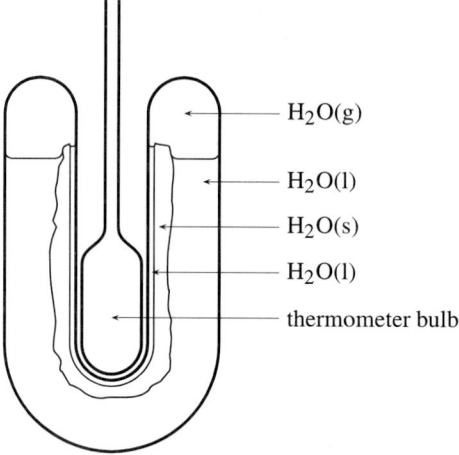

Figure 2.7 Cross-section of a water triple-point cell. The cell has cylindrical symmetry. Pure water of the same isotopic composition as H_2O in ocean water is distilled into the cell. The air is pumped out and the cell is sealed. A freezing mixture is placed in the inner well to cause a thin layer of ice to form next to the inner wall. The freezing mixture is removed, and some of the ice is allowed to melt to a film of very pure water between the ice and inner wall. The thermometer bulb is placed in the inner well as shown, together with ice water (not shown) for good thermal contact.

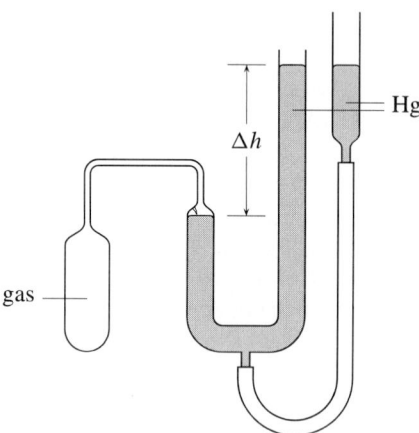

Figure 2.8 Simple version of a constant-volume gas thermometer.

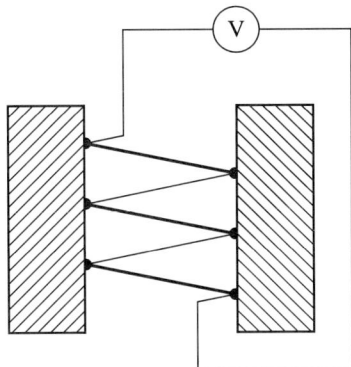

Figure 2.9 A thermopile used to measure the difference in the temperatures of two bodies.

against p_1 or another appropriate extrapolating function. Note that values of n and R are not needed.

2.6.5 Practical thermometers

Liquid-in-glass thermometers use indicating liquids whose volume change with temperature is much greater than that of the glass. A mercury-in-glass thermometer can be used in the range 234 K (the freezing point of mercury) to 600 K, and typically can be read to 0.01 K. A Beckmann thermometer covers a range of only a few kelvins but can be read to 0.001 K.

A *resistance thermometer* is included in a circuit that measures the electric resistance. Platinum resistance thermometers are widely used because of their high sensitivity (0.0001 K) and stability. Thermistors use metal oxides and can be made very small; they have greater sensitivity than platinum thermometers but are not as stable over time.

A *thermocouple* consists of wires of two dissimilar metals (e.g., constantan alloy and copper) connected in series at soldered or welded junctions. A many-junction thermocouple is called a *thermopile* (Fig. 2.9). When adjacent junctions are placed in thermal contact with bodies of different temperatures, an electric potential develops that is a function of the two temperatures.

Finally, two other temperature-measuring devices are the *quartz crystal thermometer*, incorporating a quartz crystal whose resonance frequency is temperature dependent, and *optical pyrometers*, which are useful above about 1300 K to measure the radiant intensity of a black body emitter.

PROBLEMS

2.1 Let X represent the quantity V^2 with dimensions length^6. Give a reason that X is or is not an extensive property. Give a reason that X is or is not an intensive property.

2.2 Calculate the *relative uncertainty* (the uncertainty divided by the value) for each of the measurement methods listed in Table 2.1 on page 18 using the typical values shown. For each of the five physical quantities listed, which measurement method has the least relative uncertainty?

Table 2.3 Pressure measurements in a constant-volume gas thermometer

p_1/Pa	p_2/Pa
13 490.00	18 426.58
13 490.00	18 426.43
13 490.00	18 426.50
74 410.00	101 638.87
74 410.00	101 638.73

2.3 L. A. Guildner and R. E. Edsinger (*J. Res. Natl. Bur. Stand. (U.S.)* **1973**, *77A*, 383–389) describe measurements of the steam point made using a constant-volume helium-filled gas thermometer at the U.S. National Bureau of Standards (now the National Institute of Science and Technology). The data in Table 2.3 are based on their results with an assumption that no correction is needed for the thermal expansion of the thermometer bulb. The values of p_1 were measured at the triple point of water; p_2 was measured at the steam point. Use these data to find the value of the steam point and its approximate uncertainty.

Chapter 3

THE FIRST LAW

In science a **law** is a statement or equation that concisely describes reproducible experimental observations. Classical thermodynamics is built on a foundation of three laws, none of which can be derived from principles that are more fundamental. This chapter explains some basic terminology of thermodynamics; discusses theoretical aspects of the first law; gives examples of spontaneous and reversible processes and the work that occurs in them; and introduces two extensive state functions, enthalpy and heat capacity.

3.1 TERMINOLOGY

3.1.1 The system, surroundings, and boundary

As explained in chap. 2, a thermodynamic **system** is any three-dimensional region of physical space on which we wish to focus our attention. Usually we consider only one system at a time and call it simply "the system." The rest of the physical universe constitutes the **surroundings** of the system. The **boundary** is the closed surface that encloses the system and separates it from the surroundings. The boundary may (and usually does) coincide with real physical surfaces: the interface between two phases, the inner or outer surface of the wall of a flask or other vessel, and so on. Alternatively, part or all of the boundary may be an imagined surface in space, unrelated to any physical structure. The size and shape of the system, as defined by its boundary, may change in time. In short, our choice of the region that constitutes the system is arbitrary—but it is essential that we know exactly what this choice is.

We usually think of the system as a part of the physical universe that we are able to influence only indirectly through its interaction with the surroundings, and the surroundings as the part of the universe that we are able to directly manipulate with various physical devices under our control. That is, we (the experimenters) are part of the surroundings, not the system.

For some purposes we may wish to treat the system as being divided into subsystems, or two or more systems as being combined into a supersystem.

If matter is transferred in either direction across the boundary, the system is **open**; otherwise it is **closed**. In an open system, matter may pass through a fixed boundary, or the boundary may move through fixed matter.

If the boundary allows heat transfer between the system and surroundings, the boundary is **diathermal**. An **adiabatic**[1] **boundary**, on the other hand, is a boundary that does not allow heat transfer. We can, in principle, construct an adiabatic boundary with a thermally insulated wall—a wall containing perfect thermal insulation and a perfect radiation shield.

An **isolated** system is one that is constrained to exchange no matter or energy with the surroundings at any portion of the boundary and cannot change its energy as a whole. A closed system with an adiabatic boundary, constrained to do no work and have no work done on it, is an isolated system.

> The constraints required to prevent work usually involve opposing forces between the system and surroundings. In that sense an isolated system may interact with the surroundings. For instance, a gas contained by rigid, thermally insulated walls is an isolated system; the gas exerts a force on each wall, and the wall exerts an equal and opposite force on the gas. An isolated system may also experience a constant external field, such as a gravitational field.

3.1.2 The state of the system

The thermodynamic **state** of a system[2] is an important and subtle concept. Here is the basic idea. At each instant the system is in some definite state that we may describe with information about properties we consider relevant. At any given instant the properties have values that define the state at that time. If any of these properties subsequently changes, the system has changed its state. If at a later time each property returns to its original value, then the system has returned to the original state.

Various factors determine what states of a system are physically possible. If a phase has an equation of state, property values must be consistent with it. The system may have certain built-in or externally imposed conditions or constraints that keep some properties from changing. For instance, a closed system has constant mass; a system with a rigid boundary has constant volume. We may know about other conditions that affect the properties during the time the system is under observation.

In most cases we can define the state of the system at any instant by specifying values of a certain minimum number of variable properties, the **independent variables**. The number of independent variables that is sufficient to describe the state depends on how complete for our purposes we require the description to be. Once we have selected a set of independent variables, consistent with the physical nature of the system and any constraints or conditions, we can treat all other variable properties that are affected by the state as **dependent variables**. All of the independent and dependent variables are called **state functions** (or state variables or state parameters) because their values at one instant depend only on the state of the system at that instant, not on the system's prior history.[3]

---

[1] Greek: *impassable*.

[2] Do not confuse the *state* of the system with the kind of *physical state* or state of aggregation of a phase discussed in Sec. 2.1.4. A *change* of state refers to a change in the state of the system, not necessarily to a phase transition.

[3] A property of a solid phase that we treat as a dependent state function can cease to be a state function when we stress the solid beyond its elastic limit because then hysteresis in the solid causes the property to depend on the

Table 3.1 Example of the description of the state of an aqueous sucrose solution (A = water, B = sucrose)

temperature	$T = 293.15\,\text{K}$
pressure	$p = 1.00\,\text{atm}$
amount of water	$n_A = 39.18\,\text{mol}$
amount of sucrose	$n_B = 1.375\,\text{mol}$
volume	$V = 1000\,\text{cm}^3$
mass	$m = 1176.5\,\text{g}$
density	$\rho = 1.1765\,\text{g cm}^{-3}$
mole fraction of sucrose	$x_B = 0.03390$
osmotic pressure	$\Pi = 58.2\,\text{bar}$
refractive index, sodium D line	$n_D = 1.400$

Usually we can choose as independent variables the temperature and pressure of each phase and the amount of each substance in each phase (Sec. 2.1.7). Consider the example in Table 3.1, which lists the values of 10 properties (state functions) of an aqueous sucrose solution in a particular state. The first four properties (T, p, n_A, n_B) are ones that we can vary independently, and their values suffice to define the state for most purposes. Experimental measurements will convince us that, whenever the system has the particular values listed for these four variables, each of the other properties necessarily has the one definite value listed; we cannot alter any of the other properties without changing one or more of the first four variables. Thus, if T, p, n_A, and n_B are taken as independent variables, the other listed properties are *dependent* variables. The other properties include one (V) that is determined by the equation of state; three (m, ρ, and x_B) that can be calculated from the independent variables and the equation of state; a solution property (Π) treated by thermodynamics (Sec. 9.4.4); and an optical property (n_D), which, while having no direct connection to thermodynamics, is nevertheless a state function. In addition to these independent variables, this system has innumerable others: isothermal compressibility, heat capacity at constant pressure, and so on.

We could make other choices of independent variables for the aqueous sucrose system. For instance, we could choose the set T, p, V, and x_B, or the set p, V, ρ, and x_B. The choice is arbitrary, but the *number* of independent variables remains four. (Note that we could not arbitrarily choose just any four variables; for instance p, V, m, and ρ are not independent of one another because of the relation $\rho = m/V$.)

If the system has imposed conditions, there will be fewer independent variables. Suppose we know the sucrose solution is in a closed system of known fixed mass and known fixed mole fraction composition; then there are only two independent variables and we could describe the state by the values of just T and p.

A system with more than one phase could in principle have $2 + s$ independent variables

---

phase's prior history (Sec. 2.1.9). The discussion in the rest of this section assumes the stress on any solid in the system is kept below the elastic limit.

for each phase, where s is the number of substances in the phase. There are various equilibria and other conditions that can reduce the number of independent variables needed for the system as a whole; for instance, each phase may have the same temperature and the same pressure, equilibrium may exist with respect to chemical reaction and transfer between phases (Sec. 3.1.3), and the system may be closed. (While these various conditions do not *have* to be present, the relations among T, p, V, and amounts described by the equation of state of a phase are *always* present.) On the other hand, additional independent variables are required if properties such as the shapes and surface areas of phases, which do not depend on the usual independent variables, are relevant to the description of the state.[4]

> We must be careful to choose a set of independent variables that defines the state without ambiguity. The set p and V might be a poor choice for a closed system of liquid *water* because the molar volume of water passes through a minimum as T is varied at constant p. Thus, the values $p = 1.000$ bar and $V = 18.016$ cm^3 would describe one mole of water at both 2 °C and 6 °C, so these values would not uniquely define the state. Better choices of independent variables in this case would be either T and p or T and V.

How may we describe the state of a system with nonuniform regions? In this case we may envision the regions to be divided into many small volume elements, each small enough to be essentially uniform but large enough to contain many molecules; we then describe the state by specifying values of independent variables for each volume element. If there is internal macroscopic motion (e.g., flow), the independent variables will have to include velocity components. Obviously, the quantity of information needed to describe a complicated state may be enormous.

> We can imagine situations in which classical thermodynamics would be completely incapable of describing the state. For instance, turbulent flow in a fluid or a shock wave in a gas may involve inhomogeneities all the way down to the molecular scale. Macroscopic variables would not suffice to define the states in these cases.

Whatever our choice of independent variables, all we need to know to be sure a system is in the same state at two different times is that *the value of each independent variable is the same at both times.*

3.1.3 Equilibrium states

An **equilibrium state** is a state that can exist in an isolated system (a closed system without heat or work) and that remains unchanged indefinitely as long as the system remains isolated. An equilibrium state of an isolated system has no natural tendency to change over time. If changes *do* occur in an isolated system, they continue until an equilibrium state is reached. Obviously a system with macroscopic internal motion, or one containing a fluid subjected to shear stress, whether or not isolated is *not* in an equilibrium state.

A system in an equilibrium state may have some or all of the following four kinds of internal equilibria:

---

[4] The important question of how many independent intensive variables a particular kind of system has is answered by the Gibbs phase rule, which is taken up in Secs. 6.1.7 and 10.1.

thermal equilibrium: Each phase has the same uniform temperature.

mechanical equilibrium: Each fluid phase has the same uniform pressure.

transfer equilibrium: There is equilibrium with respect to the transfer of each species from one phase to another.

reaction equilibrium: Every possible chemical reaction is at equilibrium.

> The meaning of reaction equilibrium in the context of an equilibrium state is that no perceptible reaction occurs during the period we keep the isolated system under observation. For instance, a system containing a homogeneous mixture of gaseous H_2 and O_2 at 25 °C and 1 bar is in an equilibrium state on a time scale of hours or days; but if a measurable amount of H_2O forms over a longer period, the state is not an equilibrium state on this longer time scale. This consideration of time scale is similar to the one we have to apply to the persistence of deformation in distinguishing a solid from a fluid (Sec. 2.1.4).

A *homogeneous* system (one phase) must by definition have a uniform temperature and pressure, and so has thermal and mechanical equilibrium. It is in an equilibrium state if it also has reaction equilibrium.

A *heterogeneous* system is in an equilibrium state if each of the four kinds of internal equilibrium is present.

A state lacking one or more kinds of internal equilibrium may still be an equilibrium state if the effects of an internal equilibrium are replaced by a constraint or the influence of an external field. Here are five examples.

1. An isolated system with an internal adiabatic partition separating two phases can have an equilibrium state lacking thermal equilibrium. The adiabatic partition allows the two phases to remain indefinitely at different temperatures.

2. An experimental system used to measure osmotic pressure (Fig. 9.2 on page 312) has a semipermeable membrane separating a liquid solution phase and a pure solvent phase. The membrane prevents the transfer of solute from the solution to the pure solvent and allows the liquid phases to have different pressures. In its equilibrium state, this system lacks both transfer equilibrium with respect to solute transfer and mechanical equilibrium.

3. In the equilibrium state of an isolated galvanic cell (Sec. 3.8.2), the separation of the reactants and products effectively prevents the cell reaction from coming to reaction equilibrium.

4. A system containing mixed reactants of a reaction might have an equilibrium state without reaction equilibrium if we withhold a catalyst or initiator or introduce a hypothetical "anticatalyst" that prevents reaction.

5. An example of a system influenced by an external field is a tall column of gas in a gravitational field. In the equilibrium state, the pressure decreases continuously with increasing altitude (Sec. 6.1.4).

Keep in mind that regardless of the presence or absence of constraints or external fields, the essential feature of an equilibrium state is that *the properties of the isolated system in this state do not change over time.*

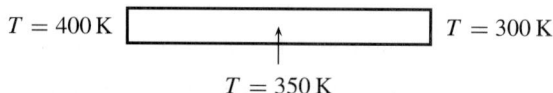

Figure 3.1 Steady state in a metal rod with heat conduction.

Strictly speaking, a system can be in an equilibrium state only if there is no exchange of energy with the surroundings. However, at a given instant the state of a system exchanging heat or work can very closely approximate an equilibrium state if the rate of energy transfer across the boundary is very slow. Energy transfer at a finite rate causes nonuniform temperature and pressure and prevents thermal and mechanical equilibrium. The sequence of equilibrium states approached in the limit of infinitely slow changes is called a *reversible* process, a hypothetical kind of process to be discussed at length beginning in Sec. 3.4.

It is important not to confuse an equilibrium state with a **steady state**, a state that is constant over time even though the system exchanges matter or energy with the surroundings. The heat-conducting metal rod shown in Fig. 3.1 is a system in such a steady state. Each end of the rod is in thermal contact with a **heat reservoir**, a body or system whose temperature remains constant and uniform when heat is transferred to or from it.[5] The two heat reservoirs have different temperatures, causing a temperature gradient to form along the length of the rod and heat to be transferred from the warmer reservoir to the rod and from the rod to the cooler reservoir. Although the properties of the rod remain constant, the system is clearly not in an equilibrium state because the temperature gradient will disappear if we isolate the system. Another kind of steady-state system is a stirred reaction system that has reactants continuously supplied and products continuously withdrawn.

3.1.4 Processes and paths

A **process** is a change in the state of the system over time, starting with a definite initial state and ending with a definite final state. The process is defined by a **path**, which is the sequence of consecutive states through which the system passes including the initial state, the intermediate states, and the final state. We could describe a path by a curve in an N-dimensional space in which each axis represents one of the N independent variables (Fig. 3.2 on the next page). The process has a direction along the path. The complete specification of a process includes the path, the time rate of progress along the path, the quantities of heat transferred across each diathermal portion of the boundary, any external fields acting on the system, and forces exerted on the system by the surroundings (including those responsible for work).

Expansion is a process in which the system volume increases; in **compression**, the volume decreases.

An **isothermal** process is one in which the temperature of the system remains constant

---

[5]A heat reservoir can be a body that is so large its temperature changes only imperceptibly, a thermostat with a heater and cooler to control the temperature, or a system of two coexisting phases of a pure substance.

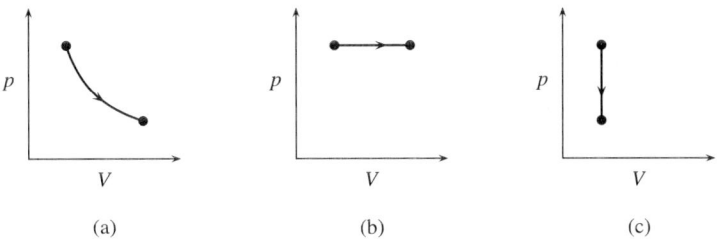

Figure 3.2 Paths of three processes of a closed ideal-gas system, with p and V as the independent variables: (a) isothermal expansion, (b) isobaric expansion, (c) isochoric pressure reduction.

and uniform. An **isobaric** or newtermisopiestic process refers to constant uniform pressure, and an **isochoric** process to constant volume.

An **adiabatic** process is one in which no heat is transferred across any portion of the boundary. We may ensure that a process is adiabatic either by using an adiabatic boundary or, if the boundary is diathermal, by continuously adjusting the external temperature to eliminate a temperature gradient at the boundary.

Recall that a state function is a property whose value at each instant depends only on the state of the system at that instant. The finite change of a state function X in a process is written ΔX. The symbol ΔX always has the meaning $X_2 - X_1$, where X_1 is the value in the initial state and X_2 is the value in the final state. Therefore, the change ΔX depends only on the values of X_1 and X_2. *The change of a state function during a process depends only on the initial and final states of the system, not on the path of the process.*

We write dX for an infinitesimal change of the state function X. The mathematical operation of summing an infinite number of infinitesimal changes is integration, and the sum is an integral (see the brief calculus review in Appendix E). If we take the sum of the changes in X along a path, we obtain a definite integral that is equal to ΔX:

$$\int_{X_1}^{X_2} dX = X_2 - X_1 = \Delta X$$

If dX obeys this relation—that is, if its integral for given limits has the same value regardless of the path—it is called an **exact differential**.

A **cyclic process** is a process in which the path returns the system to the initial state. In this case the integral of dX is written with a cyclic integral sign: $\oint dX$. Since a state function X has the same initial and final values in a cyclic process, X_2 is equal to X_1 and the cyclic integral of dX is zero:

$$\oint dX = 0$$

Heat (q) and work (w) are examples of quantities that are *not* state functions, and it would be incorrect to write "Δq" or "Δw." Instead, q and w depend on the path (except in special circumstances such as an adiabatic process) and are called **path functions**.

We use the same notation for infinitesimal quantities of a path function as for infinitesimal changes of a state function (e.g., dq and dw). However, the sum of many infinitesimal quantities of a path function is not the difference of two values of the path function; instead, the sum is the *net* quantity:

$$\int dq = q \qquad \int dw = w$$

The infinitesimal quantities dq and dw, because their integrals depend on the path, are **inexact differentials**.

It is important to keep in mind a fundamental distinction between a state function (such as temperature or volume) and a path function (such as heat or work): *The value of a state function refers to one instant of time; the value of a path function refers to an interval of time.*

> The distinction between a state function and a path function in thermodynamics is analogous to the distinction between altitude and trail length in mountain hiking. The hiker at each instant is at a definite altitude. During a climb from the base of a mountain to the summit, the hiker's change of altitude is independent of the trail used; but the distance traveled during the climb can be different for different trails.

3.2 ENERGY

Energy is sometimes described as "the capacity to do work." This is hardly a complete (or even correct) definition and raises the question of what is meant by "work."

Appendix G gives a detailed analysis of energy and work from the viewpoint of the classical mechanics of microscopic particles. The analysis leads to the following conclusions regarding different kinds of energy. The energy of the system at any instant, E_{sys}, is the sum of the kinetic energy and the potential energy of the system. The kinetic energy is the sum of the kinetic energies of the constituent particles. There are two kinds of contributions to the potential energy: those that are functions of interparticle distances, and those that are functions of the positions of particles in a conservative time-independent external field, such as a gravitational or electric field. The energy of the surroundings, like that of the system, is also a sum of kinetic and potential energies. The various potential energies are defined in such a way that a certain quantity—the sum of the energy of the system, the energy of the surroundings, and any potential energy shared by both system and surroundings—is constant in time.

3.2.1 The energy of the system

It is the energy of the system, E_{sys}, that remains constant in an isolated system. From the macroscopic viewpoint taken by thermodynamics, we may consider E_{sys} to be the sum of three contributions:

$$E_{sys} = E_k + E_p + U \qquad (3.2.1)$$

where E_k is the kinetic energy of translation and rotation of the system as a whole, E_p is the potential energy of the system as a whole due to its position in a conservative time-independent external field, and U is the **internal energy**. The manner in which we divide E_{sys} into these three contributions is to some extent arbitrary. We can think of the coordinates that specify the position of the system as a whole (and whose rates of change determine E_k) as *external* coordinates; these coordinates are irrelevant to the state of the system, and so E_k and E_p are not state functions. The internal energy U, however, *is* a state function, and its value depends on the state of the system. We consider state functions to be *internal* coordinates.

As a simple illustration of the difference between the three contributions to E_{sys}, consider a fixed mass m of water in a beaker. Suppose that initially the beaker is at rest on a laboratory bench. The system (the water) is in an equilibrium state defined by two independent variables, which we may choose to be the temperature T and the pressure p. If we slide the beaker along the bench with velocity v, E_{sys} increases by $(1/2)mv^2$ but T and p do not change. Since this is a kinetic energy increment that does not affect the state of the system, it is the contribution E_k that increases.[6] If we lift the beaker a distance Δh above the bench, the potential energy of the water in the earth's gravitational field increases by $mg\Delta h$,[7] again without affecting the system state; thus this potential energy change is an increase in the contribution E_p. If, however, something happens to change either T or p, the concurrent change in E_{sys} is a change in the internal energy U—the part of E_{sys} that depends on the state of the system.

> The total energy of a body of mass m when it is at rest is given by the Einstein relation $E = mc_0^2$, where c_0 is the velocity of light in a vacuum. In principle, then, we could calculate the internal energy of a system at rest from its mass, and we could determine ΔU for a process from the change in mass. In practice, an absolute value of U calculated from a measured mass has too much uncertainty to be of any practical use. For example, the typical uncertainty of mass measured with a microbalance, about $0.1\,\mu g$ (Table 2.1), would introduce the enormous uncertainty in energy of about 10^{10} joules. Only values of the *change* ΔU are useful, and we do not calculate these values from Δm because the change in mass during an ordinary chemical process is too small to be detected.

3.2.2 The internal energy of an ideal gas

The concept of an ideal gas is used in many places in the development of thermodynamics. For examples to follow, we need the following definition: An ideal gas is a gas

1. whose equation of state is the ideal gas equation, $pV = nRT$; and

2. whose internal energy in a closed system is a function only of temperature.[8]

---

[6] Systems in motion, including the modifications needed to take special relativity into account at high velocities, are discussed by Guggenheim, Chapter 14 (reference in Appendix I).

[7] This formula assumes the intensity of the gravitational field does not change appreciably in distance Δh.

[8] A gas with this second property is sometimes called a "perfect gas." In Sec. 5.4, we show that if a gas has the first property, it must also have the second.

On the molecular level, a gas with negligible intermolecular interactions fulfills both of these requirements. Kinetic-molecular theory predicts that a gas containing noninteracting molecules obeys the ideal gas equation. If intermolecular forces (the only forces that depend on intermolecular distance) are negligible, the internal energy is simply the sum of the energies of the individual molecules. These energies are independent of V but depend on T. We can make any real gas approach ideal-gas behavior by expanding it isothermally. As V_m becomes large (and p becomes small), the average distance between molecules becomes large and intermolecular forces become negligible.

Since U for a fixed amount of an ideal gas depends only on T, an infinitesimal change of U depends only on T and its infinitesimal change: $dU = f(T)\,dT$, where $f(T)$ is a function only of T. This function is the heat capacity at constant volume C_V of the ideal gas discussed in Sec. 3.12. Accordingly, we have the relation

$$dU = C_V\,dT \tag{3.2.2}$$

(closed system, ideal gas)

3.3 HEAT, WORK, AND THE FIRST LAW

The **first law of thermodynamics** may be stated as follows.

> In a closed system:
> $dU = dq + dw$
> $\Delta U = q + w$
> where U is a state function, the internal energy of the system;
> q is heat transferred across the boundary; and
> w is thermodynamic work transferred across the boundary.

(3.3.1)

The first law in effect sets up a balance sheet for the internal energy of a closed system by equating the internal energy change during a process to the net quantity of energy transferred across the boundary by means of heat and work. (If the system is open, energy can also be brought across the boundary by the transport of matter.) The equation $dU = dq + dw$ is the *differential* form of the first law, and $\Delta U = q + w$ is the *integrated* form.

We define the heat and work appearing in the first law in a general way as follows.

Heat refers to the transfer of energy across the boundary on account of a temperature difference[9] or temperature gradient at the boundary.

Work refers to the transfer of energy across the boundary on account of the displacement of an internal coordinate affecting a part of the system on which the surroundings exert a force.

---

[9]We may treat the temperature as being different on either side of the boundary, but the temperature is not actually discontinuous there. Instead there is a thin zone with a temperature gradient. We can ignore the presence of this zone for most purposes.

An infinitesimal quantity of heat is written dq,[10] and a finite quantity is written q (Sec. 3.1.4). The corresponding quantities of work are dw and w. Any of these quantities is *positive* if the effect is to *increase* the internal energy, and *negative* if the effect is to *decrease* the internal energy. Thus, positive heat is energy entering the system, and negative heat is energy leaving the system. Positive work is work done by the surroundings on the system, and negative work is work done by the system on the surroundings.

Energy transfer in the form of heat may occur by conduction, convection, or radiation. We can reduce conduction with good thermal insulation, we can eliminate conduction and convection with a vacuum gap, and we can minimize radiation with highly reflective surfaces at both sides of the vacuum gap. However, the only way to completely prevent heat is to arrange conditions in the surroundings so there is no temperature difference or temperature gradient at any part of the boundary. Under these conditions, a process is adiabatic, and any energy transfer in a closed system is in the form of work.

3.3.1 The concept of thermodynamic work

Recall that the system's total energy includes the kinetic and potential energies of the system as a whole and the internal energy: $E_{sys} = E_k + E_p + U$ (Eq. 3.2.1 on page 34). Changes in E_k and E_p during a process do not affect the state of the system. A change in the sum of E_k and E_p can only occur through work performed on or by the system. The total work w_{sys} performed on the system includes the changes in E_k and E_p as well as the thermodynamic work w, which affects the state:

$$dw_{sys} = dE_k + dE_p + dw$$

The terms dE_k and dE_p are for work done on the system involving the displacement of *external* coordinates (those giving the position of the system as a whole or of its center of gravity), whereas dw is work involving the displacement of *internal* coordinates—state functions whose change affects the state of the system.

The detailed analysis in Appendix G shows that the infinitesimal quantity of work dw_{sys} done on a closed system by the surroundings can be written[11]

$$dw_{sys} = \sum_i F_i \, dx_i \qquad (3.3.2)$$
(closed system)

Each term in the sum is associated with a position coordinate x_i. Then dx_i is an infinitesimal displacement in the system at the point a force is exerted by the surroundings, and F_i is the component of the force in the $+x_i$ direction. If the displacement is in the same direction as the applied force, $F_i \, dx_i$ is positive; if the displacement is in the opposite direction, $F_i \, dx_i$ is negative. Appendix G shows moreover that any force exerted by a conservative time-independent *external field* should be omitted from the sum in Eq. 3.3.2; this is because

---

[10] It should be understood that dq may refer to heat at an infinitesimal surface element of the boundary. The integral $\int dq = q$ for a process is obtained by integrating over the total boundary surface and the entire path of the process.

[11] This is Eq. G.17 with a change of notation.

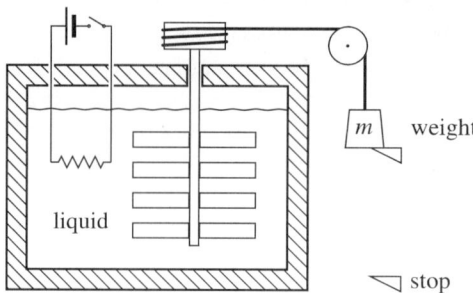

Figure 3.3 System containing a Joule paddle wheel and an electrical resistor immersed in a liquid.

displacement with this kind of force causes a change of potential energy in the field that is equal and opposite in sign to the kinetic energy change, with no net change in E_{sys}.

An alternative to using Eq. 3.3.2 for evaluating w_{sys} is to imagine that the only effect of the work on the *surroundings* is to cause a weight to rise or sink in the surroundings; then w_{sys} is equal in magnitude and opposite in sign to the change in the weight's energy $(m \Delta v^2/2 + mg \Delta h)$. This viewpoint can be helpful for seeing whether work occurs and for deciding on its sign, but of course cannot be used to determine its *value* if the actual surroundings include no such weight.

Since the thermodynamic work dw appearing in the first law involves displacements of internal coordinates only, we can write

$$dw = \sum_{\text{internal}} F_i \, dx_i \qquad (3.3.3)$$

where the sum includes displacements only of *internal* coordinates. The thermodynamic work of a given system can frequently be written in the more general form

$$dw = \sum Y_i \, dy_i$$

where Y_i is a *generalized* force called a **work coefficient** and y_i is a generalized internal coordinate called a **work coordinate**. The work coefficients and work coordinates are independent state functions that are not necessarily forces and position coordinates. For example, we see in Sec. 3.6.2 that expansion work is given by $dw = -p \, dV$; in this case, the work coefficient is $-p$ and the work coordinate is V.

Hereafter in this book, thermodynamic work is called simply *work*.

3.3.2 Heat and work as path functions

Consider the apparatus depicted in Fig. 3.3. The *system* consists of the liquid plus the immersed parts shown: stirring paddles attached to a shaft (a Joule paddle wheel) and an electrical resistor attached to wires. Assume the pressure and the amount of liquid are constant. We may describe the equilibrium states of this system by two independent variables: the temperature T and the angular position of the shaft. We shall suppose the initial state is described by $T_1 = 300.0$ K and a certain angular position of the shaft. We can carry out three experiments, each having the same initial state and the same final state.

For the first experiment, we surround the system with thermal insulation and release the external weight, which is linked mechanically to the paddle wheel. The resulting paddle-wheel rotation causes turbulent churning of the liquid and an increase in its temperature until the weight hits a stop and comes to rest. We can calculate the work done on the system from the potential energy lost by the weight, with a correction for the kinetic energy gained by the weight just before it hits the stop. We wait until the liquid comes to rest and the system comes to thermal equilibrium, then measure the final temperature. Assume the final temperature is $T_2 = 300.1$ K and the shaft comes to rest in the same angular position as the initial position.

In the second experiment, we start with the system in the same initial state and again surround it with thermal insulation. This time, instead of releasing the weight we complete the electrical circuit to the resistor and allow the same quantity of electrical work to be done on the system as the mechanical work done in the first experiment. We discover the final temperature is exactly the same as in the first experiment (300.1 K). The process and path are different from the first experiment, but the work and the initial and final states are the same.

In the third experiment, we return the system to the initial state, remove the thermal insulation, and place the system in thermal contact with a heat reservoir of temperature 300.1 K. Energy can now enter the system in the form of heat, and does so on account of the temperature gradient at the boundary. By a substitution of heat for mechanical or electrical work, the system changes to the same final state as in the first two experiments.

Although the paths in the three experiments are entirely different, the change in state is the same. In fact, a person who observes only the initial and final states and has no knowledge of changes in the surroundings will be ignorant of the path. Did the paddle wheel turn? Did an electric current pass through the resistor? How much energy was transferred as work and how much as heat? The observer cannot know. In other words, *heat and work are not state functions* because neither considered individually is uniquely related to properties of the initial and final states. Only the *sum* of heat and work is related to the initial and final states; this sum is the change in the state function U, as expressed by the integrated form of the first law, $\Delta U = q + w$.

It follows from this discussion that neither heat nor work are quantities possessed by the system. A system at a given instant does not *have* or *contain* a particular quantity of heat or a particular quantity of work. Instead, heat and work are always associated with a *process*. They are *path* functions.

When we speak of heating a body, we mean that its temperature increases. Here are three statements that have similar meanings but that could be misleading:

"Heat is transferred to the system."
"Heat flows from a warmer body to a cooler body."
"To remove heat from a hot body, place it in cold water."

Statements such as these may give the false impression that heat is like a substance that retains its identity as it moves from one body to another.[12] Actually, heat and work do not exist once a process is completed. Nevertheless, the wording of statements such as these is imbedded in our everyday language, and no harm is done if we interpret them correctly. This book, for conciseness, often refers to "heat transfer" and "heat flow," instead of using the technically more correct phrase "energy transfer in the form of heat."

It is sometimes useful to speak of **thermal energy**, the energy possessed by matter on account of its temperature. Thermal energy has no exact definition, but roughly speaking it is the average kinetic energy of the translational, vibrational, and rotational motions of microscopic particles. This average kinetic energy increases when the temperature increases. Thermal energy and heat are not directly related, as demonstrated by the experiments described earlier in which the increase of the system's thermal energy in the first experiment was caused by mechanical work, in the second by electrical work, and in the third by heat. Furthermore, heat does not always cause a change of thermal energy; for example, when heat is slowly transferred to an equilibrium system of coexisting ice and water, the ice melts *without* a change in temperature.

3.4 SPONTANEOUS AND REVERSIBLE PROCESSES

A **spontaneous process** is a process that can actually occur in a finite length of time under the existing conditions. Any change in the state of a system that we observe experimentally is a spontaneous process.

The concept of a **reversible process** is important for defining certain quantities, notably work and changes of entropy, in terms of changes of state functions. This concept is not easy to describe or grasp, one reason being that a reversible process is not a process that a real system ever actually undergoes. However, we can make certain kinds of spontaneous processes approach reversible processes as closely as we wish by carrying out the changes sufficiently slowly. Thus, a reversible process is an *idealized* process that may be approached by a spontaneous process in the *limit* of infinite slowness.

A reversible process has the following characteristics:

1. It is an imaginary process in which the system passes through a continuous sequence of equilibrium states;[13] that is, the state at each instant is one that in an isolated system would persist with no tendency to change.

2. This sequence of equilibrium states can be approximated, as closely as desired, by the states of a spontaneous process carried out sufficiently slowly. The reverse sequence of equilibrium states can also be approximated, as closely as desired, by the states of another spontaneous process carried out sufficiently slowly.[14]

---

[12] The concept of heat as a substance is the essence of the caloric theory, which was finally disproved by Joule's experiments on the mechanical equivalent of heat in the 1840s.

[13] A process with this property is sometimes called a *quasistatic* process.

[14] This requirement prevents any process with hysteresis, such as the stretching of a metal wire beyond its elastic limit, from being a reversible process.

3. A spontaneous process that approaches a reversible process in the limit of infinite slowness must be a process with nonzero values of heat, work, or both—the system cannot be an isolated one. It must be possible to use conditions in the surroundings to control the rate at which energy is transferred across the boundary as heat and work, and thus to make the process go as slowly as desired.

4. When a reversible process is reversed, the magnitudes of the heat and work are unchanged and their signs are reversed. Thus, heat transferred in one direction across the boundary during a reversible process is fully recovered in the reverse process as heat transferred in the opposite direction; likewise work is fully recovered in the reverse process.[15]

We must imagine the reversible process to proceed at a finite rate or there would be no change of state over time. The precise rate of the change, however, is not important. Imagine a gas whose volume, temperature, and pressure are changing at some finite rate while the temperature and pressure magically stay perfectly uniform throughout the system. During these imaginary changes, there is no temperature or pressure gradient—no physical "driving force" tending to make the changes occur in a particular direction. This imaginary process is a reversible process—one whose states of uniform temperature and pressure are approached in a real process as the rate of change of the real process becomes slower and slower.

The following five sections, 3.5–3.9, describe some examples of spontaneous processes, give formulas for work, and illustrate the concept of a reversible limit in situations where such a limit exists.

3.5 HEAT TRANSFER

Consider the spontaneous heating of a solid body depicted schematically in Fig. 3.4(a) on the next page. The *system* is the body, which has a diathermal boundary and is immersed in a well-stirred water bath. Assume we can control the temperature T_{ext} of the water bath (the subscript "ext" stands for *external*). Suppose the bath and the body are initially equilibrated at temperature T_1, and we wish to raise the temperature of the body to a higher final temperature T_2. One way to do this is to rapidly increase the bath temperature to T_2 and keep it there. It takes time for the interior of the body to reach this higher temperature, so a temperature *gradient* is established within the body. The gradient at the boundary causes energy in the form of heat to enter the system; this of course is a spontaneous process because heat flows spontaneously from a higher to a lower temperature. Eventually the temperature in the body becomes uniform and equal to T_2.

The initial and final states of this process are equilibrium states. The intermediate states have temperature gradients, and no one single temperature characterizes these states. If we were to isolate the system with an adiabatic boundary while it is in an intermediate state, the

---

[15] This characteristic of a reversible process is sometimes described by saying that it is possible to restore both the system and the surroundings to their original states, with no further changes anywhere.

Figure 3.4 Sequences of states (left to right) for three processes of a solid body. Darker shading indicates higher temperature. (a) Spontaneous heating. (b) Reversible heating. (c) Spontaneous cooling.

temperature gradient would disappear and the state would change. Thus, the intermediate states are *not* equilibrium states.

To make the intermediate states more nearly uniform in temperature, we can warm the bath at a slower rate. The more slowly we increase T_{ext}, the smaller is the temperature gradient in the solid. The sequence of states approached as we use progressively slower warming rates is indicated in Fig. 3.4(b). In each state of this limiting sequence, the temperature is perfectly uniform throughout the body and is equal to T_{ext}; that is, each state has thermal equilibrium both internally and with respect to the surroundings. A single temperature now suffices to define each state. Each state is an *equilibrium* state because it would have no tendency to change if we isolated the system with an adiabatic boundary. This limiting sequence of states is a *reversible* heating process.

The reverse of the reversible heating process is a cooling process in which the temperature again is uniform in each state. The sequence of states of this reverse process is the limit of the spontaneous cooling process depicted in Fig. 3.4(c), as we reduce T_{ext} more and more slowly.

In any real heating or cooling process occurring at a finite rate, the body's temperature could not remain perfectly uniform. However, if we raise the bath temperature very slowly the temperature in all parts of the system will be very close to that of the bath. At any point in this very slow heating process, it would then take only a small decrease in the bath temperature to begin a slow *cooling* process; that is, the practically reversible heating process would be reversed. The important thing to notice about the intermediate states of the spontaneous cooling process shown in Fig. 3.4(c) is that none resembles an intermediate state of the spontaneous heating process shown in Fig. 3.4(a); the temperature gradients are in opposite directions. Only in the reversible limits of the heating and cooling processes are the intermediate states the same, except for their order of occurrence.

Now consider a different kind of system, one consisting of two coexisting phases of a

Section 3.6 Expansion and Compression

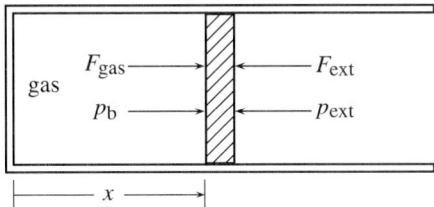

Figure 3.5 Forces and pressures in a cylinder-and-piston apparatus.

pure substance. At a given pressure, this kind of system can be in transfer equilibrium at only one temperature. The transfer of heat into or out of the system causes a phase transition (Sec. 2.1.5). If the external temperature is slightly greater than the equilibrium temperature, heat is transferred into the system and the substance is transferred from one phase to the other; a slightly lower external temperature causes heat transfer out of the system and transfer of the substance in the opposite direction. The closer the external temperature is to the equilibrium temperature, the smaller are the temperature gradients and the closer is the system to an equilibrium state. In the limit as the temperature difference approaches zero, the system passes through a sequence of equilibrium states in which the temperature is uniform and constant, heat is transferred into or out of the system, and the substance is transferred from one phase to the other. This idealized process is an *equilibrium* phase transition, and it is a reversible process.

3.6 EXPANSION AND COMPRESSION

As a model for the volume change of a fluid, consider the arrangement shown in Fig. 3.5 in which gas is confined in a cylinder by a piston. The *system* is the gas; the piston and cylinder walls are in the *surroundings*. The piston position is given by the internal coordinate x. Movement of the piston to the right, in the $+x$ direction, expands the gas; movement to the left, in the $-x$ direction, compresses it.

We do not need to specify at this point whether the boundary is diathermal or adiabatic.

3.6.1 Conditions for spontaneity and equilibrium

The piston is a body with finite mass. Three forces can act on it: the force F_{gas} exerted in the $+x$ direction by the gas; an external force F_{ext} in the $-x$ direction, which we can control in the surroundings; and a frictional force F_{fric} in the direction opposite to the piston's velocity when the piston moves. (The friction occurs at the seal between the edge of the piston and the cylinder wall.) Suppose p_b is the average gas pressure in the cylinder *at the piston*; that is to say, at the moving system boundary (the subscript of p_b stands for *boundary*). Then the force exerted by the gas on the piston is given by

$$F_{\text{gas}} = p_b A_s \qquad (3.6.1)$$

where A_s is the cross-section area of the cylinder.

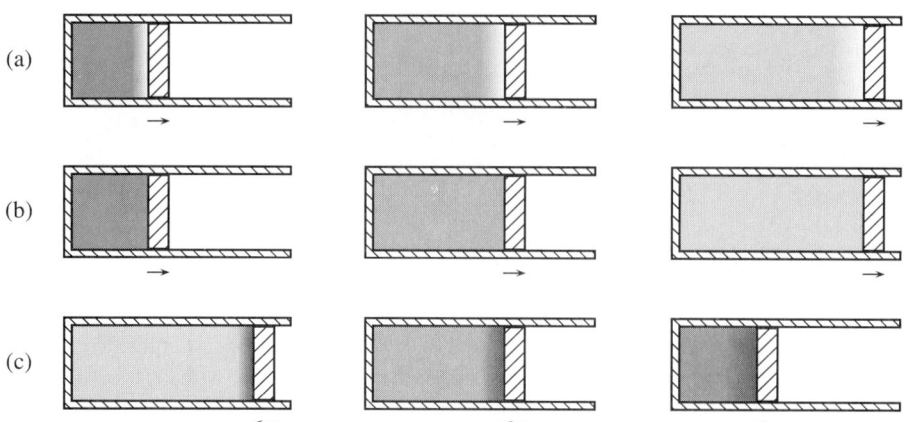

Figure 3.6 Sequences of states (left to right) for three processes of a gas in a cylinder-and-piston apparatus. Darker shading indicates higher pressure. (a) Spontaneous expansion. (b) Reversible expansion. (c) Spontaneous compression.

The net force F_{net} acting on the piston in the $+x$ direction is given by

$$F_{\text{net}} = F_{\text{gas}} - F_{\text{ext}} + F_{\text{fric}} \tag{3.6.2}$$

Here, F_{gas} and F_{ext} are taken as positive; F_{fric} is negative if the piston moves to the right, positive if the piston moves to the left, and zero if the piston is stationary.

Suppose that initially the system (the gas) is in an equilibrium state of uniform and constant temperature T and pressure p and the piston is stationary, so that F_{fric} is zero. According to Newton's second law of motion, the net force F_{net} on the piston is also zero since otherwise the piston would accelerate. Then, from Eqs. 3.6.1 and 3.6.2, F_{ext} is equal to pA_s.

By reducing F_{ext} from its initial value, we cause a spontaneous expansion to begin. As the piston moves to the right at a finite rate, the gas just behind it flows to the right and becomes attenuated, making p_b less than the pressure further from the piston; see Fig. 3.6(a). The pressure and temperature are nonuniform, and we cannot describe intermediate states of this spontaneous process with single values of p and T. These states are not equilibrium states since the pressure and temperature gradients would disappear in an isolated system.

The more slowly we allow the piston to move, the more nearly uniform is the gas. Each state of the sequence in the limit of infinite slowness, Fig. 3.6(b), is an equilibrium state in which the gas is perfectly uniform in both pressure and temperature.

By *increasing* F_{ext} instead of decreasing it, we cause a spontaneous compression. A pressure gradient develops in the opposite direction from that in expansion, as shown in Fig. 3.6(c). The states approached in the limit as we carry out the compression more and more slowly are equilibrium states, occurring in the reverse sequence from that shown in Fig. 3.6(b). The sequence of equilibrium states, taken in either direction, is a *reversible* process.

3.6.2 Expansion work

Now let us consider the work involved in expansion and compression of the gas shown in Fig. 3.5 on page 43. The general formula for an infinitesimal quantity of work is $dw = \sum F_i \, dx_i$ (Eq. 3.3.3), where dx_i is the displacement of an internal position coordinate and F_i is the component in the $+x$ direction of a force exerted by the surroundings on the system at the point where x_i changes. The appropriate internal coordinate in this process is the piston position x. The gas exerts a force F_{gas} on the piston directed in the $+x$ direction. Newton's third law of action and reaction tells us the piston exerts an opposing force of equal magnitude on the gas. Thus, the component in the $+x$ direction of the force exerted by the surroundings on the system is $-F_{\text{gas}}$. When the piston moves an infinitesimal distance dx, the resulting infinitesimal quantity of work from Eq. 3.3.3 is given by the relation

$$dw = -F_{\text{gas}} \, dx \tag{3.6.3}$$

which, with the substitution $F_{\text{gas}} = p_b A_s$, becomes

$$dw = -p_b A_s \, dx \tag{3.6.4}$$

During expansion (positive dx), dw is negative and the system does work on the surroundings. During compression (negative dx), dw is positive and the surroundings do work on the system.

The system volume is given by $V = A_s x$ so that an infinitesimal change dx changes the volume by $dV = A_s \, dx$. The infinitesimal quantity of work for an infinitesimal volume change is then given by

$$dw = -p_b \, dV \tag{3.6.5}$$
(expansion work, closed system)

and the finite work for a finite volume change, found by integrating from the initial to the final volume, is

$$w = -\int_{V_1}^{V_2} p_b \, dV \tag{3.6.6}$$
(expansion work, closed system)

Equations 3.6.5 and 3.6.6 are valid whether or not the gas is uniform or the process is reversible (provided p_b is the average pressure at the piston), and they apply to both expansion ($dV > 0$) and compression ($dV < 0$). We use the term "expansion work" in this book to include the work of both expansion and compression. Both equations are also valid for expansion and compression of a *liquid*; that is, for any fluid.

> When carrying out dimensional analysis, you will find it helpful to remember that the product of two quantities with dimensions of pressure and volume (such as $p_b \, dV$) has dimensions of energy and $1 \, \text{Pa} \, \text{m}^3$ is equal to $1 \, \text{J}$.

If the piston motion is sufficiently slow for the gas to be effectively uniform (an essentially reversible process), with practically the same pressure p throughout, Eq. 3.6.5 becomes

$$\mathrm{d}w = -p\,\mathrm{d}V \qquad (3.6.7)$$
(reversible expansion work, closed system)

and Eq. 3.6.6 becomes

$$w = -\int_{V_1}^{V_2} p\,\mathrm{d}V \qquad (3.6.8)$$
(reversible expansion work, closed system)

The appearance of the symbol p in these equations implies that the system has at any given instant a single uniform pressure. As a general rule, *an equation containing the symbol of an intensive property not assigned to a specific phase is valid only if that property is uniform throughout the system*, and this will not be explicitly indicated as a condition of validity.

Suppose expansion occurs in a system that is an isotropic fluid of arbitrary shape or an isotropic fluid with solid phases immersed in it. We then must apply the relation $\mathrm{d}w = -F_{\text{fluid}}\,\mathrm{d}x$ (analogous to $\mathrm{d}w = -F_{\text{gas}}\,\mathrm{d}x$) to each surface element of the boundary. If the fluid pressure is uniform, we obtain the same equation as before for reversible expansion work: $\mathrm{d}w = -p\,\mathrm{d}V$.[16] This equation is also valid for a boundary that coincides with the outer surface of a solid immersed in an isotropic fluid of pressure p in the surroundings. The equations $\mathrm{d}w = -p\,\mathrm{d}V$ and $w = \int p\,\mathrm{d}V$, then, are formulas for expansion work of *any* closed system that has an isotropic fluid phase of uniform pressure p on one or both sides of each surface element of the boundary.

Equations 3.6.3, 3.6.5, and 3.6.7 give three different expressions for the infinitesimal quantity of work during an infinitesimal displacement of the piston shown in Fig. 3.5: $\mathrm{d}w = -F_{\text{gas}}\,\mathrm{d}x$, $\mathrm{d}w = -p_{\text{b}}\,\mathrm{d}V$, and $\mathrm{d}w = -p\,\mathrm{d}V$. Each expression is the product of a work coefficient and the infinitesimal change of a work coordinate. The first two expressions are general, but the third expression applies only to expansion work in the reversible limit. In this limit the system has two independent variables (e.g., p and V). The number of independent variables is one greater than the number of work coordinates, a statement that will turn out to be true for any reversible process of a closed system.

> Some texts state that expansion work for a cylinder-and-piston apparatus like that shown in Fig. 3.5 should be calculated from $w = -\int p_{\text{ext}}\,\mathrm{d}V$, where p_{ext} is the pressure in the *surroundings* that exerts an external force on the piston. However, if the system is the gas, the correct general expression is $w = -\int p_{\text{b}}\,\mathrm{d}V$ as given in Eq. 3.6.6 because it is p_{b} and not p_{ext} that is the pressure at the boundary of the system. Both expressions for w give the same result if p_{b} and p_{ext} are equal, as they must be for expansion or compression in the reversible limit. The two expressions do not, however, necessarily

---

[16] We can reason that if the state of the system is independent of the fluid shape, the work must be the same as if the volume change occurred in a cylinder-and-piston apparatus.

give the same result for *spontaneous* expansion or compression. The integral $-\int p_b \, dV$ is the work done by the piston on the system, whereas the integral $-\int p_{ext} \, dV$ is the work done on the piston by the rest of the surroundings. The two integrals will have different values if, for instance, the kinetic energy of the piston changes during the process or there is sliding friction at the edge of the piston.[17]

If the only kind of work occurring in a closed system is expansion work, and the pressure is essentially uniform, the first law becomes $dU = dq - p\, dV$. Since we will often want to distinguish between expansion work and other kinds of work, we will frequently write the first law in the form

$$dU = dq - p\, dV + dw' \qquad (3.6.9)$$
(closed system)

where dw' is nonexpansion work (any work that is not expansion work).

3.6.3 Reversible isothermal expansion of an ideal gas

During a reversible expansion, the temperature and pressure remain uniform. If we substitute $p = nRT/V$ from the ideal gas equation into Eq. 3.6.8 and take T as a constant, we obtain

$$w = -nRT \int_{V_1}^{V_2} \frac{dV}{V} = -nRT \ln \frac{V_2}{V_1} \qquad (3.6.10)$$
(reversible isothermal expansion, ideal gas)

The appearance of the amount n as a constant implies that the system is closed during the process, without that having to be stated explicitly as a condition of validity.

3.6.4 Reversible adiabatic expansion of an ideal gas

The internal energy change of a fixed amount of an ideal gas is given by $dU = C_V \, dT$, where C_V is the heat capacity of the gas at constant volume (Eq. 3.2.2). In a reversible adiabatic expansion, the heat is zero and the first law becomes $dU = dw = -p\, dV$. We equate these two expressions for dU

$$C_V \, dT = -p\, dV$$

and substitute $p = nRT/V$ from the ideal gas equation:

$$C_V \, dT = -\frac{nRT}{V} dV$$

When we divide both sides of this equation by T to separate the variables T and V, and integrate on the assumption that C_V is independent of T, we obtain

$$C_V \int_{T_1}^{T_2} \frac{dT}{T} = -nR \int_{V_1}^{V_2} \frac{dV}{V}$$

---

[17] For an informative discussion of this point, see Bauman, R. P. *J. Chem. Educ.* **1964**, *41*, 102–104; also comments in *J. Chem. Educ.* **1964**, *41*, 674–677; also Kivelson, D.; Oppenheim, I. *J. Chem. Educ.* **1966**, *43*, 233–235.

$$C_V \ln \frac{T_2}{T_1} = -nR \ln \frac{V_2}{V_1}$$

We can rearrange this result into the form

$$\ln \frac{T_2}{T_1} = -\frac{nR}{C_V} \ln \frac{V_2}{V_1} = \ln \left(\frac{V_1}{V_2}\right)^{nR/C_V}$$

and take the exponential of both sides:

$$\frac{T_2}{T_1} = \left(\frac{V_1}{V_2}\right)^{nR/C_V}$$

The final temperature is then given as a function of the initial and final volumes by

$$T_2 = T_1 \left(\frac{V_1}{V_2}\right)^{nR/C_V} \tag{3.6.11}$$
(reversible adiabatic expansion, ideal gas)

This relation shows that the temperature decreases during an adiabatic expansion and increases during an adiabatic compression, as expected from the effect of the expansion work on the internal energy.

To find the work from the temperature change, we can use the relation

$$w = \Delta U = \int dU = C_V \int_{T_1}^{T_2} dT$$
$$= C_V (T_2 - T_1)$$

To express the final *pressure* as a function of the initial and final volumes, we make the substitutions $T_1 = p_1 V_1/nR$ and $T_2 = p_2 V_2/nR$ in Eq. 3.6.11

$$\frac{p_2 V_2}{nR} = \frac{p_1 V_1}{nR} \left(\frac{V_1}{V_2}\right)^{nR/C_V}$$

and solve for p_2:

$$p_2 = p_1 \left(\frac{V_1}{V_2}\right)^{1+nR/C_V} \tag{3.6.12}$$
(reversible adiabatic expansion, ideal gas)

3.6.5 Indicator diagrams

An **indicator diagram** (or pressure–volume diagram) is the curve of a plot of p as a function of V. The curve describes the path of an expansion or compression process of a gas that is essentially uniform. The area under the curve is the integral $\int p \, dV$, which is the negative of the reversible expansion work given by $w = -\int p \, dV$ (Eq. 3.6.8).

Section 3.6 Expansion and Compression

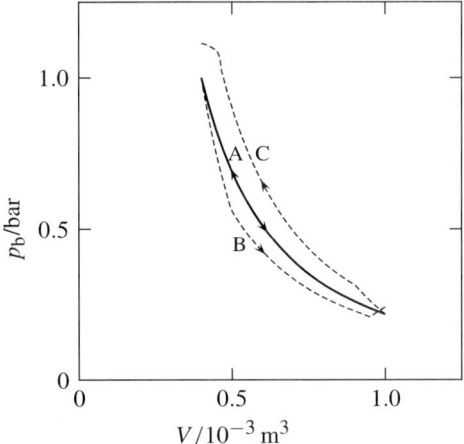

Figure 3.7 Indicator diagram for adiabatic processes in a closed system of a gas. Path A: reversible expansion or compression. Path B: spontaneous expansion starting with the same initial state as for expansion along Path A. The final state is an equilibrium state with a slightly higher temperature than the final state of the reversible expansion because ΔU is less negative. Path C: spontaneous compression starting with the same initial state as for compression along Path A. The final state is an equilibrium state with a higher temperature than the final state of the reversible compression because ΔU is more positive.

Historically, an indicator diagram was a diagram drawn by an "indicator," an instrument invented by James Watt in the late 1700s to monitor the performance of steam engines. The steam engine indicator was a simple pressure gauge: a piston moving in a small cylinder, with the steam pressure of the main cylinder on one side and a compressed spring opposing this pressure on the other side. A pointer attached to the small piston indicated the steam pressure. In later versions, the pointer was replaced with a pencil moving along a paper-covered roll, which in turn was mechanically linked to the piston of the main cylinder. During a cycle of the engine, the pencil moved back and forth along the length of the roll and the roll rotated in a reciprocating motion, causing the pencil to trace a closed curve whose area was proportional to the net work performed by the engine.

3.6.6 Spontaneous adiabatic expansion

The indicator diagram of Fig. 3.7 shows three paths for the adiabatic expansion or compression of a gas. Path A is the reversible path,[18] which is the same in both directions. Path B for a spontaneous adiabatic expansion with the same volume change shows how the pressure at the piston is less, at most intermediate values of V, than for the reversible adiabatic expansion. This is because of the attenuation of the gas just behind the piston

---

[18] The path is calculated from Eq. 3.6.12 assuming an ideal gas with $C_V/n = (3/2)R$.

Figure 3.8 Free expansion into a vacuum.

described in Sec. 3.6.1. Since the area under the curve is the negative of the work, w is less negative for the spontaneous expansion compared with the reversible expansion with the same change of the work coordinate V.

For a spontaneous adiabatic *compression* (Path C), we see that the increased pressure at the piston makes w more positive than for the reversible adiabatic compression with the same change of V.

We can summarize these observations by the statement that, for an adiabatic process of a closed system with a given change of a work coordinate, the work in the reversible limit is smaller (less positive or more negative) than when the change is spontaneous. This statement will turn out to be true for adiabatic processes in general.

3.6.7 Expansion into a vacuum

When we open the stopcock of the apparatus shown in Fig. 3.8, the gas expands from one vessel into a second evacuated vessel. This process is called **free expansion**. We assume that both vessels are perfectly rigid and that the *system* is the gas. The surroundings exert a force on the gas only at the walls, where there is no macroscopic displacement. Thus, in free expansion, there is *no* work: $w = 0$.

3.7 WORK WITH DISSIPATED ENERGY

The piston-and-cylinder apparatus shown in Fig. 3.9 is contrived to illustrate work performed with internal friction. A rod R attached to the piston P passes through a bushing B fixed to the cylinder, and we can control the piston motion with an external force exerted on the piston. The *system* is the combination of gas, rod, and bushing; the piston and cylinder walls are in the surroundings.

The work in this system is given by $dw = F\,dx$ (Eq. 3.3.3), where F is the component

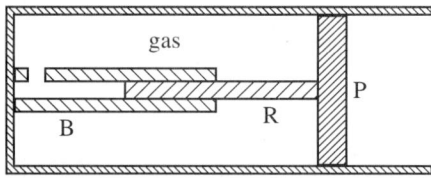

Figure 3.9 Work with internal friction.

Section 3.7 Work with Dissipated Energy

in the $+x$ direction of the force exerted by the surroundings on the system at the moving boundary (the left-hand surface of the piston). F is equal in magnitude and opposite in sign to the force exerted in the $+x$ direction by the system, which is the sum of the force F_{gas} exerted by the gas and the frictional drag F_{frict} at the bushing. The work is thus given by

$$dw = -\left(F_{\text{gas}} + F_{\text{frict}}\right) dx$$

Suppose the piston moves slowly to the right or left. This is not a reversible process, even if the gas pressure is essentially uniform, because the internal friction causes a temperature gradient in the bushing.

Does this process have a reversible limit? It does not if the surface contact between the rod and bushing has static friction that prevents motion at velocities below a certain minimum value. If, however, the bushing is lubricated with a liquid film, we can make the velocity be as close to zero as desired.[19] Under these conditions, a reversible limit does exist because F_{frict} approaches zero as the velocity approaches zero. In this limit, the gas becomes uniform with a single pressure p, F_{gas} becomes equal to pA_s, and the work is given by $dw = -pA_s dx = -p dV$ just as for a system *without* internal friction.

Suppose the system has an adiabatic boundary, so that all energy transfer across the boundary is by work. If we allow the piston to move spontaneously to the left at a *finite* rate, F_{frict} is positive and therefore the work is more positive than for the same change of x in the *reversible* limit. Spontaneous motion of the piston to the right at a finite rate makes F_{frict} negative; the work is then less negative for the same change of x than in the reversible limit. We know that in the reversible limit, changing the direction of a process changes the sign but not the magnitude of the work. Thus, the energy transferred to the system by spontaneous movement of the piston to the left is greater than the energy that the system can return to the surroundings when the piston returns either spontaneously or reversibly to its initial position. In other words, energy transferred as work in a spontaneous process cannot be fully recovered as work.

When work, performed on a closed system at a finite rate in an adiabatic process, cannot be fully recovered by returning the work coordinates adiabatically to their initial values, energy is said to be **dissipated**. In the illustration here, the dissipation occurs because the internal friction converts mechanical energy, the kinetic energy of *concerted* motion, to thermal energy, the kinetic energy of *random* motion. The dissipated energy is the portion of the energy transferred across the boundary by work that could be replaced by heat with the same net result in the system.

If we remove all the gas from the cylinder so that F_{gas} is zero, we can control the piston motion by pushing or pulling it from the surroundings. The work performed on the system (the rod and bushing) is then $dw = -F_{\text{frict}} dx$. With liquid lubrication at the bushing, $-F_{\text{frict}}$ approaches zero in the limit of zero piston velocity. It is a characteristic of work like this that is completely dissipated to thermal energy that the work coefficient approaches zero as the rate of change of the work coordinate approaches zero. The reversible limit in this case has *no* work, even though there is a change in the work coordinate during the reversible process.

---

[19] Since a liquid flows under shear stress of any magnitude (Sec. 2.1.4), a lubricating film separating two solid surfaces allows sliding motion with a relative velocity of any magnitude.

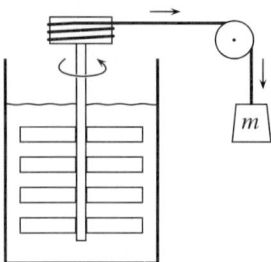

Figure 3.10 Joule's apparatus for the measurement of the mechanical equivalent of heat (schematic).

Energy dissipation can be caused by effects other than friction. The pressure gradients in the spontaneous expansion or compression of a gas described in Sec. 3.6.1 also dissipate energy. Gas flow along a pressure gradient converts mechanical to thermal energy within the system, allowing less work to be done by the system on the surroundings in spontaneous expansion, or requiring more work to be done by the surroundings on the system in spontaneous compression, compared to the reversible process with the same volume change. However, when mechanical energy is converted to thermal energy in a *reversible* process, as for example in the adiabatic compression of a gas described in Sec. 3.6.4, the energy is fully recovered when the reversible process is reversed. There is no dissipation of energy in a reversible process.

Another example of energy dissipation is operation of the paddle wheel devised by James Joule in the 1840s to determine the "mechanical equivalent of heat." The apparatus is shown schematically in Fig. 3.10; it is the same as the Joule paddle wheel shown earlier in Fig. 3.3 on page 38. A sinking weight of mass m in the surroundings is linked mechanically to a paddle wheel that rotates and churns a liquid (Joule used both water and mercury) in a thermally insulated container. The paddle arms move through openings in vanes (not shown) fixed inside the container; the vanes prevent the liquid from simply moving around in a circle. The frictional drag at the paddles keeps the weight from accelerating. The result is turbulent motion (shearing flow or viscous flow) in the liquid and an increase in the liquid's temperature.

Joule evaluated the work from the change of position of the external weight. If the weight descends at constant speed and the pulley bearings have negligible friction, the work done on the system (the liquid and the immersed portion of the paddle wheel and shaft) equals the change of the weight's potential energy: $dw = -mg\,dh$.

> Joule's apparatus had two weights wound up with a crank to a height about 1.6 meters above the floor. He made corrections for heat losses through the container wall, friction in the external bearings, and the kinetic energy of the weights when they reached the floor. His results for the mechanical equivalent of heat, expressed as the work needed to heat one gram of water from 14.5 °C to 15.5 °C (the "15 °C calorie"), was 4.17 J, remarkably close to the modern value of 4.186 J.

To analyze the process in a different way, consider the portion of the boundary where the shaft passes into the liquid. Here energy is transferred across the boundary as rotational mechanical work. When the shaft rotates, the frictional drag at the paddles causes a torque in the shaft. The portion of the shaft outside the boundary then exerts a force (a torsional stress) on the portion inside, and work is done given by

$$dw = \tau \, d\theta$$

where τ is the torque at the boundary and θ is the angle of rotation. In this equation, τ is a work coefficient and θ is a work coordinate. While the shaft rotates and the liquid is churned, the states of the system are not equilibrium states; many independent variables are needed to describe them. If the rate of rotation is made slower and slower, τ approaches zero. In the reversible limit, the paddle wheel rotates without work.

In addition to rotational mechanical work, there is a small quantity of expansion work in Joule's apparatus. The system has two work coordinates (V and θ), and they are both state functions. The number of independent variables in equilibrium states is three, which we could choose as T, p, and θ or as T, V, and θ. Thus, the number of independent variables of a reversible process of this system is one greater than the number of work coordinates, in agreement with the general rule mentioned on page 46.

3.8 ELECTRICAL WORK

The electric potential energy of a charge Q_{el} at a point where the electric potential is ϕ is equal to ϕQ_{el}. If an infinitesimal charge dQ_{el} enters the system, the energy transferred as electrical work is $dw_{el} = \phi \, dQ_{el}$.

Electric current is usually conducted in an electrical circuit. Consider a system that is part of a circuit: Electrons enter the system through one wire, and an equal number of electrons simultaneously leave through a second wire. To simplify the description, we refer to the wires as the right-hand conductor and the left-hand conductor. If an infinitesimal charge dQ_{el} enters the system at the right-hand conductor, where the electric potential is ϕ_R, and an equal charge leaves at the left-hand conductor, where the electric potential is ϕ_L, the work is given by the energy difference $(\phi_R - \phi_L) \, dQ_{el}$. Thus, the general formula for an infinitesimal quantity of electrical work, valid whenever the system is part of an electrical circuit, is:

$$dw_{el} = \Delta\phi \, dQ_{el} \qquad (3.8.1)$$
(electrical work in a circuit)

where dQ_{el} is defined as the charge entering at the right-hand conductor and $\Delta\phi$ is defined as the electric potential difference

$$\Delta\phi = \phi_R - \phi_L$$

The electric current I is the rate at which charges pass a point in the circuit: $I = dQ_{el}/dt$ (where t is time). Here I is taken as positive if the charge entering at the right conductor is

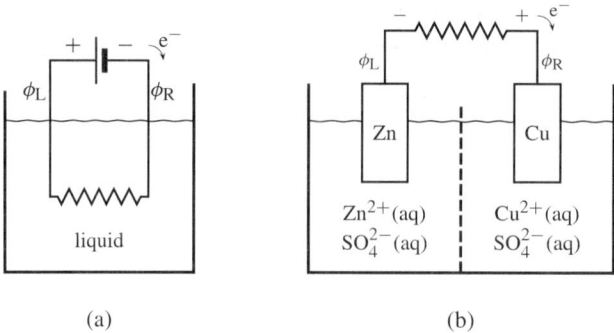

Figure 3.11 Two systems to illustrate electrical work. (a) An electrical resistor immersed in a liquid. (b) A galvanic cell with a porous barrier separating two half-cells.

positive. This relation provides an alternative form of Eq. 3.8.1:

$$dw_{el} = I \Delta\phi \, dt \tag{3.8.2}$$
(electrical work in a circuit)

Equations 3.8.1 and 3.8.2 are general equations for electrical work in a circuit. The next two sections discuss an example of a circuit in which the source of the electric potential difference is in the surroundings and an example in which the source is in the system.

3.8.1 Electrical heating

Figure 3.11(a) shows an electrical resistor immersed in a liquid. The *system* consists of the liquid, the resistor, and the immersed portions of the wires. An external galvanic cell provides an electric potential difference (the "voltage"), $\Delta\phi$, across the wires. When electric current flows in the circuit, the resistor and the rest of the system become warmer due to the ohmic resistance of the resistor, a phenomenon variously called electrical heating, Joule heating, ohmic heating, and resistive heating. The heating is caused by inelastic collisions of the moving electrons with the stationary atoms of the resistor and is yet another example of work with energy dissipation.

While current flows, $\Delta\phi$ is nonzero and there is an electric potential gradient within the resistor. The gradient affects the state of the resistor by polarizing its material. The state is not an equilibrium state since isolating the system by opening the electrical circuit would immediately remove the gradient.

The work performed on the system is given by Eq. 3.8.1 or Eq. 3.8.2. The portion of the electrical circuit inside the system has an electric resistance R_{el} defined by $R_{el} = \Delta\phi/I$ (Ohm's law). Substitution in Eq. 3.8.2 gives

$$dw_{el} = I^2 R_{el} \, dt \tag{3.8.3}$$
(electrical work on a resistor)

We now have three equivalent expressions for the electrical work in this system: $dw_{el} = \Delta\phi \, dQ_{el}$, $dw_{el} = I\Delta\phi \, dt$, and $dw_{el} = I^2 R_{el} \, dt$. The work coefficients in these expressions are state functions, but the work coordinates (Q_{el} and t) are not. The work is completely dissipated to thermal energy, the work coefficients approach zero as the rate of work approaches zero, and in the limit of zero rate (zero current) there is no work. Since in this limit there is no state function that changes as equal numbers of electrons enter and leave the system, we cannot speak of a limiting reversible process.

Suppose we redefine the system to be only the liquid, leaving the electrical circuit entirely in the surroundings. Then the boundary is at the interface where the wires and resistor contact the liquid. Operation of the circuit will cause a temperature gradient at this interface. With this choice of the system, there is *heat* instead of work.

3.8.2 Electrical work with a galvanic cell

Figure 3.11(b) shows the performance of electrical work involving a zinc-copper galvanic cell (a Daniell cell), which is the system. The electric potential difference between the wires, $\Delta\phi$, arises at the interfaces between phases within the system. The galvanic cell does electrical work on the resistor in the surroundings. The energy for this work comes from the spontaneous cell reaction

$$Zn(s) + Cu^{2+}(aq) \rightarrow Zn^{2+}(aq) + Cu(s)$$

The formula for the electrical work is again given by Eq. 3.8.1: $dw_{el} = \Delta\phi \, dQ_{el}$. The figure shows $\Delta\phi$ as positive and dQ_{el} as negative, so dw_{el} is negative. The work coefficient $\Delta\phi$ is a state function. In this system, the work coordinate Q_{el} is also a state function because it is directly related to the amount of each reactant and product of the cell reaction: for each copper(II) ion that becomes reduced to a copper atom, two electrons must enter the system at the right conductor.

If we isolate the system by cutting the wires, the system immediately reaches an equilibrium state in which no further reaction occurs (at least in the short term before ions have time to diffuse through the porous barrier). In this state, the electric potential difference E between the wires (called the electromotive force, or emf) is greater than it was when current was being drawn. The system remains isolated if we attach each wire to a lead of an adjustable voltage source (a potentiometer) with exactly the same electric potential. Then, by making a small change in the adjustable voltage, we can control the rate and direction of the process. As the adjustable voltage approaches the open-circuit value E, the rate and current approach zero and the work approaches $dw_{el} = E \, dQ_{el}$. A reversible process exists in this limit because the state function Q_{el} changes.

The internal resistance of the cell causes dissipation of energy for finite current. Consequently, the electrical work for a given change in the work coordinate Q_{el} (or in the cell reaction), when the process is adiabatic, is the most negative or least positive in the reversible limit.

The thermodynamics of galvanic cells is treated in detail in chap. 11.

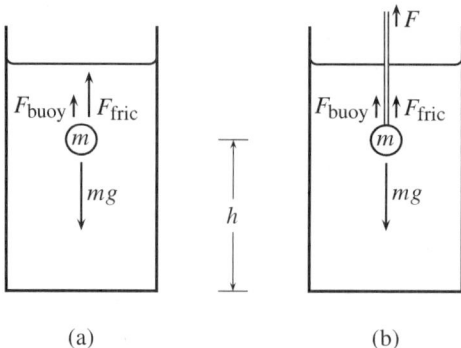

Figure 3.12 Two systems to illustrate gravitational work. (a) A body falling freely through liquid. (b) The same, with a string attached.

3.9 GRAVITATIONAL WORK

Suppose that a body of mass m falls through a liquid in a system with an adiabatic boundary on account of an external gravitational field as shown in Fig. 3.12(a). This process is clearly spontaneous. However, there is *no* work[20] and no change in the system's internal energy. (As explained in Sec. 3.3.1, we do not use forces exerted by a time-independent external field to calculate work.) As the system loses potential energy, it gains an equal quantity of kinetic and thermal energy.

Now suppose we use tension in a thin string to exert an upward force F on the body, as shown in Fig. 3.12(b). Since the string provides a means for the surroundings to exert a force on the system, vertical displacement of the body causes gravitational work given by $\mathrm{d}w = F\,\mathrm{d}h$, where h is the body's elevation. The displaced internal coordinate is h, and F is the force exerted by the surroundings. If the body sinks ($\mathrm{d}h < 0$), the work is negative, as we could see by the fact that with a suitable pulley arrangement the string could raise a weight in the surroundings. If the body rises, the work is positive.

The components in the $+x$ direction of the forces on the body are (1) a gravitational force $-mg$ (g is the acceleration of free fall); (2) a buoyant force $F_{\text{buoy}} = +\rho V'g$, where ρ is the density of the liquid and V' is the volume of the body;[21] (3) a frictional force F_{fric} of opposite sign from the velocity $\mathrm{d}h/\mathrm{d}t$; and (4) the force F exerted by the string. According to Newton's second law, if the body moves up or down without acceleration, the net force on the body is zero, so that F is given by

$$F = mg - F_{\text{buoy}} - F_{\text{fric}}$$

As the velocity in either direction approaches zero, F_{fric} approaches zero and the work approaches the reversible limit $\mathrm{d}w = (mg - F_{\text{buoy}})\,\mathrm{d}h$.[22] As in the other examples of work

---

[20] We are ignoring expansion work caused by the small temperature increase.

[21] The buoyant force is a consequence of the pressure gradient that exists in the equilibrium state of a fluid in a gravitational field (see Sec. 6.1.4). We ignore this gradient when we treat the fluid as a uniform phase.

[22] In this expression for reversible work, $mg\,\mathrm{d}h$ is the change in the potential energy of the body in the gravitational

with energy dissipation, the work in the reversible limit is most negative or least positive for a given change of the work coordinate.

3.10 SPONTANEOUS AND REVERSIBLE PROCESSES: GENERALITIES

In this section, we list some general characteristics of spontaneous and reversible processes in closed systems, as illustrated by the examples of Secs. 3.5–3.9. We need these generalities to develop aspects of the second law in chap. 4.

A *spontaneous* process is a real process that takes place in a finite time period and whose intermediate states are not equilibrium states.

A *reversible* process is an imaginary process taking place in a finite, unspecified time period whose states are equilibrium states. The sequence of states can be approached by a spontaneous process in the limit of infinite slowness, and so can the reverse sequence of states. A spontaneous process whose rate we can control so as to be very slow is practically a reversible process if, when we change conditions in the surroundings only slightly at any point along the path, the system begins to retrace practically the same path in the reverse direction.

We calculate infinitesimal quantities of work during a process from an expression of the form $dw = \sum Y_i \, dy_i$, where Y_i is a work coefficient and y_i is a work coordinate.

There is dissipation of energy when work occurs spontaneously. The dissipation consists of the conversion of energy of concerted motion in whole or part to thermal energy, which cannot be completely converted back to work. The effect on the system of a spontaneous process in which work is completely dissipated, as in the Joule paddle-wheel experiment and electrical heating, could equally well be accomplished with heat. In a process with such completely dissipated work, the work coefficient approaches zero as the rate of change of the work coordinate approaches zero.

If a process is carried out adiabatically, the work for given changes in the work coordinates is smallest (the least positive or most negative) in the reversible limit.

When work occurs in a reversible process, the work coefficients and work coordinates are state functions, and each work coefficient is nonzero in the reversible limit.

When a reversible process occurs in the reverse direction, each work coefficient remains unchanged and the changes in work coordinates change sign. Consequently, energy transferred across the boundary as work in a reversible process is fully recovered as work of the opposite sign in the reverse process. It follows from the first law that heat is also fully recovered in the reverse process.

In a reversible process of a closed system, the number of independent variables of equilibrium states of uniform temperature (with no constraints other than those needed to prevent transport of matter across the boundary) is one greater than the number of work coordinates.[23] Thus, we could choose the independent variables for a reversible process to

---

field, and $-F_{\text{buoy}} \, dh$ is the change in the potential energy of the liquid due to displacement by the body.

[23] If the system has internal adiabatic partitions that allow different phases to have different temperatures in equilibrium states, then the number of independent variables is equal to the number of work coordinates plus the

be each of the work coordinates and the temperature.[24] The number of independent variables needed to describe a nonequilibrium state of a *spontaneous* process is greater (often *much* greater) than this.

Table 3.2 lists general formulas for various kinds of work, including those illustrated in Sections 3.6–3.9.

Table 3.2 Some kinds of work

Kind	Formula	Definitions
Linear mechanical work	$dw = F\,dx$	F = x-component of force exerted by the surroundings dx = displacement in x direction
Rotational mechanical work	$dw = \tau\,d\theta$	τ = torque θ = angle of rotation
Expansion work	$dw = -p_b\,dV$	p_b = average pressure at moving boundary
Surface work	$dw = \gamma\,dA_s$	γ = surface tension, A_s = surface area
Stretching or compression of a rod or spring	$dw = F\,dl$	F = stress (positive for tension, negative for compression) l = length
Gravitational work	$dw = mg\,dh$	m = mass, h = height g = acceleration of free fall
Electrical work	$dw = \phi\,dQ_{el}$	ϕ = electric potential Q_{el} = charge
Electrical work in a circuit	$dw = \Delta\phi\,dQ_{el}$	$\Delta\phi$ = electric potential difference $= \phi_R - \phi_L$ dQ_{el} = charge entering system at right
Electric polarization	$dw = V\boldsymbol{E}\cdot d\boldsymbol{P}$	$\boldsymbol{E}$ = electric field strength $\boldsymbol{P}$ = dielectric polarization
Magnetization	$dw = \left(\frac{\mu_0}{4\pi}\right)V\boldsymbol{H}\cdot d\boldsymbol{M}$	μ_0 = permeability of vacuum $\boldsymbol{H}$ = magnetic field strength $\boldsymbol{M}$ = magnetization

---

number of independent temperatures.

[24] There may be exceptions to this statement in special cases. An example is along the triple line of a pure substance where values of V and T are not sufficient to determine the amounts in each of the three possible phases.

3.11 ENTHALPY

The first law shows that in a process of a closed system without work, the internal energy change equals the heat. The process might, for instance, be the free expansion of a gas (Sec. 3.6.7), or it might be a process at constant volume without electrical work or other nonexpansion work:

$$dU = dq \tag{3.11.1}$$

(closed system, constant V, $dw' = 0$)

Many experiments are carried out at constant pressure instead of constant volume. It will prove useful to define a new state function whose change equals the heat under these conditions, analogous to the way the change of internal energy equals the heat at constant volume. This function is the **enthalpy** H defined by

$$H \equiv U + pV \tag{3.11.2}$$

Enthalpy has dimensions of energy. It is a state function because it is a function only of state functions, and it is extensive because U and V are extensive.

For a fixed amount of an *ideal gas* of uniform temperature, U depends only on T (Sec. 3.2.2) and so does pV since it is equal to nRT. Thus, the enthalpy of a fixed amount of an ideal gas, like the internal energy, depends only on T.

The general definition of enthalpy assumes the system has uniform pressure p. We can define the enthalpy of a nonuniform fluid by applying $H = U + pV$ to constituent phases, or to subsystems small enough to be essentially uniform, and summing.

The enthalpy change is equal to the heat if the system is closed, the pressure is uniform and constant, and the only work is expansion work. We can see this property of H by taking its differential

$$dH = dU + p\,dV + V\,dp \tag{3.11.3}$$

and making the substitution $dU = dq - p\,dV + dw'$ (Eq. 3.6.9) for a closed system to give (after cancellation of the $p\,dV$ terms) $dH = dq + V\,dp + dw'$. Thus, in a process at constant pressure ($dp = 0$) with expansion work only ($dw' = 0$), the infinitesimal change of enthalpy is equal to the heat:

$$dH = dq \tag{3.11.4}$$

(closed system, constant p, $dw' = 0$)

The integrated form of this relation is $\int dH = \int dq$ or

$$\Delta H = q \tag{3.11.5}$$

(closed system, constant p, $w' = 0$)

We make extensive use of the enthalpy function in later chapters, particularly in chap. 8 where the enthalpy changes of chemical processes are discussed.

3.12 HEAT CAPACITY

Transfer of energy into a closed system in the form of heat usually causes a temperature increase. The **heat capacity** of a closed system is defined as the infinitesimal quantity of heat divided by the infinitesimal temperature increase under specified conditions:

$$\text{heat capacity} \equiv \frac{dq}{dT} \qquad (3.12.1)$$
(closed system)

The specified conditions usually include the requirements that nonexpansion work be absent (e.g., that there be no electrical heating) and that no phase change occur. Since q is a path function, the heat capacity depends on what other changes occur in the system as T increases. The heat capacities of isochoric and isobaric processes are of particular interest.

The **heat capacity at constant volume**, C_V, is the ratio dq/dT for a process in a closed constant-volume system with no nonexpansion work—thus no work at all. Under these conditions, we can replace dq by dU (Eq. 3.11.1) and write the heat capacity as a partial derivative:

$$C_V = \left(\frac{\partial U}{\partial T}\right)_V \qquad (3.12.2)$$
(closed system)

If the closed system has more than two independent variables, additional conditions will be needed to define C_V unambiguously. For instance, if the system is a gas mixture in which reaction can occur, we might specify that the system remains in reaction equilibrium as T changes at constant V. However, it is not necessary to specify that the nonexpansion work, dw', be zero to use Eq. 3.12.2 to evaluate C_V because all quantities appearing in the equation are *state* functions whose relations to one another are fixed by the nature of the system. Thus, if the transfer of heat into the system at constant V causes U to increase at a certain rate with respect to T, and this rate is defined as C_V, the performance of electrical work on the system at constant V will cause the same rate of increase of U with respect to T and can equally well be used to evaluate C_V.

Note that C_V is a state function whose value depends in general on the state of the system—that is, on T, V, and any additional independent variables. C_V is an *extensive* property. The combination of two identical phases has twice the value of C_V that one of the phases has by itself. For a phase containing a pure substance, we define the **molar heat capacity at constant volume** by $C_{V,\text{m}} = C_V/n$. $C_{V,\text{m}}$ is an *intensive* property.

If the system is an ideal gas, its internal energy depends only on T, regardless of whether V is constant, and Eq. 3.12.2 can be simplified to

$$C_V = \frac{dU}{dT} \qquad (3.12.3)$$
(closed system, ideal gas)

Thus, the internal energy change of an ideal gas is $dU = C_V \, dT$, a relation already given as Eq. 3.2.2.

The **heat capacity at constant pressure**, C_p, is the ratio dq/dT for a process in a closed system carried out at constant pressure with expansion work only. From Eq. 3.11.4, we obtain a relation analogous to Eq. 3.12.2:

$$C_p = \left(\frac{\partial H}{\partial T}\right)_p \quad (3.12.4)$$
(closed system)

C_p is an extensive state function. For a phase containing a pure substance, the **molar heat capacity at constant pressure** is $C_{p,m} = C_p/n$, an intensive property.

Since the enthalpy of a fixed amount of an ideal gas depends only on T (Sec. 3.11), we can write a relation analogous to Eq. 3.12.3:

$$C_p = \frac{dH}{dT} \quad (3.12.5)$$
(closed system, ideal gas)

PROBLEMS

3.1 The value of ΔU for the formation of one mole of crystalline potassium iodide from its elements at 25 °C and 1 bar is -327.9 kJ. Calculate Δm for this process. Comment on the feasibility of measuring this mass change.

3.2 Refer to the apparatus depicted in Fig. 3.3 on page 38. Suppose the mass of the external weight is $m = 1.50$ kg, the resistance of the electrical resistor is $R_{el} = 5.50$ kΩ, and the acceleration of free fall is $g = 9.81$ m s^{-2}. How long will the external cell have to operate, providing an electric potential difference $|\Delta\phi| = 1.30$ V, to cause the same change in the state of the system as when the weight sinks 20.0 cm? Assume both processes occur adiabatically.

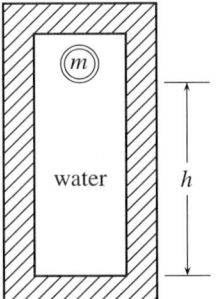

Figure 3.13

3.3 The apparatus shown in Fig. 3.13 consists of a fixed amount of water and a glass sphere (a marble) of mass m. Assume the marble has an adiabatic outer layer so that its temperature cannot change. The apparatus is enclosed with a rigid adiabatic wall. Initially the marble is suspended at height h_0 above the bottom of the vessel. When released it falls through the water and comes to rest at the bottom of the vessel ($h = 0$), causing the temperature and pressure of the water to increase. For each of the following definitions of the system, give formulas for q, w, ΔU, ΔE_p, and ΔE_{sys}; also describe what, if any, change occurs in the state of the system. (Ignore the expansion work done by the water on the marble.)

(a) The system is the marble; h is an external coordinate.

(b) The system is the water; h is an internal coordinate.

(c) The system is the combination of the water and the marble; h is an internal coordinate.

3.4 Consider a horizontal piston-and-cylinder apparatus similar to the one shown in Fig. 3.5 on page 43. The piston has mass m. The cylinder is diathermal and in thermal contact with a heat reservoir of temperature T_{ext}. The *system* is an amount n of an ideal gas confined by the piston. The initial state of the system is an equilibrium state described by p_1 and $T = T_{ext}$. There is a constant external pressure p_{ext}, equal to twice p_1, that supplies a constant external force on the piston. When the piston is released, it begins to move to the left. Make the idealized assumptions that (1) the piston moves with negligible friction; and (2) the gas remains uniform (because the piston is massive and its motion is slow) and has constant temperature $T = T_{ext}$ (because temperature equilibration is rapid).

(a) Describe the resulting process.

(b) Describe how you could calculate w and q during the period needed for the piston velocity to become zero again.

3.5 Discuss the proposition that, to a certain degree of approximation, a living organism is a steady-state system.

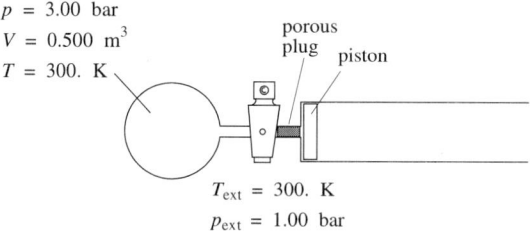

Figure 3.14

3.6 Figure 3.14 shows the initial state of an apparatus consisting of an ideal gas in a bulb, a stopcock, a porous plug, and a cylinder containing a frictionless piston. The walls are diathermal, and the surroundings are at a constant temperature of 300. K.

When the stopcock is opened, the gas diffuses slowly through the porous plug, and the piston moves slowly to the right. The process ends when the pressures are equalized and the piston stops moving. The *system* is the gas. Assume that during the process the temperature throughout the system differs only infinitesimally from 300. K and the pressure on both sides of the piston differs only infinitesimally from 1.00 bar.

(a) Which of these terms correctly describes the process: isothermal, isobaric, isochoric, spontaneous, reversible?

(b) Calculate q and w.

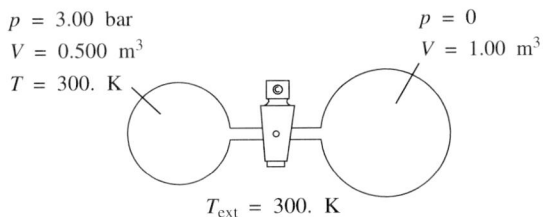

Figure 3.15

3.7 Figure 3.15 shows the initial state of an apparatus containing an ideal gas. When the stopcock is opened, gas passes into the evacuated vessel. The *system* is the gas. Find q, w, and ΔU for the following conditions.

  (a) The vessels have adiabatic walls.

  (b) The vessels have diathermal walls in thermal contact with a heat reservoir at 300. K, and the temperature is $T = 300.$ K in the final state.

3.8 Assume you have a metal spring that obeys Hooke's law: $F = c(l - l_0)$, where F is the force exerted on the spring of length l, l_0 is the length of the unstressed spring, and c is the spring constant.

  (a) Find an expression for the work done on the spring when you reversibly compress it from length l_0 to length l.

  (b) Suppose you place the spring and a length of nylon cord in a beaker of concentrated acid at 25 °C. The acid completely oxidizes the metal, and you end up with hydrogen gas, an aqueous solution of metal sulfates, and the intact cord. You allow the temperature to return to 25 °C. Will the final state at 25 °C be the same if before placing the spring in the acid you compress it to length l and bind it tightly with the cord, as compared with the final state when you place the unstressed spring and the untied cord in the acid? What difference, if any, will there be in the heat during the two ways of carrying out this process?

3.9 From concepts in this chapter, show that the heat capacities C_V and C_p of a fixed amount of an ideal gas are functions only of T.

Chapter 4

THE SECOND AND THIRD LAWS

The second law of thermodynamics concerns entropy and the spontaneity of processes in the universe, and the third law concerns the entropy of perfectly ordered crystals at zero kelvins. This chapter discusses theoretical aspects and practical applications of these laws.

We have seen that the first law allows us to set up a balance sheet for energy changes during a process, but has nothing to say about why some processes occur spontaneously and others are impossible. The laws of physics explain some spontaneous processes. For instance, unbalanced forces on matter at the system boundary cause spontaneous work, and a temperature gradient at a diathermal boundary causes spontaneous heat transfer. But what causes a phase change, a chemical reaction, or the mixing of two gases to be spontaneous? The second law provides the principle we need to answer these and other questions—a general criterion for spontaneity in a closed system.

4.1 TYPES OF PROCESSES

Any conceivable process is either spontaneous, reversible, or impossible. Which of these three possibilities applies to a particular process depends on conditions in the system and at the boundary during the course of the process.

A **spontaneous** process, as described in Secs. 3.4 and 3.10, is a change in the state of the system that can actually occur in a finite length of time. (A spontaneous process is sometimes called a natural, feasible, possible, allowed, or real process.)

A **reversible** process, also described in Secs. 3.4 and 3.10, is an idealized change consisting of a continuous sequence of equilibrium states. This sequence, and the reverse sequence, can be approached by a spontaneous process in the limit of infinite slowness.

An **impossible** process is a change that cannot occur under the existing conditions, even in a limiting sense. It is also known as an unnatural or disallowed process. Sometimes it is useful to describe a hypothetical impossible process that we can imagine but that does not occur in reality. We presently introduce the second law of thermodynamics by describing two such impossible processes.

The spontaneous processes relevant to chemistry are irreversible. An **irreversible** pro-

Section 4.1 Types of Processes

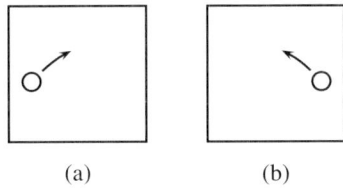

(a)   (b)

Figure 4.1 Two purely mechanical processes that are the reverse of one another: a thrown ball moving through a vacuum (a) to the right; (b) to the left. The ball moves spontaneously in either direction.

cess is a spontaneous process whose reverse is an impossible process.

To visualize the reverse of a spontaneous process, imagine making a movie film of the events of the process. Each frame of the film is a "snapshot" picture of the state at one instant. If we run the film backward through a movie projector, we see the reverse process: The values of system properties such as p and V appear to change in reverse chronological order, and every velocity changes sign. If the time sequence of the events depicted by the film when it is run backward could not occur in reality, the spontaneous process is irreversible.

A good example of a spontaneous, irreversible process is the Joule paddle-wheel experiment shown in Fig. 3.10 on page 52, in which a weight sinks, a paddle wheel rotates, and the temperature of the liquid (water, say) increases. Mechanical energy is dissipated into thermal energy (Sec. 3.7). If you make a movie film of the paddle-wheel experiment with a thermometer inserted in the water, when you run the film backward you will see the paddle wheel rotating in the direction that raises the weight, and the thermometer indicating the water becoming cooler. Clearly, this reverse process is impossible in the real physical world. It is not difficult to understand why when we consider events on the microscopic level: It is extremely unlikely that the H_2O molecules next to the paddle would happen to move simultaneously over a period of time in the concerted motion needed to raise the weight.

There is a class of spontaneous processes that are also spontaneous in reverse; that is, spontaneous but not irreversible. These are purely *mechanical* processes involving the motion of perfectly elastic macroscopic bodies without friction, temperature gradients, viscous flow, or other dissipative effects. Such conditions are idealizations and of little interest in chemistry. As a simple example of a purely mechanical process, consider the situation in Fig. 4.1: The ball can move spontaneously in either direction. Another example is a flywheel with frictionless bearings rotating in a vacuum.

It is true that completely reversible processes and purely mechanical processes are idealized processes that cannot be carried out in practice; but a spontaneous process can be *practically* reversible if carried out slowly, or *practically* purely mechanical if friction is negligible. In that sense, they are not impossible processes. We reserve the term "impossible" for a process that cannot be approached by any spontaneous process no matter how slowly or carefully carried out.

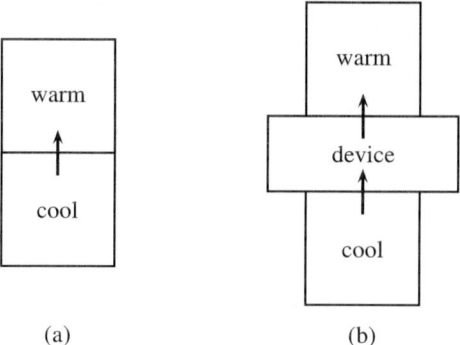

Figure 4.2 Two impossible processes. (a) Transfer of heat from a cool to a warm body. (b) The same, with a device that operates in a cycle.

4.2 STATEMENTS OF THE SECOND LAW

The **mathematical form of the second law** is

$$\begin{aligned}
&\mathrm{d}S = \mathrm{d}q/T_\mathrm{b} \text{ for a reversible process of a closed system;}\\
&\mathrm{d}S > \mathrm{d}q/T_\mathrm{b} \text{ for an irreversible process of a closed system;}\\
&\text{where } S \text{ is an extensive state function, the \textbf{entropy}; and}\\
&\quad \mathrm{d}q \text{ is an infinitesimal quantity of heat transferred at a}\\
&\quad \text{portion of the boundary where the thermodynamic}\\
&\quad \text{temperature is } T_\mathrm{b}.
\end{aligned} \tag{4.2.1}$$

The box includes three distinct statements. First, there is the assertion that the entropy S is a state function. Second, there is an equation for calculating the entropy change of a closed system: $\mathrm{d}S$ is equal to $\mathrm{d}q/T_\mathrm{b}$ during a reversible process.[1] Third, there is a criterion for irreversibility: $\mathrm{d}S$ is greater than $\mathrm{d}q/T_\mathrm{b}$ during an irreversible process. There is also mention of thermodynamic temperature, which is explained in Sec. 4.3.4.

Each of the three statements is an essential element of the second law but is somewhat abstract. What fundamental principle, based on experimental observation, may we take as the starting point to obtain these statements? Two principles are available, one associated with Clausius and the other with Kelvin and Planck. Both principles are equivalent statements of the second law. Each states that a certain kind of process is impossible, in agreement with common experience.

Consider the process depicted in Fig. 4.2(a). The system is isolated and consists of a cool body in thermal contact with a warm body. During the process, energy in the form of heat is transferred from the cool to the warm body; the temperature of the cool body falls and

---

[1] During a reversible process, the temperature usually has the same value T throughout the system, in which case we can simply write $\mathrm{d}S = \mathrm{d}q/T$. The equation $\mathrm{d}S = \mathrm{d}q/T_\mathrm{b}$ allows for the possibility that an equilibrium state has phases of different temperatures separated by internal adiabatic partitions.

Section 4.2 Statements of the Second Law 67

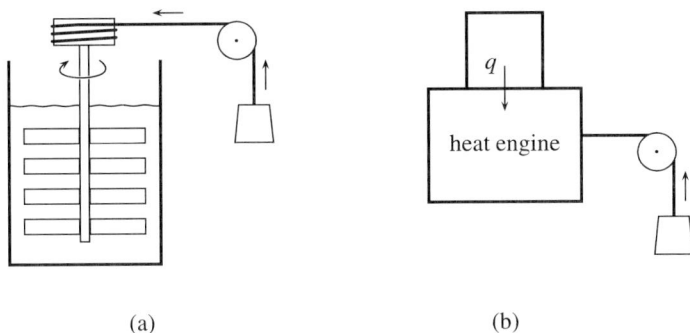

(a) (b)

Figure 4.3 Two more impossible processes. (a) A weight rises as a liquid becomes cooler. (b) The same, with a heat engine.

that of the warm body increases. Of course, this process is impossible; we never observe heat flowing directly from a cooler to a warmer body. (However, the reverse process, a transfer of heat from the warmer to the cooler body, is spontaneous and irreversible.) Note that this impossible process does not violate the first law since energy is conserved.

Suppose we attempt to bring about the same changes in the two bodies by interposing a device of some sort between them, as depicted in Fig. 4.2(b). Here is how we would like the device to operate in the isolated system: Heat should flow from the cool body to the device, an equal quantity of heat should flow from the device to the warm body, and the final state of the device should be the same as the initial state. In other words, we want the device to transfer heat quantitatively from the cool to the warm body while operating in a *cycle*. If the device could do this, there would be no limit to the quantity of heat it could transfer because after each cycle the device would be ready to repeat the process. But experience shows that *it is impossible to build such a device!* The proposed process of Fig. 4.2(b) is impossible even in the limit of infinite slowness.

The general principle was expressed by Rudolph Clausius[2] in the words: "Heat can never pass from a colder to a warmer body without some other change, connected therewith, occurring at the same time." For use in the derivation to follow, we reword the statement as follows.

The Clausius statement of the second law: It is impossible to construct a device that will operate in a cycle, and produce no effect except the transfer of heat from a body to the device and the transfer of an equal quantity of heat from the device to a warmer body.

Next consider the impossible process shown in Fig. 4.3(a). A Joule paddle wheel rotates at constant angular velocity in a container of water as a weight rises. As the weight gains potential energy, the water loses an equal quantity of thermal energy and its temperature decreases. As mentioned in Sec. 4.1, this process is just the reverse of the Joule paddle-

---

[2]Clausius, R. *The Mechanical Theory of Heat, with Its Applications to the Steam-Engine and to the Physical Properties of Bodies*; John Van Voorst: London, 1867.

wheel experiment depicted in Fig. 3.10 on page 52. We might again attempt to use some sort of device operating in a cycle to accomplish the same overall process, as shown in Fig. 4.3(b). A closed system that has heat and work transferred across its boundary, does net work on the surroundings, and operates in a cycle is called a **heat engine**. The heat engine shown in Fig. 4.3(b) is a special one. During one cycle, a quantity of heat is transferred from a heat reservoir to the engine, and the engine performs an *equal* quantity of work on a weight causing it to rise. At the end of the cycle, the engine has returned to its initial state. This would be a very desirable engine because it could convert thermal energy into an equal quantity of useful mechanical work with no other effect on the surroundings.[3] This engine could power a ship; it would use the ocean as a heat reservoir and require no fuel. Unfortunately, it is impossible to construct such a heat engine!

The principle was expressed by William Thomson (Lord Kelvin) in 1852 as follows: "It is impossible by means of inanimate material agency to derive mechanical effect from any portion of matter by cooling it below the temperature of the coldest of the surrounding objects." Max Planck[4] gave this statement: "It is impossible to construct an engine which will work in a complete cycle, and produce no effect except the raising of a weight and the cooling of a heat-reservoir." For our purposes, we word the principle as follows.

The Kelvin–Planck statement of the second law: It is impossible to construct a heat engine that will operate in a cycle and produce no effect except the transfer of heat from a heat reservoir to the engine and the performance of an equal quantity of work on the surroundings.

Both the Clausius and Kelvin–Planck statements assert that certain processes, although they do not violate the first law, are nevertheless *impossible*.

> These processes would not be impossible if we could control the trajectories of large numbers of individual particles. Newton's laws of motion are invariant to time reversal. Suppose we could measure the position and velocity of each molecule of a macroscopic system in the final state of an irreversible process. Then, if we could somehow arrange at one instant to place each molecule in the same position with its velocity reversed and the molecules behave classically, they would retrace their trajectories in reverse and we would observe the reverse "impossible" process.

The plan of the next few sections of this chapter is as follows. In Sec. 4.3, a hypothetical device called a Carnot engine is introduced and used to prove that the two physical statements of the second law (the Clausius statement and the Kelvin–Planck statement) are equivalent, in the sense that if one is true so is the other. An expression is also derived for the efficiency of a Carnot engine for the purpose of defining thermodynamic temperature. Section 4.4 combines Carnot cycles and the Kelvin–Planck statement to derive the mathematical statements given in Box 4.2.1. Section 4.5 shows the relevance of the second law to various adiabatic processes. Section 4.6 summarizes the results of these sections.

> Carnot engines and Carnot cycles are admittedly outside the normal experience of chemists, and using them to derive the mathematical statement of the second law may

---

[3]This hypothetical process is called "perpetual motion of the second kind."
[4]Planck, p. 89 (reference in Appendix I).

Section 4.3   Concepts Developed with Carnot Engines

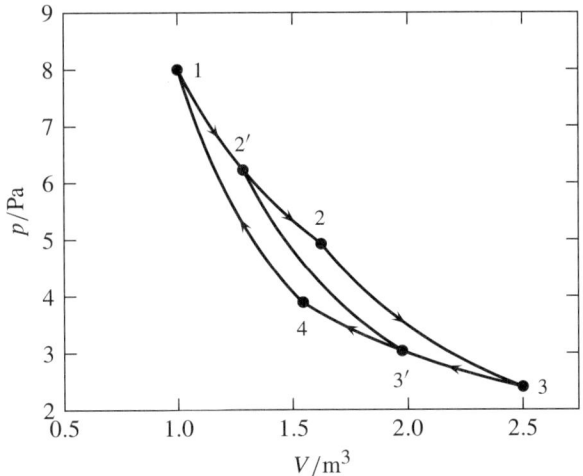

Figure 4.4 Indicator diagram for a Carnot engine using an ideal gas as the working substance. In this example, $T_h = 400\,\text{K}$, $T_c = 300\,\text{K}$, $\epsilon = 1/4$, $C_{V,m} = (3/2)R$, $n = 2.41\,\text{mmol}$. The processes of Paths $1 \to 2$ and $3 \to 4$ are isothermal; those of Paths $2 \to 3$, $2' \to 3'$, and $4 \to 1$ are adiabatic. The cycle $1 \to 2 \to 3 \to 4 \to 1$ has net work $w = -1\,\text{J}$; the cycle $1 \to 2' \to 3' \to 4 \to 1$ has net work $w = -0.5\,\text{J}$.

seem to be completely arcane. G. N. Lewis and M. Randall, in the classic text *Thermodynamics and the Free Energy of Chemical Substances* (McGraw-Hill: New York, 1923), complained of the presentation of " 'cyclical processes' limping about eccentric and not quite completed cycles." However, there seems to be no way to carry out a rigorous *general* derivation without invoking thermodynamic cycles. You may avoid the details by skipping Secs. 4.3 and 4.4. (Incidently, the cycles described in these sections are complete!)

4.3   CONCEPTS DEVELOPED WITH CARNOT ENGINES

4.3.1   Carnot engines and Carnot cycles

A heat engine, as mentioned in Sec. 4.2, is a closed system that converts heat to work and operates in a cycle. A **Carnot engine** is a particular kind of heat engine, one that performs **Carnot cycles** with a working substance. A Carnot cycle has four reversible steps, alternating isothermal and adiabatic; see the examples in Figs. 4.4 and 4.5 in which the working substances are an ideal gas and H_2O, respectively. The steps of a Carnot cycle are as follows (the *system* is the working substance).

Path $1 \to 2$: A quantity of heat q_h is transferred reversibly and isothermally from a heat reservoir (the "hot" reservoir) at temperature T_h to the system, also at T_h.

Path $2 \to 3$: The system undergoes a reversible adiabatic change that does work on the surroundings and reduces the temperature to T_c.

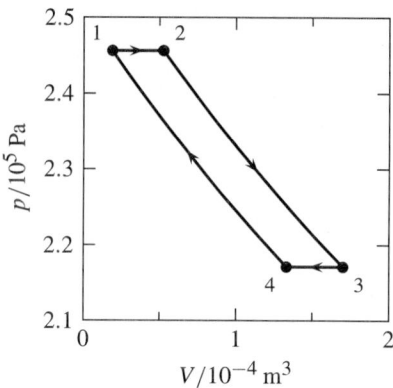

Figure 4.5 Indicator diagram for a Carnot engine using H_2O as the working substance. In this example, $T_h = 400$ K, $T_c = 396$ K, $\epsilon = 1/100$, $w = -1$ J. In State 1, the system consists of one mol of $H_2O(l)$. The processes (all carried out reversibly) are: $1 \to 2$, vaporization of 2.54 mmol H_2O at 400 K; $2 \to 3$, adiabatic expansion, causing vaporization of an additional 7.68 mmol; $3 \to 4$, condensation of 2.50 mmol at 396 K; $4 \to 1$, adiabatic compression returning the system to the initial state.

Path $3 \to 4$: A quantity of heat q_c is transferred reversibly and isothermally from the system to a second heat reservoir (the "cold" reservoir) at temperature T_c.

Path $4 \to 1$: The system undergoes a reversible adiabatic change in which work is done on the system, the temperature returns to T_h, and the system returns to the initial state to complete the cycle.

There is work in each step.[5] The net work w in one cycle is negative. In one cycle a quantity of energy is transferred as heat from the hot reservoir to the system, a portion of this energy is transferred as heat to the cold reservoir, and the remainder of the energy is the net work done on the surroundings. (It is the transfer of heat to the cold reservoir that keeps the Carnot engine from being an impossible Kelvin–Planck engine.) Adjustment of the length of Path $1 \to 2$ makes the magnitude of w as large or small as desired (note the two cycles with different values of w described in the caption of Fig. 4.4).

> The Carnot engine is an idealized heat engine because its paths are reversible processes. It does not resemble the design of any practical steam engine. A cylinder of a typical working steam engine (such as those once used for motive power in train locomotives and steamships) is an *open* system undergoing the following irreversible steps in each cycle: (1) high-pressure steam enters the cylinder from a boiler and pushes the piston from the closed end toward the open end of the cylinder; (2) the supply valve closes and the steam expands in the cylinder until its pressure decreases to atmospheric pressure; (3) an exhaust valve opens to release the steam either to the atmosphere or to a condenser;

---

[5] There could, however, be an isothermal step in which pressure changes at constant volume; this step would have no work.

Section 4.3 Concepts Developed with Carnot Engines

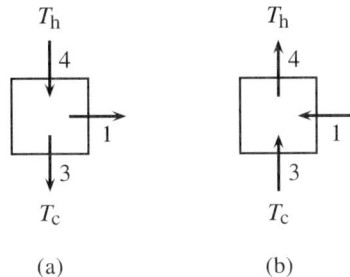

Figure 4.6 (a) One cycle of a Carnot engine that does work on the surroundings. (b) The same system run in reverse as a Carnot heat pump. Figures 4.6–4.8 use the following symbols: A square box represents a system (a Carnot engine or Carnot heat pump). Vertical arrows indicate heat and horizontal arrows indicate work; the arrow direction shows whether energy is transferred into or out of the system. The number next to each arrow is the absolute value of the number of joules of energy transferred in one cycle. For example, (a) shows 4 joules of heat transferred to the system from the hot reservoir, 3 joules of heat transferred from the system to the cold reservoir, and 1 joule of work done by the system on the surroundings.

(4) the piston returns to its initial position, driven either by an external force or by suction created by steam condensation.

The energy transfers involved in one cycle of a Carnot engine are shown schematically in Fig. 4.6(a). When the cycle is reversed, as shown in Fig. 4.6(b), the device is called a **Carnot heat pump**. Since each step of a Carnot engine or Carnot heat pump is a reversible process, neither device is an impossible device.

4.3.2 The equivalence of the Clausius and Kelvin–Planck statements

We can use the logical tool of *reductio ad absurdum* to prove the equivalence of the Clausius and Kelvin–Planck statements of the second law. Let us assume for the moment that the Clausius statement is incorrect and the device the Clausius statement claims is impossible (a "Clausius device") is actually possible. If the Clausius device is possible, then we can combine one of these devices with a Carnot engine as shown in Fig. 4.7(a) on the next page. We adjust the cycles of the Clausius device and Carnot engine to transfer equal quantities of heat from and to the cold reservoir. The combination of the Clausius device and Carnot engine is a system. When the Clausius device and Carnot engine each performs one cycle, the system has performed one cycle as shown in Fig. 4.7(b). There has been a transfer of heat into the system and the performance of an equal quantity of work on the surroundings, with no other change. This system is a heat engine that according to the Kelvin–Planck statement is impossible. Thus, if the Kelvin–Planck statement is correct, it is impossible to operate the Clausius device as shown, and our provisional assumption that the Clausius statement is incorrect must be wrong. In conclusion, if the Kelvin–Planck statement is

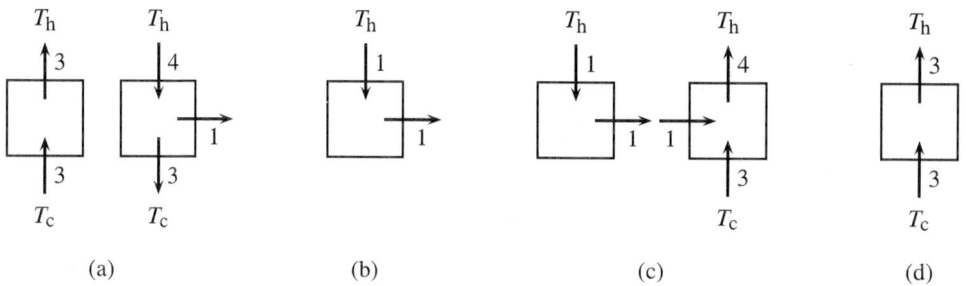

Figure 4.7 (a) A Clausius device combined with the Carnot engine shown in Fig. 4.6(a). (b) The resulting impossible Kelvin–Planck engine. (c) A Kelvin–Planck engine combined with the Carnot heat pump shown in Fig. 4.6(b). (d) The resulting impossible Clausius device.

correct, then the Clausius statement must also be correct.

We can use a similar line of reasoning for the heat engine that the Kelvin–Planck statement claims is impossible (a "Kelvin–Planck engine") by seeing what happens if we assume this engine is actually possible. We combine a Kelvin–Planck engine with a Carnot heat pump, such that the work performed on the Carnot heat pump in one cycle is equal to the work performed by the Kelvin–Planck engine in one cycle, as shown in Fig. 4.7(c). One cycle of the combined system, shown in Fig. 4.7(d), shows the system to be a device that the Clausius statement says is impossible. So if the Clausius statement is correct, then the Kelvin–Planck statement must also be correct.

This conclusion completes the proof that the Clausius and Kelvin–Planck statements are equivalent; the truth of one implies the truth of the other. We may take either statement as the fundamental physical principle of the second law and use it as the starting point for deriving the mathematical form of the second law. This derivation is taken up in Sec. 4.4.

4.3.3 The efficiency of a Carnot engine

Integrating the first law, $dU = dq + dw$, for one cycle of a Carnot engine, we obtain

$$0 = q_h + q_c + w \qquad (4.3.1)$$
(one cycle of a Carnot engine)

The **efficiency**, ϵ, of a heat engine is defined as the fraction of the heat input q_h that is returned as net work $-w$ done on the surroundings: $\epsilon = -w/q_h$. By substituting for w from Eq. 4.3.1, we obtain

$$\epsilon = 1 + \frac{q_c}{q_h} \qquad (4.3.2)$$
(Carnot engine)

The example shown in Fig. 4.6(a) is a Carnot engine with $\epsilon = 1/4$.

We will be able to reach an important conclusion regarding efficiency by considering a Carnot engine operating between the temperatures T_h and T_c, combined with a Carnot heat

Section 4.3 Concepts Developed with Carnot Engines   73

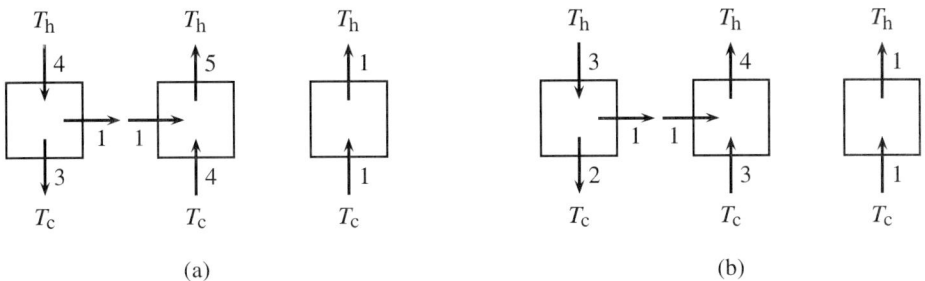

Figure 4.8 (a) A Carnot engine ($\epsilon = 1/4$) combined with a Carnot engine of different efficiency ($\epsilon = 1/5$) run in reverse and the resulting impossible Clausius device. (b) A Carnot engine ($\epsilon = 1/3$) combined with the Carnot engine of (a) ($\epsilon = 1/4$) run in reverse and the resulting impossible Clausius device.

pump operating between the same two temperatures. The combination is a system, and one cycle of the engine and heat pump is one cycle of this combined system. We adjust the cycles of the engine and heat pump to produce zero net work for one cycle of the combined system.

Could the efficiency of the Carnot engine be different from the efficiency the heat pump would have when run in reverse as a Carnot engine? If so, either the combined system is an impossible Clausius device as shown in Fig. 4.8(a), or the combined system operated in reverse (with the engine and heat pump switching roles) is an impossible Clausius device as shown in Fig. 4.8(b). We conclude that *all Carnot engines operating between the same two temperatures have the same efficiency.*

Thus, the efficiency of a Carnot engine must depend only on T_c and T_h regardless of the nature of the working substance. Since the efficiency is given by $\epsilon = 1 + q_c/q_h$, the ratio q_c/q_h must be a unique function of T_c and T_h only. To find this function for temperatures on the ideal-gas temperature scale, it is simplest to choose an ideal gas for the working substance. An ideal gas has the equation of state $pV = nRT$; its internal energy change in a closed system is given by $dU = C_V\,dT$ (Eq. 3.2.2), where C_V (a function only of T) is the heat capacity at constant volume. The work in a Carnot engine is reversible expansion work given in general by $dw = -p\,dV$ and for an ideal gas by $dw = -(nRT/V)\,dV$. Substituting these expressions for dU and dw in the first law, $dU = dq + dw$, and solving for dq gives

$$dq = C_V\,dT + \frac{nRT}{V}\,dV \qquad (4.3.3)$$
(ideal gas, reversible expansion work only)

In the two adiabatic steps of the Carnot cycle, dq is zero. We obtain a relation among the volumes of the four labeled states shown in Fig. 4.4 by integrating dq/T in these steps and setting the integrals equal to zero:

$$\text{Path } 2 \to 3: \qquad \int \frac{dq}{T} = \int_{T_h}^{T_c} \frac{C_V\, dT}{T} + nR \ln \frac{V_3}{V_2} = 0$$

$$\text{Path } 4 \to 1: \qquad \int \frac{dq}{T} = \int_{T_c}^{T_h} \frac{C_V\, dT}{T} + nR \ln \frac{V_1}{V_4} = 0$$

Adding these two equations (the integrals shown with limits cancel) gives the relation

$$nR \ln \frac{V_1 V_3}{V_2 V_4} = 0$$

which we can rearrange to

$$\ln(V_2/V_1) = -\ln(V_4/V_3) \qquad (4.3.4)$$
$$\text{(ideal gas, Carnot cycle)}$$

We obtain expressions for the heat in the two isothermal steps by integrating Eq. 4.3.3 with dT set equal to 0:

$$\text{Path } 1 \to 2: \qquad q_h = nRT_h \ln(V_2/V_1)$$
$$\text{Path } 3 \to 4: \qquad q_c = nRT_c \ln(V_4/V_3)$$

The ratio of q_c and q_h obtained from these expressions is

$$\frac{q_c}{q_h} = \frac{T_c}{T_h} \times \frac{\ln(V_4/V_3)}{\ln(V_2/V_1)}$$

By means of Eq. 4.3.4, this ratio becomes

$$\frac{q_c}{q_h} = -\frac{T_c}{T_h} \qquad (4.3.5)$$
$$\text{(Carnot cycle)}$$

Accordingly, the unique function of T_c and T_h we seek for q_c/q_h is $-T_c/T_h$. The efficiency, from Eq. 4.3.2, is then given by

$$\epsilon = 1 - \frac{T_c}{T_h} \qquad (4.3.6)$$
$$\text{(Carnot engine)}$$

In Eqs. 4.3.5 and 4.3.6, T_c and T_h are temperatures on the ideal-gas scale. As we have seen, these equations must be valid for *any* working substance; we do not need to specify as a condition of validity that the system is an ideal gas. The ratio T_c/T_h is positive but less than one, so the efficiency is less than one. We did not use the second law to reach this conclusion, but the Carnot engine does illustrate the Kelvin–Planck statement of the second law: A heat engine cannot have an efficiency of unity—that is, cannot in one cycle convert all of the heat transferred from a single heat reservoir into work. The example shown in Fig. 4.6 on page 71, with $\epsilon = 1/4$, must have $T_c/T_h = 3/4$ (e.g., $T_c = 300\,\text{K}$, $T_h = 400\,\text{K}$).

Keep in mind that a Carnot engine operates *reversibly* between two heat reservoirs. The expression of Eq. 4.3.6 gives the efficiency of this kind of heat engine only. If any part of the cycle is carried out spontaneously, dissipation of mechanical energy will cause the efficiency to be *lower* than the theoretical value given by Eq. 4.3.6.

4.3.4 Thermodynamic temperature

The ratio q_c/q_h for a Carnot cycle depends only on the temperatures of the two heat reservoirs. Kelvin (1848) proposed that this ratio be used to establish an "absolute" temperature scale. The physical quantity now called **thermodynamic temperature** is defined by the relation

$$\frac{T_c}{T_h} = -\frac{q_c}{q_h} \qquad (4.3.7)$$
(Carnot cycle)

That is, the ratio of the thermodynamic temperatures of two heat reservoirs is equal, by definition, to the ratio of the absolute quantities of heat transferred in the isothermal steps of a Carnot cycle operating between these two temperatures. In principle, a measurement of q_c/q_h during a Carnot cycle, combined with a defined value of the thermodynamic temperature of one of the heat reservoirs, would establish the thermodynamic temperature of the other heat reservoir. The thermodynamic temperature of the triple point of H_2O is defined as exactly 273.16 kelvins (Sec. 2.6.1). Just as measurements with a gas thermometer in the limit of zero pressure establish the ideal-gas temperature scale (Sec. 2.6.4), the behavior of a Carnot engine in the limit of reversible conditions establishes the thermodynamic temperature scale. However, a reversible Carnot engine as a "thermometer" to measure thermodynamic temperature is a concept and not a practical instrument since a completely reversible process cannot occur in practice.

We are now able to justify the statement in Sec. 2.6.1 that the ideal-gas temperature scale is proportional to the thermodynamic temperature scale. Both Eq. 4.3.5 and Eq. 4.3.7 equate the ratio T_c/T_h to $-q_c/q_h$; but whereas T_c and T_h refer in Eq. 4.3.5 to the *ideal-gas* temperatures of the heat reservoirs, in Eq. 4.3.7 they refer to the *thermodynamic* temperatures. This means the ratio of the ideal-gas temperatures of two bodies is equal to the ratio of the thermodynamic temperatures of the same bodies, and therefore the two scales are proportional to one another. The proportionality factor is arbitrary, but must be unity if the same unit (e.g., kelvins) is used in both scales. Thus, as discussed in Sec. 2.6.1, the two scales expressed in kelvins are identical.

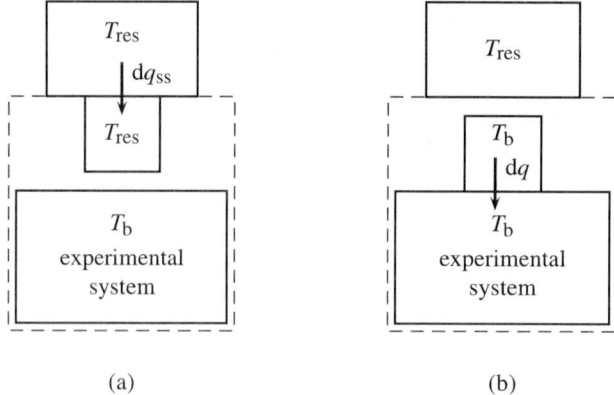

Figure 4.9 Experimental system, Carnot engine (represented by a small square box), and heat reservoir. The dashed lines indicate the boundary of the supersystem. (a) Reversible heat transfer from heat reservoir to Carnot engine. (b) Reversible heat transfer from Carnot engine to experimental system.

4.4 DERIVATION OF THE MATHEMATICAL FORM OF THE SECOND LAW

4.4.1 The existence of the entropy function

We derive the existence and properties of the state function called entropy[6] by considering various processes of a closed system, the "experimental" system. The experimental system may have any degree of complexity: Phase changes and reactions may be possible, constraints and external fields may be present, and so on.

Suppose we wish to study a particular nonadiabatic experimental process A → B that is either irreversible or reversible (but not impossible). We begin by imagining that the experimental system follows the path of interest in combination with a hypothetical Carnot engine whose function is to exchange the required quantities of heat with the experimental system. The combination of the experimental system and the Carnot engine is a closed *supersystem* (see Fig. 4.9). As the process proceeds, the Carnot engine undergoes many small Carnot cycles (cycles with infinitesimal net work). We treat the four steps of each cycle, described in Sec. 4.3.1, as infinitesimal changes. In one of the isothermal steps of a cycle the Carnot engine is in thermal contact with a heat reservoir of arbitrary fixed temperature T_{res} in the surroundings of the supersystem; heat dq_{ss} is transferred from the heat reservoir to the Carnot engine; see Fig. 4.9(a). In the other isothermal step, the Carnot engine is in thermal contact with the experimental system at a portion of its boundary where the temperature at the moment is T_b; heat dq is transferred from the Carnot engine to the experimental system[7] ; see Fig. 4.9(b). The heat dq is the heat required by the experimental

---

[6] So named by Clausius, from the Greek word *trope* meaning *transformation*.

[7] A negative value of either dq_{ss} or dq corresponds to heat transfer in the opposite direction.

Section 4.4 Derivation of the Mathematical Form of the Second Law

system in the current infinitesimal path element of its process.

The relation between the temperatures and heats in the isothermal steps of a Carnot cycle is $T_c/T_h = -q_c/q_h$ (Eq. 4.3.7), where the subscripts c and h stand for "cold" and "hot." If T_b is less than T_{res}, heat flow is in the directions of the arrows shown in Fig. 4.9; then T_c in Eq. 4.3.7 should be replaced by T_b, T_h by T_{res}, q_c by $-dq$, and q_h by dq_{ss}. If T_b is greater than T_{res}, then T_c is replaced by T_{res}, T_h by T_b, q_c by $-dq_{ss}$, and q_h by dq. In either case, the relation becomes, for one small Carnot cycle,

$$\frac{T_b}{T_{res}} = \frac{dq}{dq_{ss}}$$

which we rearrange to

$$dq_{ss} = T_{res} \frac{dq}{T_b} \qquad (4.4.1)$$

After many small Carnot cycles, the experimental process is complete, the experimental system is in its final state B, and the Carnot engine has returned to its initial state in thermal contact with the heat reservoir. Integration of dq_{ss} in Eq. 4.4.1 gives, for the total heat entering the supersystem during the process:

$$q_{ss} = T_{res} \int_A^B \frac{dq}{T_b} \qquad (4.4.2)$$

In the integral $\int dq/T_b$, the integration is to be taken over each path element of the experimental process and each surface element of the boundary.

Suppose the integral $\int dq/T_b$ is positive so that q_{ss} is positive. Would it be possible in this case for the experimental system to return from its final state B to its initial state A by a further *adiabatic* process? (The Carnot engine would be inactive during the adiabatic process.) If so, the combination of the two processes would be a cycle of the supersystem in which the net heat is positive and the net work is negative (to make the internal energy change zero). In other words, in the cycle the supersystem would convert heat from a single heat reservoir completely into work, a process the Kelvin–Planck statement of the second law says is impossible. Since the initial nonadiabatic change of the experimental system from A to B is not impossible, and the Carnot engine undergoes only reversible changes, it must be the *adiabatic* change of the experimental system that is impossible.

Thus, we have shown from the Kelvin–Planck statement that if the integral $\int dq/T_b$ is positive for any reversible or irreversible process A → B in a closed system, it is impossible for the change B → A to occur by any adiabatic process, whether reversible or irreversible.

Next let us investigate a completely *reversible* nonadiabatic process of the closed experimental system that changes it from a particular equilibrium state A to another equilibrium state B. Let us assume heat flows continuously into the system—that is, dq is positive in each path element. If each infinitesimal quantity of heat dq is positive, then $\int dq/T_b$ must be positive; then, according to our earlier conclusion, it is impossible to carry out the reverse change B → A by any *adiabatic* process, including a reversible adiabatic process. Any reversible process can be carried out in reverse. Thus, by reversing the reversible nonadiabatic process, it is possible to change the state from B to A by a reversible process with a continuous flow of heat out of the system. In contrast, the lack of a reversible *adiabatic* path from B to A means that it is impossible to carry out the change A → B by a reversible adiabatic process.

The general rule, then, is that whenever one equilibrium state of a closed system can be changed to another by a reversible nonadiabatic process with "one-way" heat (i.e., the flow of heat is either entirely into the system or else entirely out of it), it is impossible for the system to change from either of these states to the other by a reversible *adiabatic* process.

> A simple example will relate this rule to experience. We can increase the temperature of a metal body at constant pressure by allowing heat to flow reversibly into the body. It is impossible to duplicate this change of state by a reversible process without heat. There is nothing in the rule that says we can't increase the temperature *irreversibly* without heat, as in fact we can with electrical work (electrical heating).

States A and B can be arbitrarily close. We conclude that *every equilibrium state of a closed system has other equilibrium states infinitesimally close to it that are inaccessible by a reversible adiabatic process*. This is Carathéodory's principle of adiabatic inaccessibility.[8]

Next let us consider the reversible adiabatic processes that *are* possible. To carry out a reversible adiabatic process, starting at an initial equilibrium state, we use an adiabatic boundary and vary one or more of the work coordinates. A certain final temperature will result. It is helpful in visualizing this process to think of an N-dimensional space in which each axis represents one of the N independent variables needed to describe an equilibrium state. A point in this space represents an equilibrium state, and the path of a reversible process can be represented as a curve in this space. A suitable set of independent variables for equilibrium states of a closed system of uniform temperature consists of the temperature T and each of the work coordinates (Sec. 3.10). Since, in our reversible adiabatic process, we can vary the work coordinates independently while keeping the boundary adiabatic, the paths for possible reversible adiabatic processes can connect any arbitrary combinations of work coordinate values. But there is the additional dimension of temperature in the N-dimensional space. Do the paths for possible reversible adiabatic processes, starting from a common initial point, lie in a *volume* in the N-dimensional space? Or do they fall on a *surface* described by T as a function of the work coordinates? If the paths lie in a volume, then every point in a volume element surrounding the initial point must be accessible from the initial point by a reversible adiabatic path. This accessibility is precisely what Carathéodory's principle of adiabatic inaccessibility denies. Therefore, the paths for all possible reversible adiabatic processes with a common initial state must lie on a unique *surface* (or a curve if N is 2). We refer to one of these surfaces or curves as a **reversible adiabatic surface**.

Now consider the initial and final states of a reversible process with one-way heat (i.e., each infinitesimal quantity of heat dq has the same sign). Since we have seen that it is impossible for there to be a reversible *adiabatic* path between these states, the points for

---

[8] Constantin Carathéodory in 1909 combined this principle with a purely mathematical theorem (Carathéodory's theorem) to deduce the existence of the entropy function. The argument outlined here avoids the complexities of that mathematical treatment and leads to the same results.

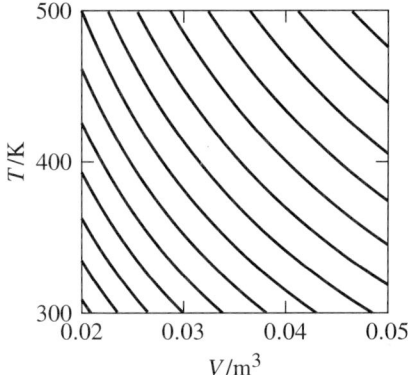

Figure 4.10 A family of adiabatic curves of an ideal gas with V and T as independent variables. A reversible adiabatic process moves the state of the system along a curve, whereas a reversible diathermal process with positive heat moves the state from one curve to another above and to the right. The curves are calculated for $n = 1$ mol and $C_{V,\mathrm{m}} = (3/2)R$. Adjacent curves differ in entropy by $1\,\mathrm{J\,K^{-1}}$.

these states must lie on different reversible adiabatic surfaces that do not intersect anywhere in the N-dimensional space. Consequently, there is an infinite number of nonintersecting reversible adiabatic surfaces filling the N-dimensional space. (To imagine what this looks like in a three-dimensional space, think of distorted layers of sedimentary rock or the pages of a flexed magazine.) A reversible, nonadiabatic process with one-way heat is represented by a path beginning at a point on one reversible adiabatic surface and ending at a point on a different surface. If q is positive, the final surface lies on one side of the initial surface, and if q is negative, the final surface is on the opposite side.

4.4.2 Using reversible processes to define the entropy

The existence of reversible adiabatic surfaces is the justification for defining a new state function S, the **entropy**. We specify that S has the same value everywhere on one of these surfaces, and a different, unique value on each different surface. In other words, the reversible adiabatic surfaces are surfaces of *constant entropy* in the N-dimensional space. The fact that the surfaces fill this space without intersecting ensures that S is a state function for equilibrium states because any point in this space represents an equilibrium state and also lies on a single reversible adiabatic surface with a definite value of S.

We know the entropy function must exist since the reversible adiabatic surfaces exist. For instance, Fig. 4.10 shows a family of these surfaces for a closed system of a pure substance in one phase. In this system, N is equal to 2, and the surfaces are two-dimensional curves. Each curve is a contour of constant S. At this stage in the derivation, our assignment of values of S to the different curves is entirely arbitrary, except that S should increase monotonically in the direction of positive heat.

How can we assign a unique value of S to each reversible adiabatic surface? We can order the values by letting a reversible process with *positive* one-way heat, which moves the point for the state to a new surface, correspond to an *increase* in the value of S. Negative one-way heat will then correspond to decreasing S. We can assign an arbitrary value to the entropy on one particular reversible adiabatic surface. (The third law of thermodynamics is used for this purpose—see Sec. 4.10.) Then all that is needed to assign a value of S to each equilibrium state is a formula for evaluating the *difference* in the entropies of any two surfaces.

Consider a reversible process with positive one-way heat that changes the system from State A to State B. The path for this process must move the system from a reversible adiabatic surface of lower entropy to a different surface of greater entropy. We combine the experimental system with a Carnot engine to form a supersystem that exchanges heat with a single heat reservoir of constant temperature T_{res}. The net heat entering the supersystem is, from Eq. 4.4.2,

$$q_{\text{ss}} = T_{\text{res}} \int_A^B \frac{dq}{T_b}$$

and it is positive. Now consider a second reversible process of the experimental system, not necessarily involving one-way heat, along a second path from State C on the same lower entropy surface to State D on the same higher entropy surface; the net heat entering the supersystem is q'_{ss}. We can then devise a *cycle* of the supersystem in which the experimental system undergoes the reversible Steps A → B → D → C → A. Step A → B is the first process described above, Step D → C is the reverse of the second process described above, and Steps B → D and C → A each have a path on one of the reversible adiabatic surfaces. The net heat entering the supersystem in the cycle is $q_{\text{ss}} - q'_{\text{ss}}$. Now unless q'_{ss} and q_{ss} are equal, either the net heat in the cycle is positive or the net heat in the reverse cycle is positive, with the heat drawn from a single heat reservoir. Either situation is impossible according to the Kelvin–Planck statement. So q_{ss} and q'_{ss} must be equal, and the integral $\int dq/T_b$ must be the same when evaluated along either of the reversible paths from the lower to the higher entropy surface. Note that since the second path does not necessarily have one-way heat, it could have taken the experimental system through any sequence of intermediate entropy values as long as the path started at the lower entropy surface and ended at the higher. Furthermore, since the path is reversible, it can be carried out in reverse resulting in reversal of the sign of ΔS and $\int dq/T_b$.

It should now be clear that a satisfactory formula for defining the entropy change of a reversible process in a closed system is

$$\Delta S = \int \frac{dq}{T_b} \qquad (4.4.3)$$

(reversible process, closed system)

This formula satisfies the necessary requirements: It makes ΔS for a reversible process positive if the process has positive one-way heat, negative if the process has negative one-way heat, and zero if the process is adiabatic. It gives the same value of ΔS for any reversible change between the same two reversible adiabatic surfaces, and it makes the sum of the ΔS values of several consecutive reversible processes equal to ΔS for the overall process.

For an infinitesimal change, Eq. 4.4.3 becomes

$$dS = \frac{dq}{T_b} \qquad (4.4.4)$$
(reversible process, closed system)

It is common to see this equation written in the form $dS = dq_{\text{rev}}/T$, where dq_{rev} denotes an infinitesimal quantity of heat in a reversible process.

> In Eq. 4.4.4, the quantity $1/T_b$ is called an *integrating factor* for dq, a factor that makes the product dq/T_b be the infinitesimal change of a state function. The quantity $1/cT_b$, where c is any nonzero constant, would also be a satisfactory integrating factor; so actually the definition of entropy, using $c = 1$, is one of an infinite number of possible choices for assigning values to the reversible adiabatic surfaces.

4.4.3 Some properties of the entropy

It is not difficult to show that S is an *extensive* property of a closed system in an equilibrium state. If one imagines a system of uniform temperature T to be divided into two closed subsystems A and B and a reversible process to occur, the entropy changes of the subsystems are $\Delta S_A = \int dq_A/T$ and $\Delta S_B = \int dq_B/T$ and of the system $\Delta S = \int dq/T$. But q is the sum of q_A and q_B, which gives $\Delta S = \Delta S_A + \Delta S_B$. Thus, the entropy changes are additive, and so entropy must be extensive.[9]

How can we evaluate S for a particular equilibrium state of the system? We must assign an arbitrary value to one state and then evaluate the change ΔS along a reversible path from this state to the state of interest using Eq. 4.4.3.

We need to be able to evaluate S for *non*equilibrium states as well. The general procedure is to apply Eq. 4.4.3 to a reversible path that changes the system from an equilibrium state of known entropy to the nonequilibrium state of interest. If the nonequilibrium state is nonuniform, it may be necessary conceptually to divide the system into many regions of essentially uniform intensive properties and carry out a reversible process on each region individually. Since S is extensive, the entropy changes for the separate portions are additive. It may also be possible to maintain equilibrium states during the process with the use of constraints such as semipermeable membranes, internal adiabatic partitions, and reaction anticatalysts (Sec. 3.1.3).

An additional strategy is needed for a nonequilibrium state that has macroscopic internal motion since such a state cannot be part of a reversible process. To handle this situation, we simply define the entropy of a state with macroscopic motion to be the same as the entropy of the state with the same internal structure but no internal motion—the same state frozen in time. The entropy of the frozen state can be evaluated with the help of Eq. 4.4.3.

---

[9]The argument is not quite complete because we have not shown that when each subsystem has an entropy of zero so does the entire system. The zero of entropy is discussed in Sec. 4.10.

4.4.4 Irreversible processes

Consider a reversible process of a closed system that increases the entropy: The integral $\int dq/T_b$ is positive. In Sec. 4.4.1, we used the Kelvin–Planck statement to deduce that the system in this case cannot return to its initial state by any adiabatic process. Thus, *it is impossible for the entropy to decrease during an adiabatic process of a closed system.* If this is true for an arbitrary finite change, it must be true for each infinitesimal path element: dS is either zero or greater than zero in an adiabatic process of a closed system. We know that dS is zero in each infinitesimal path element of a *reversible* adiabatic process, but what about an *irreversible* adiabatic process?

The reverse of an irreversible process is impossible. We would expect the Kelvin–Planck statement to show that this reverse adiabatic process of a closed system is impossible by causing a decrease in entropy. Thus, an irreversible adiabatic process of a closed system should cause an entropy *increase*.[10] If this is true for an arbitrary finite change, it must be true for each infinitesimal path element:

$$dS > 0 \qquad (4.4.5)$$
(irreversible adiabatic process, closed system)

What about an irreversible *non*adiabatic process? As in Sec. 4.4.1, we combine the experimental system with a Carnot engine and a heat reservoir of constant temperature T_{res}, but this time the supersystem includes the heat reservoir in addition to the experimental system and the Carnot engine as shown in Fig. 4.11 on the next page. During the experimental process, the supersystem exchanges work but no heat with its surroundings. During one small cycle of the Carnot engine, the entropy change of the Carnot engine is zero, the net entropy change of the experimental system is dS, the heat entering the experimental system from the Carnot engine is dq, and the heat entering the Carnot engine from the heat reservoir is $T_{res}\, dq/T_b$ (Eq. 4.4.1). (Any of these infinitesimal quantities of heat is negative if heat leaves instead of enters.) Heat leaves the heat reservoir reversibly, so the entropy change of the heat reservoir is $dS_{res} = dq_{res}/T_{res} = (-T_{res}\, dq/T_b)/T_{res} = -dq/T_b$. The entropy change of the supersystem, which includes the heat reservoir, is

$$dS_{ss} = dS + dS_{res} = dS - dq/T_b$$

Since the supersystem is changing adiabatically, its entropy must increase if the change within the experimental system is irreversible: $dS_{ss} > 0$. In this case, the entropy change

---

[10]This argument is plausible but not rigorous. While we know it is impossible for S to decrease in an adiabatic process of a closed system, we cannot prove that S must decrease in every impossible adiabatic process. Invariably, however, we find that S *increases* in an irreversible adiabatic process. Some examples to illustrate this fact are given in Sec. 4.5.

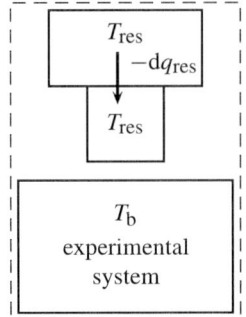 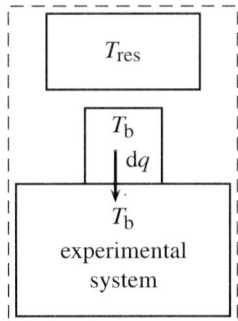

Figure 4.11 Adiabatic supersystem including the experimental system, Carnot engine, and heat reservoir.

in the experimental system must be greater than dq/T_b:[11]

$$dS > \frac{dq}{T_b} \qquad (4.4.6)$$
(irreversible process, closed system)

This relation is known as the **Clausius inequality**. It is equivalent to Eq. 4.4.5 if the irreversible process is adiabatic.

We have now derived, from the Kelvin–Planck statement, all parts of the mathematical statement of the second law shown in Box 4.2.1.

The mathematical statement includes an equality, $dS = dq/T_b$, that is true for a *reversible* process and an inequality, $dS > dq/T_b$, that is true for an *irreversible* process. The integrated forms are $\Delta S = \int dq/T_b$ for a reversible process and $\Delta S > \int dq/T_b$ for an irreversible process. Consider two particular equilibrium states 1 and 2 of a closed system. The system can change from State 1 to State 2 either reversibly or irreversibly. Since S is a state function, it is the integral $\int dq/T_b$ that is different in the two cases, not ΔS. Our general principle for evaluating the entropy difference between the two states must be to equate it to the integral evaluated for a *reversible* change from one state to the other. For an irreversible change between the same states, the integral has a smaller value.

> Note that Box 4.2.1 makes no statement about an impossible process. In fact, the infinitesimal entropy change dS of an impossible process could be less than, equal to, or greater than dq/T_b. Consider a system containing three bodies that have different temperatures and are in thermal contact; an adiabatic process in which heat flows internally to both the warmest and coolest bodies from the body of intermediate temperature is impossible, and dS could be either less than, equal to, or greater than zero.

---

[11] In the derivation T_b is the temperature of the Carnot engine as it exchanges heat with the experimental system. If we want to treat the temperature as being *discontinuous* across a surface element where heat is transferred, T_b must represent the temperature in the experimental system's *surroundings*.

4.5 ADIABATIC PROCESSES: EXAMPLES

An adiabatic process is a process without heat: $dq = 0$. The second law says the entropy of a closed system is constant if the adiabatic process is reversible and increases if the process is irreversible. The examples of irreversible adiabatic processes in this section illustrate alternative reversible paths that allow us to evaluate the entropy change and confirm that the change is positive.

4.5.1 Adiabatic processes with work

The generalization was made in Sec. 3.10 that, for a given change in a work coordinate, the work is smallest (i.e., the least positive or most negative) when the adiabatic process is carried out reversibly. That is, ΔU for a reversible adiabatic process is algebraically less than ΔU for an irreversible adiabatic process with the same change of work coordinate.

We can use this result to show that, in the case of an irreversible adiabatic process with work, the entropy must increase as stated by the second law. This is a process with energy dissipation as discussed in Sec. 3.7. In the irreversible process, there is a certain change in a work coordinate. Suppose we make the same change in the work coordinate reversibly and adiabatically; by definition, the entropy is constant. If we then increase U by another reversible process while keeping the work coordinate constant, we can bring the system to the final state of the irreversible process. This last reversible step requires positive heat, which increases the entropy. Since entropy is a state function, it must be greater in the final state of the irreversible process than in the initial state.

4.5.2 Isolated systems

A closed system undergoing an adiabatic process without work is an *isolated* system. Any spontaneous change in an isolated system must arise solely from conditions within the system, uninfluenced by the surroundings; the change occurs "by itself." Furthermore, it can occur only if the isolated system starts in a nonequilibrium state because by definition an equilibrium state of an isolated system does not change over time. Unless the spontaneous change is purely mechanical, it is irreversible and we may describe it as a drive toward an equilibrium state; that is, as a drive toward thermal, mechanical, transfer, and reaction equilibrium. The second law states that the entropy increases. We find this principle to be one of the most useful for deriving conditions for spontaneity and equilibrium in chemical systems: *The entropy of an isolated system increases during a spontaneous process and reaches a maximum at equilibrium.*

If we treat the universe as an isolated system (although cosmology provides no assurance that this is valid), we can say that as long as anything happens in the universe its entropy continually increases. Clausius summarized the first and second laws in the famous statement: "Die Energie der Welt ist constant; die Entropie der Welt strebt einem Maximum zu" (The energy of the universe is constant; the entropy of the universe strives for a maximum).

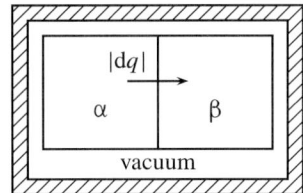

Figure 4.12 Internal heat flow in an isolated system.

4.5.3 Internal heat flow

Figure 4.12 shows a simple setup that illustrates the principle of entropy increase in an isolated system. The system consists of two bodies, α and β, which initially are separated and have uniform but different temperatures T^α and T^β, T^α being greater. We place the bodies in thermal contact and enclose them with an adiabatic boundary. We know the spontaneous process is the irreversible transfer of heat from the warm body to the cool body, in this case from α to β. Eventually the system reaches an equilibrium state in which both bodies have the same uniform temperature, intermediate in value between T^α and T^β.

What is the entropy change in this irreversible process? We can calculate it by devising a *reversible* path between the same initial and final states. We treat each body as a subsystem. The same change of state as before will take place if we place each body in thermal contact with separate heat baths and use reversible processes to simultaneously transfer heat from body α to its surroundings and heat to body β from its surroundings. We carry out these reversible heat exchanges at equal rates but so slowly that temperature gradients are negligible (which was not true of the irreversible process). The quantities of heat transferred from body α and to body β must be equal in magnitude and the same as the heat transferred between the two bodies during the irreversible process. In an infinitesimal path element of the reversible process body α (treated as a subsystem) has negative heat of magnitude $|dq|$ and body β has positive heat of the same magnitude. Then according to Eq. 4.4.4, the infinitesimal entropy change is

$$dS = dS^\alpha + dS^\beta = -\frac{|dq|}{T^\alpha} + \frac{|dq|}{T^\beta} = |dq|\left(-\frac{1}{T^\alpha} + \frac{1}{T^\beta}\right)$$

which is *positive* as long as T^α is greater than T^β. Therefore, the overall entropy change ΔS in the reversible process is positive. Since entropy is a state function, its change depends only on the initial and final states; thus, the entropy change in the irreversible process in the isolated system must also have been positive.

4.5.4 Free expansion of an ideal gas

For the final example of a spontaneous process in an isolated system, consider the free expansion shown in Fig. 3.8 on page 50. At the instant the stopcock is opened, the system consists of a uniform gas phase in one portion and a vacuum in the other. When the gas expands into the vacuum, there is no work. If the walls are adiabatic, there is no heat and so

Table 4.1 Relations for dS in closed systems

	reversible	irreversible	purely mechanical
adiabatic process	dS = 0	dS > 0	dS = 0
nonadiabatic process	dS = dq/T_b	dS > dq/T_b	

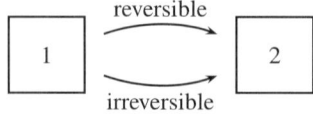

Figure 4.13 Two paths between an initial and a final state. ΔS is the same for both paths, whereas $\int \mathrm{d}q/T_b$ is in general different.

no change in internal energy. Then if the gas is ideal, its temperature remains constant. We can carry out the same change of state by a reversible isothermal expansion with negative work (see Eq. 3.6.10). Since ΔU is zero, the heat in the reversible process is positive, and therefore the entropy change is positive.

4.6 IN RETROSPECT: PROCESSES AND ENTROPY CHANGES

Any conceivable process is either spontaneous, reversible, or impossible.

The lengthy derivation in Sec. 4.4 was based on the Kelvin–Planck statement describing the impossibility of converting energy, transferred into the system as heat from a single heat reservoir, completely into work. Eventually the derivation showed that during a reversible process of a closed system the infinitesimal quantity dq/T_b (dq is heat transferred at the boundary where the temperature is T_b) equals the infinitesimal change of a state function called the entropy, S. Furthermore, a change in a closed system that makes dS less than dq/T_b is an impossible change.

A *reversible* process proceeds by a continuous sequence of equilibrium states. Note that the equality of dS and dq/T_b is a necessary, but not sufficient, condition for a process of a closed system to be reversible.

A *spontaneous* process is one that proceeds naturally at a finite rate. An *irreversible* process is a spontaneous process that cannot take place in reverse. In each path element of an irreversible process in a closed system, dS is greater than dq/T_b. A *purely mechanical process* is a process that is spontaneous in either direction; the entropy is constant during a purely mechanical process of a closed system. Except for the idealized purely mechanical process, we can use the terms *spontaneous* and *irreversible* interchangeably. Table 4.1 summarizes the relations for dS in these various kinds of processes.

Since entropy is a state function, its change during a process depends only on the initial and final states, not on the path (see Fig. 4.13). Heat, being a path function, depends in

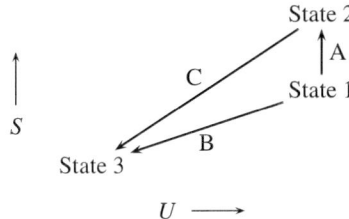

Figure 4.14 Paths to illustrate a change in unavailable energy.

general on the path. Thus, our basic strategy for determining the entropy change of a process in a closed system is to find a reversible path from the initial to the final state and then to evaluate the integral $\int dq/T_b$ along this reversible path. The entropy change is equal to the value of this integral.

The second law establishes no general relation between entropy changes and heat in an open system or for an impossible process. The entropy of an open system may increase or decrease depending on whether matter enters or leaves. It is possible to find examples of impossible processes in which dS is less than, equal to, or greater than dq/T_b.

4.7 WHAT IS THE "MEANING" OF ENTROPY?

We know that, by definition, the difference in the entropy of two states of a closed system is equal to the integral $\int dq/T$ integrated along a reversible path between the states. But what is the physical significance of entropy? There are two short answers: Entropy is a measure of energy that is unavailable to do work on the surroundings, and entropy is a measure of the disorder of the system.

4.7.1 Entropy as a measure of unavailable energy

Consider the paths between states of a closed system indicated in Fig. 4.14. State 1 is an arbitrary state of the system. Path A is for a spontaneous, *irreversible* process that changes State 1 to State 2 while the system is *isolated*. The internal energy change in this process, $\Delta_A U$, must be zero and the entropy change $\Delta_A S$ must be positive.

We now want to compare the maximum quantities of work the system in these two states can perform on the surroundings while changing to a reference state (State 3) without withdrawing heat from the surroundings. (If heat were allowed to enter the system, the system could perform an indefinite quantity of work.)

We shall assume that the surroundings contain a heat reservoir whose temperature T_0 is no greater than the temperature of any part of the system in States 1 or 2. State 1 changes to State 3 by Path B, and State 2 changes to State 3 by Path C. In both of these paths, we want the net effect on the surroundings to be the transfer of energy into the surroundings in the form of work and the transfer of heat into the heat reservoir. The maximum work the system can perform on the surroundings (i.e., the most negative value of w) occurs when the heat is transferred reversibly by a hypothetical Carnot engine operating in small cycles.

The concept of the reversible heat transfer is the same as shown in Fig. 4.9 on page 76, where the experimental system and the Carnot engine constitute a supersystem. In Paths B and C, the heat transfers are in the opposite directions from those indicated by the arrows in Fig. 4.9. The heat transferred into the supersystem is given by Eq. 4.4.2 on page 77, which for Paths B and C is $q_{ss} = T_0 \int dq/T = T_0 \Delta S$. The work in Paths B and C is the work of the supersystem given according to the first law by $\Delta U - q_{ss}$. Accordingly, we write $w_B = \Delta_B U - T_0 \Delta_B S$ and $w_C = \Delta_C U - T_0 \Delta_C S$. The difference between the work in Paths B and C, considering that $\Delta_B U$ and $\Delta_C U$ are equal, is then

$$w_C - w_B = -T_0(\Delta_C S - \Delta_B S) = T_0 \Delta_A S$$

Since $\Delta_A S$ is positive, w_C is less negative than w_B.

In other words, an irreversible process in an isolated system reduces the quantity of work that the system can subsequently perform on the surroundings. When an isolated system undergoes an irreversible change, its energy remains constant but the energy is degraded—unable to perform as much work on the surroundings. This is the *principle of the degradation of energy*.

Energy transferred as work is sometimes called "available energy" because in principle, in the absence of friction, it can be completely converted to useful mechanical work such as the raising of a weight or the elastic deformation of a spring. The decrease of available energy, or the increase of unavailable energy, during an irreversible process of an isolated system is equal to $T_0 \Delta S$, where ΔS is the entropy increase and T_0 is the lowest temperature available in the surroundings. Thus, for a given value of T_0, the entropy is a measure of the unavailable energy.

4.7.2 Entropy as a measure of disorder

Entropy is often said to be a measure of "disorder." Consider the kinds of processes that increase the entropy of a closed system. The entropy change of any reversible process in which heat flows into the system ($dq > 0$) is positive. This includes the reversible isothermal expansion of a gas and an endothermic equilibrium phase transition such as melting or vaporization. In these processes, we can think of the microscopic particles (e.g., molecules) as gaining spatial freedom, either because of increased volume in which to move or decreased restrictions from interparticle forces.

The entropy change is also positive if heat causes the temperature to increase at constant volume. Thus, the "disorder" in question has both spatial and energetic aspects.

The discipline of statistical mechanics provides a precise definition of entropy based on the behavior of macroscopic amounts of microscopic particles. The fundamental assumption of statistical mechanics is that different quantum states of a closed system that have equal energies are equally probable. The entropy then turns out to be given by Boltzmann's equation

$$S = k \ln W$$

where k is the Boltzmann constant $k = R/N_A$ and W is the number of quantum states of the system that have energies close to the average energy of the system.

The second law is a statement of behavior that on the macroscopic level is inevitable but on the microscopic level can be violated locally for a short time. It is all a matter of probability. For example, consider the free expansion of an ideal gas from one vessel into an evacuated second vessel in an isolated system (Fig. 3.8 and Sec. 4.5.4). The process is irreversible and the entropy change is positive. The second law states that the reverse process, the movement of the gas entirely back into the original vessel while the system is isolated, is impossible because that would require ΔS (negative) to be less than $\int dq/T_b$ (zero).

The individual molecules of an ideal gas, however, move independently of one another. We can easily calculate the probability that each of the molecules is in just one of the vessels when the connection between the vessels is open. Suppose both vessels have the same volume; then each molecule has an equal probability of being in either vessel, independent of the positions of any other molecules. If the system has only one molecule, the probability of its being in the left-hand vessel is 1/2; this is the fraction of time that it spends in the left-hand vessel. If there are two molecules, the first spends half its time in the left-hand vessel and during one half of this time the second molecule is also in the same vessel; the probability P that both molecules are in the left-hand vessel simultaneously is thus $P = (1/2)(1/2) = 1/4$. Thus, if the system contains only two gas molecules, it is far from impossible for the gas to move from both vessels into one.

Now consider the general case of N molecules. The probability that each is in the left-hand vessel simultaneously is $P = 1/2^N$. If N equals 100 (not really a macroscopic system), P is equal to 8×10^{-31}, which means that the probability is vanishingly small that a gas of only 100 molecules will move spontaneously into one vessel.[12] The odds become smaller still for the much larger values of N, perhaps of the order of 10^{23}, in a truly macroscopic gas phase.

The essence of the second law on the microscopic level is the statistical certainty that the energy of any concerted motion of particles, such as that associated with mechanical and electrical work, is dissipated over time into the random motion of thermal energy. The entropy of the universe is thereby increased.

4.8 THE HELMHOLTZ ENERGY AND GIBBS ENERGY

Two useful extensive state functions are defined as follows:

Helmholtz energy $\quad A \equiv U - TS \quad$ (4.8.1)

Gibbs energy $\quad G \equiv H - TS \quad$ (4.8.2)

These quantities, and especially the Gibbs energy, are used extensively in thermodynamics. The next section shows how they provide criteria for spontaneity under conditions of constant volume or constant pressure.

---

[12] To put the magnitude of this value of P into perspective, consider a time period of 10^{10} years. This is an estimate of the age of the universe according to the "big bang" theory of cosmology. The fraction 8×10^{-31} of this time period amounts to only 2×10^{-13} seconds, which is the total length of time during this period that we would expect to find all 100 molecules in the same vessel.

Both quantities have dimensions of energy. Both are state functions (because the quantities used to define them are state functions) and are extensive (because U, H, and S are extensive). If temperature is not uniform in the system, we can apply the definitions to constituent phases, or to subsystems small enough to be essentially uniform, and sum.

> Alternative names for the Helmholtz energy are Helmholtz function, Helmholtz free energy, and work function; and alternative names for the Gibbs energy are Gibbs function and Gibbs free energy. Both the Helmholtz energy and Gibbs energy have been called simply free energy, and the symbol F has been used for both. The nomenclature in this text follows the recommendations of the IUPAC Green Book.

We find expressions for infinitesimal changes of A and G by applying the rules of differentiation to their defining equations:

$$dA = dU - T\,dS - S\,dT \tag{4.8.3}$$

$$\begin{aligned}dG &= dH - T\,dS - S\,dT \\ &= dU + p\,dV + V\,dP - T\,dS - S\,dT\end{aligned} \tag{4.8.4}$$

4.9 COMBINING THE FIRST AND SECOND LAWS

Here we use the first and second laws to derive some general relations for changes in reversible and irreversible processes of closed systems. The processes may have any desired degree of complexity: Temperature and pressure may change, species may move between phases, reactions may occur, and so on.

We shall assume that during the process the system has a practically uniform temperature T and pressure p even if the process is irreversible. For example, a spontaneous homogeneous reaction might be occurring that causes the temperature and pressure to change uniformly in all parts of the system.

The second law states that dS is equal to dq/T_b in a reversible process and is greater than dq/T_b in an irreversible process. We replace T_b, the temperature at the portion of the boundary where heat is transferred, by T because the temperature is uniform.[13] It is convenient to combine the equality and inequality on the same line:

$$dS \geq dq/T$$

or

$$dq \leq T\,dS \tag{4.9.1}$$

$$(\tfrac{\text{irrev}}{\text{rev}}, \text{closed system})$$

The *equality* applies to a reversible process and the *inequality* to an irreversible process. This convention is indicated by the notation $\tfrac{\text{irrev}}{\text{rev}}$ in the equations of this section.

---

[13] Recall (page 46) that the use of a symbol such as T or p to refer to an intensive property of a macroscopic system implies that the property is uniform throughout the system unless otherwise stated.

When we substitute $\mathrm{d}q$ from Eq. 4.9.1 into the first law in the form $\mathrm{d}U = \mathrm{d}q - p\,\mathrm{d}V + \mathrm{d}w'$ (Eq. 3.6.9), where $\mathrm{d}w'$ is nonexpansion work, we obtain

$$\mathrm{d}U \leq T\,\mathrm{d}S - p\,\mathrm{d}V + \mathrm{d}w' \qquad (4.9.2)$$
$$(\substack{\text{irrev} \\ \text{rev}}, \text{closed system})$$

Substitution of $\mathrm{d}U$ from Eq. 4.9.2 into the differential of enthalpy, $\mathrm{d}H = \mathrm{d}U + p\,\mathrm{d}V + V\,\mathrm{d}p$ (Eq. 3.11.3), gives us

$$\mathrm{d}H \leq T\,\mathrm{d}S + V\,\mathrm{d}p + \mathrm{d}w' \qquad (4.9.3)$$
$$(\substack{\text{irrev} \\ \text{rev}}, \text{closed system})$$

and substitution of $\mathrm{d}U$ into $\mathrm{d}A = \mathrm{d}U - T\,\mathrm{d}S - S\,\mathrm{d}T$ (Eq. 4.8.3) yields

$$\mathrm{d}A \leq -S\,\mathrm{d}T - p\,\mathrm{d}V + \mathrm{d}w' \qquad (4.9.4)$$
$$(\substack{\text{irrev} \\ \text{rev}}, \text{closed system})$$

Finally, substitution of $\mathrm{d}H$ from Eq. 4.9.3 into $\mathrm{d}G = \mathrm{d}H - T\,\mathrm{d}S - S\,\mathrm{d}T$ (Eq. 4.8.4) gives us

$$\mathrm{d}G \leq -S\,\mathrm{d}T + V\,\mathrm{d}p + \mathrm{d}w' \qquad (4.9.5)$$
$$(\substack{\text{irrev} \\ \text{rev}}, \text{closed system})$$

Equation 4.9.4 provides a useful criterion for spontaneity in a process constrained to take place at constant temperature and volume ($\mathrm{d}T = 0$, $\mathrm{d}V = 0$). Under these conditions, $\mathrm{d}A$ is less than $\mathrm{d}w'$ during an irreversible change. If in addition to these conditions there is no nonexpansion work such as electrical work, the Helmholtz energy decreases during an irreversible process. If A changes along the path of a process carried out with these constraints, the spontaneous direction must be the direction of decreasing A and the value of A at equilibrium must be the minimum value.

Equation 4.9.5 provides an even more useful criterion for spontaneity, for we can apply it to the common situation of a reaction or other process carried out at uniform and constant temperature and pressure. Under these conditions, $\mathrm{d}G$ is less than $\mathrm{d}w'$ in an irreversible change. If G changes along the path of an isothermal, isobaric process of a closed system doing expansion work only, the spontaneous direction is the direction of decreasing G and the value of G at equilibrium is the minimum value.

4.10 THE THIRD LAW

When a reaction or phase transition with pure, perfectly ordered crystals is studied at low temperatures, the entropy change is found to approach zero as the temperature approaches zero kelvins:

$$\lim_{T \to 0} \Delta_\mathrm{r} S = 0 \qquad (4.10.1)$$
$$(\text{pure, perfectly ordered crystals})$$

Equation 4.10.1 is the mathematical statement of the *Nernst heat theorem*[14] or **third law** of thermodynamics. It is true in general only if each reactant and product is a *pure, perfectly ordered* crystal.

There is no theoretical relation between the entropies of different elements. We can arbitrarily choose the entropy of every pure crystalline element to be zero at zero kelvins. Then the experimental observation expressed by Eq. 4.10.1 requires that the entropy of every pure crystalline *compound* also be zero at zero kelvins so that the entropy change for the formation of a compound from its elements be zero. According to this convention, every substance (element or compound) in a pure, perfectly ordered crystal at 0 K and any pressure[15] has a molar entropy of zero:

$$S_m(0\,\text{K}) = 0 \qquad (4.10.2)$$
(pure, perfectly ordered crystal)

This convention establishes a scale of absolute entropies called **third-law entropies** at temperatures above zero kelvins, as explained in the next section.

4.11 MOLAR ENTROPIES

With the convention that the entropy of a pure, perfectly ordered crystalline solid at zero kelvins and any pressure is zero, we can establish the third-law value of the molar entropy of a pure substance at any temperature and pressure. These molar entropy values are needed to calculate changes of various thermodynamic quantities in various kinds of processes. While absolute values are not required in these calculations, since only entropy differences appear in the expressions, absolute values are what are usually tabulated for calculational use.

4.11.1 Molar entropy in a pure phase

Consider a closed system containing a pure substance in a single phase at a given temperature T' and pressure p. The same substance, in a perfectly ordered crystal at zero kelvins and any pressure, has an entropy of zero. The entropy of the state of interest at T' and p, then, is simply the entropy change ΔS of a heating process at constant pressure p that converts the perfectly ordered crystal at zero kelvins to this state of interest.

If the heating process is carried out reversibly, the entropy change is given by $\Delta S = \int dq/T$ (Eq. 4.4.3). If the phase in the final state (the state of interest) is a liquid or gas, or is a solid of different crystal structure than at zero kelvins, the process must include one or more equilibrium phase transitions under conditions where two phases are in equilibrium at the same temperature and pressure (Sec. 2.1.5). For example, transforming the solid at 0 K

---

[14] Nernst preferred to avoid the use of the entropy function and to use in its place the partial derivative $-(\partial A/\partial T)_V$ (Eq. 5.3.10). The original 1906 version of his heat theorem was in the form $\lim_{T \to 0}(\partial \Delta_r A/\partial T)_V = 0$ (Cropper, W. H. *J. Chem. Educ.* **1987**, *64*, 3–8).

[15] The entropy becomes independent of pressure as T approaches zero kelvins. This behavior can be deduced from the relation $(\partial S/\partial p)_T = -\alpha V$ (Table 5.1 on page 122) and the experimental observation that the cubic expansion coefficient α approaches zero as T approaches zero kelvins.

Section 4.11 Molar Entropies

to a gas by heating, at a pressure above the triple point, requires melting and vaporization transitions, and possibly one or more transitions from one crystal form to another.

For a temperature range of the reversible isobaric heating process in which the temperature of a single phase is increasing, the definition of heat capacity as dq/dT (Eq. 3.12.1) allows us to substitute $C_p \, dT$ for dq (where C_p is the heat capacity of the phase at constant pressure). When the temperature of the heating process reaches a value T_{trs} at which an equilibrium phase transition occurs, there must be a transfer of heat into the system for one phase to change to the other. At this temperature, the phase transition is a reversible process. The quantity of heat is called the heat of the transition, $\Delta_{\text{trs}} H$, and is an enthalpy change (see Sec. 6.3.1). The entropy change of the heating process is then given by

$$\Delta S = \int_0^{T'} \frac{C_p}{T} \, dT + \sum \frac{\Delta_{\text{trs}} H}{T_{\text{trs}}}$$

and the resulting operational equation for the calculation of the *molar* entropy of the substance at temperature T' and pressure p is

$$S_{\text{m}}(T') = \frac{\Delta S}{n} = \int_0^{T'} \frac{C_{p,\text{m}}}{T} \, dT + \sum \frac{\Delta_{\text{trs}} H_{\text{m}}}{T_{\text{trs}}} \qquad (4.11.1)$$

(pure substance, constant p)

where $C_{p,\text{m}} = C_p/n$ is the molar heat capacity at constant pressure. The summation is over each equilibrium phase transition occurring during the heating process, and $\Delta_{\text{trs}} H_{\text{m}}$ is the molar heat of the transition.

Since $C_{p,\text{m}}$ and each $\Delta_{\text{trs}} H_{\text{m}}$ are positive quantities, the molar entropy of a substance is *positive* at any temperature above zero kelvins.

The heat capacity and transition enthalpy data required to use Eq. 4.11.1 come from calorimetry. The calorimeter can be cooled to about 10 K with liquid hydrogen, but it is difficult to make measurements below this temperature. Statistical mechanical theory may be used to approximate the part of the integral in Eq. 4.11.1 between zero kelvins and the lowest temperature at which a value of $C_{p,\text{m}}$ is measured. The appropriate formula for nonmagnetic nonmetals comes from the Debye theory for the lattice vibration of a monatomic crystal, which predicts the molar heat capacity at constant volume to be proportional to the cube of T at low temperature (from 0 K to about 30 K): $C_{V,\text{m}} = aT^3$, where a is a constant. For a solid, the molar heat capacities at constant volume and at constant pressure are practically equal. Thus, for the integral in Eq. 4.11.1, we write, to a good approximation,

$$\int_0^{T'} \frac{C_{p,\text{m}}}{T} \, dT = a \int_0^{T''} T^2 \, dT + \int_{T''}^{T'} \frac{C_{p,\text{m}}}{T} \, dT$$

where T'' is the lowest experimental temperature. The first integral on the right side is

$$(1/3) a \, T^3 \Big|_0^{T''} = (1/3) a (T'')^3$$

But $a(T'')^3$ is the value of $C_{p,m}$ at T'', so Eq. 4.11.1 becomes

$$S_m(T') = \frac{C_{p,m}(T'')}{3} + \int_{T''}^{T'} \frac{C_{p,m}}{T}\, dT + \sum \frac{\Delta_{\text{trs}} H_m}{T_{\text{trs}}} \qquad (4.11.2)$$
(pure substance, constant p)

> In the case of a metal, statistical mechanical theory predicts an electronic contribution to the molar heat capacity, proportional to T at low temperature, that should be added to the Debye T^3 term: $C_{p,m} = aT^3 + bT$. The error in using Eq. 4.11.2, which ignores the electronic term, is usually negligible if the heat capacity measurements are made down to about 10 K.

We may evaluate the integral on the right side of Eq. 4.11.2 by numerical integration. We need the area under the curve of $C_{p,m}/T$ plotted as a function of T between some low temperature, T'', and the temperature at which the molar entropy is to be evaluated, T'. Since the integral may be written in the form

$$\int_{T''}^{T'} \frac{C_{p,m}}{T}\, dT = \int_{T=T''}^{T=T'} C_{p,m}\, d\ln(T/K)$$

we may also use the area under the curve of $C_{p,m}$ plotted as a function of $\ln(T/K)$.

The procedure of evaluating the entropy from the heat capacity is illustrated for the case of H_2O in Fig. 4.15 on the next page. The areas under the curves of $C_{p,m}/T$ versus T, and of $C_{p,m}$ versus $\ln T/K$, in a given temperature range are numerically identical. In a case like that shown in the figure, in which the $C_{p,m}/T$ curve displays a relative maximum, it may be more convenient to integrate the $C_{p,m}$ curve. From the integration comes the curve of S_m versus T shown in the figure. Note how the molar entropy constantly increases with increasing T and has a discontinuity at each phase transition; this behavior is a consequence of Eq. 4.11.1 and the fact that $C_{p,m}$, $\Delta_{\text{fus}} H_m$, and $\Delta_{\text{vap}} H_m$ are all positive.

> We have stated the convention that the zero of entropy of any substance refers to the pure, perfectly ordered crystal. In practice, experimental entropy values depart from this convention in two respects. First, an element is usually a mixture of two or more isotopes, and so the substance is not isotopically pure. Second, if any of the nuclei have spins, weak interactions between the nuclear spins in the crystal would cause the spin orientations to become ordered at a very low temperature; above 1 K, however, nuclear spins are randomly oriented, and this change is not included in the Debye T^3 formula. The neglect of these two effects results in a *practical entropy scale* or conventional entropy scale on which the crystal that is assigned an entropy of zero has randomly mixed isotopes and randomly oriented nuclear spins, but is pure and ordered in other respects. This is the scale that is used for published values of absolute "third-law" molar entropies. The shift of the zero away from a completely pure and perfectly ordered crystal introduces no inaccuracies into the calculated value of ΔS for any process occurring above 1 K because the shift is the same in the initial and final states; that is, isotopes remain randomly mixed and nuclear spins remain randomly oriented.

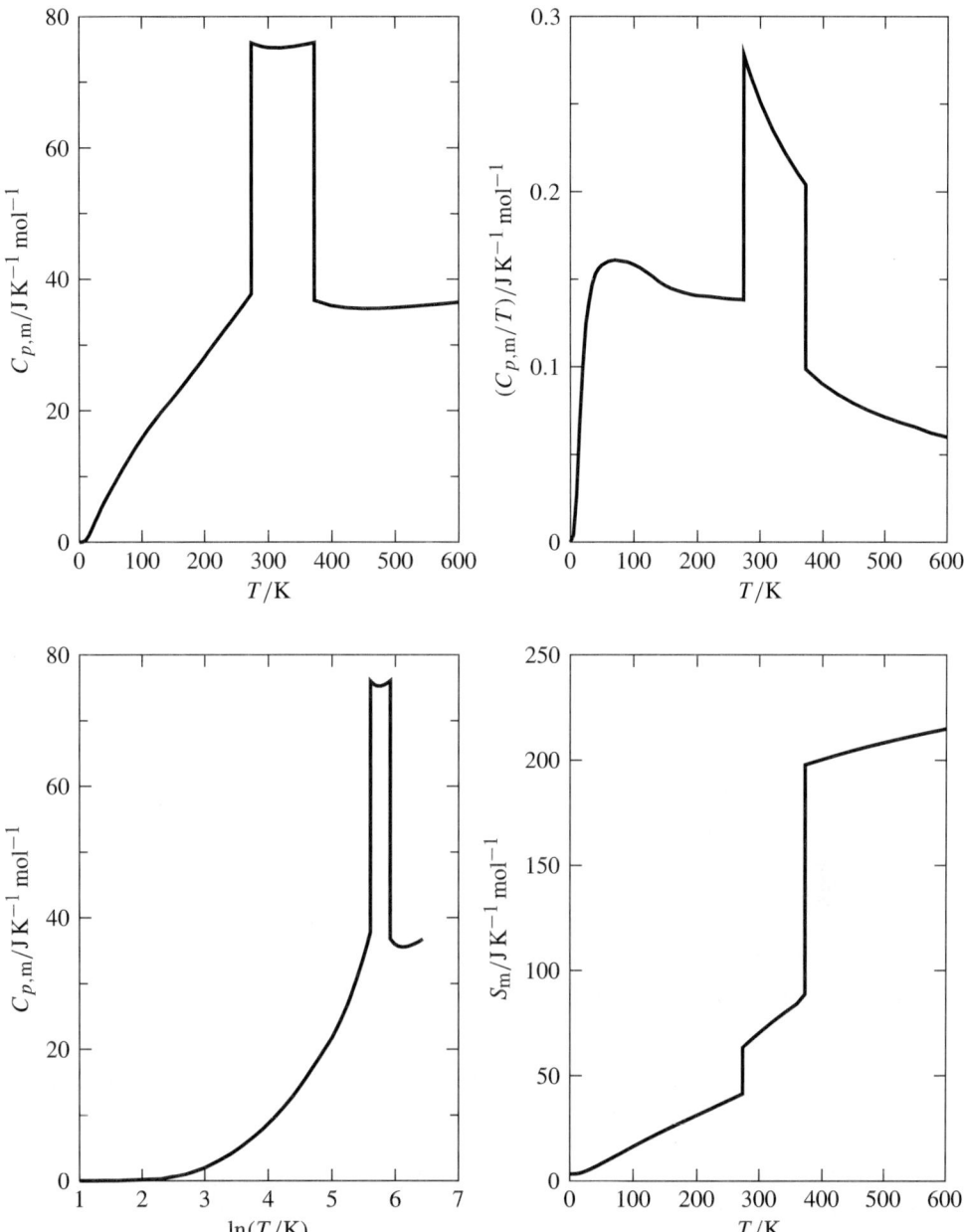

Figure 4.15 Data for H_2O: the dependence of $C_{p,m}$, $C_{p,m}/T$, and S_m on temperature at a pressure of 1 bar. The discontinuities are at the melting and boiling temperatures. The molar entropy includes a residual entropy of $3\,\mathrm{J\,K^{-1}\,mol^{-1}}$ (see Sec. 4.11.2).

4.11.2 The molar entropy of an ideal gas from statistical mechanics

Statistical mechanical theory provides an accurate method to evaluate the molar entropy of a pure ideal gas at any temperature and pressure from experimental molecular properties. This is often the preferred method of evaluating S_m for a gas. The zero of entropy is the same as the practical entropy scale; that is, isotope mixing and nuclear spin interactions are ignored. Intermolecular interactions are also ignored, which is why the results apply only to an ideal gas.

The statistical mechanics formula writes the molar entropy as the sum of a translational contribution and an internal contribution: $S_m = S_{m,\text{trans}} + S_{m,\text{int}}$. The translational contribution is given by the Sackur–Tetrode equation

$$S_{m,\text{trans}} = R \ln \frac{(2\pi M)^{3/2} (RT)^{5/2}}{ph^3 N_A^4} + (5/2) R$$

Here h is the Planck constant and N_A is the Avogadro constant. The internal contribution is given by

$$S_{m,\text{int}} = R \ln q_{\text{int}} + RT (d \ln q_{\text{int}}/dT)$$

Here q_{int} is the molecular partition function defined by

$$q_{\text{int}} = \sum_i \exp(-\epsilon_i / kT)$$

where ϵ_i is the energy of a molecular quantum state relative to the lowest energy level, k is the Boltzmann constant, and the sum is over the quantum states of one molecule (with appropriate averaging for natural isotopic abundance). The experimental data needed to evaluate q_{int} consist of the energies of low-lying electronic energy levels, values of electronic degeneracies, fundamental vibrational frequencies, rotational constants, and other spectroscopic parameters.

When S_m is evaluated by this method with p set equal to the standard pressure $p° = 1$ bar, the value is the *standard* molar entropy, $S_m°$. This is because the standard state of a gas is the hypothetical state in which the gas is at the standard pressure and behaves as an ideal gas.

Molar entropy values calculated by the statistical mechanics method should agree with those obtained by the calorimetric (third-law) method. For certain substances, however, the S_m value obtained from the statistical mechanics calculation is greater than that from calorimetric data by more than can be explained by the uncertainty of the data. The difference is called *residual entropy*. Residual entropy is found in crystalline CO, NO, N_2O, and H_2O. It is caused by molecular disorder that is frozen into the crystals and is present at the lowest temperature of the heat capacity measurements. In a crystal of solid H_2O, for instance, each oxygen atom has about it four bonds arranged approximately tetrahedrally: covalent bonds to the two hydrogen atoms in the same molecule, and hydrogen bonds to two hydrogen atoms in neighboring molecules. The arrangements of the two kinds of bonds throughout

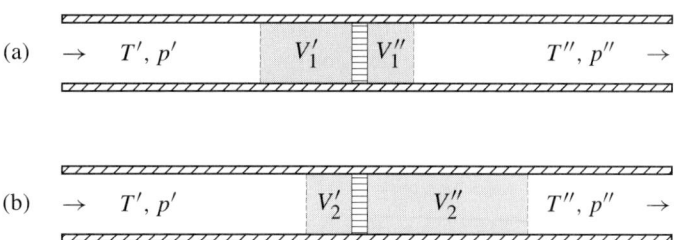

Figure 4.16 Joule–Thomson expansion of a gas through a porous plug. The shaded area represents a fixed-amount sample of the gas (a) at Time 1 and (b) at a later Time 2.

the crystal are practically random, even at the lowest temperatures at which calorimetric measurements are made, but the arrangements would not be random in a perfectly ordered crystal. The resulting residual entropy is found to be $3.41 \, \text{J K}^{-1} \, \text{mol}^{-1}$.

4.12 CRYOGENICS

The field of cryogenics involves the production of very low temperatures and the study of the behavior of matter at these temperatures. We have seen that low temperatures are needed to evaluate third-law entropies. There are some additional interesting thermodynamics applications.

4.12.1 Joule–Thomson expansion

A gas may be cooled by allowing it to expand adiabatically against a piston (Sec. 3.6.4), and a liquid may be cooled by pumping on its vapor to cause evaporation (vaporization). An evaporation procedure with a refrigerant fluid is what produces the cooling in an ordinary kitchen refrigerator.

For further cooling of a fluid, a common procedure is to use a continuous **throttling process** in which the fluid is forced to flow through a porous plug, valve, or other constriction that causes an abrupt drop in pressure. A slow continuous adiabatic throttling of a gas is called the **Joule–Thomson experiment**, or Joule–Kelvin experiment, after the two scientists who collaborated between 1852 and 1862 to design and analyze this procedure.[16]

The principle of the Joule–Thomson experiment is shown in Fig. 4.16. A tube with thermally insulated walls contains a gas maintained at a constant pressure p' at the left side of a porous plug and at a constant lower pressure p'' at the right side. Because of the pressure difference, the gas flows from left to right through the plug. The flow is slow, and the pressure is essentially uniform throughout the portion of the tube at each side of the plug but has a large gradient within the pores of the plug.

After the gas has been allowed to flow for a period of time, a steady state develops in the tube. In this steady state, the gas is assumed to have a uniform temperature T' at the left side of the plug and a uniform temperature T'' (not necessarily equal to T') at the right

---
[16] William Thomson later became Lord Kelvin.

side of the plug.

Consider the segment of gas whose position at Times 1 and 2 is indicated by shading in Fig. 4.16. This segment contains a fixed amount of gas and expands as it moves through the porous plug. We can treat this portion of the gas as a *closed system*. During the interval between Times 1 and 2, the system passes through a sequence of different states, none of which is an equilibrium state since the process is irreversible. The energy transferred across the boundary as heat is *zero* because the tube wall is insulated and there is no temperature gradient at either end of the gas segment. We calculate the energy transferred as work at each end of the gas segment from $dw = -p_b A_s\, dx$ (Eq. 3.6.4 on page 45), where p_b is the pressure (either p' or p'') at the moving boundary, A_s is the cross-section area of the tube, and x is the distance along the tube. The result is

$$w = -p'(V_2' - V_1') - p''(V_2'' - V_1'')$$

where the meaning of the volumes V_2' and so on is indicated in the figure.

The internal energy change ΔU of the system must be equal to w since q is zero. Now let us find the enthalpy change ΔH. At each instant, a portion of the gas is in the pores of the plug, but this portion contributes an unchanging contribution to both U and H because of the steady state. The rest of the gas is in uniform portions of the system, and its enthalpy is the sum of $U + pV$ for each uniform portion. The overall enthalpy change must be

$$\Delta H = \Delta U + (p'V_2' + p''V_2'') - (p'V_1' + p''V_1'')$$

which, when combined with the expression for $w = \Delta U$, shows that ΔH is *zero*. In other words, a given amount of gas has the same enthalpy before and after it passes through the plug; the adiabatic throttling process is *isenthalpic*.

The temperatures T' and T'' can be measured directly. When values of T'' versus p'' are plotted for a series of Joule–Thomson experiments having the same values of T' and p' and different values of p'', the curve drawn through the points is a curve of constant enthalpy. The slope at any point on this curve is equal to the **Joule–Thomson coefficient** (or Joule–Kelvin coefficient) defined by

$$\mu_{\text{JT}} \equiv \left(\frac{\partial T}{\partial p}\right)_H \tag{4.12.1}$$

For an ideal gas, μ_{JT} is zero because the enthalpy of an ideal gas depends only on T (Sec. 3.11); T cannot change if H is constant. For a *nonideal* gas, μ_{JT} is a function of T and p and the kind of gas.[17] For most gases, μ_{JK} is positive at low to moderate pressures unless the temperature is much greater than room temperature. A Joule–Thomson expansion of a gas under these conditions, to any lower final pressure, causes a cooling effect because T will decrease as p decreases at constant H. Hydrogen and helium, however, have negative values of μ_{JK} at room temperature and must be cooled by other means to about 200 K and 40 K, respectively, for a Joule–Thomson expansion to cause further cooling.

---

[17] See Sec. 5.7.2 for the relation of the Joule–Thomson coefficient to other properties of a gas.

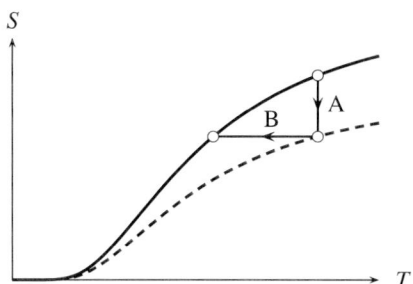

Figure 4.17 Adiabatic demagnetization to achieve a low temperature in a paramagnetic solid.

The cooling effect of a Joule–Thomson expansion is often used to cool a gas down to its condensation temperature. This procedure can be carried out continuously by pumping the gas through the throttle and recirculating the cooler gas on the low-pressure side through a heat exchanger to help cool the gas on the high-pressure side. Starting at room temperature, gaseous nitrogen can be condensed by this means to liquid nitrogen at 77.4 K. The liquid nitrogen can then be used as a cooling bath for gaseous hydrogen. At 77.4 K, hydrogen has a positive Joule–Thomson coefficient, so it in turn can be converted by a throttling process to liquid hydrogen at 20.3 K. Finally, gaseous helium, whose Joule–Thomson coefficient is positive at 20.3 K, can be converted to liquid helium at 4.2 K. Further cooling of the liquid helium to about 1 K can be carried out by pumping to cause rapid evaporation.

4.12.2 Adiabatic demagnetization

To obtain temperatures below 1 K, the technique of **adiabatic demagnetization** is used. This method, suggested by Peter Debye in 1926 and independently by William Giaque in 1927, requires a paramagnetic solid in which ions with unpaired electrons are sufficiently separated that at 1 K the orientations of the magnetic dipoles are almost completely random. Gadolinium(III) sulfate octahydrate, $Gd_2(SO_4)_3 \cdot 8H_2O$, is commonly used. Figure 4.17 illustrates the principle of the technique. The solid curve shows the temperature dependence of the entropy of a paramagnetic solid in the absence of an applied magnetic field, and the dashed curve is for the solid in a constant, nonzero magnetic field. The temperature range shown is from 0 K to approximately 1 K. At 0 K, the magnetic dipoles are perfectly ordered. The increase of S shown by the solid curve between 0 K and 1 K is due almost entirely to increasing disorder in the orientations of the magnetic dipoles as heat enters the system, rather than to the heat capacity of the crystal lattice.

Path A represents the process that occurs when the paramagnetic solid, surrounded by gaseous helium in thermal contact with liquid helium that has been cooled to about 1 K, is slowly moved into a strong magnetic field. The process is isothermal magnetization, which partially orients the magnetic dipoles and reduces the entropy. Heat is transferred to the liquid helium during this process, and some of the liquid boils away. In Path B, the thermal contact between the solid and the liquid helium has been broken by pumping away the gas surrounding the solid, and the sample is slowly moved away from the magnetic field. This step is a reversible adiabatic demagnetization. Because the process is reversible and adiabatic, the entropy change is zero, which brings the state of the solid to a lower temperature as shown.

Section 5.12.2 gives a more rigorous derivation of the signs of the entropy and temperature changes occurring in the paths shown in the figure.

By repeatedly carrying out the procedure of isothermal magnetization and adiabatic demagnetization, starting each stage at the temperature produced by the previous stage, it has been possible to attain a temperature as low as 0.0015 K. The temperature can be reduced still further, down to 16 microkelvins, by using adiabatic nuclear demagnetization. However, as is evident from the figure, if in accordance with the third law both of the entropy curves come together at the absolute zero of the kelvin scale, then it is not possible to attain a temperature of zero kelvins in a finite number of stages of adiabatic demagnetization. This conclusion is called the *principle of the unattainability of absolute zero*.

PROBLEMS

4.1 Explain why an electric refrigerator, which transfers heat from the cold food storage compartment to the warmer air in the room, is not an impossible "Clausius device."

4.2 During the reversible expansion of a fixed amount of an ideal gas, the heat is given by the expression $dq = C_V \, dT + (nRT/V) \, dV$ (Eq. 4.3.3).

(a) A necessary and sufficient condition for dq to be an exact differential is that the reciprocity relation must be satisfied for the independent variables T and V (see Appendix F). Apply this test to the preceding expression to show that dq is *not* an exact differential and therefore that q is not a state function.

(b) By the same method, show that $dS = dq/T$ is an exact differential so that S is a state function.

4.3 A system consisting of a fixed amount of an ideal gas is maintained in thermal equilibrium with a heat reservoir at temperature T. The system is subjected to the following isothermal cycle:

1. The gas, initially in an equilibrium state with volume V_0, is allowed to expand into a vacuum and reach a new equilibrium state of volume V'.

2. The gas is reversibly compressed from V' to V_0.

For this cycle, find expressions or values for w, $\oint dq/T$, and $\oint dS$.

4.4 In an irreversible isothermal process of a closed system:

(a) Is it possible for ΔS to be negative?

(b) Is it possible for ΔS to be less than q/T?

4.5 Refer to the apparatus shown in Figs. 3.14 and 3.15 and described in Probs. 3.6 and 3.7. For each system, evaluate ΔS for the process that results from opening the stopcock. Also evaluate $\int dq/T_{\text{ext}}$ for both processes (for the apparatus in Fig. 3.15, assume the vessels have adiabatic walls). Are your results consistent with the mathematical form of the second law?

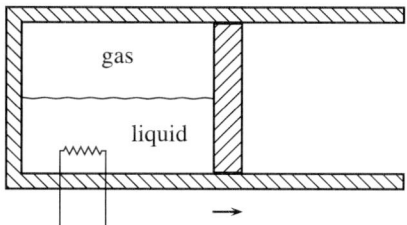

Figure 4.18

4.6 Figure 4.18 represents a hypothetical liquid in equilibrium with its vapor and an electrical resistor in a cylinder with a piston. The *system* is the interior of the cylinder (gas, liquid, and resistor). The initial state of the system is described by

$$V_1 = 0.2200 \text{ m}^3 \qquad T_1 = 300.0 \text{ K} \qquad p_1 = 2.50 \times 10^5 \text{ Pa}$$

A constant current $I = 0.5000$ A is passed for 1600 s through the resistor, which has electric resistance $R_{el} = 50.00 \; \Omega$. The piston moves slowly to the right against a constant external pressure equal to the vapor pressure of the liquid, 2.50×10^5 Pa, and some of the liquid vaporizes. Assume that the process is adiabatic and that T and p remain uniform and constant. The final state is described by

$$V_2 = 0.2400 \text{ m}^3 \qquad T_2 = 300.0 \text{ K} \qquad p_2 = 2.50 \times 10^5 \text{ Pa}$$

(a) Calculate q, w, ΔU, and ΔH.

(b) Is the process reversible? Explain.

(c) Devise a reversible process that accomplishes the same change of state, and use it to calculate ΔS.

(d) Compare q for the reversible process with ΔH; does your result agree with Eq. 3.11.5?

4.7 Calculate the molar entropy of carbon disulfide at 25.00 °C and 1 bar from the heat capacity data for the solid in Table 4.2 on the next page and the following information. At the standard melting point, 161.11 K, the molar enthalpy of fusion is $\Delta_{fus} H_m^\circ = 4.39 \times 10^3$ J mol^{-1}. The molar heat capacity of the liquid in the range 161–300 K is described by $C_{p,m}^\circ = a + bT$, where the constants have the values $a = 74.6$ J K^{-1} mol^{-1} and $b = 0.0034$ J K^{-2} mol^{-1}.

Table 4.2 Molar heat capacity of $CS_2(s)$ (Brown, L. I.; Manov, G. G. *J. Am. Chem. Soc.* **1937**, *59*, 500–502)

T/K	$C_{p,m}^\circ/\text{J K}^{-1}\text{ mol}^{-1}$
15.05	6.9
20.15	12.0
29.76	20.8
42.22	29.2
57.52	35.6
75.54	40.0
94.21	45.0
108.93	48.5
131.54	52.6
156.83	56.6

Chapter 5

PURE SUBSTANCES IN SINGLE PHASES

This chapter applies the first and second laws of thermodynamics to the simplest kind of system, a single substance in a single phase. The importance of the independent variables of the system is stressed, and relations between various properties of a single phase are derived. Among the processes discussed are changes of temperature, pressure, and volume. The important concepts of standard states and chemical potential are introduced.

5.1 SYSTEM VARIABLES

With the usual neglect of surface effects and external fields, a uniform phase containing one substance has *three* independent variables. We could choose these to be the temperature, the pressure, and the amount of the substance. There are two ways to think of the significance of the independent variables:

1. We are able to vary any one of the three independent variables independently of the other two. For example, we can take an arbitrary amount of the substance and, while the substance remains in a liquid phase, change its temperature over a range and at each temperature change its pressure over a range.

2. Whenever we adjust the three independent variables to given values, every other property of the phase can have only one definite, reproducible value. For example, a one-substance phase with given values of T, p, and n has a volume determined by an equation of state, a mass equal to nM, a molar volume given by $V_m = V/n$, and a density given by $\rho = nM/V$.

   A single-phase system containing substances in reaction equilibrium, or a fixed-composition mixture of nonreacting substances, also has three independent variables. Examples are liquid H_2O in equilibrium with H^+(aq) and OH^-(aq); and air, a mixture of gases in fixed proportions. All of the results in this chapter (except for molar properties) apply to such systems. This is indicated in numbered equations by the condition $C = 1$, where C, the number of components, is the minimum number of substances or mixtures of fixed composition needed to form the phase. If a chemical reaction could occur in the phase, C would be greater than 1 unless we specify that the reactants and products remain in reaction equilibrium at all times, or else that they are constrained from reacting at all. The concept of the number of components is discussed in more detail in chap. 10.

If the one-substance (or one-component), single-phase system is *closed*, then the amount is fixed and the number of independent variables is reduced to two. We could choose these to be T and p.

Strictly speaking, the phase will not remain uniform in intermediate states of a spontaneous process; that is, one occurring at a nonzero rate. However, if we carry out the process sufficiently slowly, the system will be uniform for all practical purposes, and we can describe the state at each instant with single values of intensive properties such as T and p.

5.2 VOLUME PROPERTIES

Two *mechanical coefficients* of a closed system are defined as follows:

$$\textbf{cubic expansion coefficient} \quad \alpha \equiv \frac{1}{V}\left(\frac{\partial V}{\partial T}\right)_p \qquad (5.2.1)$$

$$\textbf{isothermal compressibility} \quad \kappa_T \equiv -\frac{1}{V}\left(\frac{\partial V}{\partial p}\right)_T \qquad (5.2.2)$$

The cubic expansion coefficient is also called the coefficient of thermal expansion and the expansivity coefficient. Other symbols for the isothermal compressibility are β and γ_T.

These definitions show that α is the fractional volume increase per unit temperature increase at constant pressure, and κ_T is the fractional volume decrease per unit pressure increase at constant temperature. Both quantities are *intensive* properties. Most substances have positive values of α (an exception is liquid water below its temperature of maximum density, 3.98 °C),[1] and all substances have positive values of κ_T because a pressure increase at constant temperature requires a volume decrease.

If we divide the numerator and denominator of the right sides of Eqs. 5.2.1 and 5.2.2 by the amount n, in the case of a pure substance in a single phase, we obtain the alternative expressions

$$\alpha = \frac{1}{V_m}\left(\frac{\partial V_m}{\partial T}\right)_p \qquad (5.2.3)$$
$$(P = 1, \text{ pure substance})$$

$$\kappa_T = -\frac{1}{V_m}\left(\frac{\partial V_m}{\partial p}\right)_T \qquad (5.2.4)$$
$$(P = 1, \text{ pure substance})$$

---

[1] The crystalline ceramics zirconium tungstate (ZrW_2O_8) and hafnium tungstate (HfW_2O_8) have the remarkable behavior of contracting uniformly and continuously in all three dimensions when they are heated from 0.3 K to about 1050 K; α is negative throughout this very wide temperature range (Mary, T. A. et al *Science* **1996**, *272*, 90–92).

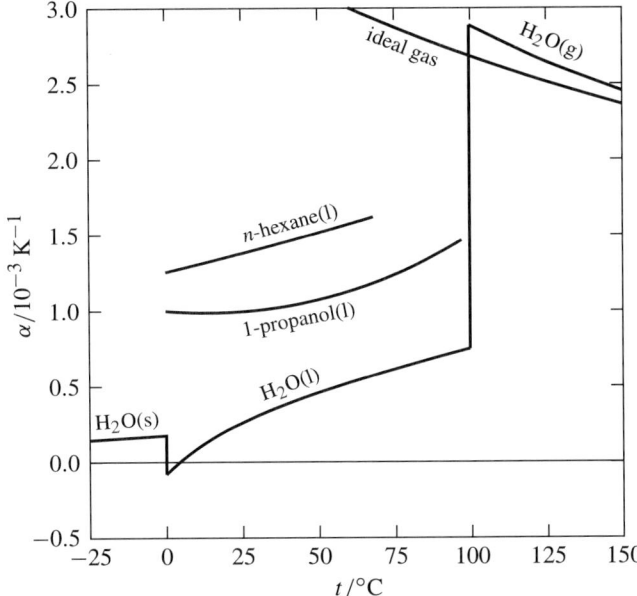

Figure 5.1 The cubic expansion coefficient of several substances and an ideal gas as functions of temperature at $p = 1$ bar. Note that because liquid water has a density maximum at $4\,°C$, α is zero at that temperature.

where V_m is the molar volume (Eq. 2.1.1). P in the conditions of validity is the number of phases. Only intensive properties appear in Eqs. 5.2.3 and 5.2.4, so the amount of the substance is irrelevant; thus, these equations are valid regardless of whether the system is closed. Figures 5.1 and 5.2 show the temperature variation of α and κ_T for several substances.

The **total differential** of a dependent state function is an expression for the infinitesimal change of the function in terms of the infinitesimal changes of the independent variables. As explained in Appendix F, the expression has one term for each independent variable. Each term is the product of a partial derivative and the infinitesimal change of the independent variable. Thus, if we choose T and p as the independent variables of the closed system, the total differential of V is given by

$$dV = \left(\frac{\partial V}{\partial T}\right)_p dT + \left(\frac{\partial V}{\partial p}\right)_T dp$$

With the substitution of $(\partial V/\partial T)_p = \alpha V$ (from Eq. 5.2.1) and $(\partial V/\partial p)_T = -\kappa_T V$ (from Eq. 5.2.2), the expression for the total differential of V becomes

$$dV = \alpha V\, dT - \kappa_T V\, dp \qquad (5.2.5)$$

(closed system, $P = 1, C = 1$)

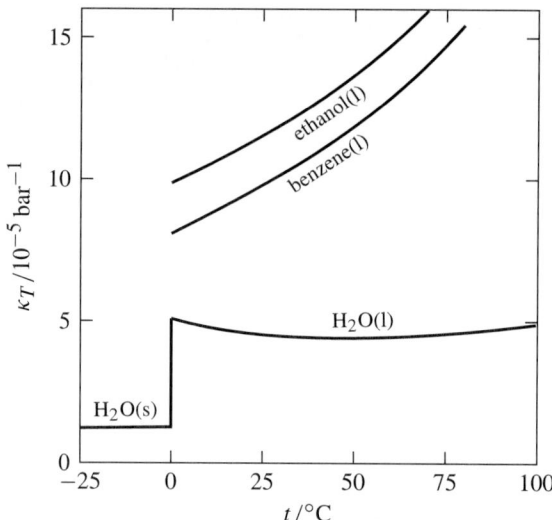

Figure 5.2 The isothermal compressibility of several substances as a function of temperature at $p = 1$ bar.

To find how p varies with T in a closed system kept at constant volume, we set dV equal to zero in Eq. 5.2.5: $0 = \alpha V \, dT - \kappa_T V \, dp$ or $dp/dT = \alpha/\kappa_T$. Since dp/dT under the condition of constant volume is the partial derivative $(\partial p/\partial T)_V$, we have the general relation

$$\left(\frac{\partial p}{\partial T}\right)_V = \frac{\alpha}{\kappa_T} \qquad (5.2.6)$$
(closed system, $P = 1, C = 1$)

5.3 ENERGY FUNCTIONS

Suppose we apply the relation $dU \leq T \, dS - p \, dV + dw'$ (Eq. 4.9.2) to a change in the state of a fixed amount of the substance. The relation shows that *if* the process is carried out reversibly and there is expansion work only ($dw' = 0$), the internal energy change is given by

$$dU = T \, dS - p \, dV \qquad (5.3.1)$$
(closed system, reversible process, $dw' = 0$)

The term $T \, dS$ is the reversible heat, and $-p \, dV$ is the reversible expansion work. This equation relates the change in one state function, U, to changes in two others, S and V. It is therefore valid for *any* process of a closed system with two independent variables

Section 5.3 Energy Functions

because U must have a unique dependence on S and V if S and V are the only independent variables. Equation 5.3.1 is valid, for instance, for a system like that shown in Fig. 3.11(a) that is heated electrically and nonreversibly, provided the temperature and pressure remain practically uniform.[2] The appearance of the intensive variables T and p in the equation imply, of course, that T and p are uniform throughout the system. The general validity of this equation for systems with two independent variables is simply the consequence of its being an expression for the *total differential* of U, with a term for each independent variable.

With appropriate changes in conditions for validity, Eq. 5.3.1 becomes

$$dU = T\,dS - p\,dV \qquad (5.3.2)$$
(closed system, $P = 1, C = 1$)

From this, using the definitions of H, A, and G (Eqs. 3.11.2, 4.8.1, and 4.8.2), we obtain

$$dH = T\,dS + V\,dp \qquad (5.3.3)$$
(closed system, $P = 1, C = 1$)

$$dA = -S\,dT - p\,dV \qquad (5.3.4)$$
(closed system, $P = 1, C = 1$)

$$dG = -S\,dT + V\,dp \qquad (5.3.5)$$
(closed system, $P = 1, C = 1$)

Equations 5.3.2–5.3.5 are sometimes called the **Gibbs equations**. They are expressions for the total differentials of the energy functions U, H, A, and G in closed systems of two independent variables. Each equation shows how the dependent variable on the left side varies as a function of changes in two independent variables (the "natural variables" of the dependent variable) on the right side. Since these are expressions for total differentials, they must apply to *any* process of a system of two independent variables, regardless of whether the process is reversible and whether there is nonexpansion work.

As explained in Appendix F, we may identify the coefficient of each term in the expression for the total differential of a function as a partial derivative of the function. From Eqs. 5.3.2–5.3.5, we then obtain the following relations (which are valid for a closed system of one uniform phase and one component):

from Eq. 5.3.2
$$\left(\frac{\partial U}{\partial S}\right)_V = T \qquad (5.3.6)$$

$$\left(\frac{\partial U}{\partial V}\right)_S = -p \qquad (5.3.7)$$

---

[2] This statement is consistent with the inequality $dU < T\,dS - p\,dV + dw'$ applied to an irreversible process with positive nonexpansion work dw'. In the irreversible heating process, dU is equal to $T\,dS - p\,dV$, which is less than $T\,dS - p\,dV + dw'$.

from Eq. 5.3.3 $\left(\dfrac{\partial H}{\partial S}\right)_p = T$ (5.3.8)

$\left(\dfrac{\partial H}{\partial p}\right)_S = V$ (5.3.9)

from Eq. 5.3.4 $\left(\dfrac{\partial A}{\partial T}\right)_V = -S$ (5.3.10)

$\left(\dfrac{\partial A}{\partial V}\right)_T = -p$ (5.3.11)

from Eq. 5.3.5 $\left(\dfrac{\partial G}{\partial T}\right)_p = -S$ (5.3.12)

$\left(\dfrac{\partial G}{\partial p}\right)_T = V$ (5.3.13)

At this point, we use for the first time an extremely useful mathematical tool called the **reciprocity relation** of a total differential. If the independent variables are x and y and the total differential of a dependent state function f is given by

$$df = M\, dx + N\, dy$$

where M and N are functions of x and y, then the reciprocity relation is

$$\left(\dfrac{\partial M}{\partial y}\right)_x = \left(\dfrac{\partial N}{\partial x}\right)_y \qquad (5.3.14)$$

The reciprocity relation is justified in Appendix F.

The reciprocity relations obtained from Eqs. 5.3.2–5.3.5 are called **Maxwell relations** (again valid for a closed system of one uniform phase and one component):

from Eq. 5.3.2 $\left(\dfrac{\partial T}{\partial V}\right)_S = -\left(\dfrac{\partial p}{\partial S}\right)_V$ (5.3.15)

from Eq. 5.3.3 $\left(\dfrac{\partial T}{\partial p}\right)_S = \left(\dfrac{\partial V}{\partial S}\right)_p$ (5.3.16)

from Eq. 5.3.4 $\left(\dfrac{\partial S}{\partial V}\right)_T = \left(\dfrac{\partial p}{\partial T}\right)_V$ (5.3.17)

from Eq. 5.3.5 $-\left(\dfrac{\partial S}{\partial p}\right)_T = \left(\dfrac{\partial V}{\partial T}\right)_p$ (5.3.18)

5.4 INTERNAL PRESSURE

The partial derivative $(\partial U/\partial V)_T$ applied to a fluid phase in a closed system is called the **internal pressure**. (Note that U and pV both have dimensions of energy; therefore, U/V has dimensions of pressure.)

Section 5.4 Internal Pressure

To relate the internal pressure to other properties, we divide Eq. 5.3.2 by dV: $dU/dV = T(dS/dV) - p$. Then we impose a condition of constant T: $(\partial U/\partial V)_T = T(\partial S/\partial V)_T - p$. When we make a substitution for $(\partial S/\partial V)_T$ from the Maxwell relation of Eq. 5.3.17, we obtain

$$\left(\frac{\partial U}{\partial V}\right)_T = T\left(\frac{\partial p}{\partial T}\right)_V - p \qquad (5.4.1)$$
(closed system, fluid phase, $C = 1$)

This equation is sometimes called the "thermodynamic equation of state" of the fluid.

For an ideal-gas phase, we can write $p = nRT/V$ and then

$$\left(\frac{\partial p}{\partial T}\right)_V = \frac{nR}{V} = \frac{p}{T}$$

Making this substitution in Eq. 5.4.1 gives us

$$\left(\frac{\partial U}{\partial V}\right)_T = 0 \qquad (5.4.2)$$
(closed system of an ideal gas)

showing that the internal pressure of an ideal gas is zero.

> In Sec. 3.2.2, we defined an ideal gas as a gas (1) that obeys the ideal gas equation, and (2) for which U in a closed system depends only on T. Equation 5.4.2, derived from the first part of this definition, expresses the second part. It thus appears that the second part of the definition is redundant, and that we could define an ideal gas simply as a gas obeying the ideal gas equation. This argument is valid provided we assume the ideal-gas temperature is the same as the thermodynamic temperature (Secs. 2.6.1 and 4.3.4) since this assumption is required to derive Eq. 5.4.2. Without this assumption, it is insufficient to define an ideal gas solely by $pV = nRT$, where T is the ideal gas temperature.

Here is a simplified interpretation of the significance of the internal pressure. When the volume of a fluid increases, the average distance between molecules increases and the potential energy due to intermolecular forces changes. If attractive forces dominate, as they usually do unless the fluid is highly compressed, expansion causes the potential energy to *increase*. The internal energy is the sum of the potential energy and thermal energy. The internal pressure, $(\partial U/\partial V)_T$, is the rate at which the internal energy changes with volume at constant temperature. At constant temperature, the thermal energy is constant; so the internal pressure is the rate at which just the potential energy changes with volume. Thus, the internal pressure is a measure of the strength of the intermolecular forces and is positive if attractive forces dominate.[3] In an ideal gas, intermolecular forces are absent and therefore the internal pressure is zero.

---

[3] These attractive intermolecular forces are the cohesive forces that can allow a negative pressure to exist in a liquid; see page 19.

With the substitution $(\partial p/\partial T)_V = \alpha/\kappa_T$ (Eq. 5.2.6), Eq. 5.4.1 becomes

$$\left(\frac{\partial U}{\partial V}\right)_T = \frac{\alpha T}{\kappa_T} - p \qquad (5.4.3)$$
(closed system, fluid phase, $C = 1$)

The internal pressure of a liquid at $p = 1$ bar is typically much larger than 1 bar (see Prob. 5.6). Equation 5.4.3 shows that, in this situation, the internal pressure is approximately equal to $\alpha T/\kappa_T$.

5.5 THERMAL PROPERTIES

Equations 3.12.2 and 3.12.4 give the following expressions for the heat capacities of a closed system at constant volume and at constant pressure:

$$C_V = \left(\frac{\partial U}{\partial T}\right)_V \qquad C_p = \left(\frac{\partial H}{\partial T}\right)_p$$

A closed system of one component in a single phase has only two independent variables. In such a system, these equations are complete and unambiguous definitions of C_V and C_p because the partial derivatives are expressed with two independent variables—T and V in one case and T and p in the other. As mentioned on page 60, additional conditions would have to be specified to define C_V for a more complicated system; the same is true for C_p.

5.5.1 The relation between constant-volume and constant-pressure heat capacities

The value of $C_{p,\text{m}}$ for a substance is greater than $C_{V,\text{m}}$. The derivation is simple in the case of a fixed amount of an *ideal gas*. Substituting $H = U + pV = U + nRT$ in $C_p = \text{d}H/\text{d}T$ (Eq. 3.12.5) and performing the differentiation, we obtain $C_p = \text{d}U/\text{d}T + nR$. The derivative $\text{d}U/\text{d}T$ in the case of an ideal gas is equal to C_V (Eq. 3.12.3). Making this substitution and dividing by n to obtain molar quantities gives

$$C_{p,\text{m}} = C_{V,\text{m}} + R \qquad (5.5.1)$$
(ideal gas, pure substance)

For any phase in general, we proceed as follows. First we write Eq. 3.12.4 as

$$C_p = \left(\frac{\partial H}{\partial T}\right)_p = \left[\frac{\partial (U + pV)}{\partial T}\right]_p = \left(\frac{\partial U}{\partial T}\right)_p + p\left(\frac{\partial V}{\partial T}\right)_p$$

Then we write the total differential of U, with T and V as independent variables, and identify one of the coefficients as C_V:

$$\text{d}U = \left(\frac{\partial U}{\partial T}\right)_V \text{d}T + \left(\frac{\partial U}{\partial V}\right)_T \text{d}V = C_V \text{d}T + \left(\frac{\partial U}{\partial V}\right)_T \text{d}V$$

When we divide both sides of the preceding equation by dT and impose a condition of constant p, we obtain

$$\left(\frac{\partial U}{\partial T}\right)_p = C_V + \left(\frac{\partial U}{\partial V}\right)_T \left(\frac{\partial V}{\partial T}\right)_p$$

Substitution of this expression for $(\partial U/\partial T)_p$ in the equation for C_p yields

$$C_p = C_V + \left[\left(\frac{\partial U}{\partial V}\right)_T + p\right]\left(\frac{\partial V}{\partial T}\right)_p$$

Finally we set the partial derivative $(\partial U/\partial V)_T$ (the internal pressure) equal to $(\alpha T/\kappa_T) - p$ (Eq. 5.4.3) and $(\partial V/\partial T)_p$ equal to αV to obtain

$$C_p = C_V + \frac{\alpha^2 T V}{\kappa_T}$$

and divide by n to obtain molar quantities:

$$C_{p,m} = C_{V,m} + \frac{\alpha^2 T V_m}{\kappa_T}$$

Since the quantity $\alpha^2 T V_m/\kappa_T$ must be positive, $C_{p,m}$ is greater than $C_{V,m}$.

5.5.2 The measurement of heat capacities

The most accurate method of evaluating the heat capacity of a phase is by measuring the temperature change resulting from heating with electrical work. The procedure in general is called calorimetry, and the apparatus containing the phase of interest and the electric heater is a **calorimeter**. The principles of three commonly used types of calorimeters with electrical heating are now described.

Adiabatic calorimeters

An adiabatic calorimeter is designed to have negligible heat flow to or from its surroundings. The calorimeter contains the phase of interest, kept at either constant volume or constant pressure, and also an electric heater and a temperature-measuring device such as a platinum resistance thermometer, thermistor, or quartz crystal oscillator. The contents may be stirred to ensure temperature uniformity.

To minimize conduction and convection, the calorimeter usually is surrounded by a jacket separated by an air gap or an evacuated space. The outer surface of the calorimeter and inner surface of the jacket may be polished to minimize radiation emission from these surfaces. These measures, however, are not sufficient to ensure a completely adiabatic boundary because heat can be transferred along the mounting hardware and through the electrical leads. Therefore, the temperature of the jacket, or of an outer metal shield, is adjusted throughout the course of the experiment so as to be as close as possible to

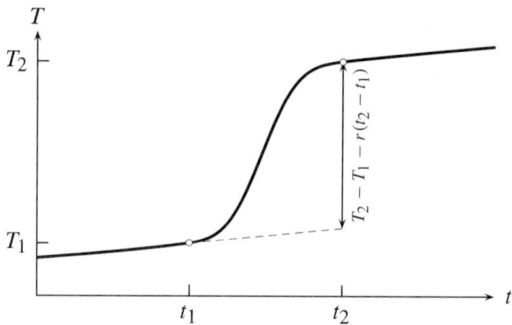

Figure 5.3 Typical heating curve of an adiabatic calorimeter.

the varying temperature of the calorimeter. This goal is most easily achieved when the temperature change is slow.

For a heat capacity measurement, a constant electric current is passed through the heater circuit for a known period of time. The *system* is the calorimeter and its contents. The electrical work w_{el} performed on the system by the heater circuit is calculated from the integrated form of Eq. 3.8.3 on page 54: $w_{el} = I^2 R_{el} \Delta t$ (where I is the electric current, R_{el} is the electric resistance, and Δt is the time interval). We assume the boundary is adiabatic and write the first law in the form

$$dU = -p\,dV + dw_{el} + dw_{cont} \tag{5.5.2}$$

where $-p\,dV$ is expansion work and w_{cont} is any continuous mechanical work from stirring (the subscript "cont" stands for continuous). If electrical work is done on the system by a thermometer using an external electrical circuit, such as a platinum resistance thermometer, this work is included in w_{cont}.

Consider an adiabatic calorimeter in which the heating process is carried out at constant *volume*. There is no expansion work, and Eq. 5.5.2 becomes

$$dU = dw_{el} + dw_{cont} \tag{5.5.3}$$
$$(\text{constant } V)$$

An example of a measured heating curve (temperature T as a function of time t) is shown in Fig. 5.3. We select two points on the heating curve, indicated in the figure by open circles. The point at time t_1 is at or shortly before the time the heater circuit is closed and electrical heating begins, and the point at t_2 is after the heater circuit has been opened and the slope of the curve has become essentially constant.

In the time periods before t_1 and after t_2, the temperature may exhibit a slow rate of increase due to the continuous work w_{cont} from stirring and temperature measurement. If this work is performed at a constant rate throughout the course of the experiment, the slope is constant and the same in both time periods as shown in the figure.

The relation between the slope and the rate of work is given by a quantity called the **energy equivalent**, ϵ. The energy equivalent is the heat capacity of the calorimeter under

Section 5.5 Thermal Properties

the conditions of an experiment. The heat capacity of a constant-volume calorimeter is given by $\epsilon = (\partial U/\partial T)_V$ (Eq. 3.12.2). Thus, at a time before t_1 or after t_2, when dw_{el} is zero and dU equals dw_{cont}, the slope r of the heating curve is given by

$$r = \frac{dT}{dt} = \frac{dT}{dU}\frac{dU}{dt} = \frac{1}{\epsilon}\frac{dw_{cont}}{dt}$$

The rate of the continuous work is therefore $dw_{cont}/dt = \epsilon r$. This rate is constant throughout the experiment. In the time interval from t_1 to t_2, the total quantity of continuous work is $w_{cont} = \epsilon r(t_2 - t_1)$, where r is the slope of the heating curve measured *outside* this time interval.

To find the energy equivalent, we integrate Eq. 5.5.3 between the two points on the curve:

$$\Delta U = w_{el} + w_{cont} = w_{el} + \epsilon r(t_2 - t_1) \qquad (5.5.4)$$
$$\text{(constant } V)$$

Then the average heat capacity between temperatures T_1 and T_2 is

$$\epsilon = \frac{\Delta U}{T_2 - T_1} = \frac{w_{el} + \epsilon r(t_2 - t_1)}{T_2 - T_1}$$

Solving for ϵ, we obtain

$$\epsilon = \frac{w_{el}}{T_2 - T_1 - r(t_2 - t_1)} \qquad (5.5.5)$$

The value of the denominator on the right side is indicated by the vertical line in Fig. 5.3. It is the temperature change that would have been observed if w_{cont} had been zero.

We can also consider the heating process in a constant *pressure* calorimeter. In this case, the enthalpy change is $dH = dU + pdV$, and in place of Eq. 5.5.3 we use

$$dH = dw_{el} + dw_{cont}$$

By the same procedure, replacing U by H and C_V by C_p, we obtain

$$\Delta H = w_{el} + w_{cont} = w_{el} + \epsilon r(t_2 - t_1) \qquad (5.5.6)$$
$$\text{(constant } p)$$

in place of Eq. 5.5.4 and end up again with the expression of Eq. 5.5.5 for ϵ.

The value of ϵ calculated from Eq. 5.5.5 is an *average* value for the temperature interval from T_1 to T_2, and we can identify it with the heat capacity at the temperature of the midpoint of the interval. By taking the difference of values of ϵ measured with and without the phase of interest present in the calorimeter, we obtain C_V or C_p for the phase alone.

It may seem paradoxical that we can use an adiabatic process, one without heat, to evaluate a quantity defined by heat (heat capacity $\equiv dq/dT$). The explanation is that energy transferred into the adiabatic calorimeter as electrical work, and dissipated completely to thermal energy, substitutes for the heat that would be needed for the same change of state without electrical work.

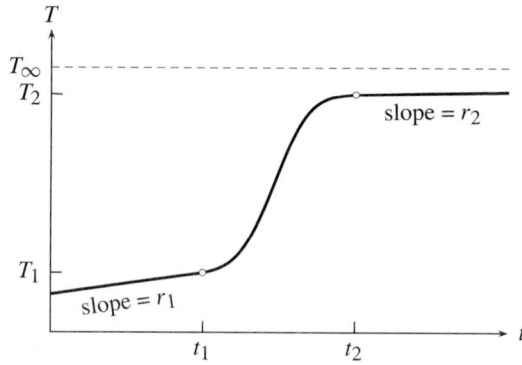

Figure 5.4 Typical heating curve of an isothermal-jacket calorimeter.

Isothermal-jacket calorimeters

Another common type of calorimeter is similar in construction to an adiabatic calorimeter, except that the surrounding jacket is maintained at constant temperature. It is sometimes called an *isoperibol calorimeter*. A correction is made for heat transfer resulting from the difference in temperature across the gap separating the jacket from the outer surface of the calorimeter. It is important in making this correction that the outer surface have a uniform temperature without "hot spots."

If the outer surface of the calorimeter has a uniform temperature T, the jacket temperature is T_{ext}, and convection is eliminated by evacuating the gap, heat transfer is by conduction and radiation. The rate of heat transfer is given by Newton's law of cooling:

$$\frac{dq}{dt} = -k(T - T_{\text{ext}}) \tag{5.5.7}$$

where k is a constant (the thermal conductance). Heat flows from a warmer to a cooler body, so dq/dt is positive if T is less than T_{ext} and negative if T is greater than T_{ext}.

The possible kinds of work are the same as for the adiabatic calorimeter: expansion work $-p\,dV$, intermittent work w_{el} done by the heater circuit, and continuous work w_{cont}. By combining the first law and Eq. 5.5.7, we obtain the following relation for the rate at which the internal energy changes:

$$\frac{dU}{dt} = \frac{dq}{dt} + \frac{dw}{dt} = -k(T - T_{\text{ext}}) - p\frac{dV}{dt} + \frac{dw_{\text{el}}}{dt} + \frac{dw_{\text{cont}}}{dt}$$

For heating at constant *volume* ($dV/dt = 0$), this relation becomes

$$\frac{dU}{dt} = -k(T - T_{\text{ext}}) + \frac{dw_{\text{el}}}{dt} + \frac{dw_{\text{cont}}}{dt} \tag{5.5.8}$$
$$\text{(constant } V\text{)}$$

An example of a heating curve is shown in Fig. 5.4. In contrast to the curve of Fig. 5.3, the slopes are different before and after the heating interval due to the changed rate of heat flow. The points at times t_1 and t_2, indicated in the figure by open circles, are at times before and after the heater circuit is closed. In any time interval before time t_1 or after time t_2, the system behaves as if it is approaching a steady state of constant temperature T_∞ (called the convergence temperature), which it would eventually reach if the experiment were continued without closing the heater circuit. T_∞ is greater than T_ext because of the energy transferred to the system by stirring and electrical temperature measurement. By setting dU/dt and dw_el/dt equal to zero and T equal to T_∞ in Eq. 5.5.8, we obtain $dw_\text{cont}/dt = k(T_\infty - T_\text{ext})$. We assume dw_cont/dt is constant. Substituting this expression into Eq. 5.5.8 gives us a general expression for the rate at which U changes in terms of the unknown quantities k and T_∞:

$$\frac{dU}{dt} = -k(T - T_\infty) + \frac{dw_\text{el}}{dt} \qquad (5.5.9)$$
$$\text{(constant } V\text{)}$$

This relation is valid throughout the experiment, not only while the heater circuit is closed. If we multiply by dt and integrate from t_1 to t_2, we obtain the internal energy change in the time interval from t_1 to t_2:

$$\Delta U = -k \int_{t_1}^{t_2} (T - T_\infty)\, dt + w_\text{el} \qquad (5.5.10)$$
$$\text{(constant } V\text{)}$$

All the intermittent work w_el is performed in this time interval.

> There is nothing in the derivation that prevents Eq. 5.5.10 from being applied to a isothermal-jacket calorimeter in which a reaction is occurring. In Sec. 8.5.2 we will mention the use of this equation for an internal energy correction of a reaction calorimeter with an isothermal jacket.

The average value of the energy equivalent in the temperature range T_1 to T_2 is

$$\epsilon = \frac{\Delta U}{T_2 - T_1} = \frac{-\epsilon(k/\epsilon)\int_{t_1}^{t_2}(T - T_\infty)\,dt + w_\text{el}}{T_2 - T_1}$$

Solving for ϵ, we obtain

$$\epsilon = \frac{w_\text{el}}{(T_2 - T_1) + (k/\epsilon)\int_{t_1}^{t_2}(T - T_\infty)\,dt} \qquad (5.5.11)$$

The value of w_el is known from $w_\text{el} = I^2 R_\text{el} \Delta t$, where Δt is the time interval the heater circuit is closed; and the integral can be evaluated numerically once T_∞ is known.

For heating at constant *pressure*, dH is equal to $dU + p\,dV$, and we can write

$$\frac{dH}{dt} = \frac{dU}{dt} + p\frac{dV}{dt} = -k(T - T_\text{ext}) + \frac{dw_\text{el}}{dt} + \frac{dw_\text{cont}}{dt}$$

which is analogous to Eq. 5.5.8. By the same procedure described before for the case of constant V, we obtain

$$\Delta H = -k \int_{t_1}^{t_2} (T - T_\infty)\, dt + w_{el} \tag{5.5.12}$$
(constant p)

At constant p, the energy equivalent is equal to $C_p = \Delta H/(T_2 - T_1)$, and the final expression for ϵ is the same as Eq. 5.5.11.

To obtain values of k/ϵ and T_∞ for use in Eq. 5.5.11, we need the slopes of the heating curve in time intervals (rating periods) just before t_1 and just after t_2. Consider the case of constant *volume*. In these intervals, dw_{el}/dt is zero and dU/dt equals $-k(T - T_\infty)$ (from Eq. 5.5.9). The heat capacity at constant volume is $C_V = dU/dT$. The slope r in general is then given by

$$r = \frac{dT}{dt} = \frac{dT}{dU}\frac{dU}{dt} = -(k/\epsilon)(T - T_\infty)$$

Applying this relation to the points at times t_1 and t_2, we have the following simultaneous equations in the unknowns k/ϵ and T_∞:

$$r_1 = -(k/\epsilon)(T_1 - T_\infty) \qquad r_2 = -(k/\epsilon)(T_2 - T_\infty)$$

The solutions are

$$(k/\epsilon) = \frac{r_1 - r_2}{T_2 - T_1} \qquad T_\infty = \frac{r_1 T_2 - r_2 T_1}{r_1 - r_2}$$

Finally, k is given by

$$k = (k/\epsilon)\epsilon = \left(\frac{r_1 - r_2}{T_2 - T_1}\right)\epsilon \tag{5.5.13}$$

By the same procedure, we obtain the same relations for k/ϵ, T_∞, and k when the *pressure* is constant.

Continuous-flow calorimeters

In a flow calorimeter used to measure the heat capacity of a fluid phase of a substance, the gas or liquid flows through a tube at a known constant rate past an electrical heater of known constant power input. After a steady state is achieved in the tube, the fluid temperature is measured upstream and downstream from the heater.

If dw_{el}/dt is the rate at which electrical work is performed (the electric power) and dm/dt is the mass flow rate, then in time interval Δt a quantity $w = (dw_{el}/dt)\Delta t$ of work is performed on an amount $n = (dm/dt)\Delta t/M$ of the fluid (where M is the molar mass). If heat flow is negligible, the molar heat capacity of the substance is given by

$$C_{p,m} = \frac{w}{n\Delta T} = \frac{M(dw_{el}/dt)}{\Delta T (dm/dt)}$$

where ΔT is the measured temperature increase at the heater. To correct for the effects of heat flow, ΔT is usually measured over a range of flow rates and the results extrapolated to infinite flow rate.

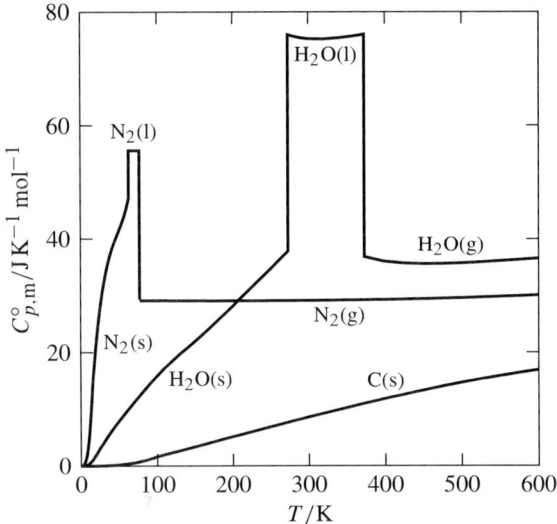

Figure 5.5 Standard molar heat capacities at constant pressure of H_2O, N_2, and C (graphite).

5.5.3 Typical values

The thermal energy of an ideal monatomic gas in which each molecule is in its ground electronic state is simply the kinetic energy of the translational motion of the atoms. According to statistical mechanical theory, this energy is equal to $(3/2)nRT$. Thus, from Eq. 3.12.3, the heat capacity at constant volume of this gas is $(3/2)nR$ and the molar heat capacity at constant volume is $C_{V,m} = (3/2)R$. The molar heat capacity at constant *pressure*, from Eq. 5.5.1, is $C_{p,m} = C_{V,m} + R = (5/2)R = 20.786 \, \text{J K}^{-1} \, \text{mol}^{-1}$. At high temperature, electronic excitation occurs and $C_{p,m}$ increases. In more complicated systems, the thermal energy includes, in addition to translational and electronic energy, contributions from rotational energy, vibrational energy, and intermolecular potentials; the value of $C_{p,m}$ is accordingly greater than $(5/2)R$. Thermodynamics is unable to predict this value.

Figure 5.5 shows the temperature dependence of $C_{p,m}$ for several substances. The discontinuities seen at certain temperatures occur at equilibrium phase transitions. We might say that at these temperatures the heat capacity is infinite since the phase transition of a pure substance has finite heat with zero temperature change.

5.6 HEATING AT CONSTANT VOLUME OR PRESSURE

Consider the process of changing the temperature of a phase from T_1 to T_2 while keeping the *volume* constant.[4] Since the rate of change of internal energy with T under these conditions

---

[4]Keeping the volume exactly constant while increasing the temperature is not as simple as it sounds. Most solids expand when heated unless we arrange to increase the external pressure at the same time. If we use solid

is $(\partial U/\partial T)_V$, the infinitesimal change of U is $dU = (\partial U/\partial T)_V\, dT$. But $(\partial U/\partial T)_V$ is equal to C_V. Accordingly, the infinitesimal change is

$$dU = C_V\, dT \qquad (5.6.1)$$
$$\text{(closed system, } P=1, C=1, \text{ constant } V)$$

Integration then gives the finite change

$$\Delta U = \int_{T_1}^{T_2} C_V\, dT \qquad (5.6.2)$$
$$\text{(closed system, } P=1, C=1, \text{ constant } V)$$

Three comments, relevant to these and other equations in this chapter, are in order:

1. Equation 5.6.2 allows us to calculate the finite change of a state function, U, by integrating C_V over T. We derived the equation under the condition that V is constant during the process, and the use of the integration variable T implies that the system has a single, uniform temperature at each instant during the process. The integrand C_V may depend on both V and T, and we should integrate with V held constant and C_V treated as a function only of T.

2. Suppose we want to evaluate ΔU for a process in which the volume is the same in the initial and final states but is different in some intermediate states, and the temperature is *not* uniform in some of the intermediate states. We know the change of a state function depends only on the initial and final states, so we can still use Eq. 5.6.2 to evaluate ΔU for this process. We integrate with V held constant, although V was not constant during the actual process. To generalize, a finite change ΔX of a state function, evaluated under the condition that another state function Y is constant, is the same as ΔX under the less stringent condition $Y_2 = Y_1$.

3. We may convert an expression for an infinitesimal or finite change of an extensive property of a pure substance to an expression for the change of the corresponding *molar* property by dividing by n. For instance, Eq. 5.6.1 becomes

$$dU_m = C_{V,m}\, dT$$

and Eq. 5.6.2 becomes

$$\Delta U_m = \int_{T_1}^{T_2} C_{V,m}\, dT$$

---

walls to contain a fluid phase, the container volume will change with temperature. For practical purposes, these volume changes are usually negligible.

Section 5.6 Heating at Constant Volume or Pressure

If, at a fixed volume and over the temperature range T_1 to T_2, the value of C_V is essentially constant (i.e., independent of T), Eq. 5.6.2 becomes

$$\Delta U = C_V(T_2 - T_1) \tag{5.6.3}$$

(closed system, $P = 1$, $C = 1$, constant V and C_V)

The entropy change of a reversible process in a closed system is given according to the second law by $dS = dq/T$ (Box 4.2.1). The heat dq, in the absence of nonexpansion work w', is equal to the heat capacity times dT (Eq. 3.12.1). The appropriate heat capacity for heating at constant volume is C_V. Accordingly, when an isochoric heating process is carried out reversibly and with dw' equal to zero, the entropy change is given by the relation $dS = (C_V/T) dT$. Since there is only one independent variable when V is kept constant, there is only one possible infinitesimal change of S for a given infinitesimal change of T. Therefore, the relation is valid regardless of whether the process is reversible or dw' is zero; the only conditions of validity are that the system be a single phase of one component and that the volume be constant:

$$dS = \frac{C_V}{T} dT \tag{5.6.4}$$

(closed system, $P = 1, C = 1$, constant V)

Integration of this equation yields the finite change

$$\Delta S = \int_{T_1}^{T_2} \frac{C_V}{T} dT \tag{5.6.5}$$

(closed system, $P = 1, C = 1$, constant V)

If C_V is treated as constant, Eq. 5.6.5 becomes

$$\Delta S = C_V \ln \frac{T_2}{T_1} \tag{5.6.6}$$

(closed system, $P = 1$, $C = 1$, constant V and C_V)

Since C_V is positive, we see from Eqs. 5.6.2 and 5.6.5 that heating a phase at constant volume causes both U and S to increase.

We may derive relations for a temperature change at constant *pressure* by the same methods. From $C_p = (\partial H/\partial T)_p$ (Eq. 3.12.4), we obtain

$$\Delta H = \int_{T_1}^{T_2} C_p \, dT \tag{5.6.7}$$

(closed system, $P = 1, C = 1$, constant p)

If C_p is treated as constant, Eq. 5.6.7 becomes

$$\Delta H = C_p(T_2 - T_1) \tag{5.6.8}$$
(closed system, $P = 1$,
$C = 1$, constant p and C_p)

A reversible, isobaric process with dw' equal to zero has $dS = dq/T = (C_p/T)\,dT$. This relation applies to any isobaric process in a closed system of one component in one phase:

$$dS = \frac{C_p}{T}\,dT \tag{5.6.9}$$
(closed system,
$P = 1$, $C = 1$, constant p)

Integration of Eq. 5.6.9 gives

$$\Delta S = \int_{T_1}^{T_2} \frac{C_p}{T}\,dT \tag{5.6.10}$$
(closed system,
$P = 1$, $C = 1$, constant p)

or, with C_p treated as constant,

$$\Delta S = C_p \ln \frac{T_2}{T_1} \tag{5.6.11}$$
(closed system, $P = 1$,
$C = 1$, constant p and C_p)

C_p is positive, so heating a phase at constant pressure causes H and S to increase. The Gibbs energy changes according to $(\partial G/\partial T)_p = -S$ (Eq. 5.3.12), so heating at constant pressure causes G to decrease.

5.7 PARTIAL DERIVATIVES WITH RESPECT TO TEMPERATURE, PRESSURE, AND VOLUME

5.7.1 Tables of partial derivatives

The tables in this section collect useful expressions for partial derivatives of the seven quantities T, p, V, U, H, S, and G in a closed, single-phase system. Each derivative is taken with respect to one of the three easily controlled variables T, p, or V while another of these variables is held constant. We have already seen some of these expressions, and the derivations of the others are indicated in this section.

We can use these partial derivatives (1) for writing an expression for the total differential of any of the seven quantities listed earlier, and (2) for expressing the finite change in one of these quantities as an integral under conditions of constant T, p, or V. For instance, given the expressions

$$\left(\frac{\partial S}{\partial T}\right)_p = \frac{C_p}{T}$$

Section 5.7 Partial Derivatives with Respect to Temperature, Pressure, and Volume

and

$$\left(\frac{\partial S}{\partial p}\right)_T = -\alpha V$$

we may write the total differential of S, taking T and p as the independent variables, as

$$dS = \frac{C_p}{T} dT - \alpha V \, dp$$

Furthermore, the first expression is equivalent to the differential form

$$dS = \frac{C_p}{T} dT$$

provided p is constant; we can integrate this equation to obtain the finite change ΔS under isobaric conditions as shown in Eq. 5.6.10.

The expressions are given in Tables 5.1, 5.2, and 5.3. The third column of each table shows an expression valid for an ideal gas.

We may derive these expressions as follows. We are considering differentiation with respect only to T, p, and V. Expressions for $(\partial V/\partial T)_p$, $(\partial V/\partial p)_T$, and $(\partial p/\partial T)_V$ come from Eqs. 5.2.1, 5.2.2, and 5.2.6 and are shown as functions of the mechanical coefficients α and κ_T. The reciprocal of each of these three expressions provides the expression for another partial derivative from the general relation

$$(\partial y/\partial x)_z = \frac{1}{(\partial x/\partial y)_z}$$

This procedure gives us expressions for the six partial derivatives of T, p, and V.

The remaining expressions are for partial derivatives of U, H, S, and G. We obtain the expression for $(\partial U/\partial T)_V$ from Eq. 3.12.2, for $(\partial U/\partial V)_T$ from Eq. 5.4.3, for $(\partial H/\partial T)_p$ from Eq. 3.12.4, for $(\partial S/\partial T)_V$ from Eq. 5.6.4, for $(\partial S/\partial T)_p$ from Eq. 5.6.9, for $(\partial S/\partial p)_T$ from Eq. 5.3.18, for $(\partial G/\partial p)_T$ from Eq. 5.3.13, and for $(\partial G/\partial T)_p$ from Eq. 5.3.12. We find an expression for $(\partial U/\partial T)_p$ by differentiating $H - pV$ (which is equal to U) and making appropriate substitutions: $(\partial U/\partial T)_p = (\partial H/\partial T)_p - p(\partial V/\partial T)_p = C_p - pV\alpha$. Similarly, we obtain expressions for $(\partial H/\partial V)_T$ and $(\partial H/\partial T)_V$ by differentiating $U + pV$ and an expression for $(\partial G/\partial T)_V$ by differentiating $H - TS$. Whenever a partial derivative appears in a derived expression, we replace it with an expression derived in an earlier step.

Finally, we use each of these partial derivatives of U, H, S, and G to obtain the expression for a partial derivative in which the variable of differentiation is different and the constant variable is the same. This transformation involves multiplying by an appropriate partial derivative of T, p, or V. For instance, from the partial derivative $(\partial U/\partial V)_T = (\alpha T/\kappa_T) - p$, we obtain

$$\left(\frac{\partial U}{\partial p}\right)_T = \left(\frac{\partial U}{\partial V}\right)_T \left(\frac{\partial V}{\partial p}\right)_T = \left(\frac{\alpha T}{\kappa_T} - p\right)(-\kappa_T V) = (-\alpha T + \kappa_T p) V$$

The expressions derived by these steps constitute the full set shown in Tables 5.1, 5.2, and 5.3.

Table 5.1 *Constant temperature*: expressions for partial derivatives of state functions with respect to pressure and volume in a closed, single-phase system

Partial derivative	General expression	Ideal gas
$\left(\dfrac{\partial p}{\partial V}\right)_T$	$-\dfrac{1}{\kappa_T V}$	$-\dfrac{p}{V}$
$\left(\dfrac{\partial V}{\partial p}\right)_T$	$-\kappa_T V$	$-\dfrac{V}{p}$
$\left(\dfrac{\partial U}{\partial p}\right)_T$	$(-\alpha T + \kappa_T p)V$	0
$\left(\dfrac{\partial U}{\partial V}\right)_T$	$\dfrac{\alpha T}{\kappa_T} - p$	0
$\left(\dfrac{\partial H}{\partial p}\right)_T$	$(1 - \alpha T)V$	0
$\left(\dfrac{\partial H}{\partial V}\right)_T$	$\dfrac{\alpha T - 1}{\kappa_T}$	0
$\left(\dfrac{\partial S}{\partial p}\right)_T$	$-\alpha V$	$-\dfrac{V}{T}$
$\left(\dfrac{\partial S}{\partial V}\right)_T$	$\dfrac{\alpha}{\kappa_T}$	$\dfrac{p}{T}$
$\left(\dfrac{\partial G}{\partial p}\right)_T$	V	V
$\left(\dfrac{\partial G}{\partial V}\right)_T$	$-\dfrac{1}{\kappa_T}$	$-p$

Bridgman[5] devised a simple method to obtain expressions for these and many other partial derivatives from a relatively small set of formulas.

5.7.2 The Joule–Thomson coefficient

The Joule–Thomson coefficient of a gas was defined in Eq. 4.12.1 on page 98 by $\mu_{JT} = (\partial T/\partial p)_H$. It can be evaluated with measurements of T and p during adiabatic throttling processes as described in Sec. 4.12.1.

To relate μ_{JT} to other properties of the gas, we write the total differential of the enthalpy

---

[5] Bridgman, P. W. *Phys. Rev.* **1914**, *3*, 273; Bridgman, P. W. *The Thermodynamics of Electrical Phenomena in Metals and a Condensed Collection of Thermodynamic Formulas*; Dover Publications: New York, 1961, p. 199-241.

Section 5.7 Partial Derivatives with Respect to Temperature, Pressure, and Volume

Table 5.2 *Constant pressure*: expressions for partial derivatives of state functions with respect to temperature and volume in a closed, single-phase system

Partial derivative	General expression	Ideal gas
$\left(\dfrac{\partial T}{\partial V}\right)_p$	$\dfrac{1}{\alpha V}$	$\dfrac{T}{V}$
$\left(\dfrac{\partial V}{\partial T}\right)_p$	αV	$\dfrac{V}{T}$
$\left(\dfrac{\partial U}{\partial T}\right)_p$	$C_p - \alpha p V$	C_V
$\left(\dfrac{\partial U}{\partial V}\right)_p$	$\dfrac{C_p}{\alpha V} - p$	$\dfrac{C_V T}{V}$
$\left(\dfrac{\partial H}{\partial T}\right)_p$	C_p	C_p
$\left(\dfrac{\partial H}{\partial V}\right)_p$	$\dfrac{C_p}{\alpha V}$	$\dfrac{C_p T}{V}$
$\left(\dfrac{\partial S}{\partial T}\right)_p$	$\dfrac{C_p}{T}$	$\dfrac{C_p}{T}$
$\left(\dfrac{\partial S}{\partial V}\right)_p$	$\dfrac{C_p}{\alpha T V}$	$\dfrac{C_p}{V}$
$\left(\dfrac{\partial G}{\partial T}\right)_p$	$-S$	$-S$
$\left(\dfrac{\partial G}{\partial V}\right)_p$	$-\dfrac{S}{\alpha V}$	$-\dfrac{TS}{V}$

of a closed, one-phase system in the form

$$dH = \left(\frac{\partial H}{\partial T}\right)_p dT + \left(\frac{\partial H}{\partial p}\right)_T dp$$

and divide both sides by dp:

$$\frac{dH}{dp} = \left(\frac{\partial H}{\partial T}\right)_p \frac{dT}{dp} + \left(\frac{\partial H}{\partial p}\right)_T$$

Next we impose a condition of constant H; the ratio dT/dp becomes a partial derivative:

$$0 = \left(\frac{\partial H}{\partial T}\right)_p \left(\frac{\partial T}{\partial p}\right)_H + \left(\frac{\partial H}{\partial p}\right)_T$$

Table 5.3 *Constant volume*: expressions for partial derivatives of state functions with respect to temperature and pressure in a closed, single-phase system

Partial derivative	General expression	Ideal gas
$\left(\dfrac{\partial T}{\partial p}\right)_V$	$\dfrac{\kappa_T}{\alpha}$	$\dfrac{T}{p}$
$\left(\dfrac{\partial p}{\partial T}\right)_V$	$\dfrac{\alpha}{\kappa_T}$	$\dfrac{p}{T}$
$\left(\dfrac{\partial U}{\partial T}\right)_V$	C_V	C_V
$\left(\dfrac{\partial U}{\partial p}\right)_V$	$\dfrac{\kappa_T C_p}{\alpha} - \alpha T V$	$\dfrac{T C_V}{p}$
$\left(\dfrac{\partial H}{\partial T}\right)_V$	$C_p + \dfrac{\alpha V}{\kappa_T}(1 - \alpha T)$	C_p
$\left(\dfrac{\partial H}{\partial p}\right)_V$	$\dfrac{\kappa_T C_p}{\alpha} + V(1 - \alpha T)$	$\dfrac{C_p T}{p}$
$\left(\dfrac{\partial S}{\partial T}\right)_V$	$\dfrac{C_V}{T}$	$\dfrac{C_V}{T}$
$\left(\dfrac{\partial S}{\partial p}\right)_V$	$\dfrac{\kappa_T C_p}{\alpha T} - \alpha V$	$\dfrac{C_V}{p}$
$\left(\dfrac{\partial G}{\partial T}\right)_V$	$\dfrac{\alpha V}{\kappa_T} - S$	$\dfrac{pV}{T} - S$
$\left(\dfrac{\partial G}{\partial p}\right)_V$	$V - \dfrac{\kappa_T S}{\alpha}$	$V - \dfrac{T S}{p}$

Rearrangement gives

$$\left(\frac{\partial T}{\partial p}\right)_H = -\frac{(\partial H/\partial p)_T}{(\partial H/\partial T)_p}$$

The left side of this equation is the Joule–Thomson coefficient. An expression for the partial derivative $(\partial H/\partial p)_T$ is given in Table 5.1, and the partial derivative $(\partial H/\partial T)_p$ is the heat capacity at constant pressure (Eq. 3.12.4). These substitutions give us the desired relation

$$\mu_{\mathrm{JT}} = \frac{(\alpha T - 1)V}{C_p} = \frac{(\alpha T - 1)V_{\mathrm{m}}}{C_{p,\mathrm{m}}}$$

5.8 ISOTHERMAL PRESSURE AND VOLUME CHANGES

In various applications, we need formulas for the effect of changing the pressure at constant temperature on the internal energy, enthalpy, entropy, or Gibbs energy of a phase. We obtain the formulas by integrating expressions found in Table 5.1; for example, ΔU is given by $\int (\partial U/\partial p)_T \, dp$.

The resulting formulas, for an isothermal change of pressure from p_1 to p_2, are

$$\Delta U = \int_{p_1}^{p_2} (-\alpha T + \kappa_T p) V \, dp \qquad (5.8.1)$$
(closed system, $P = 1, C = 1$, constant T)

$$\Delta H = \int_{p_1}^{p_2} (1 - \alpha T) V \, dp \qquad (5.8.2)$$
(closed system, $P = 1, C = 1$, constant T)

$$\Delta S = -\int_{p_1}^{p_2} \alpha V \, dp \qquad (5.8.3)$$
(closed system, $P = 1, C = 1$, constant T)

$$\Delta G = \int_{p_1}^{p_2} V \, dp \qquad (5.8.4)$$
(closed system, $P = 1, C = 1$, constant T)

5.8.1 Ideal gases

Simplifications result when the phase is an ideal gas. In this case, we can make the substitutions $V = nRT/p$, $\alpha = 1/T$, and $\kappa_T = 1/p$ resulting in the formulas

$$\Delta U = 0 \qquad (5.8.5)$$
(closed system, ideal gas, $C = 1$, $T_2 = T_1$)

$$\Delta H = 0 \qquad (5.8.6)$$
(closed system, ideal gas, $C = 1$, $T_2 = T_1$)

$$\Delta S = -nR \ln \frac{p_2}{p_1} = nR \ln \frac{V_2}{V_1} \qquad (5.8.7)$$
(closed system, ideal gas, $C = 1$, $T_2 = T_1$)

$$\Delta G = nRT \ln \frac{p_2}{p_1} = -nRT \ln \frac{V_2}{V_1} \qquad (5.8.8)$$
$$\text{(closed system,}$$
$$\text{ideal gas, } C = 1, T_2 = T_1)$$

(Equation 5.8.7 is derived in fewer steps from $dS = dq/T = (dU - dw)/T$; for an ideal gas at constant T, dU is zero and dw is $-p\,dV = -(nRT/V)\,dV$.)

The Helmholtz energy is related to the Gibbs energy by $A = G - pV$. Since pV is constant for isothermal changes of an ideal gas, the expressions for ΔG in Eq. 5.8.8 apply also to ΔA.

We may summarize Eqs. 5.8.5–5.8.8 by saying that, when an ideal gas expands isothermally, the internal energy and enthalpy stay constant, the entropy increases, and the Helmholtz energy and Gibbs energy decrease.

5.8.2 Condensed phases

Liquids and solids are much less compressible than gases. Typically the isothermal compressibility, κ_T, of a liquid or solid at room temperature and $p = 1$ bar is no greater than 1×10^{-4} bar^{-1} (see Fig. 5.2 on page 106), whereas an ideal gas under these conditions has $\kappa_T = 1/p = 1$ bar^{-1}. Consequently, it is frequently valid to assume that V for a liquid or solid is essentially constant during a pressure change at constant temperature. Because κ_T is small, the quantity $\kappa_T p$ for a liquid or solid is usually much smaller than αT. Furthermore, κ_T for liquids and solids does not change rapidly with p as it does for gases; neither does α.

With the approximations that V, α, and κ_T are constant during an isothermal pressure change and that $\kappa_T p$ is negligible compared with αT, Eqs. 5.8.1–5.8.4 become

$$\Delta U \approx -\alpha T V \Delta p \qquad (5.8.9)$$
$$\text{(closed system, liquid or}$$
$$\text{solid, } C = 1, \text{constant } T)$$

$$\Delta H \approx (1 - \alpha T) V \Delta p \qquad (5.8.10)$$
$$\text{(closed system, liquid or}$$
$$\text{solid, } C = 1, \text{constant } T)$$

$$\Delta S \approx -\alpha V \Delta p \qquad (5.8.11)$$
$$\text{(closed system, liquid or}$$
$$\text{solid, } C = 1, \text{constant } T)$$

$$\Delta G \approx V \Delta p \qquad (5.8.12)$$
$$\text{(closed system, liquid or}$$
$$\text{solid, } C = 1, \text{constant } T)$$

5.9 STANDARD STATES OF PURE SUBSTANCES

It is often useful to refer to a reference pressure, the **standard pressure**, denoted $p°$, which has an arbitrary but constant value in any given application. Until 1982, chemists used a standard pressure of 1 atm (1.01325×10^5 Pa). The IUPAC now recommends the value $p° = 1$ bar (exactly 10^5 Pa). This book uses the latter value unless stated otherwise.

> A degree symbol (°) denotes a standard quantity or standard-state conditions. An alternative symbol for this purpose, used extensively outside the U.S., is a superscript plimsoll mark ($^{\ominus}$).[6]

A **standard state** of a pure substance is a particular reference state appropriate for the kind of phase and is described by intensive variables. We follow the recommendations of the IUPAC Green Book for various standard states.

The standard state of a *pure gas* is the hypothetical state in which the gas is at pressure $p°$ and the temperature of interest, and the gas behaves as an ideal gas. The molar volume of a gas at 1 bar may have a measurable deviation from the molar volume predicted by the ideal gas equation due to intermolecular forces. We must imagine the standard state in this case to consist of the gas with the intermolecular forces magically "turned off" and the molar volume adjusted to the ideal-gas value $RT/p°$.

The standard state of a *pure liquid or solid* is the liquid or solid at pressure $p°$ and the temperature of interest.[7] If the liquid or solid is stable under these conditions, this is a real (not hypothetical) state. Any standard molar quantity of a substance in a liquid or solid phase is simply the molar quantity under these conditions; for instance, $S_m°(T) = S_m(T, p°)$.

Section 7.7 introduces additional standard states for constituents of mixtures.

5.10 CHEMICAL POTENTIAL AND FUGACITY

The **chemical potential**, μ, of a pure substance is defined as the molar Gibbs energy:

$$\mu \equiv G_m = \frac{G}{n} \qquad (5.10.1)$$
(pure substance)

(Section 7.3 introduces a more general definition of chemical potential that applies also to a constituent of a mixture.) The chemical potential is an intensive state function.

The total differential of the Gibbs energy of a fixed amount of a pure substance in a single phase, with T and p as independent variables, is $dG = -S\,dT + V\,dp$ (Eq. 5.3.5). Dividing both sides of this equation by n gives the total differential of the chemical potential with these same independent variables:

$$d\mu = -S_m\,dT + V_m\,dp \qquad (5.10.2)$$
(pure substance, $P = 1$)

---

[6]The plimsoll mark is named after the British merchant Samuel Plimsoll, at whose instigation Parliament passed an act in 1875 requiring the symbol to be placed on the hulls of cargo ships to indicate the maximum depth for safe loading.

[7]To be in its standard state, a solid must not be stressed. For instance, the solid must have no internal cavities containing a vacuum or fluid at a pressure different from $p°$.

(Since all quantities in this equation are intensive, it is not necessary to specify a closed system; the amount of the substance in the system is irrelevant.)

We identify the coefficients of the terms on the right side of Eq. 5.10.2 as

$$\left(\frac{\partial \mu}{\partial T}\right)_p = -S_m \qquad (5.10.3)$$
(pure substance, $P=1$)

and

$$\left(\frac{\partial \mu}{\partial p}\right)_T = V_m \qquad (5.10.4)$$
(pure substance, $P=1$)

The last equation shows that the chemical potential increases with increasing pressure in an isothermal process.

The **standard chemical potential**, $\mu°$, of a pure substance in a given phase and at a given temperature is the chemical potential of the substance when it is in the standard state of the phase at this temperature. Whereas μ has a definite fixed value (G_m) under given conditions, the value of $\mu°$ depends on the choice of standard state. There is no way we can evaluate[8] the absolute value of either μ or $\mu°$, but we can measure or calculate the *difference* $\mu - \mu°$.

5.10.1 Gases

For the standard chemical potential of a gas, we usually use the symbol $\mu°(g)$ to emphasize the choice of a *gas* standard state.

An ideal gas is in its standard state at a given temperature when its pressure is the standard pressure. To relate the chemical potential of an ideal gas to its pressure and its standard chemical potential at the same temperature, we integrate $d\mu = (RT/p)\,dp$ (Eq. 5.10.2 with dT set equal to zero and V_m set equal to RT/p) from the standard state to the state at pressure p:

$$\int_{\mu°}^{\mu} d\mu = RT \int_{p°}^{p} \frac{dp}{p}$$

The result is μ as a function of p:

$$\mu = \mu°(g) + RT \ln \frac{p}{p°} \qquad (5.10.5)$$
(pure ideal gas, constant T)

This function is shown as the dashed curve in Fig. 5.6 on the next page.

If a gas is not an ideal gas, its standard state is a hypothetical state. The **fugacity**, f, of a real gas (a gas that is not necessarily an ideal gas) is defined by an equation with the same

---

[8] At least not to any useful degree of precision. The values of μ and $\mu°$ include the molar internal energy whose absolute value can only be calculated from the Einstein relation; see Sec. 3.2.

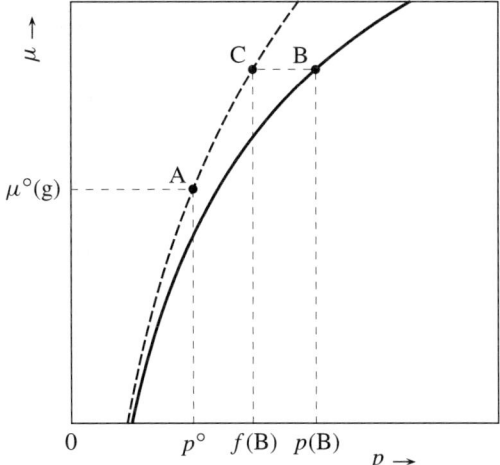

Figure 5.6 Chemical potential as a function of pressure at constant temperature, for a real gas (solid curve) and the same gas behaving ideally (dashed curve). Point A is the gas standard state and Point B is a state of the real gas. The fugacity of the gas at Point B is equal to the pressure of the ideal gas having the same chemical potential (Point C).

form as Eq. 5.10.5:

$$\mu = \mu^\circ(g) + RT \ln \frac{f}{p^\circ} \tag{5.10.6}$$
(pure gas, constant T)

Note that fugacity has the dimensions of pressure. Fugacity is a kind of effective pressure. Specifically, it is the pressure that the hypothetical ideal gas (the gas with intermolecular forces "turned off") would need to have for its chemical potential at the same temperature to be the same as the chemical potential of the real gas (see Point C in Fig. 5.6). The fugacity of an ideal gas is the same as its pressure.

To evaluate the fugacity of a real gas at a given T and p, we must relate the chemical potential to the pressure–volume behavior. Let μ be the chemical potential and f be the fugacity at the pressure p of interest; let μ' be the chemical potential and f' be the fugacity at some low pressure p' (all at the same temperature). Then from $\mu = \mu^\circ(g) + RT \ln(f/p^\circ)$ and $\mu' = \mu^\circ(g) + RT \ln(f'/p^\circ)$ (Eq. 5.10.6), we obtain

$$\mu - \mu' = RT \ln \frac{f}{f'}$$

From $d\mu = V_m \, dp$ (Eq. 5.10.2), we obtain

$$\mu - \mu' = \int_{\mu'}^{\mu} d\mu = \int_{p'}^{p} V_m \, dp$$

Equating the two expressions for $\mu - \mu'$ gives $RT \ln(f/f') = \int_{p'}^{p} V_m \, dp$ or

$$\ln \frac{f}{f'} = \int_{p'}^{p} \frac{V_m}{RT} \, dp$$

In principle, we could use the equation in this form to evaluate f by choosing p' to be a low pressure at which the gas behaves as an ideal gas so that f' would equal p'. However, the integral would then be very large and difficult to evaluate. We avoid this difficulty by subtracting from the preceding equation the identity

$$\ln \frac{p}{p'} = \int_{p'}^{p} \frac{dp}{p}$$

which is simply the result of integrating the function $1/p$ from p' to p. The result is

$$\ln \frac{fp'}{f'p} = \int_{p'}^{p} \left(\frac{V_m}{RT} - \frac{1}{p} \right) dp$$

Now we take the limit of both sides of the equation as p' approaches 0; in this limit, the gas at pressure p' approaches ideal-gas behavior, f' approaches p', and $fp'/f'p$ approaches f/p:

$$\ln \frac{f}{p} = \int_{0}^{p} \left(\frac{V_m}{RT} - \frac{1}{p} \right) dp$$

The **fugacity coefficient** ϕ of a gas is defined by

$$f = \phi p \qquad (5.10.7)$$
$$\text{(pure gas)}$$

With this definition, the equation for $\ln(f/p)$ becomes

$$\ln \phi = \int_{0}^{p} \left(\frac{V_m}{RT} - \frac{1}{p} \right) dp \qquad (5.10.8)$$
$$\text{(pure gas, constant } T\text{)}$$

For a real gas at temperature T and pressure p, Eq. 5.10.8 allows us to evaluate the fugacity coefficient from the experimental equation of state (V_m as a function of T and p). Then we can find the fugacity from Eq. 5.10.7.

The dimensionless ratio $\phi = f/p$ is an example of an *activity coefficient*, and the dimensionless ratio $f/p°$ is an example of an *activity*; see Sec. 7.7.

5.10.2 Liquids and solids

The dependence of the chemical potential on pressure at constant temperature, from Eq. 5.10.2, is $d\mu = V_m dp$. Integrating from $p°$ to p we obtain

$$\mu = \mu° + \int_{p°}^{p} V_m \, dp \qquad (5.10.9)$$
$$\text{(pure liquid or solid,}$$
$$\text{constant } T\text{)}$$

With an approximation of zero compressibility, this becomes
$$\mu \approx \mu° + V_m(p - p°) \tag{5.10.10}$$
(pure liquid or solid, constant T)

5.11 STANDARD MOLAR FUNCTIONS OF A GAS

We have seen that the standard state of a gas is a hypothetical state in which the gas behaves ideally at pressure $p°$ without influence of intermolecular forces. We need to relate molar properties of the gas in this state, at a given temperature, to properties of the real gas at the same temperature.

We begin by combining Eqs. 5.10.6–5.10.8 to obtain a relation between the chemical potential and the standard chemical potential at the same temperature:

$$\mu = \mu°(g) + RT \ln \frac{p}{p°} + \int_0^p \left(V_m - \frac{RT}{p} \right) dp \tag{5.11.1}$$
(pure gas)

Note that this expression for μ is not what we would obtain by simply integrating $d\mu = V_m \, dp$ from $p°$ to p because the real gas is not necessarily in its standard state when its pressure is $p°$.

Substitution of μ from Eq. 5.11.1 in $S_m = -(\partial \mu/\partial T)_p$ (from Eq. 5.10.3) gives

$$S_m = S_m°(g) - R \ln \frac{p}{p°} - \int_0^p \left[\left(\frac{\partial V_m}{\partial T} \right)_p - \frac{R}{p} \right] dp \tag{5.11.2}$$
(pure gas)

where the quantity $S_m°(g)$ is given by

$$S_m°(g) = - \left(\frac{\partial \mu°(g)}{\partial T} \right)_p$$

$S_m°(g)$ is the **standard molar entropy** of the gas—that is, the molar entropy when the gas is in its standard state at the same temperature as the real gas—as we may readily deduce by noting that the two other terms on the right side of Eq. 5.11.2 become zero when the pressure is $p°$ and the gas behaves ideally.[9]

If we are interested in evaluating $S_m°(g)$ experimentally, we can combine Eq. 5.11.2 with the value of S_m obtained by the calorimetric method described in Sec. 4.11.1.

Now we use these formulas for μ and S_m to find expressions for the molar enthalpy, internal energy, and heat capacity relative to standard molar quantities. Dividing both sides of the equation $G = H - TS$ by n, we obtain

$$\mu = H_m - T S_m \tag{5.11.3}$$
(pure substance, $P = 1$)

---

[9]For an ideal gas, we have $V_m = RT/p$ and $(\partial V_m/\partial T)_p = R/p$; thus, the integrand of the integral in Eq. 5.11.2 is zero for an ideal gas. We see furthermore that the second partial derivative $(\partial^2 V_m/\partial T^2)_p$ appearing in Eq. 5.11.6 is given for an ideal gas by $[\partial(R/p)/\partial T]_p = 0$.

or $H_m = \mu + TS_m$. Substitution from Eqs. 5.11.1 and 5.11.2 gives

$$H_m = H_m^\circ(g) + \int_0^p \left[V_m - T \left(\frac{\partial V_m}{\partial T} \right)_p \right] dp \qquad (5.11.4)$$
(pure gas)

where $H_m^\circ(g)$ is given by

$$H_m^\circ(g) = \mu^\circ(g) + TS_m^\circ(g)$$

We see that $H_m^\circ(g)$ is the **standard molar enthalpy**, the molar enthalpy of the gas in its standard state, because the integrand of the integral in Eq. 5.11.4 is zero for an ideal gas (at any pressure).

From the relation $H = U + pV$, we have $H_m = U_m + pV_m$ or $U_m = H_m - pV_m$. Substitution from Eq. 5.11.4 gives

$$U_m = U_m^\circ(g) + \int_0^p \left[V_m - T \left(\frac{\partial V_m}{\partial T} \right)_p \right] dp + RT - pV_m \qquad (5.11.5)$$
(pure gas)

where $U_m^\circ(g)$ is given by

$$U_m^\circ(g) = H^\circ(g) - p^\circ V_m^\circ = H^\circ(g) - RT$$

Dividing both sides of $C_p = (\partial H/\partial T)_p$ (Eq. 3.12.4) by n gives $C_{p,m} = (\partial H_m/\partial T)_p$. Substitution from Eq. 5.11.4 then gives

$$C_{p,m} = C_{p,m}^\circ(g) - \int_0^p T \left(\frac{\partial^2 V_m}{\partial T^2} \right)_p dp \qquad (5.11.6)$$
(pure gas)

where $C_{p,m}^\circ(g)$ is given by

$$C_{p,m}^\circ(g) = \left(\frac{\partial H_m^\circ(g)}{\partial T} \right)_p$$

$C_{p,m}^\circ(g)$ is the **standard molar heat capacity at constant pressure** because the integrand of the integral in Eq. 5.11.6 is zero for an ideal gas.

The isothermal behavior of real gases at low to moderate pressures (up to at least 1 bar) is usually adequately described by a two-term equation of state of the form given in Eq. 2.1.7:

$$V_m \approx \frac{RT}{p} + B$$

where B is the second virial coefficient, a function of T. With this equation of state, Eqs. 5.10.8 and 5.11.1–5.11.6 become

$$\ln \phi \approx \frac{Bp}{RT} \qquad (5.11.7)$$
(pure gas)

$$\mu \approx \mu°(g) + RT \ln \frac{p}{p°} + Bp \qquad (5.11.8)$$
$$\text{(pure gas)}$$

$$S_m \approx S_m°(g) - R \ln \frac{p}{p°} - p \frac{dB}{dT} \qquad (5.11.9)$$
$$\text{(pure gas)}$$

$$H_m \approx H_m°(g) + p \left(B - T \frac{dB}{dT} \right) \qquad (5.11.10)$$
$$\text{(pure gas)}$$

$$U_m \approx U_m°(g) - pT \frac{dB}{dT} \qquad (5.11.11)$$
$$\text{(pure gas)}$$

$$C_{p,m} \approx C_{p,m}°(g) - pT \frac{d^2 B}{dT^2} \qquad (5.11.12)$$
$$\text{(pure gas)}$$

We can see what the quantities on the left sides of these equations look like if the gas is ideal simply by setting B equal to zero. We find that an ideal gas has a fugacity coefficient $\phi = 1$ and a chemical potential given by $\mu = \mu°(g) + RT \ln(p/p°)$, which we already knew. The molar entropy of an ideal gas is given by $S_m = S_m°(g) - R \ln(p/p°)$. The molar enthalpy, molar internal energy, and molar heat capacity are the same as the corresponding standard molar quantities.

In summary, when the pressure of an ideal gas increases at constant temperature, the chemical potential increases, the entropy decreases, and the enthalpy, internal energy, and heat capacity are unaffected.

5.12 ADDITIONAL WORK COORDINATES

Sometimes we need more than the usual two independent variables to describe an equilibrium state of a closed system of one substance in one phase. This is the case when, in addition to expansion work, other kinds of reversible work are possible. As examples we consider systems with surface work and magnetization work; in these systems, surface area or magnetic moment are relevant to the description of the state.

5.12.1 Surface tension

A liquid–gas interface behaves somewhat like a stretched membrane. The upper and lower surfaces of the liquid film in the device depicted in Fig. 5.7 on the next page exert a force F on the sliding rod, tending to pull it in the direction that would reduce the surface area.[10]

---

[10] The force occurs because the liquid has a lower internal energy when its surface area is reduced. On the microscopic scale, a molecule at the surface has fewer attractive interactions with other molecules than does a molecule in the bulk liquid.

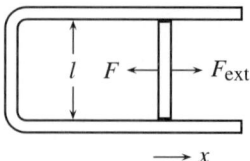

Figure 5.7 Device to measure the surface tension of a liquid film. The film is stretched between a bent wire and a sliding rod.

We can measure the force by determining the opposing force F_{ext} needed to prevent the rod from moving. This force is found to be proportional to the length of the rod, but independent of the total surface area. The force also depends on the temperature and pressure.

The **surface tension**, γ, is the force exerted by each surface per unit length in an equilibrium state. In the case of the film shown in Fig. 5.7, there are two surfaces, so we have $\gamma = F/2l$ where l is the rod length. To increase the surface area of the film by a practically reversible process, we make F_{ext} very slightly greater than F and allow the rod to slowly move a distance dx. The surface work is the product of force and distance: $dw' = F\,dx = 2\gamma l\,dx$. The increase in surface area, dA_s, is $2l\,dx$, so the surface work is $dw' = \gamma\,dA_s$. In this expression, γ is a work coefficient and A_s is a work coordinate.

If the system (the liquid) is closed and any changes are practically reversible, the first-law expression $dU = T\,dS - p\,dV + dw'$ (Eq. 4.9.2) becomes

$$dU = T\,dS - p\,dV + \gamma\,dA_s \tag{5.12.1}$$

(closed system, pure substance, $P = 1$)

Substitution into $dG = dU + p\,dV + V\,dP - T\,dS - S\,dT$ (Eq. 4.8.4) gives

$$dG = -S\,dT + V\,dp + \gamma\,dA_s \tag{5.12.2}$$

(closed system, pure substance, $P = 1$)

which is the total differential of G with T, p, and A_s as independent variables. Identifying the coefficient of the last term on the right side, we find the following expression for the surface tension:

$$\gamma = \left(\frac{\partial G}{\partial A_s}\right)_{T,p}$$

That is, the surface tension is not only a force per unit length, but also a Gibbs energy per unit area.

From Eq. 5.12.2, we obtain the reciprocity relation

$$\left(\frac{\partial \gamma}{\partial T}\right)_{p,A_s} = -\left(\frac{\partial S}{\partial A_s}\right)_{T,p} \tag{5.12.3}$$

Thus, the variation of surface tension with temperature tells us how the entropy of the liquid varies with surface area. It is valid to replace the partial derivative on the left side by $(\partial \gamma / \partial T)_p$ because γ is not a function of A_s.

5.12.2 Magnetization

A formula for the work of magnetization is given in Table 3.2 on page 58. For an isotropic phase, it can be written $dw_{mag} = (\mu_0/4\pi) V H_{mag} dM$, where μ_0 is the permeability of vacuum, H_{mag} is the magnitude of the magnetic field strength, and M is the magnitude of the magnetization (the magnetic moment per unit volume).

For convenience we write $dw_{mag} = F\, dM$, where the function F is defined by

$$F \equiv (\mu_0/4\pi) V H_{mag}$$

Then the first-law expression $dU = T\, dS - p\, dV + dw'$ is

$$dU = T\, dS - p\, dV + F\, dM$$

We can define a modified enthalpy function H' (sometimes called the *magnetic enthalpy*) by $H' \equiv U + pV - FM$. This function has the useful property that two terms of its differential, $dH' = dU + p\, dV + V\, dp - F\, dM - M\, dF$, drop out when combined with the expression for dU:

$$dH' = T\, dS + V\, dp - M\, dF$$

The closed system has three independent variables, so this is an expression for the total differential of H'. From it we obtain the reciprocity relation

$$\left(\frac{\partial T}{\partial F}\right)_{S,p} = -\left(\frac{\partial M}{\partial S}\right)_{p,F} \tag{5.12.4}$$

We are interested in the sign of $(\partial T/\partial F)_{S,p}$ because this will tell us the sign of the temperature change during a reversible adiabatic demagnetization (Path B of Fig. 4.17 on page 99). Curie's law states that the magnetization M at constant magnetic field strength H_{mag} is proportional to $1/T$; this law applies when H_{mag} is small, but even if H_{mag} is not small M decreases with increasing T. As T increases, heat enters the system and S increases. Thus, $(\partial M/\partial S)_{p,F}$ is negative and, according to Eq. 5.12.4, $(\partial T/\partial F)_{S,p}$ must be positive. Adiabatic demagnetization is a constant-entropy process in which F decreases, and therefore the temperature also *decreases*.

We can find the sign of the entropy change during the isothermal magnetization process shown as Path A in Fig. 4.17 on page 99. We need to use T, p, and F as the independent variables, and for this purpose we define a modified Gibbs energy function by $G' \equiv H' - TS$. Its differential is $dG' = dH' - T\, dS - S\, dT$ or

$$dG' = -S\, dT + V\, dp - M\, dF$$

From this total differential, we obtain the reciprocal relation

$$\left(\frac{\partial S}{\partial F}\right)_{T,p} = \left(\frac{\partial M}{\partial T}\right)_{p,F}$$

Since M at constant F decreases with increasing T, as explained earlier, we see that the entropy change during isothermal magnetization is *negative*.

5.13 OPEN SYSTEMS

To this point in the chapter, we have considered only closed systems. If matter can pass through the boundary, we need an additional independent variable to describe the state. An open system with one substance in a single phase usually has *three* independent variables.

Let us write the formal expression for the total differential of U, with S, V, and n as the three independent variables:

$$dU = \left(\frac{\partial U}{\partial S}\right)_{V,n} dS + \left(\frac{\partial U}{\partial V}\right)_{S,n} dV + \left(\frac{\partial U}{\partial n}\right)_{S,V} dn \qquad (5.13.1)$$
(pure substance, $P = 1$)

We know the total differential of U in a *closed* system may be written $dU = T\, dS - p\, dV$ (Eq. 5.3.2). We identify the coefficients here as $T = (\partial U/\partial S)_V$ and $-p = (\partial U/\partial V)_S$. Since both of these partial derivatives are for closed systems in which n is constant, they are the same as the first two partial derivatives on the right side of Eq. 5.13.1. We represent the third partial derivative by the symbol μ and show that it is the same as the chemical potential introduced in Sec. 5.10. Then Eq. 5.13.1 becomes

$$dU = T\, dS - p\, dV + \mu\, dn \qquad (5.13.2)$$
(pure substance, $P = 1$)

where μ is defined by

$$\mu = \left(\frac{\partial U}{\partial n}\right)_{S,V} \qquad (5.13.3)$$
(pure substance, $P = 1$)

Next we substitute the expression for dU into expressions for dH, dA, and dG (Eqs. 3.11.3, 4.8.3, and 4.8.4) to obtain

$$dH = T\, dS + V\, dp + \mu\, dn \qquad (5.13.4)$$
(pure substance, $P = 1$)

$$dA = -S\, dT - p\, dV + \mu\, dn \qquad (5.13.5)$$
(pure substance, $P = 1$)

$$dG = -S\,dT + V\,dp + \mu\,dn \qquad (5.13.6)$$
$$\text{(pure substance, } P = 1\text{)}$$

Equations 5.13.2 and 5.13.4–5.13.6 are expressions for total differentials since they have as many terms (three) as the number of independent variables. Note that these equations are the same as the four Gibbs equations (Eqs. 5.3.2–5.3.5) with the addition of a term $\mu\,dn$ to allow for a change in the amount of substance.

Identification of the coefficient of the third term on the right side of Eq. 5.13.6 shows that μ is equal to $(\partial G/\partial n)_{T,p}$. That is, μ is the rate at which the Gibbs energy increases with the amount of substance added to a phase whose intensive properties remain constant. Thus, μ is the same as the Gibbs energy divided by the amount or the molar Gibbs energy of the substance. This is the intensive property that was called the *chemical potential* in Sec. 5.10.

In addition to being equal to the partial derivatives $(\partial U/\partial n)_{S,V}$ and $(\partial G/\partial n)_{T,p}$, μ is also equal (from Eqs. 5.13.4 and 5.13.5) to $(\partial H/\partial n)_{S,p}$ and $(\partial A/\partial n)_{T,V}$. Thus, all four of these partial derivatives have the same value for a given state of the system; the value, of course, depends on what that state is.

With additional work coordinates, more terms are needed in the total differentials. For example, if surface area is needed to describe the state of an open system, the total differential of U becomes

$$dU = T\,dS - p\,dV + \gamma\,dA_s + \mu\,dn \qquad (5.13.7)$$
$$\text{(pure substance, } P = 1\text{)}$$

(Compare Eqs. 5.12.1 and 5.13.2.)

PROBLEMS

5.1 Derive the following relations from the definitions of α, κ_T, and ρ:

$$\alpha = -\frac{1}{\rho}\left(\frac{\partial \rho}{\partial T}\right)_p \qquad \kappa_T = \frac{1}{\rho}\left(\frac{\partial \rho}{\partial p}\right)_T$$

5.2 Use equations in this chapter to derive the following expressions for an ideal gas:

$$\alpha = 1/T \qquad \kappa_T = 1/p$$

5.3 For a gas with the simple equation of state

$$V_m = \frac{RT}{p} + B$$

(Eq. 2.1.7), where B is the second virial coefficient (a function of T), find expressions for α, κ_T, and $(\partial U_m/\partial V)_T$ in terms of dB/dT and other state functions.

5.4 Show that when the virial equation $pV_m = RT(1 + B'p + C'p^2 + \cdots)$ (Eq. 2.1.4) adequately represents the equation of state of a real gas, the Joule–Thomson coefficient is given by

$$\mu_{JT} = \frac{RT^2[dB'/dT + (dC'/dT)p + \cdots]}{C_{p,m}}$$

Note that the limiting value at low pressure, $RT^2(dB'/dT)/C_{p,\mathrm{m}}$, is not necessarily equal to zero even though the equation of state approaches that of an ideal gas in this limit.

5.5 The quantity $(\partial T/\partial V)_U$ is called the *Joule coefficient*. James Joule attempted to evaluate this quantity by measuring the temperature change accompanying the expansion of air into a vacuum—the "Joule experiment." Write an expression for the total differential of U with T and V as independent variables, and by a procedure similar to that used in Sec. 5.7.2 show that the Joule coefficient is equal to

$$\frac{p - \alpha T/\kappa_T}{C_V}$$

5.6 p-V-T data for several organic liquids were measured by Gibson and Loeffler (*J. Amer. Chem. Soc.* **1939**, *61*, 2515–2522). The following formulas describe the results for aniline:

Molar volume as a function of temperature at $p = 1$ bar (298–358 K):

$$V_{\mathrm{m}} = a + bT + cT^2 + dT^3$$

where the parameters have the values

$a = 69.287 \text{ cm}^3 \text{ mol}^{-1}$ $\quad c = -1.0443 \times 10^{-4} \text{ cm}^3 \text{ K}^{-2} \text{ mol}^{-1}$

$b = 0.08852 \text{ cm}^3 \text{ K}^{-1} \text{ mol}^{-1}$ $\quad d = 1.940 \times 10^{-7} \text{ cm}^3 \text{ K}^{-3} \text{ mol}^{-1}$

Molar volume as a function of pressure at $T = 298.15$ K (1–1000 bar):

$$V_{\mathrm{m}} = e - f \ln(g + p/\text{bar})$$

where the parameter values are

$e = 156.812 \text{ cm}^3 \text{ mol}^{-1}$ $\quad f = 8.5834 \text{ cm}^3 \text{ mol}^{-1}$ $\quad g = 2006.6$

(a) Use these equations to calculate values of α, κ_T, $(\partial p/\partial T)_V$, and $(\partial U/\partial V)_T$ (the internal pressure) for aniline at $T = 298.15$ K and $p = 1.000$ bar.

(b) Estimate the pressure increase if the temperature of a fixed amount of aniline is increased by 0.10 K at constant volume.

5.7 (a) From the total differential of H with T and p as independent variables, derive the relation $(\partial C_{p,\mathrm{m}}/\partial p)_T = -T(\partial^2 V_{\mathrm{m}}/\partial T^2)_p$.

(b) Evaluate $(\partial C_{p,\mathrm{m}}/\partial p)_T$ for liquid aniline at 300.0 K and 1 bar using data in Prob. 5.6.

5.8 (a) From the total differential of V with T and p as independent variables, derive the relation $(\partial \alpha/\partial p)_T = -(\partial \kappa_T/\partial T)_p$.

(b) Use this relation to estimate the value of α for benzene at 25 °C and 500 bar given that the value of α is 1.2×10^{-3} K^{-1} at 25 °C and 1 bar. (Use information from Fig. 5.2 on page 106.)

5.9 Certain equations of state supposed to be applicable to nonpolar liquids and gases are of the form $p = Tf(V_{\mathrm{m}}) - a/V_{\mathrm{m}}^2$, where $f(V_{\mathrm{m}})$ is a function of the molar volume only and a is a constant.

(a) Show that the van der Waals equation of state $(p + a/V_{\mathrm{m}}^2)(V_{\mathrm{m}} - b) = RT$ (where a and b are constants) is of this form.

(b) Show that any fluid with an equation of state of this form has an internal pressure equal to a/V_m^2.

5.10 Suppose that the molar heat capacity at constant pressure of a substance has a temperature dependence given by $C_{p,m} = a + bT + cT^2$, where a, b, and c are constants. Consider the heating of an amount n of the substance from T_1 to T_2 at constant pressure. Find expressions for ΔH and ΔS for this process in terms of a, b, c, n, T_1, and T_2.

5.11 The temperature dependence of the standard molar heat capacity of gaseous carbon dioxide in the temperature range 298–2000 K is given by

$$C_{p,m}^\circ = a + bT + \frac{c}{T^2}$$

where the constants have the values

$a = 44.2 \, \text{J K}^{-1} \, \text{mol}^{-1} \quad b = 8.8 \times 10^{-3} \, \text{J K}^{-2} \, \text{mol}^{-1} \quad c = -8.6 \times 10^5 \, \text{J K mol}^{-1}$

Calculate the enthalpy and entropy changes when one mole of CO_2 is heated at 1 bar from 300.00 K to 800.00 K.

5.12 At $p = 1$ atm, the molar heat capacity at constant pressure of aluminum is given by

$$C_{p,m} = a + bT$$

where the constants have the values

$a = 20.67 \, \text{J K}^{-1} \, \text{mol}^{-1} \quad b = 0.01238 \, \text{J K}^{-2} \, \text{mol}^{-1}$

Calculate the quantity of electrical work needed to heat 2.000 mol of aluminum from 300.00 K to 400.00 K at 1 atm in an adiabatic enclosure.

5.13 A chemist, needing to determine the specific heat capacity of a certain liquid but not having an electrically heated calorimeter at her disposal, used the following simple procedure known as *drop calorimetry*. She placed 500.0 g of the liquid in a thermally insulated container equipped with a lid and a thermometer. After recording the initial temperature of the liquid, $24.80\,°\text{C}$, she removed a 60.17-g block of aluminum metal from a boiling water bath at $100.00\,°\text{C}$ and quickly immersed it in the liquid in the container. When the contents of the container were thermally equilibrated, she recorded a final temperature of $27.92\,°\text{C}$. She calculated the specific heat capacity C_p/m of the liquid from these data, making use of the molar mass of aluminum ($M = 26.9815 \, \text{g mol}^{-1}$) and the formula for the molar heat capacity of aluminum given in Prob. 5.12.

(a) From these data, find the specific heat capacity of the liquid under the assumption that its value does not vary with temperature. Hint: Treat the temperature equilibration process as adiabatic and isobaric ($\Delta H = 0$), and equate ΔH to the sum of the enthalpy changes in the two phases.

(b) Show that the value obtained in Part (a) is actually an average value of C_p/m over the temperature range between the initial and final temperatures of the liquid given by

$$\frac{\int_{T_1}^{T_2} (C_p/m) \, dT}{T_2 - T_1}$$

5.14 Suppose a gas has the virial equation of state $pV_m = RT(1 + B'p + C'p^2)$, where B' and C' depend only on T, and higher powers of p can be ignored.

(a) Derive an expression for the fugacity coefficient, ϕ, of this gas.

(b) For $CO_2(g)$ at $0.00\,°C$, the virial coefficients have the values $B' = -6.67 \times 10^{-3}\,\text{bar}^{-1}$ and $C' = -3.4 \times 10^{-5}\,\text{bar}^{-2}$. Evaluate the fugacity f at $0.00\,°C$ and $p = 20.0\,\text{bar}$.

Table 5.4 Molar volume of $H_2O(g)$ at $400.00\,°C$

$p/10^5$ Pa	$V_m/10^{-3}\,\text{m}^3\,\text{mol}^{-1}$	$p/10^5$ Pa	$V_m/10^{-3}\,\text{m}^3\,\text{mol}^{-1}$
1	55.896	100	0.47575
10	5.5231	120	0.37976
20	2.7237	140	0.31020
40	1.3224	160	0.25699
60	0.85374	180	0.21447
80	0.61817	200	0.17918

5.15 Table 5.4 lists values of the molar volume of gaseous H_2O at $400.00\,°C$ at 12 pressures.

(a) Evaluate the fugacity coefficient and fugacity of $H_2O(g)$ at $400.00\,°C$ and 200 bar.

(b) Show that the second virial coefficient B in the virial equation of state, $pV_m = RT(1 + B/V_m + C/V_m^2 + \cdots)$, is given by

$$B = RT \lim_{p \to 0} \left(\frac{V_m}{RT} - \frac{1}{p} \right)$$

where the limit is taken at constant T. Then evaluate B for $H_2O(g)$ at $400.00\,°C$.

5.16 Write expressions for the six possible partial derivatives of the Helmholtz energy A in a closed system with T, p, and V as the variables. Two of the derivatives are found in Sec. 5.3; you may obtain the others by differentiating $A = U - TS$ and making substitutions from Tables 5.1–5.3.

5.17 From statistical mechanical theory, a simple model for a hypothetical "hard-sphere" liquid (spherical molecules of finite size without attractive forces) gives the following expression for the Helmholtz energy with T, V, and n as independent variables:

$$A = -nRT \ln\left[cT^{3/2}\left(\frac{V}{n} - b\right)\right] - nRT + nA_0$$

Here b, c, and A_0 are constants. Derive expressions for the following intensive properties of this hypothetical liquid, all as functions of T and p only. (You should find the expressions in the order listed below, because some depend on others.)

(a) p (Hint: Eq. 5.3.11)

(b) V_m

(c) S_m (Hint: Eq. 5.3.10)

(d) α and κ_T

(e) U_m and H_m

(f) μ

(g) $C_{p,m}$

Table 5.5 Surface tension of water at 1 atm (International Critical Tables, Vol. 4)

$t/°C$	$\gamma/10^{-6}\,\text{J cm}^{-2}$
15	7.349
20	7.275
25	7.197
30	7.118
35	7.038

5.18 Use the data in Table 5.5 to evaluate $(\partial S/\partial A_s)_{T,p}$ at 25 °C, which is the rate at which the entropy changes with the area of the air–water interface at this temperature. If increased entropy represents increased molecular disorder, are water molecules at the surface more or less "disordered" than those in bulk water?

5.19 When an ordinary rubber band is hung from a clamp and stretched with constant force (F) by attaching a weight at the bottom end, gentle heating is observed to cause the rubber band to contract in length (l). If the length of the rubber band is to remain constant during heating, F must be increased.

(a) From these observations, evaluate the signs of $(\partial S/\partial l)_{T,p}$ and $(\partial S/\partial F)_{T,p}$. (Hint: The formula for stretching work is $dw' = F\,dl$. Find the total differential of G with T, p, and l as independent variables; find the total differential of the state function $X = G - Fl$ with T, p, and F as independent variables.)

(b) Find the total differential of H with S, p, and l as independent variables. Prove the relation $(\partial T/\partial l)_{S,p} = (\partial F/\partial S)_{p,l}$. Use this relation and the preceding observations to evaluate the sign of $(\partial T/\partial l)_{S,p}$; then predict whether stretching of the rubber band will cause a heating or a cooling effect. (You can check your prediction experimentally by touching a rubber band to the side of your face before and after you rapidly stretch it.)

Chapter 6

PHASE TRANSITIONS AND EQUILIBRIA OF PURE SUBSTANCES

A system of two or more phases of a single substance, in the absence of effects of an external field or other constraints, is in an equilibrium state when each phase has the same temperature, the same pressure, and the same chemical potential. This chapter describes the derivation and consequences of this simple principle, the general appearance of phase diagrams of single-substance systems, and thermodynamic changes during the equilibrium phase transitions of such systems.

6.1 PHASE EQUILIBRIA

6.1.1 Conditions for spontaneity and equilibrium

As explained in Sec. 3.1.3, an equilibrium state of an isolated system does not change with time.

However, we expect an isolated system that is *not* in an equilibrium state to undergo a spontaneous, irreversible process and eventually to reach an equilibrium state. Just how rapidly this process occurs is a matter of kinetics, not thermodynamics. During this irreversible adiabatic process, the entropy increases (Table 4.1 on page 86) until it reaches a maximum in the equilibrium state.

We now introduce a general method, used in several places in this book, for finding conditions for equilibrium under given constraints. The method has four steps:

1. Write an expression for the total differential of the internal energy consistent with any constraints and the number of independent variables of the system.

2. Impose conditions of isolation for the system, thereby reducing the number of independent variables.

3. Rearrange to obtain an expression for the total differential of the entropy consistent with the reduced number of independent variables.

4. The conditions for an equilibrium state are those that make the infinitesimal entropy change, dS, equal to zero for all infinitesimal changes of the independent variables of the isolated system.

6.1.2 Equilibrium in a two-phase system

As our first application of this method, consider a system of one substance in two uniform phases labeled α and β. For instance, α might be a liquid phase and β a gas phase. Since internal energy is extensive, its change is given by $dU = dU^\alpha + dU^\beta$. We assume that any changes are slow enough to keep each phase practically uniform at all times. Treating each phase as an open subsystem, we use the relation $dU = T\,dS - p\,dV + \mu\,dn$ (Eq. 5.13.2) to replace dU^α and dU^β (with appropriate superscripts to indicate the phase):

$$dU = (T^\alpha\,dS^\alpha - p^\alpha\,dV^\alpha + \mu^\alpha\,dn^\alpha) + (T^\beta\,dS^\beta - p^\beta\,dV^\beta + \mu^\beta\,dn^\beta) \quad (6.1.1)$$

This is an expression for the total differential of U with S^α, V^α, n^α, S^β, V^β, and n^β as the independent variables.

We make the system an isolated system by imposing constraints that keep the system closed with constant internal energy and no work. The first law shows that any process with these constraints has no heat.

We assume there is an isotropic fluid of uniform pressure on one or both sides of each surface element of the boundary (see the discussion in Sec. 3.6.2). To ensure that the system as a whole does no expansion work on the surroundings, we need only constrain its volume to be constant. The constraints in the present application, then, are given by the relations $dU = 0$, $dV = 0$ (or $dV^\beta = -dV^\alpha$), and $dn = 0$ (or $dn^\beta = -dn^\alpha$). Each of these three relations is an independent restriction that reduces the number of independent variables by one. Using the relations to substitute for dU, dV^β, and dn^β in Eq. 6.1.1 and rearranging gives

$$0 = T^\alpha\,dS^\alpha + T^\beta\,dS^\beta - (p^\alpha - p^\beta)\,dV^\alpha + (\mu^\alpha - \mu^\beta)\,dn^\alpha$$

For the infinitesimal entropy change of the system, we write $dS = dS^\alpha + dS^\beta$ or $dS^\beta = dS - dS^\alpha$. Making this substitution for dS^β and solving for dS, we obtain

$$dS = \frac{T^\beta - T^\alpha}{T^\beta}\,dS^\alpha - \frac{p^\beta - p^\alpha}{T^\beta}\,dV^\alpha + \frac{\mu^\beta - \mu^\alpha}{T^\beta}\,dn^\alpha \quad (6.1.2)$$
(pure substance, $P = 2$, isolated system)

This is an expression for the total differential of S in the isolated system, with S^α, V^α, and n^α as the independent variables.

Spontaneous changes in the isolated system involve changes in these three independent variables that increase the entropy. When the isolated system is no longer able to change spontaneously, it has reached an equilibrium state. Once the system is in this state, an imagined finite change of any of the independent variables (a so-called *virtual displacement*) corresponds to an impossible process with an entropy decrease. Thus, the equilibrium state has the *maximum* entropy that is possible for the isolated system. For S to be a maximum, dS must be zero for an infinitesimal change of any of the independent variables of the isolated system. This requirement is satisfied only if the coefficient of each term on the right side of Eq. 6.1.2 is zero; that is, if T^α and T^β are equal, p^α and p^β are equal, and μ^α and μ^β are equal. These are, respectively, the conditions of *thermal* equilibrium, *mechanical* equilibrium, and *transfer* equilibrium described in Sec. 3.1.3.

These conditions characterize an equilibrium state regardless of the process by which the system attains that state. A particular equilibrium state reached by an isolated system could also, for instance, be reached by a process with heat and work.

We may easily generalize the derivation for a system with more than two phases. In general, for the system containing a single substance to be in an equilibrium state, each phase must have the same temperature and the same pressure, and in each phase the substance must have the same chemical potential.

6.1.3 A simple derivation of equilibrium conditions

Here is a simpler, less formal derivation of the three equilibrium conditions in a multiphase system of one substance.

It is intuitively obvious that, unless there are special constraints (such as internal partitions), an equilibrium state must have thermal and mechanical equilibrium. A temperature difference between two phases would cause a spontaneous transfer of heat from the warmer to the cooler phase; a pressure difference would cause spontaneous flow or motion of matter.

When some of the substance is transferred from one phase to another under conditions of constant T and p, the intensive properties of the phases remain the same including the chemical potentials. The chemical potential of a pure phase is the Gibbs energy per amount of substance in the phase. We know that in a closed system of constant T and p with expansion work only, the Gibbs energy decreases during a spontaneous process and is constant during a reversible process (Eq. 4.9.5). The Gibbs energy will decrease only if there is a transfer of substance from a phase of higher chemical potential to a phase of lower chemical potential, and this will be a spontaneous change. No spontaneous transfer is possible if both phases have the same chemical potential, so this is a condition for an equilibrium state.

6.1.4 Gas pressure in a gravitational field

The earth's gravitational field is an example of an external field that acts on a system placed in it. Usually we ignore its effects on the state of the system unless the system's vertical extent is considerable.

When an external field is present, we define a potential function that is zero at an arbitrary reference point in space and that allows us to calculate a potential energy. Once we select the reference point, the potential has a definite value at every other point. For instance, the potential of the earth's gravitational field is gh, where g is the acceleration of free fall and h is the elevation relative to an arbitrary reference elevation. The gravitational potential energy

Section 6.1 Phase Equilibria

of a body of mass m in this field is mgh. It is a characteristic of the earth's gravitational field that the presence of the system has a completely negligible effect on the potential function and that g is essentially constant over any range of elevation that is small compared with the earth's radius.

What are the equilibrium conditions for a gas in a tall column? Since we anticipate that some intensive properties will vary with elevation, we treat the system as a vertical stack of many slab-shaped phases, each thin enough to be treated as uniform in its intensive properties.

We follow the general method described in Sec. 6.1.1. A given phase β is at elevation h^β and is open. Its volume is constant because it has a fixed vertical dimension and the column has fixed cross-section area. The total differential of its internal energy, from Eq. 5.13.2 with dV set equal to zero, is

$$dU^\beta = T^\beta \, dS^\beta + \mu^\beta \, dn^\beta$$

where μ^β is the chemical potential at elevation h^β. One particular phase, α, is located at the reference elevation $h = 0$ at the bottom of the column.

We write the change of the internal energy of the whole system in the form

$$dU = dU^\alpha + \sum_{\beta \neq \alpha} dU^\beta$$
$$= T^\alpha \, dS^\alpha + \sum_{\beta \neq \alpha} T^\beta \, dS^\beta + \mu^\alpha \, dn^\alpha + \sum_{\beta \neq \alpha} \mu^\beta \, dn^\beta \quad (6.1.3)$$

We write the entropy change in the form $dS = dS^\alpha + \sum_{\beta \neq \alpha} dS^\beta$. The conditions needed to isolate the system are $dU = 0$, $dV = 0$, and $dn^\alpha + \sum_{\beta \neq \alpha} dn^\beta = 0$. We solve the first and last of these equations for dS^α and dn^α and substitute into Eq. 6.1.3 with dU set equal to zero. We then solve for dS and obtain, for changes in the isolated system, the expression

$$dS = \sum_{\beta \neq \alpha} \frac{T^\alpha - T^\beta}{T^\alpha} \, dS^\beta + \sum_{\beta \neq \alpha} \frac{\mu^\alpha - \mu^\beta}{T^\alpha} \, dn^\beta \quad (6.1.4)$$

In an equilibrium state, the coefficient multiplying each differential on the right side of the equation must be zero. We conclude that at equilibrium the temperature of each phase is equal to that of the reference phase, α, and that the chemical potential in each phase is equal to the chemical potential of the reference phase. That is, at equilibrium *the temperature and the chemical potential are uniform throughout the system.*

The relation of the chemical potential to the gas fugacity is given by Eq. 5.10.6 on page 129, which we write in the form

$$\mu(h) = \mu^\circ(h) + RT \ln \frac{f(h)}{p^\circ}$$

Here, $\mu^\circ(h)$ is the chemical potential of the gas in its standard state at the temperature of interest and *at the elevation of interest*—that is, at the same temperature and elevation as the gas that has chemical potential $\mu(h)$ and fugacity $f(h)$. We can see this by the fact that, since the logarithm of 1 is zero, the equation becomes $\mu(h) = \mu°(h)$ when the fugacity is $p°$.

The value of $\mu°(h)$ depends on the elevation, as we can see by the following argument. When the elevation of a body of mass m changes by dh, its potential energy changes by $mg\,dh$. If the change occurs by a reversible adiabatic process, the work is equal to this potential energy change (Sec. 3.9) and the internal energy changes by $mg\,dh$. If we have an amount n of a gas in its standard state, and we increase its elevation reversibly and adiabatically by dh, its internal energy must change by $nMg\,dh$ (where M is the molar mass). Then, since the Gibbs energy is given by $G = U + pV - TS$ and the values of p, V, T, and S remain constant for the gas in its standard state, the Gibbs energy also changes by $nMg\,dh$.[1] Consequently, the chemical potential (or molar Gibbs energy) of the gas in its standard state at elevation h, relative to the chemical potential at the reference elevation, is given by

$$\mu°(h) = \mu°(0) + Mgh \qquad (6.1.5)$$

and the chemical potential in general is given by

$$\mu(h) = \mu°(0) + Mgh + RT \ln \frac{f(h)}{p°} \qquad (6.1.6)$$

Some thermodynamicists call the expression on the right side of Eq. 6.1.6 the "total chemical potential" or "gravitochemical potential" and reserve the term "chemical potential" for the function $\mu° + RT \ln(f/p°)$, where $\mu°$ is independent of elevation. This book, for simplicity and consistency, defines the chemical potential of a pure substance as the molar Gibbs energy.

We know that in the equilibrium state the chemical potential $\mu(h)$ has the same value at each elevation h. Equation 6.1.6 shows that for this to be true the fugacity must decrease with increasing elevation. Let $f(0)$ be the fugacity in the reference phase, where h is zero. Then equating expressions from Eq. 6.1.6 for $\mu(h)$ at an arbitrary elevation h and $\mu(0)$ in the reference phase, we obtain

$$\mu°(0) + Mgh + RT \ln \frac{f(h)}{p°} = \mu°(0) + RT \ln \frac{f(0)}{p°}$$

or

$$f(h) = f(0)e^{-Mgh/RT}$$

If we treat the gas as ideal, so that the fugacity equals the pressure, this equation becomes

$$p(h) = p(0)e^{-Mgh/RT} \qquad (6.1.7)$$
(pure ideal gas)

---

[1] We can also deduce this from the fact that the change in Gibbs energy during a reversible process at constant temperature and pressure is equal to the nonexpansion work (Eq. 4.9.5).

Equation 6.1.7 is the *barometric formula* for a pure ideal gas. It shows that in the equilibrium state of a tall column of gas the pressure decreases exponentially with increasing elevation.

This derivation of the barometric formula has introduced a method that is used in Sec. 7.9 for dealing with *mixtures* in a gravitational field. However, there is a shorter derivation based on Newton's second law of motion and not involving the chemical potential. Consider one of the thin slab-shaped phases. Let its density be ρ, the area of each face be A_s, and its thickness be dh. Its mass is then $m = \rho A_s\, dh$. Three forces act on this phase: an upward force pA_s on its lower face, a downward force $-(p+dp)A_s$ on its upper face, and a downward gravitational force $-mg = -\rho A_s g\, dh$. If the phase is at rest, the net force is zero:

$$pA_s - (p+dp)A_s - \rho A_s g\, dh = 0$$

We can rearrange this equation to

$$dp = -g\rho\, dh \tag{6.1.8}$$

This is a general relation between changes in elevation and hydrostatic pressure in *any* fluid. To apply it to an ideal gas, we replace the density by $\rho = nM/V = M/V_m = Mp/RT$ and rearrange to $dp/p = -(gM/RT)\, dh$. Treating g and T as constants, we integrate from $h=0$ to an arbitrary elevation h and obtain the same result as Eq. 6.1.7.

6.1.5 The pressure in a liquid droplet

To describe the state of a system, we may need to include variables that are usually ignored. For instance, consider a system containing one substance in both a liquid and a gas phase. If the internal energy of this system depends significantly on the surface area A_s of the liquid–gas interface, we must include A_s or a property dependent on it as an independent variable to describe the state. For the total differential of the internal energy of this system, we may write

$$dU = T^l\, dS^l - p^l\, dV^l + \mu^l\, dn^l + T^g\, dS^g - p^g\, dV^g + \mu^g\, dn^g + \gamma\, dA_s$$

where γ is the surface tension of the liquid–gas interface (Sec. 5.12.1). The system has as many independent variables as terms in the total differential, seven in all.

Suppose the system consists of a spherical liquid droplet of radius r surrounded by the gas phase. The spherical constraint means the volume and surface area of the drop are not independent since both depend on r. This particular system then has six independent variables, but the expression for dU is not the same as when we neglect the surface area. A derivation by the method described in Sec. 6.1.1 shows that at equilibrium the gas and liquid have equal temperatures and equal chemical potentials, but the pressure in the droplet is greater than the gas pressure by $2\gamma/r$. The pressure differential is significant if r is small. The derivation is left as an exercise (Prob. 6.1).

6.1.6 The number of independent variables

From this point on in this book, unless stated otherwise, the discussions of multiphase systems implicitly assume the existence of thermal, mechanical, and transfer equilibrium.

Numbered equations do not explicitly show these equilibria as a condition of validity. Unless stated otherwise, we assume the variables are the usual ones such as temperature, pressure, volume, and amount and do not include elevation in a gravitational field, interface surface area, extent of stretching of a solid, and so on.

How many of the usual variables of a multiphase one-substance equilibrium system are independent? To find out, we go through the following argument. In the absence of any kind of equilibrium, we could treat each phase as having three independent variables (e.g., T^α, p^α, and n^α). A system of P phases would then have $3P$ independent variables. However, each independent relation resulting from equilibrium imposes a restriction on the system and reduces the number of independent variables by one.

> We must decide how to count the number of phases. It is usually of no thermodynamic significance whether a phase, with particular values of its intensive properties, is contiguous. For instance, splitting a crystal into several pieces is not usually considered to change the number of phases or the state of the system provided the increased surface area makes no significant contribution to properties such as internal energy. Thus, the number of phases P refers to the number of different *kinds* of phases.

A two-phase system with thermal equilibrium has the single relation $T^\beta = T^\alpha$. For a three-phase system, there are two such relations that are independent, for instance $T^\beta = T^\alpha$ and $T^\gamma = T^\alpha$. (The additional relation $T^\gamma = T^\beta$ is not independent since we may deduce it from the other two.) In general, thermal equilibrium gives $P - 1$ independent relations among temperatures.

By the same reasoning, mechanical equilibrium involves $P - 1$ independent relations among pressures, and transfer equilibrium involves $P - 1$ independent relations among chemical potentials. The total number of independent relations for equilibrium is $3(P - 1)$, which we subtract from $3P$ (the number of independent variables in the absence of equilibrium) to obtain the number of independent variables in the equilibrium system: $3P - 3(P - 1) = 3$. Thus, *an open system at equilibrium with one substance in any number of phases has three independent variables*. For example, in a two-phase system we may vary T, n^α, and n^β independently, in which case p is a dependent variable; for a given value of T, the value of p is the one that causes both phases to have the same μ.

6.1.7 The Gibbs phase rule for a pure substance

The complete description of the state of a system must include the value of an *extensive* variable of each phase (e.g., its volume, mass, or amount) to specify how much of the phase is present. For an equilibrium system of P phases with a total of 3 independent variables, we may choose the remaining $3 - P$ variables to be *intensive*. The number of these intensive independent variables is called the **number of degrees of freedom** or **variance**, F, of the system:

$$F = 3 - P \qquad (6.1.9)$$
(pure substance)

Equation 6.1.9 is a special case of the Gibbs phase rule $F = C - P + 2$ (Eq. 10.1.1) for one component ($C = 1$). The application of the phase rule to multicomponent systems is taken up in Sec. 10.1.

We may think of F either as the number of intensive variables needed to describe an equilibrium state (in addition to how much of each phase is present) or as the maximum number of intensive properties that we may vary independently (within certain ranges) while the phases remain in equilibrium.

A system with two degrees of freedom is called *bivariant*, one with one degree of freedom is called *univariant*, and one with no degrees of freedom is called *invariant*. For a system of a pure substance, these three cases correspond to one, two, and three phases respectively. For instance, a system of liquid and gaseous H_2O (and no other substances) is univariant ($F = 3 - P = 3 - 2 = 1$); we are able to independently vary only one intensive property, such as T, while the liquid and gas remain in equilibrium.

6.2 PHASE DIAGRAMS OF PURE SUBSTANCES

A **phase diagram** is a two-dimensional map showing which phase or phases are stable—able to exist in an equilibrium state in the presence of any other phase—under various conditions. This chapter describes pressure–volume and pressure–temperature phase diagrams for a single substance, and chap. 10 describes numerous types of phase diagrams for multicomponent systems. All phase diagrams have certain concepts in common that are now described.

6.2.1 Features of phase diagrams

Two-dimensional phase diagrams for a single substance can be generated as projections of a three-dimensional p-V/n-T surface. A point on the three-dimensional surface corresponds to a physically realizable combination of values, for an equilibrium state of a system containing a total amount n of the substance, of the variables p, V/n, and T.

Three-dimensional surfaces for carbon dioxide are shown at different scales in Fig. 6.1 on the next page and Fig. 6.2 on page 151. In these figures, some areas are labeled with a single physical state (solid, liquid, gas, or supercritical fluid). A point in one of these areas corresponds to an equilibrium state of the system containing a single phase of the labeled physical state. The shape of the surface in this one-phase area gives the equation of state of the phase (i.e., the dependence of one of the variables on the other two). A point in an area labeled with two physical states corresponds to two coexisting phases. The triple line is the locus of points for all possible equilibrium systems of three coexisting phases (in this case, solid, liquid, and gas); however, a point on a triple line can also correspond to just one or two phases (see the discussion on page 152).

The two-dimensional projections shown in Figs. 6.1(b) and 6.1(c) are pressure–volume and pressure–temperature phase diagrams. Because all phases of a multiphase equilibrium system have the same temperature and pressure,[2] the projection of each two-phase area

---
[2] This statement assumes there are no constraints such as internal insulated partitions.

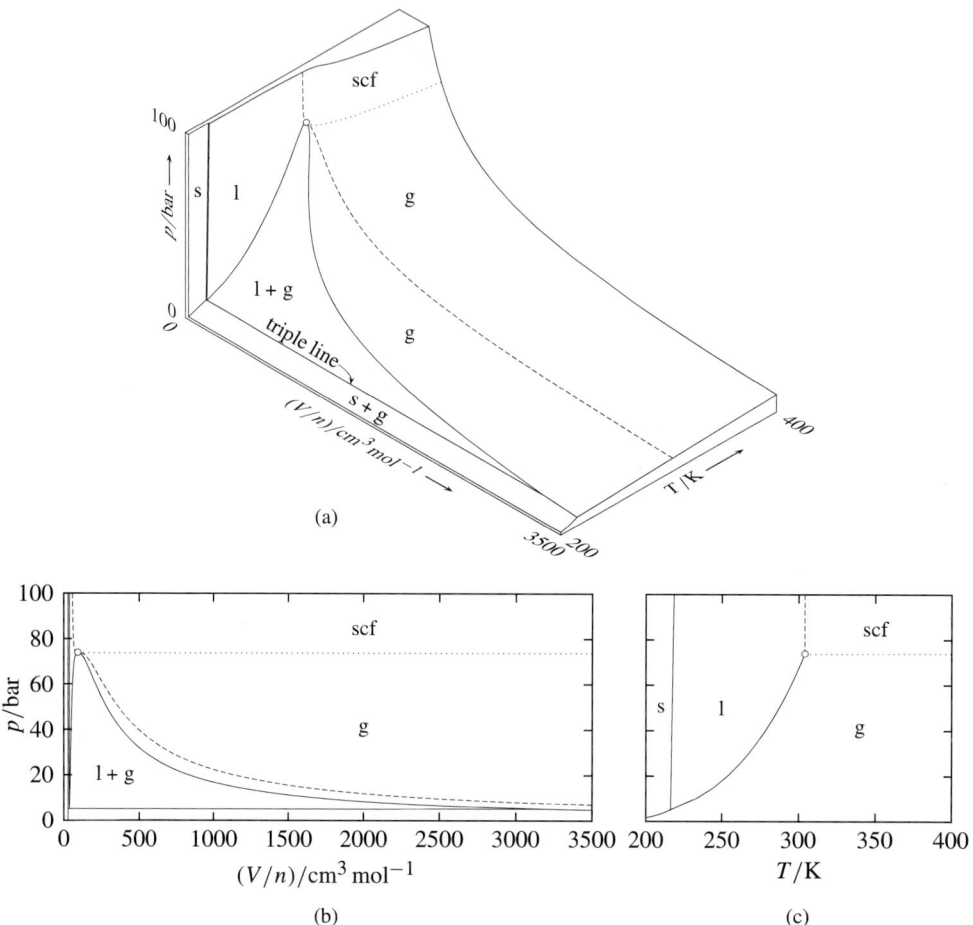

Figure 6.1 Relations among p, V/n, and T for carbon dioxide. Areas are labeled with the stable phase or phases (scf stands for supercritical fluid). The circle indicates the critical point. (a) Three-dimensional p-V/n-T surface. The dashed curve is the critical isotherm at $T = 304.21$ K, and the dotted curve is a portion of the critical isobar at $p = 73.8$ bar. (b) Pressure–volume phase diagram (projection of the surface onto the p-V/n plane). (c) Pressure–temperature phase diagram (projection of the surface onto the p-T plane).

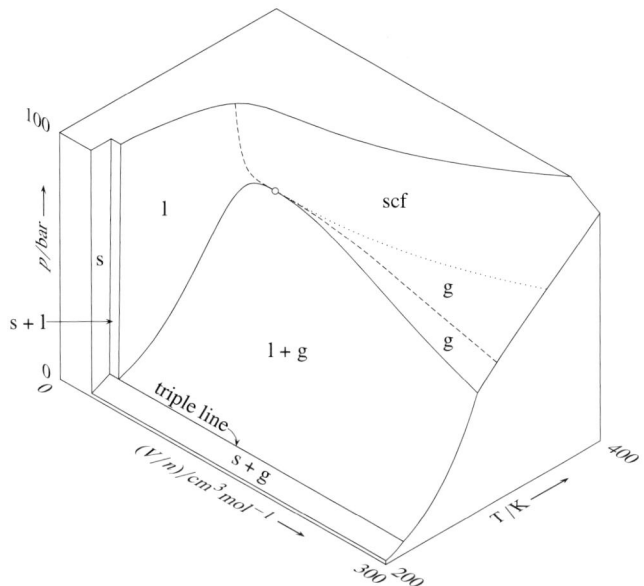

Figure 6.2 Three-dimensional p-V/n-T surface for CO_2, magnified along the V/n axis compared to Fig. 6.1. The circle is the critical point, the dashed curve is the critical isotherm, and the dotted curve is a portion of the critical isobar.

onto the pressure–temperature diagram is a curve, called a **coexistence curve** or **phase boundary**, and the projection of a triple line is a point, called a **triple point**.

How may we use a phase diagram? The two axes represent values of two independent variables, such as p and V/n or p and T. For given values of these variables, we place a point on the diagram at the intersection of the corresponding coordinates; this is the **system point**. Then depending on whether the system point falls in an area or on a coexistence curve, the diagram tells us the number and kinds of phases that can be present in the equilibrium system.

If the system point falls within an area labeled with the physical state of a *single* phase, only that one kind of phase can be present in the equilibrium system. A system containing a pure substance in a single phase is bivariant ($F = 3 - 1 = 2$), so we may vary two intensive properties independently. That is, the system point may move independently along two coordinates (p and V/n or p and T) and still remain in the one-phase area of the phase diagram. V and n refer to only one phase, and the variable V/n is the molar volume V_m of the phase.

If the system point falls in an area of the pressure–volume phase diagram labeled with symbols for *two* phases, these two phases coexist in equilibrium. The phases have the same pressure and different molar volumes. To find the molar volumes of the individual phases, we draw a horizontal line, a **tie line**, through the system point and extending from one edge of the area to the other. The horizontal position of each end of the line, where it terminates at

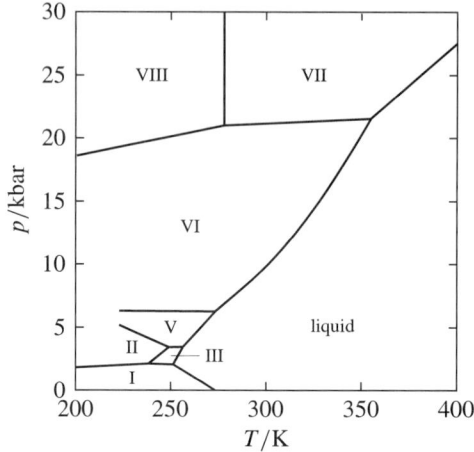

Figure 6.3 High-pressure pressure–temperature phase diagram of H_2O. The roman numerals designate seven forms of ice.

the boundary with a one-phase area, gives the molar volume of that phase in the two-phase system. For an example of a tie line, see Fig. 6.7 on page 156.

The triple line on the pressure–volume diagram represents the range of values of V/n in which three phases (solid, liquid, and gas) can coexist at equilibrium.[3] A three-phase one-component system is invariant ($F = 3 - 3 = 0$); there is only one temperature (the triple-point temperature T_{tp}) and one pressure (the triple-point pressure p_{tp}) at which the three phases can coexist. The values of T_{tp} and p_{tp}, unique to each substance, are shown by the position of the triple point on the pressure–temperature phase diagram. The molar volumes of the three coexisting phases are given by the values of V/n at the three points on the pressure–volume diagram where the triple line touches a one-phase area. These points are at the two ends and an intermediate position of the triple line. If the system point is at either end of the triple line, only the one phase of corresponding molar volume at temperature T_{tp} and pressure p_{tp} can be present. When the system point is on the triple line anywhere between the two ends, either two or three phases can be present. If the system point is at the position on the triple line corresponding to the phase of intermediate molar volume, there might be only that one phase present.

At high pressures, a substance may have additional triple points for two solid phases and the liquid, or for three solid phases. This is illustrated by the pressure–temperature phase diagram of H_2O in Fig. 6.3, which extends to pressures up to 30 kbar. (On this scale, the liquid–gas coexistence curve lies too close to the horizontal axis to be visible.) The diagram shows seven different solid phases of H_2O differing in crystal structure and designated ice I,

---

[3] Helium is the only substance lacking a solid–liquid–gas triple line. When a system containing the coexisting liquid and gas of ^{4}He is cooled to 2.17 K, a triple point is reached in which the third phase is a liquid called He-II, which has the unique property of superfluidity. It is only at high pressures (10 bar or greater) that solid helium can exist.

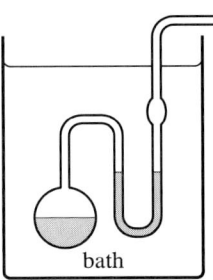

Figure 6.4 An isoteniscope. The liquid to be investigated is placed in the vessel and U-tube, as shown, and maintained at a fixed temperature in the bath. The pressure in the side tube is reduced until the liquid boils gently and its vapor sweeps out the air. The pressure is adjusted until the liquid level is the same in both limbs of the U-tube; then the vapor pressure of the liquid is equal to the pressure in the side tube, which can be measured with a manometer.

ice II, and so on. Ice I is the ordinary form of ice and is the stable form below 2 bar. On the diagram are four triple points for two solids and the liquid and three triple points for three solids. Each triple point is invariant. Note how H_2O can exist as ice VI or ice VII above its standard melting point of 273 K if the pressure is high enough ("hot ice").

6.2.2 Two-phase equilibrium

A system containing two phases of a pure substance in equilibrium is univariant. Both phases have the same values of T and of p, but these values are not independent because of the requirement that the phases have equal chemical potentials. We may vary only one intensive variable of a pure substance (such as T or p) independently while two phases coexist in equilibrium.

At a given temperature, the pressure at which solid and gas or liquid and gas are in equilibrium is called the **vapor pressure** or **saturation pressure** of the solid or liquid.[4] The vapor pressure of a solid is sometimes called the **sublimation pressure**. We may measure the vapor pressure of a liquid at a fixed temperature with a simple device called an isoteniscope (Fig. 6.4).

At a given pressure, the **melting point** or **freezing point** is the temperature at which solid and liquid are in equilibrium, the **boiling point** is the temperature at which liquid and gas are in equilibrium, and the **sublimation temperature** or **sublimation point** is the temperature at which solid and gas are in equilibrium.

The relation between temperature and pressure in a system with two phases in equilibrium is shown by the coexistence curve separating the two one-phase areas on the pressure–temperature diagram (see Fig. 6.5 on the next page). Consider the liquid–gas curve. If we

---

[4] In a system of more than one substance, *vapor pressure* can refer to the partial pressure of a substance in a gas mixture equilibrated with a solid or liquid of that substance. The effect of total pressure on vapor pressure is discussed in Sec. 9.7.1. This book refers to the *saturation pressure* of a liquid when it is necessary to indicate that it is the pure liquid and pure gas phases that are in equilibrium.

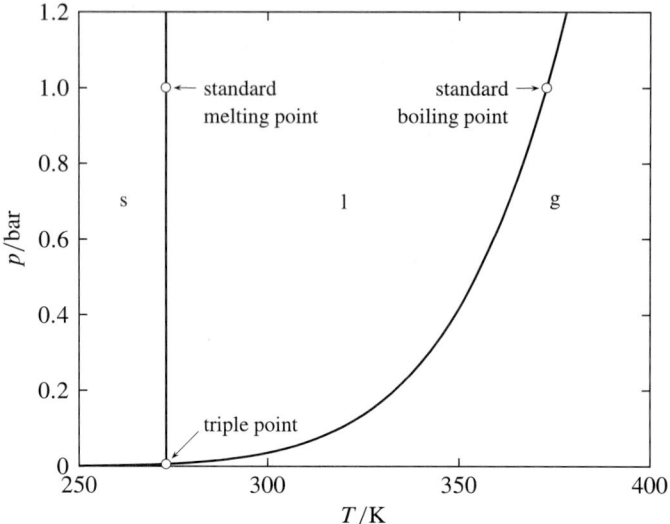

Figure 6.5 Pressure–temperature phase diagram of H$_2$O.

think of T as the independent variable, the curve is a **vapor-pressure curve** showing how the vapor pressure of the liquid varies with temperature. However, if p is the independent variable, then the curve is a **boiling-point curve** showing the dependence of the boiling point on pressure.

The *normal* melting point or boiling point refers to a pressure of one atmosphere, and the *standard* melting point or boiling point refers to the standard pressure. Thus, the normal boiling point of water (99.97 °C) is the boiling point at 1 atm; this temperature is also known as the *steam point*. The standard boiling point of water (99.61 °C) is the boiling point at the slightly lower pressure of 1 bar.

Coexistence curves are discussed further in Sec. 6.4.

6.2.3 The critical point

Every substance has a certain temperature, the **critical temperature**, above which only one fluid phase can exist at any volume and pressure (Sec. 2.1.6). The **critical point** is the point on a phase diagram corresponding to liquid–gas coexistence at the critical temperature, and the **critical pressure** is the pressure at this point.

To observe the critical point of a substance experimentally, we may evacuate a glass vessel, introduce an amount of the substance such that V/n is approximately equal to the molar volume at the critical point, seal the vessel, and raise the temperature from below the critical-point temperature. As T increases in this vessel, the density of the liquid decreases and the density of the gas increases. When T reaches the critical temperature, the two densities become equal and the meniscus disappears. Just before this point is reached, and at slightly higher temperatures, the substance exhibits a cloudy appearance from scattered light, a phenomenon called *critical opalescence*.[5] At temperatures above the critical temperature and pressures above the critical pressure, the one existing fluid phase is called a **supercritical fluid**. Thus, a supercritical fluid of a pure substance is a fluid that does not undergo a phase transition to a different fluid when we change the pressure at constant temperature or change the temperature at constant pressure.[6]

A fluid in the supercritical region can have a molar volume (and density) comparable to that of the liquid and be more compressible than the liquid (κ_T approaches infinity at the critical point). Under supercritical conditions, a substance is often an excellent solvent for solids and liquids; by varying the pressure or temperature its solvating power can be changed, and by reducing the pressure isothermally the substance can be easily removed as a gas from dissolved solutes. These properties make supercritical fluids useful for chromatography and solvent extraction.

The critical temperature of a substance can be measured quite accurately by the heating method described earlier, and the critical pressure can be measured at this temperature with a high-pressure manometer. To evaluate the density at the critical point, it is best to extrapolate the mean density of the coexisting liquid and gas phases, $(\rho^l + \rho^g)/2$, to the critical temperature as illustrated for carbon dioxide in Fig. 6.6 on the next page. The observation that the mean density closely approximates a linear function of temperature, as shown in the figure, is known as the **law of rectilinear diameters** or the law of Cailletet and Matthias. The extrapolated mean density can be used to calculate the molar volume at the critical point.

> Chemists sometimes use the term *vapor* for a gas that can be condensed to a liquid by isothermal compression. The vapor physical state of a substance is the gas at temperatures below the critical temperature.

6.2.4 The lever rule

How, when the system point is in a two-phase area of a pressure–volume phase diagram, can we determine the amounts in the two phases? As an example, consider a closed equilibrium system containing a fixed amount n of a pure substance in coexisting liquid and gas phases. When there is transfer of heat to the system at constant T and p, some of the liquid vaporizes by a liquid–gas phase transition and V increases; withdrawal of heat at constant T and p causes gas to condense and V to decrease. The molar volumes and other intensive properties of the individual liquid and gas phases, however, remain constant during these changes. On the phase diagram shown in Fig. 6.7 on the next page, the volume changes correspond to horizontal movement of the system point along the tie line (page 151) drawn in the two-phase liquid–gas area.

When all of the liquid vaporizes at the given T and p, the system point has moved to the right end of the tie line (at the boundary with the one-phase gas area) and V/n at this

---

[5] For photographs of this phenomenon in CO_2, see: Sengers, J. V.; Sengers, A. L. *Chem. Engineering News* **1968**, June 10, 104–118.

[6] However, if we increase p at constant T, the supercritical fluid will change to a solid. In the phase diagram of H_2O, the coexistence curve for ice VII and liquid shown in Fig. 6.3 on page 152 extends to a higher temperature than the critical temperature of 647 K. Thus, supercritical water can be converted to ice VII by isothermal compression.

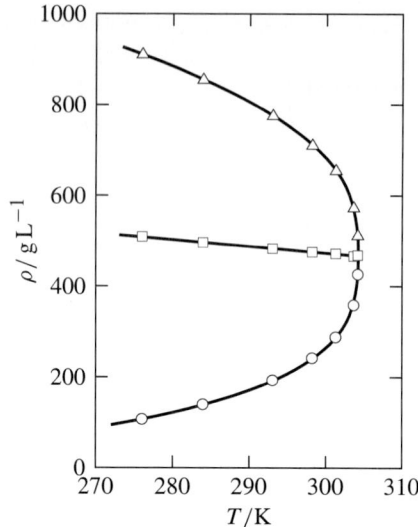

Figure 6.6 Densities of coexisting gas and liquid phases of CO_2 as functions of temperature. Gas densities are shown by circles, liquid densities by triangles, and the mean density at each experimental temperature is shown by a square. (Data of Michels, A.; Blaisse, B.; Michels, C. *Proc. Roy. Soc.* **1937**, *A160*, 358–375.)

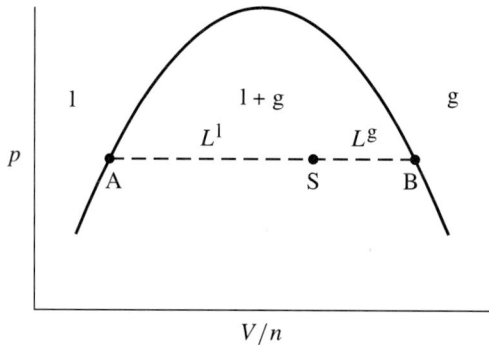

Figure 6.7 Tie line (dashed) in the liquid–gas area of a pressure–volume phase diagram. Point S is the system point and Points A and B are at the ends of the tie line. $L^{\rm l}$ and $L^{\rm g}$ are the distances AS and SB, respectively.

point must be the same as the molar volume of the gas, $V_{\rm m}^{\rm g}$. (We can see this because the system point could have moved from within the one-phase gas area to this position on the boundary, without a phase transition.) However, when all of the gas condenses, the system point is at the left end of the tie line and V/n at this point is the molar volume of the liquid, $V_{\rm m}^{\rm l}$. When both liquid and gas are present, their amounts must be such that the sum of the volumes $V^{\rm l}$ and $V^{\rm g}$ of the individual phases equals the total volume V. This requirement

Section 6.2 Phase Diagrams of Pure Substances

allows us to find the relative amounts in the two phases from the position of the system point along the tie line. Referring to Fig. 6.7, which defines the lengths L^l and L^g, we have the **lever rule** for liquid–gas equilibrium:[7]

$$n^l L^l = n^g L^g \quad \text{or} \quad \frac{n^g}{n^l} = \frac{L^l}{L^g} \tag{6.2.1}$$

(coexisting liquid and gas phases of a pure substance)

To illustrate the lever rule, suppose the amount of gas is twice the amount of liquid; the ratio of amounts is then $n^g/n^l = 2$. Of the total amount n, one third is liquid and two thirds is gas. Then the liquid-phase volume is $V^l = \frac{1}{3} n V_m^l$, the gas-phase volume is $V^g = \frac{2}{3} n V_m^g$, and the value of V/n is $(V^l + V^g)/n = \frac{1}{3} V_m^l + \frac{2}{3} V_m^g$. This places the system point on the tie line two thirds of the way from the left end and makes length L^l twice as long as L^g (as shown in Fig. 6.7). The ratio of lengths is $L^l/L^g = 2$ in agreement with the lever rule.

We cannot apply the lever rule to a point on the triple line because we need more than the value of V/n to determine the relative amounts present in three phases.

> We can derive a general form of the lever rule that applies to any two-phase area of a two-dimensional phase diagram in which a tie-line construction is valid. This general form is needed in chap. 10 for phase diagrams of systems of more than one component. Assume a function F exists that varies linearly with the position of the system point along the tie line, where F is a ratio of two *extensive* properties a and b of the system:
>
> $$F = \frac{a}{b}$$
>
> (In the pressure–volume phase diagram of Fig. 6.7, this function is $F = V/n$.) When a single uniform phase is present, the ratio of two extensive properties of the system is an intensive property of the phase. Thus, when the system point is at one end of the tie line, at the boundary with a one-phase area α, the system contains only phase α and F has a particular value F^α that is an intensive property of phase α (e.g., V_m^α). When the system point is at the other end of the tie line, at the boundary with a one-phase area β, only phase β is present and F has a different value F^β.
>
> Now consider an equilibrium state with the system point at an intermediate position on the tie line. Phases α and β are both present. Phase α has the intensive property value F^α given in terms of extensive properties of the phase by $F^\alpha = a^\alpha/b^\alpha$. Similarly, phase β has the intensive property value $F^\beta = a^\beta/b^\beta$. Since a and b are extensive, they are additive for the two phases:
>
> $$a = a^\alpha + a^\beta = b^\alpha F^\alpha + b^\beta F^\beta$$
> $$b = b^\alpha + b^\beta$$
>
> When we insert these expressions for a and b in the equation $F = a/b$, we obtain
>
> $$F = \frac{b^\alpha F^\alpha + b^\beta F^\beta}{b^\alpha + b^\beta}$$

---

[7] The relation is called the lever rule by analogy to a stationary mechanical lever, each end of which has the same value of the product of applied force and distance from the fulcrum.

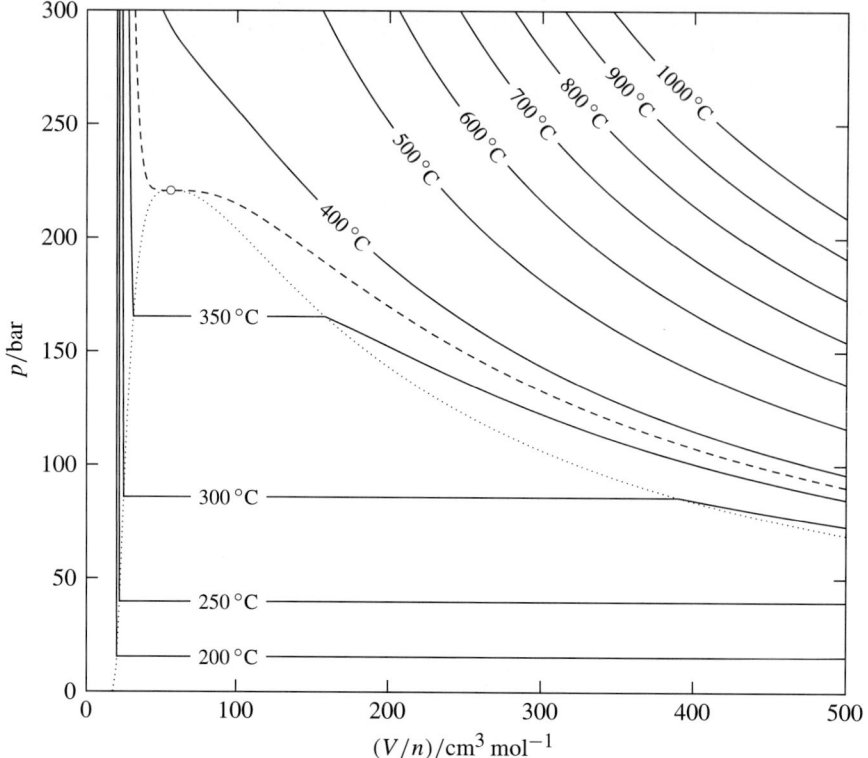

Figure 6.8 Isotherms for the fluid phases of H$_2$O. The open circle indicates the critical point, the dashed curve is the critical isotherm at 373.95 °C, and the dotted curve encloses the two-phase area of the pressure–volume phase diagram. (The triple line lies too close to the bottom of the diagram to be visible on this scale.)

which we can rearrange to

$$b^\beta (F^\beta - F) = b^\alpha (F - F^\alpha)$$

If L is a length proportional to differences in F, measured along the tie line from the system point to an end of the tie line, the preceding equation is equivalent to the general lever rule

$$b^\beta L^\beta = b^\alpha L^\alpha$$

6.2.5 Volume properties

Figure 6.8 is a pressure–volume phase diagram for H$_2$O. On the diagram are drawn *isotherms* (curves of constant T). These isotherms define the shape of the three-dimensional p-V/n-T surface like contour lines on a topographic map. The area containing the horizontal isotherm segments is the two-phase area for coexisting liquid and gas phases. The two-phase area is defined by the curve drawn through the ends of the horizontal segments. The one-phase liquid area lies to the left of this curve, the one-phase gas area lies to the right, and the critical point lies at the top.

The diagram contains the information needed to evaluate the molar volume at any temperature and pressure in the one-phase region and the derivatives of the molar volume with respect to temperature and pressure. At any point in the one-phase region, the slope of the isotherm passing through the point is the partial derivative $(\partial p/\partial V_m)_T$. Since the isothermal compressibility is given by $\kappa_T = -(1/V_m)(\partial V_m/\partial p)_T$, we have

$$\kappa_T = -\frac{1}{V_m \times \text{slope of isotherm}}$$

We see from Fig. 6.8 that the slopes of the isotherms are large and negative in the liquid region, smaller and negative in the gas and supercritical fluid regions, and approach zero at the critical point. Accordingly the isothermal compressibility of the gas and supercritical fluid is much greater than that of the liquid, approaching infinity at the critical point. The critical opalescence mentioned in Sec. 6.2.3 is caused by density inhomogeneities, which are large when κ_T is large.

6.3 PHASE TRANSITIONS

Recall (Sec. 2.1.5) that an equilibrium phase transition of a pure substance is a process in which some or all of the substance is transferred from one coexisting phase to another at constant T and p.

6.3.1 Extensive transition quantities

The quantity $\Delta_{vap}H$ is the enthalpy change for the reversible process in which liquid changes to gas *at a temperature and pressure at which the two phases coexist at equilibrium*. This quantity is called the **enthalpy of vaporization**. Since the pressure is constant during the process, $\Delta_{vap}H$ is equal to the heat q (Eq. 3.11.5). Hence, $\Delta_{vap}H$ is also called the **heat of vaporization**.

> Since $\Delta_{vap}H$ is an enthalpy *change*, it would be more accurate to call it the "enthalpy change of vaporization." This book places the subscript "vap" immediately after the Δ symbol in accordance with the recommendation of the IUPAC Green Book. An older notation places the subscript after the symbol for the property: ΔH_{vap}.

The subscript following the delta (Δ) symbol indicates the kind of phase change (Appendix D):

$\Delta_{vap}H$ = enthalpy of vaporization (a change from liquid to gas);

$\Delta_{sub}H$ = enthalpy of sublimation (a change from solid to gas);

$\Delta_{fus}H$ = enthalpy of fusion (a change from solid to liquid);

$\Delta_{\text{trs}} H$ = enthalpy of a transition between any two phases in general.

Enthalpies of vaporization, sublimation, and fusion are *positive*. The reverse processes of condensation to a liquid, condensation to a solid (or deposition), and freezing have *negative* enthalpy changes.

These subscripts are also used for the changes of other extensive properties for phase transitions under equilibrium conditions. Thus, we have the entropy of vaporization $\Delta_{\text{vap}} S$, the internal energy of sublimation $\Delta_{\text{sub}} U$, and so on. Each of these quantities is proportional to the amount transferred from one phase to the other.

6.3.2 Molar transition quantities

If we divide the change of any extensive property by the amount transferred between phases, we obtain the *molar* change for the phase transition. For example, when an amount n of a substance changes from liquid to gas, the enthalpy change is given by $\Delta_{\text{vap}} H = n H_m^g - n H_m^l$ and the **molar enthalpy of vaporization** (or molar heat of vaporization) is

$$\Delta_{\text{vap}} H_m = \frac{\Delta_{\text{vap}} H}{n} = H_m^g - H_m^l \qquad (6.3.1)$$
(pure substance)

In other words, $\Delta_{\text{vap}} H_m$ is the enthalpy change per amount vaporized and is also the difference between the molar enthalpies of the two phases.

Molar properties of a phase, being intensive, usually depend on two independent intensive variables such as T and p. However, even though $\Delta_{\text{vap}} H_m$ is the difference of two molar properties, its value depends on only *one* intensive variable because the two phases are in equilibrium and the system is univariant. Thus, we may treat $\Delta_{\text{vap}} H_m$ as a function of T only. The same is true of any other molar transition quantity.

The molar Gibbs energy of any equilibrium phase transition, $\Delta_{\text{trs}} G_m$, is a special case. For the phase transition $\alpha \to \beta$, we may write an equation analogous to Eq. 6.3.1 and equate the molar Gibbs energy in each phase to a chemical potential (see Eq. 5.10.1):

$$\Delta_{\text{trs}} G_m = G_m^\beta - G_m^\alpha = \mu^\beta - \mu^\alpha \qquad (6.3.2)$$
(pure substance)

But the transition is between two phases at equilibrium, requiring both phases to have the same chemical potential: $\mu^\beta - \mu^\alpha = 0$. Therefore, the molar Gibbs energy of *any* equilibrium phase transition is zero:

$$\Delta_{\text{trs}} G_m = 0 \qquad (6.3.3)$$
(pure substance)

Since the Gibbs energy is defined by $G = H - TS$, in phase α we have $G_m^\alpha = G^\alpha/n^\alpha = H_m^\alpha - T S_m^\alpha$. Similarly, in phase β we have $G_m^\beta = H_m^\beta - T S_m^\beta$. When we substitute these expressions in $\Delta_{\text{trs}} G_m = G_m^\beta - G_m^\alpha$ (Eq. 6.3.2) and set T equal to the transition temperature

T_{trs}, we obtain

$$\Delta_{\text{trs}} G_m = (H_m^\beta - H_m^\alpha) - T_{\text{trs}}(S_m^\beta - S_m^\alpha)$$
$$= \Delta_{\text{trs}} H_m - T_{\text{trs}} \Delta_{\text{trs}} S_m$$

Then, setting $\Delta_{\text{trs}} G_m$ equal to zero, we find a relation between the molar entropy and molar enthalpy of the equilibrium phase transition:

$$\Delta_{\text{trs}} S_m = \frac{\Delta_{\text{trs}} H_m}{T_{\text{trs}}} \qquad (6.3.4)$$
(pure substance)

where $\Delta_{\text{trs}} S_m$ and $\Delta_{\text{trs}} H_m$ are evaluated at the transition temperature T_{trs}.

> We may obtain this equation directly from the second law. With the phases in equilibrium, the transition process is reversible. The second law gives $dS = dq/T$ (box 4.2.1) or $\Delta_{\text{trs}} S = q/T_{\text{trs}} = \Delta_{\text{trs}} H / T_{\text{trs}}$. Dividing by the amount transferred between the phases gives Eq. 6.3.4.

6.3.3 Calorimetric measurement of transition enthalpies

The most precise measurement of the molar enthalpy of an equilibrium phase transition uses electrical work. A known quantity of electrical work is performed on a system containing coexisting phases, in a constant-pressure adiabatic calorimeter, and the resulting amount of substance transferred between the phases is measured. The first law shows that the electrical work $I^2 R_{\text{el}} \Delta t$ equals the heat that would be needed to cause the same change of state. This heat, at constant p, is the enthalpy change of the process.

The method is similar to that used to measure the heat capacity of a phase at constant pressure (Sec. 5.5.2), except that now the temperature remains constant and there is no need to make a correction for the heat capacity of the calorimeter.

6.3.4 Standard molar transition quantities

The *standard* molar enthalpy of vaporization, $\Delta_{\text{vap}} H_m^\circ$, is the enthalpy change when pure liquid in its standard state at a specified temperature changes to gas in its standard state at the same temperature, divided by the amount changed.

Note that the initial state of this process is a real one (the pure liquid at pressure p°), but the final state (the gas behaving ideally at pressure p°) is hypothetical. The liquid and gas are not necessarily in equilibrium with one another at pressure p° and the temperature of interest, and we cannot evaluate $\Delta_{\text{vap}} H_m^\circ$ from a calorimetric measurement with electrical work without further corrections. In contrast, $\Delta_{\text{vap}} H_m$ (without the $^\circ$ symbol) refers to a reversible transition between two *real* phases coexisting in equilibrium.

For any molar quantity X_m, $\Delta_{\text{trs}} X_m^\circ$ is the difference in the molar quantity for two phases in their standard states at a specified temperature; for example,

$$\Delta_{\text{sub}} S_m^\circ = S_m^\circ(g) - S_m^\circ(s)$$

$\Delta_{\text{trs}} X_{\text{m}}^{\circ}$ is a function of temperature only. If either of the phases is a solid or liquid, the phase must be stable at p° and the specified temperature for $\Delta_{\text{trs}} X_{\text{m}}^{\circ}$ to have meaning.

To evaluate $\Delta_{\text{sub}} X_{\text{m}}^{\circ}$ or $\Delta_{\text{vap}} X_{\text{m}}^{\circ}$ at a given temperature, we must calculate ΔX_{m} for a path that connects the standard state of the solid or liquid with that of the gas. The simplest choice of path is one of constant temperature T with the following steps:

1. Isothermal reduction of the pressure of the solid or liquid, starting with the standard state at pressure p° and ending with the pressure equal to the vapor pressure p_{vap} of the condensed phase at temperature T. The values of ΔU_{m}, ΔH_{m}, ΔS_{m}, and ΔG_{m} in this step are obtained from Eqs. 5.8.1–5.8.4 or from the approximations of Eqs. 5.8.9–5.8.12.

2. Reversible sublimation or vaporization to form the real gas at T and p_{vap}. The change of X_{m} in this step is either $\Delta_{\text{sub}} X_{\text{m}}$ or $\Delta_{\text{vap}} X_{\text{m}}$.

3. Isothermal change of the real gas at pressure p_{vap} to the hypothetical ideal gas at pressure p°. Section 5.11 has the relevant formulas relating molar quantities of a real gas to the corresponding standard molar quantities.

The sum of ΔX_{m} for these steps is the desired quantity $\Delta_{\text{sub}} X_{\text{m}}^{\circ}$ or $\Delta_{\text{vap}} X_{\text{m}}^{\circ}$.

6.4 COEXISTENCE CURVES

A coexistence curve on a pressure–temperature phase diagram shows the conditions under which two phases can coexist in equilibrium, as discussed in Sec. 6.2.2.

6.4.1 Chemical potential surfaces

We may treat the chemical potential μ of a pure substance in a single phase as a function of independent variables T and p and represent the relation by a three-dimensional surface. Since the condition for equilibrium between two phases of a pure substance is that both phases have the same T, p, and μ, equilibrium in a two-phase system can exist only along the intersection of the surfaces of the two phases as illustrated in Fig. 6.9 on the next page.

The shape of the surface for each phase is determined by the partial derivatives of the chemical potential with respect to temperature and pressure as given by Eqs. 5.10.3 and 5.10.4:

$$\left(\frac{\partial \mu}{\partial T}\right)_p = -S_{\text{m}} \qquad \left(\frac{\partial \mu}{\partial p}\right)_T = V_{\text{m}}$$

Let us explore how μ varies with T at constant p for the different physical states of a substance. The stable phase at each temperature is the one of lowest μ since, as discussed in Sec. 6.1.3, transfer of substance from a higher to a lower μ at constant T and p is spontaneous.

From the relation $(\partial \mu / \partial T)_p = -S_{\text{m}}$, we see that at constant p the slope of μ versus T is negative since molar entropy is always positive. Furthermore, the magnitude of the slope

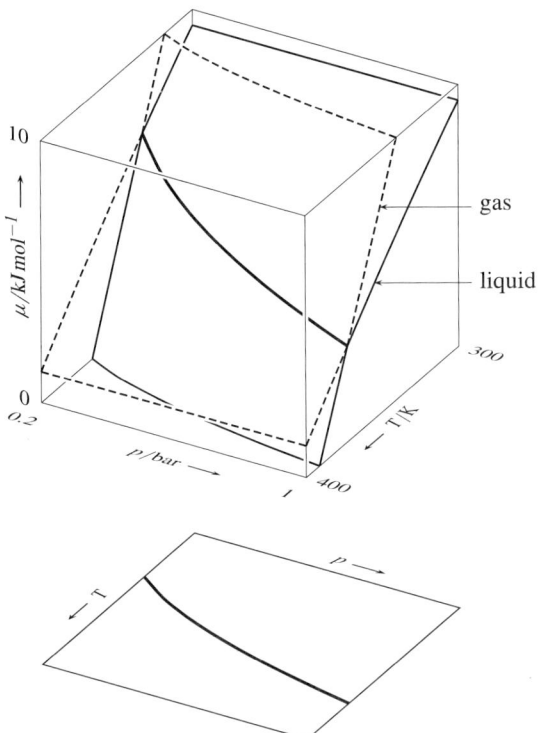

Figure 6.9 Top: chemical potential surfaces of the liquid and gas phases of H_2O; the two phases are at equilibrium along the intersection (heavy curve). (The vertical scale for μ has an arbitrary zero.) Bottom: projection of the intersection onto the p-T plane, generating the coexistence curve.

increases on going from solid to liquid and from liquid to gas since the molar entropies of sublimation and vaporization are positive. This difference in slope is illustrated by the curves for H_2O in Fig. 6.10(a) on the next page. The triple-point pressure of H_2O is 0.0062 bar. At a pressure of 0.03 bar, greater than the triple-point pressure, the curves for solid and liquid intersect at a melting point (Point A) and the curves for liquid and gas intersect at a boiling point (Point B).

From $(\partial \mu / \partial p)_T = V_m$, we see that a pressure reduction at constant temperature lowers the chemical potential of a phase. The result of a pressure reduction from 0.03 bar to 0.003 bar (below the triple-point pressure of H_2O) is a downward shift of each of the curves of Fig. 6.10(a) by a distance proportional to the molar volume of the phase. The shifts of the solid and liquid curves are too small to see ($\Delta \mu$ is only $-0.002\,\text{kJ mol}^{-1}$). Because the gas has a large molar volume, the gas curve shifts substantially to a position where it intersects with the solid curve at a sublimation point (Point C). At 0.003 bar or any other pressure below the triple-point pressure, only a solid–gas equilibrium is possible for H_2O. The liquid phase is not stable at any pressure below the triple-point pressure, as shown by

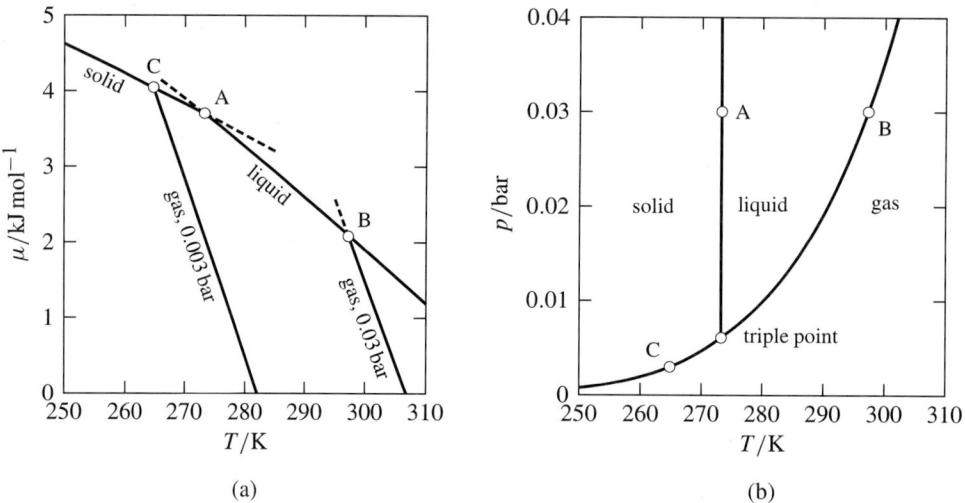

Figure 6.10 Phase stability of H_2O. (a) Chemical potentials of different physical states as functions of temperature. (The scale for μ has an arbitrary zero.) Chemical potentials of the gas are shown at 0.03 bar and 0.003 bar. The effect of pressure on the curves for the solid and liquid is negligible. At $p = 0.03$ bar, solid and liquid coexist at $T = 273.16$ K (Point A) and liquid and gas coexist at $T = 297.23$ K (Point B). At $p = 0.003$ bar, solid and gas coexist at $T = 264.77$ K (Point C). (b) Pressure–temperature phase diagram with points corresponding to those in (a).

the pressure–temperature phase diagram of H_2O in Fig. 6.10(b).

6.4.2 The Clapeyron equation

If we start with two coexisting phases, α and β, of a pure substance and make an infinitesimal change dT in their temperature without changing the pressure, the phases are no longer in equilibrium because their chemical potentials change unequally. For the phases to remain in equilibrium during the temperature change, there must be a certain simultaneous change in the pressure. The changes dT and dp must be such that the chemical potentials of both phases change equally so as to remain equal to one another: $d\mu^\alpha = d\mu^\beta$.

The infinitesimal change of μ in one phase is given by $d\mu = -S_m\, dT + V_m\, dp$ (Eq. 5.10.2). At each T and p, S_m and V_m have definite values in each phase. Thus, the two phases remain in equilibrium if dT and dp satisfy the relation

$$-S_m^\alpha\, dT + V_m^\alpha\, dp = -S_m^\beta\, dT + V_m^\beta\, dp$$

which we rearrange to

$$\frac{dp}{dT} = \frac{S_m^\beta - S_m^\alpha}{V_m^\beta - V_m^\alpha}$$

Section 6.4 Coexistence Curves

or

$$\frac{dp}{dT} = \frac{\Delta_{\text{trs}} S_m}{\Delta_{\text{trs}} V_m} \qquad (6.4.1)$$
(pure substance)

Equation 6.4.1 is one form of the **Clapeyron equation**, which contains no approximations. We find an alternative form by substituting $\Delta_{\text{trs}} S_m = \Delta_{\text{trs}} H_m / T_{\text{trs}}$ (Eq. 6.3.4):

$$\frac{dp}{dT} = \frac{\Delta_{\text{trs}} H_m}{T \Delta_{\text{trs}} V_m} \qquad (6.4.2)$$
(pure substance)

Equations 6.4.1 and 6.4.2 give the slope of the coexistence curve, dp/dT, as a function of quantities that can be measured. For the sublimation and vaporization processes, both $\Delta_{\text{trs}} H_m$ and $\Delta_{\text{trs}} V_m$ are positive. Therefore, according to Eq. 6.4.2, the solid–gas and liquid–gas coexistence curves have positive slopes. For the fusion process, however, $\Delta_{\text{fus}} H_m$ is positive, but $\Delta_{\text{fus}} V_m$ may be positive or negative depending on the substance; thus the slope of the solid–liquid coexistence curve may be either positive or negative. The absolute value of $\Delta_{\text{fus}} V_m$ is small, causing the solid–liquid coexistence curve to be relatively steep; see Fig. 6.10(b) for an example.

> Most substances expand on melting, making the slope of the solid–liquid coexistence curve positive. This is true of carbon dioxide, although in Fig. 6.1(c) on page 150 the curve is so steep that it is difficult to see the slope is positive. Exceptions at ordinary pressures are H_2O, rubidium nitrate, and the elements antimony, bismuth, and gallium.
>
> The phase diagram for H_2O in Fig. 6.3 on page 152 clearly shows that the coexistence curve for ice I and liquid has a negative slope due to ordinary ice being less dense than liquid water. The high-pressure forms of ice are more dense than the liquid, causing the slopes of the other solid–liquid coexistence curves to be positive. The ice VII–ice VIII coexistence curve is vertical because these solids have identical crystal structures except for the orientations of the H_2O molecules and equal molar volumes within experimental uncertainty.

We may rearrange Eq. 6.4.2 to give the variation of p with T along the coexistence curve:

$$dp = \frac{\Delta_{\text{trs}} H_m}{\Delta_{\text{trs}} V_m} \cdot \frac{dT}{T}$$

In the case of fusion, $\Delta_{\text{trs}} V_m$ is approximately constant for small changes of T and p since the cubic expansion coefficient and isothermal compressibility of a condensed phase are relatively small. If $\Delta_{\text{trs}} H_m$ is also practically constant, integration of the preceding equation yields

$$p_2 - p_1 \approx \frac{\Delta_{\text{fus}} H_m}{\Delta_{\text{fus}} V_m} \ln \frac{T_2}{T_1}$$

or

$$T_2 \approx T_1 \exp\left[\frac{\Delta_{\text{fus}} V_m (p_2 - p_1)}{\Delta_{\text{fus}} H_m}\right] \qquad (6.4.3)$$
(pure substance)

from which we may estimate the variation of melting point with pressure.

6.4.3 The Clausius–Clapeyron equation

The molar volume of a gas is much greater than that of a condensed phase in equilibrium with it (unless the substance is close to its critical point). Thus, we may write for the processes of sublimation and vaporization: $\Delta_{sub}V_m = V_m(g) - V_m(s) \approx V_m(g)$ and $\Delta_{vap}V_m = V_m(g) - V_m(l) \approx V_m(g)$. The further approximation that the gas behaves as an ideal gas, $V_m(g) \approx RT/p$, then changes Eq. 6.4.2 to

$$\frac{dp}{dT} \approx \frac{p \Delta_{trs} H_m}{RT^2} \tag{6.4.4}$$
(pure substance,
sublimation or vaporization)

Equation 6.4.4 is the **Clausius–Clapeyron equation** and gives an approximate expression for the slope of a solid–gas or liquid–gas coexistence curve. The expression is not valid for coexisting solid and liquid phases or for coexisting liquid and gas phases at a pressure close to the critical pressure.

At the temperature and pressure of the triple point, it is possible to carry out all three equilibrium phase transitions of fusion, vaporization, and sublimation. When fusion is followed by vaporization, the net change is sublimation. Therefore, the molar transition enthalpies at the triple point are related by

$$\Delta_{fus} H_m + \Delta_{vap} H_m = \Delta_{sub} H_m$$

Since all three transition enthalpies are positive, it follows that $\Delta_{sub} H_m$ is greater than $\Delta_{vap} H_m$ at the triple point. Therefore, according to Eq. 6.4.4, the solid–gas coexistence curve has a slightly greater slope at the triple point than the liquid–gas coexistence curve.

We may divide both sides of Eq. 6.4.4 by $p°$ and rearrange to the form

$$\frac{d(p/p°)}{p/p°} \approx \frac{\Delta_{trs} H_m}{R} \cdot \frac{dT}{T^2}$$

Then with the substitutions from calculus $d(p/p°)/(p/p°) = d\ln(p/p°)$ and $dT/T^2 = -d(1/T)$, we write the displayed equation in the three alternative forms

$$\frac{d\ln(p/p°)}{dT} \approx \frac{\Delta_{trs} H_m}{RT^2} \tag{6.4.5}$$
(pure substance,
sublimation or vaporization)

$$d\ln(p/p°) \approx -\frac{\Delta_{trs} H_m}{R} d(1/T) \tag{6.4.6}$$
(pure substance,
sublimation or vaporization)

and

$$\frac{d\ln(p/p°)}{d(1/T)} \approx -\frac{\Delta_{\text{trs}} H_{\text{m}}}{R} \qquad (6.4.7)$$
(pure substance,
sublimation or vaporization)

Equation 6.4.7 shows that the curve of a plot of $\ln(p/p°)$ versus $1/T$ (where p is the vapor pressure of a pure solid or liquid) has a slope at each temperature equal to $-\Delta_{\text{sub}} H_{\text{m}}/R$ or $-\Delta_{\text{vap}} H_{\text{m}}/R$ at that temperature. This plot provides an alternative to calorimetry for evaluating molar enthalpies of sublimation and vaporization.

If we use the recommended standard pressure of 1 bar, the ratio $p/p°$ appearing in these equations becomes p/bar; that is, $p/p°$ is simply the numerical value of p when p is expressed in bars. For the purpose of using Eq. 6.4.7 to evaluate $\Delta_{\text{trs}} H_{\text{m}}$, we can replace $p°$ by any convenient value. Thus, the curves of plots of $\ln(p/\text{bar})$ versus $1/T$, $\ln(p/\text{Pa})$ versus $1/T$, and $\ln(p/\text{Torr})$ versus $1/T$ using the same temperature and pressure data all have the same slope (but different intercepts) and yield the same value of $\Delta_{\text{trs}} H_{\text{m}}$.

If we assume $\Delta_{\text{sub}} H_{\text{m}}$ or $\Delta_{\text{vap}} H_{\text{m}}$ is essentially constant in a temperature range, we may integrate Eq. 6.4.6 from an initial to a final state along the coexistence curve to obtain

$$\ln \frac{p_2}{p_1} \approx -\frac{\Delta_{\text{trs}} H_{\text{m}}}{R}\left(\frac{1}{T_2} - \frac{1}{T_1}\right) \qquad (6.4.8)$$
(pure substance,
sublimation or vaporization)

Equation 6.4.8 allows us to estimate any one of the quantities p_1, p_2, T_1, T_2, or $\Delta_{\text{trs}} H_{\text{m}}$, given values of the other four.

PROBLEMS

6.1 Consider a small liquid droplet surrounded by the pure vapor of the liquid. The pressure inside the droplet is greater than the pressure of the vapor in equilibrium with it due to the surface tension of the liquid. The surface of the droplet acts somewhat like the stretched membrane of an inflated balloon, tending to reduce its area A_s and the liquid volume V^l until there is a balance of forces from the increased pressure inside. Assume that the droplet is a sphere of radius r; then we have the relations $V^l = (4/3)\pi r^3$, $dV^l = 4\pi r^2\, dr$, $A_s = 4\pi r^2$, and $dA_s = 8\pi r\, dr$. Find an expression for the total differential of U and then follow the method suggested in Sec. 6.1.1 to show that the equilibrium conditions are $T^g = T^l$, $\mu^g = \mu^l$, and $p^l = p^g + 2\gamma/r$, where γ is the surface tension. (The last relation is known as the *Laplace equation*.)

6.2 At 298.15 K, the standard molar entropy of gaseous H_2O calculated from spectroscopic data using statistical mechanics is $S_{\text{m}}°(\text{g}) = 188.72\,\text{J K}^{-1}\,\text{mol}^{-1}$. Liquid water at this temperature, at the standard pressure of 1 bar, has molar volume $V_{\text{m}} = 18.07\,\text{cm}^3\,\text{mol}^{-1}$ and cubic expansion coefficient $\alpha = 2.6 \times 10^{-4}\,\text{K}^{-1}$. The vapor pressure of liquid H_2O at this temperature is 0.03169 bar, and the molar enthalpy of vaporization is $\Delta_{\text{vap}} H_{\text{m}} = 4.3999 \times 10^4\,\text{J mol}^{-1}$. The values of V_{m} and α for the liquid are essentially constant between 0.03 bar and 1 bar. The second virial coefficient, B, of gaseous H_2O is $-1.25 \times 10^{-3}\,\text{m}^3\,\text{mol}^{-1}$ at 298.15 K, and $-1.07 \times 10^{-3}\,\text{m}^3\,\text{mol}^{-1}$ at 303.15 K.

(a) Use these data to calculate the standard molar entropy of *liquid* water at 298.15 K.

(b) Calculate the *standard* molar entropy and enthalpy of vaporization of H_2O at 298.15 K.

6.3 Explain why the chemical potential surfaces shown in Fig. 6.9 are concave downward; that is, why $(\partial \mu / \partial T)_p$ becomes more negative with increasing T and $(\partial \mu / \partial p)_T$ becomes less positive with increasing p.

6.4 Potassium has a standard boiling point of 773 °C and a molar enthalpy of vaporization $\Delta_{vap} H_m = 84.9 \text{ kJ mol}^{-1}$. Estimate the saturation pressure of liquid potassium at 400. °C.

6.5 Naphthalene has a melting point of 78.2 °C at 1 bar and 81.7 °C at 100 bar. The molar volume change on melting is $\Delta_{fus} V_m = 0.019 \text{ cm}^3 \text{ mol}^{-1}$. Calculate the molar enthalpy of fusion to two significant figures.

6.6 The dependence of the vapor pressure of a liquid on temperature, over a limited temperature range, is often represented by the *Antoine equation*, $\log_{10}(p/\text{Torr}) = A - B/(t+C)$, where t is the Celsius temperature and A, B, and C are constants determined by experiment. A variation of this equation, using a natural logarithm and thermodynamic temperature, is

$$\ln(p/\text{bar}) = a - \frac{b}{T+c}$$

The vapor pressure of liquid benzene at temperatures close to 298 K is adequately represented by the preceding equation with the following values of the constants:

$$a = 9.25092 \qquad b = 2771.233 \text{ K} \qquad c = -53.262 \text{ K}$$

(a) Find the standard boiling point of benzene.

(b) Use the Clausius–Clapeyron equation to evaluate the molar enthalpy of vaporization of benzene at 298.15 K.

6.7 At a pressure of one atmosphere, water and steam are in equilibrium at 99.97 °C (the normal boiling point of water). At this pressure and temperature, the water density is 0.958 g cm^{-3}, the steam density is 5.98×10^{-4} g cm^{-3}, and the molar enthalpy of vaporization is 40.66 kJ mol^{-1}.

(a) Use the Clapeyron equation to calculate the slope dp/dT of the liquid–gas coexistence curve at this point.

(b) Repeat the calculation using the Clausius–Clapeyron equation.

(c) Use your results to estimate the standard boiling point of water. (Note: The experimental value is 99.61 °C.)

6.8 At the standard pressure of 1 bar, liquid and gaseous H_2O coexist in equilibrium at 372.76 K, the standard boiling point of water.

(a) Do you expect the standard molar enthalpy of vaporization to have the same value as the molar enthalpy of vaporization at this temperature? Explain.

(b) The molar enthalpy of vaporization at 372.76 K has the value $\Delta_{vap} H_m = 40.67 \text{ kJ mol}^{-1}$. Estimate the value of $\Delta_{vap} H_m^\circ$ at this temperature with the help of Eq. 5.11.10 and the following data for the second virial coefficient of gaseous H_2O at 372.76 K:

$$B = -4.60 \times 10^{-4} \text{ m}^3 \text{ mol}^{-1} \qquad dB/dT = 3.4 \times 10^{-6} \text{ m}^3 \text{ K}^{-1} \text{ mol}^{-1}$$

(c) Would you expect the values of $\Delta_{fus}H_m$ and $\Delta_{fus}H_m^\circ$ to be equal at the standard freezing point of water? Explain.

6.9 The standard boiling point of H_2O is $99.61\,°C$, the molar enthalpy of vaporization at this temperature is $\Delta_{vap}H_m = 40.67\,kJ\,mol^{-1}$, and the molar heat capacity of the liquid at temperatures close to this value is given by $C_{p,m} = a + b(t - c)$, where the constants have the values

$$a = 75.94\,J\,K^{-1}\,mol^{-1} \qquad b = 0.022\,J\,K^{-2}\,mol^{-1} \qquad c = 99.61\,°C$$

Suppose $100.00\,mol$ of liquid H_2O is placed in a container maintained at a constant pressure of 1 bar and is carefully heated to a temperature $5.00\,°C$ above the standard boiling point, resulting in an unstable phase of superheated water. If the container is enclosed with an adiabatic boundary and the system subsequently changes spontaneously to an equilibrium state, what amount of water will vaporize? (Hint: The temperature will drop to the standard boiling point, and the enthalpy change will be zero.)

Chapter 7

MIXTURES

A homogeneous mixture is a phase of variable composition containing more than one substance. This chapter discusses composition variables and partial molar quantities of mixtures in which no chemical reaction is occurring. The ideal mixture is defined. Chemical potentials, activity coefficients, and activities of individual species in both ideal and nonideal mixtures are discussed, including solutions of electrolytes. Except for the use of fugacities to determine activity coefficients in condensed phases, a discussion of phase equilibria involving mixtures is postponed to chap. 10.

7.1 COMPOSITION VARIABLES

A **composition variable** is an intensive property that indicates the relative amount of a particular species or substance in a phase.

7.1.1 Species and substances

We sometimes need to make a distinction between a species and a substance. A **species** is any entity of definite elemental composition and charge and can be specified by a chemical formula, such as H_2O, H_3O^+, NaCl, and Na^+. A **substance** is a species that can be prepared in a pure state (e.g., N_2 and NaCl). Since we cannot prepare macroscopic amounts of one kind of ion by itself, a charged species such as H_3O^+ or Na^+ is not a substance.

7.1.2 Mixtures in general

The **mole fraction** of Species i is defined by

$$x_i \equiv \frac{n_i}{\sum_j n_j} \qquad (7.1.1)$$
$$(P = 1)$$

where n_i is the amount of Species i and the sum is taken over all species in the mixture. (If the mixture is a gas, we sometimes use the symbol y_i in place of x_i).

The **concentration** or molarity of Species i in a mixture is defined by

$$c_i \equiv \frac{n_i}{V} \qquad (7.1.2)$$
$$(P = 1)$$

The symbol M is often used to stand for units of mol L^{-1} or mol dm^{-3}; thus, a concentration of 0.5 M is 0.5 mol L^{-1} or 0.5 molar.

> Concentration is sometimes called "amount concentration" or "molar concentration" to avoid confusion with number concentration (the number of *particles* per unit volume). An alternative notation for c_A is [A].

A **binary** mixture is a mixture of *two* substances.

7.1.3 Solutions

A **solution**, strictly speaking, is a mixture in which we treat one substance, the **solvent**, in a special way. The other species comprising the mixture are then **solutes**. We denote the solvent by A and the solutes by B, C, and so on.[1] Although in principle a solution can be a gas mixture, we will consider only liquid and solid solutions.

We can prepare a solution of varying composition by mixing one or more solutes with the solvent so as to continuously decrease the mole fraction x_A of the solvent. During this process, the physical state of the solution remains the same as that of the pure solvent (either liquid or solid). As x_A decreases, we say the solution becomes more *concentrated*; if x_A increases toward its upper limit of 1, we say the solution becomes more *dilute*.

Mole fraction and concentration can be used as composition variables for both solvent and solute, just as they are for mixtures in general. A third composition variable, molality, is often used for a solute. The **molality** of solute species B is defined by

$$m_B \equiv \frac{n_B}{\text{mass of solvent}} \qquad (7.1.3)$$
$$\text{(solution)}$$

The symbol m is sometimes used to stand for units of mol kg^{-1}, although this should be discouraged because m is also the symbol for meter. For example, a solute molality of 0.6 m is 0.6 mol kg^{-1} or 0.6 molal.

7.1.4 Binary solutions

We may write simplified equations for a binary solution of two substances, Solvent A and Solute B. Equations 7.1.1–7.1.3 become

$$x_B = \frac{n_B}{n_A + n_B} \qquad (7.1.4)$$
$$\text{(binary solution)}$$

$$c_B = \frac{n_B}{V} = \frac{n_B \rho}{n_A M_A + n_B M_B} \qquad (7.1.5)$$
$$\text{(binary solution)}$$

---

[1] Some chemists denote the solvent by subscript 1 and use 2, 3, and so on for solutes.

$$m_B = \frac{n_B}{n_A M_A} \qquad (7.1.6)$$
(binary solution)

The right sides of Eqs. 7.1.4–7.1.6 express three solute composition variables in terms of the amounts and molar masses of the solvent and solute and the density ρ of the solution.

To be able to relate the values of these composition variables to one another, we solve each equation for n_B and divide by n_A to obtain an expression for the mole ratio n_B/n_A:

$$\text{from Eq. 7.1.4} \qquad \frac{n_B}{n_A} = \frac{x_B}{1 - x_B} \qquad (7.1.7)$$
(binary solution)

$$\text{from Eq. 7.1.5} \qquad \frac{n_B}{n_A} = \frac{M_A c_B}{\rho - M_B c_B} \qquad (7.1.8)$$
(binary solution)

$$\text{from Eq. 7.1.6} \qquad \frac{n_B}{n_A} = M_A m_B \qquad (7.1.9)$$
(binary solution)

These expressions for n_B/n_A allow us to find one composition variable as a function of another. For example, to find molality as a function of concentration, we can equate the expressions for n_B/n_A in Eqs. 7.1.8 and 7.1.9 and solve for m_B to obtain

$$m_B = \frac{c_B}{\rho - M_B c_B}$$

A binary solution becomes more dilute as any of the solute composition variables becomes smaller. In the limit of infinite dilution, the expressions for n_B/n_A become:

$$\frac{n_B}{n_A} = x_B$$
$$= \frac{M_A}{\rho_A^*} c_B = V_{m,A}^* c_B$$
$$= M_A m_B \qquad (7.1.10)$$
(solution at infinite dilution)

where a superscript asterisk (*) denotes a pure phase. We see that, in the limit of infinite dilution, the composition variables x_B, c_B, and m_B are proportional to one another. These expressions are also valid for a solute in a *multi*solute solution in which *each* solute is very dilute; that is, in the limit $x_A \to 1$.

> The rule of thumb that the molarity and molality values of a dilute aqueous solution are approximately equal is explained by the relation $M_A c_B/\rho_A^* = M_A m_B$ (Eq. 7.1.10), or $c_B/\rho_A^* = m_B$, and the fact that the density ρ_A^* of water is approximately $1\,\text{kg L}^{-1}$. Hence, if the solvent is water and the solution is dilute, the numerical value of c_B expressed in mol L^{-1} is approximately equal to the numerical value of m_B expressed in mol kg^{-1}.

7.1.5 The composition of a mixture

We can describe the composition of a phase with the amounts of each species or with any of the composition variables defined earlier (mole fraction, concentration, or molality). If we use mole fractions, we need the values for all but one of the species since the sum of all mole fractions is 1.

Other composition variables are sometimes used, such as mass fraction, volume fraction, mole ratio, and mole percent. To describe the composition of a gas mixture, partial pressures can be used (Sec. 7.4.1).

When the composition of a mixture is said to be *fixed* or *constant* during changes of temperature, pressure, or volume, this means there is no change in the relative *amounts* of the species. A mixture of fixed composition has fixed values of mole fractions and molalities but not necessarily of concentrations and partial pressures. Concentrations will change if the volume changes, and partial pressures in a gas mixture will change if the pressure changes.

7.2 PARTIAL MOLAR QUANTITIES

The symbol X_i, where X is an extensive property and the subscript i denotes a constituent of a mixture, stands for the **partial molar quantity** of Species i defined by

$$X_i \equiv \left(\frac{\partial X}{\partial n_i} \right)_{T,p,n_{j \neq i}} \qquad \text{(7.2.1)} \\ \text{(mixture)}$$

This is the rate at which property X changes with the amount of Species i added to the mixture when the temperature, pressure, and amount of each species other than i are kept constant. Partial molar quantities are *intensive*.

An older notation for a partial molar quantity includes an overbar: $\overline{X}_i$.

To gain insight into the significance of the rather abstract definition of a partial molar quantity given by Eq. 7.2.1, let us first apply the concept to volumes.

7.2.1 Partial molar volume

When we combine two liquid phases having the same temperatures and pressures and *identical* compositions, the new volume is the sum of the original separate volumes. This is not necessarily true when we mix two *different* liquids.

As a specific example, consider water and methanol, two liquids that mix in all proportions. Let A = water and B = methanol. At 25 °C and 1 bar, the molar volume of pure water is $V_{m,A}^* = 18.07 \text{ cm}^3 \text{ mol}^{-1}$ and that of pure methanol is $V_{m,B}^* = 40.71 \text{ cm}^3 \text{ mol}^{-1}$.

If we mix 100.0 cm³ of water with 100.0 cm³ of methanol, while maintaining the temperature at 25 °C and the pressure at 1 bar, we find the volume of the resulting mixture is not the sum 200.0 cm³, but rather the slightly smaller value 193.1 cm³. The difference is due to new intermolecular interactions in the mixture compared to the pure liquids. Let us

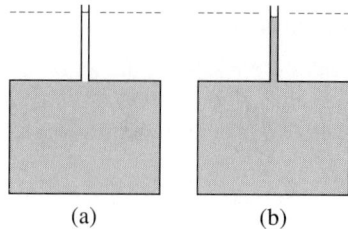

(a)   (b)

Figure 7.1 Addition of pure methanol (substance B) to a water–methanol mixture at constant T and p. (a) 40.71 cm^3 of methanol is placed in a narrow tube above a much greater volume of a mixture (shaded) of composition $x_B = 0.3074$. (b) When the two liquids mix by diffusion, the volume of the mixture increases by only 38.61 cm^3.

calculate the mole fraction composition of this binary mixture:

$$n_A = \frac{100.0 \, \text{cm}^3}{18.07 \, \text{cm}^3 \, \text{mol}^{-1}} = 5.534 \, \text{mol}$$

$$n_B = \frac{100.0 \, \text{cm}^3}{40.71 \, \text{cm}^3 \, \text{mol}^{-1}} = 2.456 \, \text{mol}$$

$$x_B = \frac{2.456}{5.534 + 2.456} = 0.3074$$

Now suppose we prepare an arbitrary but large volume of the mixture of this composition, say 10,000.00 cm^3, and add an additional 40.71 cm^3 (one mole) of pure methanol (see Fig. 7.1). We find the volume of the new mixture (at 25 °C and 1 bar) is 10,038.61 cm^3, an increase of 38.61 cm^3. The amount of methanol added is not infinitesimal, but it is small enough compared to the amount of initial mixture to cause very little change in the composition (x_B increases by only 0.5%). Treating the mixture as an open system, we see that the addition of one mole of methanol to the system at constant T, p, and n_A causes the system volume to increase by 38.61 cm^3. To a good approximation the partial molar volume of methanol in the mixture, $V_B = (\partial V/\partial n_B)_{T,p,n_A}$, is given by $\Delta V/\Delta n_B = 38.61 \, \text{cm}^3/1 \, \text{mol}$.

It does not matter what volume of mixture we add the methanol to as long as the volume is large. We would have observed the same volume increase, 38.61 cm^3, if we had mixed one mole of pure methanol with 100,000.00 cm^3 of the mixture instead of 10,000.00 cm^3.

Thus, we may interpret the partial molar volume of B as the volume change per amount of B added when B is mixed with such a large volume of mixture that the composition is not appreciably affected. We may also interpret the partial molar volume as the volume change per amount when an infinitesimal amount is mixed with a finite volume of mixture.

The partial molar volume of B is an intensive property that is a function of the composition, as well as of T and p. The limiting value of V_B as x_B approaches 1 (pure B) is $V_{m,B}^*$, the molar volume of pure B. We can see this by writing $V = nV_{m,B}^*$ for pure B, where the molar volume $V_{m,B}^*$, an intensive property, is independent of n and is constant at constant T and p. Then the partial molar volume of pure B is given by $(\partial nV_{m,B}^*/\partial n)_{T,p} = V_{m,B}^*$.

Section 7.2 Partial Molar Quantities

As x_B approaches 0, V_B approaches a certain limiting value that is the volume increase per amount of B mixed with a large amount of pure A. In the resulting mixture, each molecule of B is surrounded only by molecules of A. If we treat B as the solute of a solution, we denote this limiting value by V_B^∞, the partial molar volume of the solute at infinite dilution.

It is possible for a partial molar volume to be *negative*. Magnesium sulfate, in aqueous solutions of molality less than $0.07 \, \text{mol} \, \text{kg}^{-1}$, has a negative partial molar volume. Physically, this means that when a small amount of crystalline $MgSO_4$ dissolves in water, the liquid phase contracts. This unusual behavior is due to strong attractive water–ion interactions.

7.2.2 The total differential of the volume in an open system

Consider an open system consisting of a one-phase mixture of nonreacting substances. How many independent variables does this system have? We can prepare the mixture with various amounts of each substance, and we are able to adjust the temperature and pressure to whatever values we wish (within certain limits that prevent the formation of a second phase). Each choice of temperature, pressure, and amounts results in a definite value of every other property, such as volume, density, and mole fraction composition. Thus, an open one-phase system of C substances has $2 + C$ independent variables.[2]

For a binary mixture, the number of independent variables is $C + 2 = 2 + 2 = 4$. We may choose these variables to be T, p, n_A, and n_B and write the total differential of V as

$$dV = \left(\frac{\partial V}{\partial T}\right)_{p,n_A,n_B} dT + \left(\frac{\partial V}{\partial p}\right)_{T,n_A,n_B} dp$$

$$+ \left(\frac{\partial V}{\partial n_A}\right)_{T,p,n_B} dn_A + \left(\frac{\partial V}{\partial n_B}\right)_{T,p,n_A} dn_B \qquad (7.2.2)$$
$$\text{(binary mixture)}$$

We know the first two partial derivatives on the right side are given by[3]

$$\left(\frac{\partial V}{\partial T}\right)_{p,n_A,n_B} = \alpha V$$

and

$$\left(\frac{\partial V}{\partial p}\right)_{T,n_A,n_B} = -\kappa_T V$$

From Eq. 7.2.1, we identify the last two partial derivatives on the right side of Eq. 7.2.2 as the partial molar volumes V_A and V_B. Thus, we may write the total differential of V for this

---

[2] C is actually the number of *components*, which is usually the same as the number of substances but is less if certain constraints exist such as chemical equilibrium or a fixed composition. The precise meaning of C is discussed in Sec. 10.1.

[3] See Eqs. 5.2.1 and 5.2.2 or relations in Tables 5.1 and 5.2, which are for closed systems in agreement with the conditions of constant n_A and n_B.

open system in the compact form

$$dV = \alpha V\,dT - \kappa_T V\,dp + V_A\,dn_A + V_B\,dn_B \qquad (7.2.3)$$
(binary mixture)

If we compare this equation with the total differential of V for a one-component *closed* system, $dV = \alpha V\,dT - \kappa_T V\,dp$ (Eq. 5.2.5), we see that an additional term is required for each constituent of the mixture to allow the system to be open and the composition to vary.

When T and p are held constant, Eq. 7.2.3 becomes

$$dV = V_A\,dn_A + V_B\,dn_B \qquad (7.2.4)$$
(binary mixture, constant T and p)

We obtain an important relation between the mixture volume and the partial molar volumes by imagining the following process. Suppose we continuously pour pure water and pure methanol at fixed rates into a stirred container to form a mixture of constantly increasing volume and constant composition. For instance, equal volume rates of water and methanol would produce the mixture with the composition described in the preceding section, $x_B = 0.3074$. If this open system (the mixture) remains at constant T and p, none of its intensive properties changes during the process and the molar volumes V_A and V_B remain constant. Under these conditions, we can integrate Eq. 7.2.4 to obtain the **additivity rule**[4] for volume:

$$V = V_A n_A + V_B n_B \qquad (7.2.5)$$
(binary mixture)

This equation allows us to calculate the mixture volume from the amounts of the constituents and the appropriate partial molar volumes for the particular temperature, pressure, and composition.

For example, given that the partial molar volumes in a water–methanol mixture of composition $x_B = 0.3074$ are $V_A = 17.75\,\mathrm{cm^3\,mol^{-1}}$ and $V_B = 38.61\,\mathrm{cm^3\,mol^{-1}}$, we calculate the volume of the water–methanol mixture described at the beginning of Sec. 7.2.1 as follows:

$$V = (17.75\,\mathrm{cm^3\,mol^{-1}})(5.534\,\mathrm{mol}) + (38.61\,\mathrm{cm^3\,mol^{-1}})(2.456\,\mathrm{mol})$$
$$= 193.05\,\mathrm{cm^3}$$

We can differentiate Eq. 7.2.5 to obtain a general expression for dV under conditions of constant T and p:

$$dV = V_A\,dn_A + V_B\,dn_B + n_A\,dV_A + n_B\,dV_B$$

---

[4] Term introduced by G. W. Castellan (reference in Appendix I).

But this expression for dV is consistent with Eq. 7.2.4 only if the sum of the last two terms on the right is zero:

$$n_A \, dV_A + n_B \, dV_B = 0 \tag{7.2.6}$$
(binary mixture, constant T and p)

Equation 7.2.6 is the **Gibbs–Duhem equation** for a binary mixture, applied to partial molar volumes. (Section 7.2.4 gives a general version of this equation.) Dividing both sides of the equation by $n_A + n_B$ gives the equivalent form

$$x_A \, dV_A + x_B \, dV_B = 0 \tag{7.2.7}$$
(binary mixture, constant T and p)

The Gibbs–Duhem equation shows that changes in the values of V_A and V_B are related when the composition changes at constant T and p. If we rearrange Eq. 7.2.6 to the form

$$dV_A = -\frac{n_B}{n_A} \, dV_B \tag{7.2.8}$$
(binary mixture, constant T and p)

we see that a composition change that *increases* V_B (so that dV_B is positive) must make V_A *decrease*.

7.2.3 Measuring partial molar volumes

If an analytical formula is available for the volume of a binary mixture as a function of n_A and n_B, at a given temperature and pressure, we can evaluate the partial molar volume V_A for any composition simply by differentiating this function with respect to n_A, and we can evaluate V_B by differentiating with respect to n_B.

The **method of intercepts** is a graphical method of evaluating V_A and V_B. In this method, we plot experimental values of the quantity V/n (where n is $n_A + n_B$) versus the mole fraction x_B. (V/n is called the *mean molar volume*.) See Fig. 7.2(a) on the next page for an example. The tangent to the curve drawn at the point on the curve at the composition of interest intercepts the y axis (where x_B equals 0) at $V/n = V_A$ and intercepts the vertical line where x_B equals 1 at $V/n = V_B$.

To derive this property of a tangent line for the plot of V/n versus x_B, we use Eq. 7.2.5 to write

$$y = \frac{V}{n} = \frac{n_A V_A + n_B V_B}{n} = x_A V_A + x_B V_B$$
$$= (1 - x_B) V_A + x_B V_B = x_B(V_B - V_A) + V_A$$

Differentiation of $y = x_B(V_B - V_A) + V_A$ with respect to x_B gives

$$\frac{dy}{dx_B} = V_B - V_A + x_B \left(\frac{dV_B}{dx_B} - \frac{dV_A}{dx_B} \right) + \frac{dV_A}{dx_B}$$

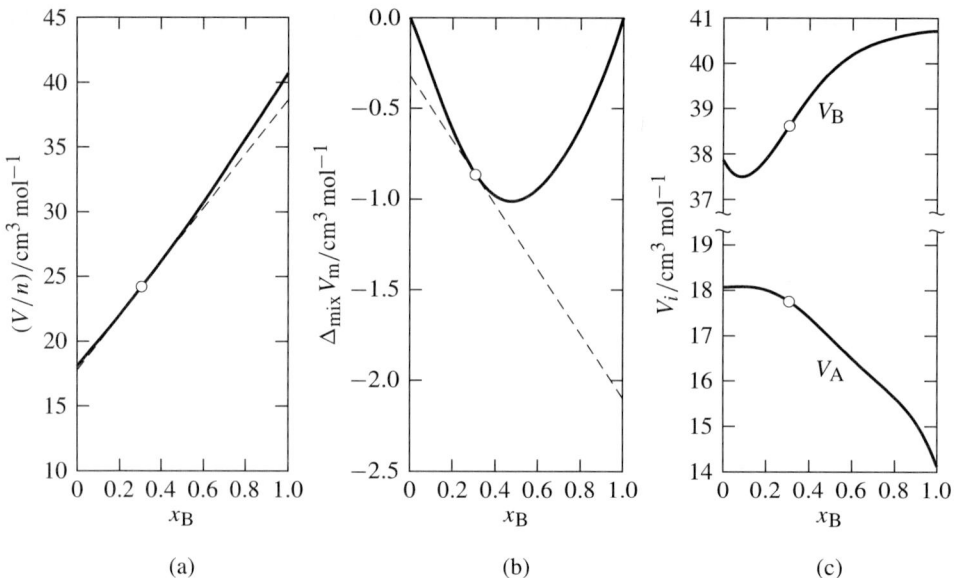

Figure 7.2 Mixtures of water (A) and methanol (B) at 25 °C and 1 bar. (a) Mean molar volume as a function of x_B. The dashed line is the tangent to the curve at $x_B = 0.3074$. (b) Molar volume of mixing as a function of x_B. The dashed line is the tangent to the curve at $x_B = 0.3074$. (c) Partial molar volumes as functions of x_B. The points at $x_B = 0.3074$ (open circles) are obtained from the intercepts of the dashed line in either (a) or (b).

$$= V_B - V_A + (1 - x_B)\frac{dV_A}{dx_B} + x_B\frac{dV_B}{dx_B}$$

$$= V_B - V_A + x_A\frac{dV_A}{dx_B} + x_B\frac{dV_B}{dx_B}$$

When we divide both sides of Eq. 7.2.7 by dx_B, the result is

$$x_A\frac{dV_A}{dx_B} + x_B\frac{dV_B}{dx_B} = 0$$

giving $dy/dx_B = V_B - V_A$. That is, the tangent to the curve of y versus x_B, at a point on the curve where the molar volumes are V_A and V_B, has a slope of $V_B - V_A$. Thus, between the point and the left intercept, the y value of the tangent changes by $-x_B(V_B - V_A)$; between the point and the right intercept, the y value changes by $(1 - x_B)(V_B - V_A)$. Adding these changes to the value $y = x_B(V_B - V_A) + V_A$ gives V_A and V_B as the left and right intercepts of the tangent.

A variation of the method of intercepts is to plot the quantity

$$\Delta_{mix}V_m = V/n - x_A V^*_{m,A} - x_B V^*_{m,B}$$

called the molar volume of mixing (see Sec. 8.2.1) versus x_B, as illustrated in Fig. 7.2(b). The tangent to the curve at the composition of interest has intercepts $V_A - V_{m,A}^*$ at $x_B = 0$ and $V_B - V_{m,B}^*$ at $x_B = 1$.

To see this, we write

$$y' = \Delta_{mix} V_m = y - x_A V_{m,A}^* - x_B V_{m,B}^* = y - x_B(V_{m,B}^* - V_{m,A}^*) - V_{m,A}^*$$

Differentiating with respect to x_B and using the relation $dy/dx_B = V_B - V_A$ from above gives

$$\frac{dy'}{dx_B} = \frac{dy}{dx_B} - (V_{m,B}^* - V_{m,A}^*) = (V_B - V_{m,B}^*) - (V_A - V_{m,A}^*)$$

By the same reasoning as before, the intercepts of the tangent to a point on the curve are at $V_A - V_{m,A}^*$ and $V_B - V_{m,B}^*$.

Figure 7.2 shows smoothed experimental data for water–methanol mixtures plotted in both kinds of graphs and the resulting partial molar volumes as functions of composition. Note how the V_A curve mirrors the V_B curve as x_B varies, as predicted by the Gibbs–Duhem equation (Eq. 7.2.6). The minimum in V_B at $x_B = 0.09$ is mirrored by a maximum in V_A in agreement with Eq. 7.2.8; the maximum is much attenuated because n_B/n_A is much less than unity.

Macroscopic measurements are unable to provide unambiguous information about molecular structure. Nevertheless, it is interesting to speculate on the implications of the minimum observed for the partial molar volume of methanol. One interpretation is that in a mostly aqueous environment methanol molecules associate, perhaps forming dimers.

7.2.4 General relations

The relations derived previously for volumes may be generalized for any extensive property X of a mixture containing any number of species.

The partial molar quantity of Species i, defined by

$$X_i \equiv \left(\frac{\partial X}{\partial n_i}\right)_{T,p,n_{j\neq i}}$$

is an intensive property that depends on T, p, and the mixture composition.

If the mole fraction of i is 1 (the phase is pure i), then the partial molar quantity X_i is the same as the molar property of pure i. For this reason, we frequently use the notation X_i^*; for example, V_A^* is the molar volume of pure A, which we could also write as $V_{m,A}^*$.

The additivity rule for property X is

$$X = \sum_i n_i X_i \qquad (7.2.9)$$
$$\text{(mixture)}$$

and the Gibbs–Duhem equation applied to X can be written in the equivalent forms

$$\sum_i n_i \, dX_i = 0 \qquad (7.2.10)$$
$$\text{(constant } T \text{ and } p\text{)}$$

and

$$\sum_i x_i \, dX_i = 0 \qquad (7.2.11)$$
$$\text{(constant } T \text{ and } p\text{)}$$

For a binary mixture of A and B, we can use the method of intercepts described in Sec. 7.2.3 to evaluate the partial molar quantities X_A and X_B at a given temperature, pressure, and composition. Usually a plot of $X/n - x_A X_A^* - x_B X_B^*$ versus x_B (at constant T and p) gives the greatest accuracy, and is the only plot possible if X is a property such as enthalpy or Gibbs energy for which we can measure only values relative to the pure liquids. On this plot, the tangent to the curve at the composition of interest has intercepts equal to $X_A - X_A^*$ at $x_B = 0$ and $X_B - X_B^*$ at $x_B = 1$.

We cannot prepare this kind of plot if the pure solute has a different physical state than the solution, as in the case of an aqueous solution of a solid salt. In this situation, if we can measure the absolute value of X, we can plot X/n as a function of x_B at constant T and p; then the tangent to the curve at the composition of interest has intercepts equal to X_A at $x_B = 0$ and X_B at $x_B = 1$.

We can obtain experimental values of such partial molar quantities of a substance as V_i, $C_{p,i}$, and S_i. However, it is not practical to evaluate the partial molar quantities U_i, H_i, and G_i because these quantities involve the internal energy brought into the system by Species i, and we cannot precisely evaluate the absolute value of an internal energy (Sec. 3.2.1). We can, however, include these quantities in useful theoretical relations.

7.3 THE CHEMICAL POTENTIAL OF A SPECIES IN A MIXTURE

Just as the molar Gibbs energy of a pure substance is given the special name *chemical potential* and the special symbol μ, the partial molar Gibbs energy G_i of Species i in a mixture is called the **chemical potential** of Species i, defined by

$$\mu_i \equiv \left(\frac{\partial G}{\partial n_i}\right)_{T,p,n_{j \neq i}} \qquad (7.3.1)$$
$$\text{(mixture)}$$

The chemical potential of a species in a phase plays a crucial role in equilibrium problems because it is a measure of the escaping tendency of the species from the phase. Although we cannot precisely determine a value of μ_i, we usually are able to evaluate the *difference* of the values of μ_i in two states or two phases.

In an open one-phase system containing a mixture of s different nonreacting species, we may independently vary T, p, and the amount of each species.[5] This is a total of $2 + s$

---

[5] However, it is not physically possible to vary the amount of a *charged* species independently of the amounts of other charged species. Any system that can be prepared in practice is electrically neutral or virtually so.

independent variables. Following the same reasoning as in Sec. 5.13, we find the total differential of the Gibbs energy is

$$dG = -S\,dT + V\,dp + \sum_{i=1}^{s} \mu_i\,dn_i \qquad (7.3.2)$$
(mixture)

which is like the corresponding expression for a pure phase, $dG = -S\,dT + V\,dp + \mu\,dn$ (Eq. 5.13.6), with a term for each species.

7.3.1 Equilibrium conditions

We now extend the derivation of equilibrium conditions in two phases described in Sec. 6.1.2 to include mixtures of nonreacting species.

The system consists of two phases, α and β, that are mixtures. Following the method of Sec. 6.1.2, we write for the total differential of the internal energy

$$\begin{aligned}dU &= dU^\alpha + dU^\beta \\ &= \left(T^\alpha\,dS^\alpha - p^\alpha\,dV^\alpha + \sum_i \mu_i^\alpha\,dn_i^\alpha\right) \\ &\quad + \left(T^\beta\,dS^\beta - p^\beta\,dV^\beta + \sum_i \mu_i^\beta\,dn_i^\beta\right)\end{aligned} \qquad (7.3.3)$$

Consider two cases.

Case 1: There is an isotropic fluid of uniform pressure on one or both sides of each surface element of the boundary. This is the same situation as in Sec. 6.1.2. The conditions of isolation are $dU = 0$, $dV^\beta = -dV^\alpha$, and $dn_i^\beta = -dn_i^\alpha$ for each species i. We also have the relation $dS^\beta = dS - dS^\alpha$. When we make these substitutions in Eq. 7.3.3 and solve for dS, we obtain

$$dS = \frac{T^\beta - T^\alpha}{T^\beta}\,dS^\alpha - \frac{p^\beta - p^\alpha}{T^\beta}\,dV^\alpha + \sum_i \frac{\mu_i^\beta - \mu_i^\alpha}{T^\beta}\,dn_i^\alpha$$

(This is like Eq. 6.1.2 with a term for each species.) When the system is in an equilibrium state, the coefficient multiplying each differential on the right side of the equation is zero. Therefore, the equilibrium conditions are $T^\alpha = T^\beta$, $p^\alpha = p^\beta$, and $\mu_i^\alpha = \mu_i^\beta$ (for each species i).

> Suppose a particular species present in phase α is completely excluded from phase β. For instance, sucrose molecules dissolved in an aqueous phase are not accommodated in the crystal structure of an ice phase. We treat this situation with additional constraints: in Eq. 7.3.3, we omit the term with the change of the amount of this species in phase β, and in the isolated system we set the change of the amount of this species in phase α equal to zero. Consequently there is no equilibrium condition involving this particular species.

Case 2: Two fluid phases are separated by a fixed, rigid semipermeable membrane. Assume some of the species can pass through the membrane and others cannot. Then the phases can have different pressures in equilibrium states of the system. In Eq. 7.3.3, the terms $\mu_i^\alpha \, dn_i^\alpha$ and $\mu_i^\beta \, dn_i^\beta$ are zero for any species that cannot pass through the membrane because its amount cannot change in either phase. The sums in Eq. 7.3.3 then include only the species that can pass through the membrane. If neither phase is to do expansion work when their pressures are different, their individual volumes must remain constant. The conditions, then, that make the system an isolated one are $dU = 0$, $dV^\alpha = 0$, $dV^\beta = 0$, and $dn_i^\beta = -dn_i^\alpha$ for each species i that can pass through the membrane. With these conditions, the total differential of S is

$$dS = \frac{T^\beta - T^\alpha}{T^\beta} \, dS^\alpha + \sum_i \frac{\mu_i^\beta - \mu_i^\alpha}{T^\beta} \, dn_i^\alpha$$

showing that the equilibrium conditions are $T^\alpha = T^\beta$ and $\mu_i^\alpha = \mu_i^\beta$ for each species i that can pass through the membrane.

We reach the following conclusions concerning a two-phase system (and by extension a system of any number of phases) that is in an equilibrium state. Each phase has the same temperature.[6] Each species present in a phase has the same chemical potential as in any other phase to which the species can transfer, regardless of whether these phases have the same pressure.

7.3.2 Relations involving partial molar quantities

Here we derive some useful relations involving partial molar quantities in a mixture. The independent variables of the phase are T, p, and the amount of each constituent species.

The Gibbs energy of the phase is defined by $G = H - TS$. Taking the partial derivatives of both sides of this equation with respect to n_i gives us

$$\left(\frac{\partial G}{\partial n_i}\right)_{T,p,n_{j\neq i}} = \left(\frac{\partial H}{\partial n_i}\right)_{T,p,n_{j\neq i}} - T\left(\frac{\partial S}{\partial n_i}\right)_{T,p,n_{j\neq i}}$$

We recognize each partial derivative as a partial molar quantity and rewrite the equation as

$$\mu_i = H_i - TS_i \qquad (7.3.4)$$
$$\text{(mixture)}$$

This is analogous to the relation $\mu = H_m - TS_m$ for a pure substance (Eq. 5.11.3).

From the total differential of the Gibbs energy, $dG = -S \, dT + V \, dp + \sum_i \mu_i \, dn_i$ (Eq. 7.3.2), we obtain the following reciprocity relations:

$$\left(\frac{\partial \mu_i}{\partial T}\right)_{p,n} = -\left(\frac{\partial S}{\partial n_i}\right)_{T,p,n_{j\neq i}} \qquad \left(\frac{\partial \mu_i}{\partial p}\right)_{T,n} = \left(\frac{\partial V}{\partial n_i}\right)_{T,p,n_{j\neq i}}$$

---

[6] Phases can have different temperatures in an equilibrium state if they are separated by internal adiabatic partitions. A semipermeable membrane, however, cannot be adiabatic.

(Subscript n on a partial derivative means the amount of *each* species is constant—that is, that the derivative is taken at constant composition.) Again we recognize partial derivatives as partial molar quantities and rewrite these relations as follows:

$$\left(\frac{\partial \mu_i}{\partial T}\right)_{p,n} = -S_i \qquad (7.3.5)$$
(mixture)

$$\left(\frac{\partial \mu_i}{\partial p}\right)_{T,n} = V_i \qquad (7.3.6)$$
(mixture)

The two preceding equations are the equivalent, for a mixture, of the relations $(\partial \mu/\partial T)_p = -S_m$ and $(\partial \mu/\partial p)_T = V_m$ for a pure phase (Eqs. 5.10.3 and 5.10.4).

By substituting from Eqs. 7.3.5 and 7.3.6 in the general expression for the total differential of μ_i, we obtain

$$d\mu_i = -S_i\, dT + V_i\, dp + \sum_i \left(\frac{\partial \mu_i}{\partial n_i}\right)_{T,p,n_{j\neq i}} dn_i \qquad (7.3.7)$$
(mixture)

Taking the partial derivatives of both sides of $U = H - pV$ with respect to n_i at constant T, p, and $n_{j\neq i}$ gives

$$U_i = H_i - pV_i \qquad (7.3.8)$$
(mixture)

Finally, we can obtain a formula for the partial molar heat capacity at constant pressure of constituent i by writing the total differential of H in the form

$$dH = \left(\frac{\partial H}{\partial T}\right)_{p,n} dT + \left(\frac{\partial H}{\partial p}\right)_{T,n} dp + \sum_i \left(\frac{\partial H}{\partial n_i}\right)_{T,p,n_{j\neq i}} dn_i$$

$$= C_p\, dT + \left(\frac{\partial H}{\partial p}\right)_{T,n} dp + \sum_i H_i\, dn_i \qquad (7.3.9)$$
(mixture)

from which we have the reciprocity relation $(\partial C_p/\partial n_i)_{T,p,n_{j\neq i}} = (\partial H_i/\partial T)_{p,n}$ or

$$C_{p,i} = \left(\frac{\partial H_i}{\partial T}\right)_{p,n} \qquad (7.3.10)$$
(mixture)

7.4 GAS MIXTURES

The gas mixtures described in this chapter are assumed to have no reactions taking place.

7.4.1 Partial pressure

The **partial pressure** p_i of substance i in a gas mixture is defined as the product of the mole fraction of the substance in the gas phase and the pressure of the phase:

$$p_i \equiv x_i p \qquad (7.4.1)$$
(gas mixture)

The sum of the partial pressures of all substances in a gas mixture is $\sum_i p_i = \sum_i x_i p = p \sum_i x_i$. Since the sum of the mole fractions of all substances in a mixture is 1, this sum becomes

$$\sum_i p_i = p \qquad (7.4.2)$$
(gas mixture)

Thus, the sum of the partial pressures equals the pressure of the gas phase. This statement is known as **Dalton's Law** and is valid for any gas mixture regardless of whether the gas obeys the ideal gas equation.

7.4.2 The ideal gas mixture

As discussed in Sec. 3.2.2, an ideal gas (whether pure or a mixture) is a gas with negligible intermolecular interactions. It obeys the ideal gas equation, $pV = nRT$ (where n in a mixture is the sum $\sum_i n_i$), and its internal energy in a closed system is a function only of temperature. The partial pressure of substance i in a mixture, from Eq. 7.4.1 and the ideal gas equation, is $p_i = x_i nRT/V$; but $x_i n$ equals n_i, giving

$$p_i = \frac{n_i RT}{V} \qquad (7.4.3)$$
(ideal gas mixture)

Equation 7.4.3 is the ideal gas equation with the partial pressure of a constituent substance substituted for the total pressure and the amount of the substance substituted for the total amount. The equation shows that the partial pressure of a substance in an ideal gas mixture is the pressure the substance by itself, with all others removed from the system, would have at the same T and V as the mixture. Note that this statement is only true for an *ideal* gas mixture. The partial pressure of a substance in a real gas mixture is often different from the pressure of the substance by itself in the same volume because the intermolecular interactions are different in the two gases.

7.4.3 Partial molar quantities in ideal gas mixtures

We need to relate the chemical potential of a constituent of a gas mixture to its partial pressure. We cannot precisely measure the absolute value of a chemical potential, but we can evaluate its value relative to the chemical potential in a particular reference state called the standard state.

The *standard state of substance i in a gas mixture* is the same as the standard state of the pure gas described in Sec. 5.9: It is the hypothetical state in which pure gaseous i has

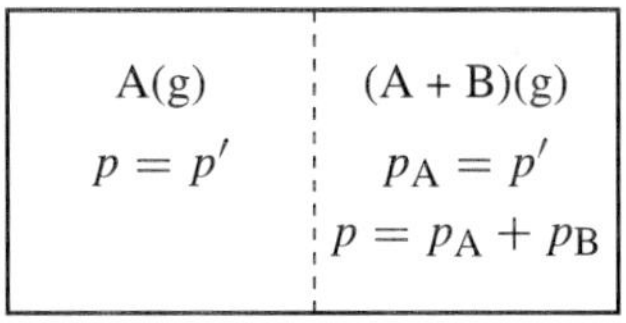

Section 7.4 Gas Mixtures

A(g)	(A + B)(g)
$p = p'$	$p_A = p'$
	$p = p_A + p_B$

Figure 7.3 System with two gas phases, pure A and a mixture, separated by a semipermeable membrane through which only A can pass. Both phases are ideal gases and have the same temperature.

the same temperature as the mixture, is at the standard pressure $p°$, and behaves as an ideal gas. The standard chemical potential $\mu_i°(g)$ of gaseous i is the chemical potential of i in this gas standard state and is a function of temperature (and possibly also of position in an external field).

To find an expression for μ_i in an ideal gas mixture relative to $\mu_i°(g)$, we make an assumption based on the following argument. Suppose we place pure A, an ideal gas, in a rigid box at pressure p'. We then slide a rigid membrane into the box so as to divide the box into two compartments. The membrane is permeable to A; that is, molecules of A pass freely through its pores. There is no reason to expect the membrane to affect the pressures on either side,[7] which remain equal to p'. Finally, without changing the volume of either compartment, we add a second gaseous substance, B, to one side of the membrane to form an ideal gas mixture, as shown in Fig. 7.3. The membrane has the property of being permeable to A but impermeable to B, so the molecules of B stay in one compartment and cause a pressure increase there. Since the mixture is an ideal gas, the molecules of A and B do not interact, and the addition of gas B causes no change in the amounts of A on either side of the membrane. Thus, the pressure of A in the pure phase and the partial pressure of A in the mixture are both equal to p'. Our assumption, then, is that the partial pressure p_A of A in an ideal gas mixture in equilibrium with pure ideal gas A is equal to the pressure of the pure gas.

Since the system shown in Fig. 7.3 is in an equilibrium state, gas A must have the same chemical potential in both phases. This is true even though the phases have different pressures (see Sec. 7.3.1). Since the chemical potential of the pure gas is given by $\mu = \mu°(g) + RT \ln(p/p°)$, and we assume that p_A in the mixture is equal to p in the pure gas, the chemical potential of A in the mixture is given by

$$\mu_A = \mu_A°(g) + RT \ln \frac{p_A}{p°}$$

In general, for any substance i in an ideal gas mixture, we have the relation

$$\mu_i = \mu_i°(g) + RT \ln \frac{p_i}{p°} \qquad (7.4.4)$$
(ideal gas mixture)

---

[7] We have to assume the gas is not adsorbed to a significant extent on the surface of the membrane or in its pores.

where $\mu_i^\circ(g)$ is the chemical potential of i in the gas standard state at the same temperature as the mixture.

Equation 7.4.4 shows that if the partial pressure of a constituent of an ideal gas mixture is equal to p°, making $\ln(p_i/p^\circ)$ equal to zero, the chemical potential is equal to the standard chemical potential. Conceptually, a standard state should be a well-defined state of the system, which in the case of a gas is the *pure* ideal gas at $p = p^\circ$. Thus, a constituent of an ideal gas mixture with a partial pressure of 1 bar is not in its standard state, although it has the same chemical potential as it does in its standard state.

We shall take Eq. 7.4.4 as the thermodynamic *definition* of an ideal gas mixture. Any gas mixture in which each constituent i obeys this relation between μ_i and p_i at all compositions is by definition an ideal gas mixture. (The nonrigorous nature of the assumption used to obtain Eq. 7.4.4 presents no difficulty if we consider the equation to be the basic definition.)

By substituting the expression for μ_i into $(\partial \mu_i/\partial T)_{p,n} = -S_i$ (Eq. 7.3.5), we obtain an expression for the partial molar entropy of substance i in an ideal gas mixture:

$$S_i = -\left[\frac{\partial \mu_i^\circ(g)}{\partial T}\right]_{p,n} - R \ln \frac{p_i}{p^\circ}$$

$$= S_i^\circ - R \ln \frac{p_i}{p^\circ} \qquad (7.4.5)$$
$$\text{(ideal gas mixture)}$$

The quantity $S_i^\circ = -[\partial \mu_i^\circ(g)/\partial T]_{p,n}$ is the **standard molar entropy** of constituent i. It is the molar entropy of i in its standard state of pure ideal gas at pressure p°, as we can deduce from Eq. 7.4.5 by noting that S_i equals S_i° if p_i is equal to p°.

Substitution of the expression for μ_i from Eq. 7.4.4 and the expression for S_i from Eq. 7.4.5 into $H_i = \mu_i + T S_i$ (from Eq. 7.3.4) yields $H_i = \mu_i^\circ(g) + T S_i^\circ$, which is equivalent to

$$H_i = H_i^\circ \qquad (7.4.6)$$
$$\text{(ideal gas mixture)}$$

This tells us that the partial molar enthalpy of a constituent of an ideal gas mixture at a given temperature is *independent* of the partial pressure.

From $(\partial \mu_i/\partial p)_{T,n} = V_i$ (Eq. 7.3.6), the partial molar volume of i in an ideal gas mixture is given by

$$V_i = \left[\frac{\partial \mu_i^\circ(g)}{\partial p}\right]_{T,n} + RT \left[\frac{\partial \ln(p_i/p^\circ)}{\partial p}\right]_{T,n}$$

The first partial derivative on the right is zero because $\mu_i^\circ(g)$ is a function only of T. For the second partial derivative, we write $p_i/p^\circ = x_i p/p^\circ$; x_i is constant when the amount of each substance is constant, so we have $[\partial \ln(x_i p/p^\circ)/\partial p]_{T,n} = 1/p$. The partial molar volume is therefore given by

$$V_i = \frac{RT}{p} \qquad (7.4.7)$$
$$\text{(ideal gas mixture)}$$

which is what we would expect simply from the ideal gas equation. The partial molar volume is not necessarily equal to the standard molar volume, which is $V_i^\circ = RT/p^\circ$ for an ideal gas.

From Eqs. 7.3.8, 7.3.10, 7.4.6, and 7.4.7 we obtain the relations

$$U_i = U_i^\circ \qquad (7.4.8)$$
(ideal gas mixture)

and

$$C_{p,i} = C_{p,i}^\circ \qquad (7.4.9)$$
(ideal gas mixture)

7.4.4 Real gas mixtures

The fugacity f of a pure gas is defined by $\mu = \mu^\circ(g) + RT \ln(f/p^\circ)$ (Eq. 5.10.6). By analogy with this equation, we define the fugacity f_i of substance i in a real gas *mixture* by the relation

$$\mu_i = \mu_i^\circ(g) + RT \ln \frac{f_i}{p^\circ} \qquad (7.4.10)$$
(gas mixture)

Just as the fugacity of a pure gas is a kind of effective pressure (Sec. 5.10.1), the fugacity of a constituent of a gas mixture is a kind of effective *partial* pressure. That is, f_i is the partial pressure substance i would have in an ideal gas mixture that is at the same temperature as the real gas mixture and in which the chemical potential of i is the same as in the real gas mixture.

To derive a relation allowing us to evaluate f_i from the pressure–volume properties of the gaseous mixture, we follow the steps described for a pure gas in Sec. 5.10.1. The temperature and composition are constant. From Eq. 7.4.10, the difference between the chemical potentials of substance i in the mixture at total pressures p and p' is

$$\mu_i - \mu_i' = RT \ln(f_i/f_i')$$

Integration of $d\mu_i = V_i \, dp$ (from Eq. 7.3.6) between these pressures yields

$$\mu_i - \mu_i' = \int_{p'}^{p} V_i \, dp$$

When we equate these two expressions for $\mu_i - \mu_i'$, divide both sides by RT, subtract the equation

$$\ln(p/p') = \int_{p'}^{p} \frac{dp}{p}$$

and take the ideal gas behavior limits $p' \to 0$ and $f_i' \to x_i p'$, we obtain

$$\ln \frac{f_i}{p_i} = \int_0^p \left(\frac{V_i}{RT} - \frac{1}{p} \right) dp \qquad (7.4.11)$$
(gas mixture, constant T)

The fugacity coefficient ϕ_i of constituent i is defined by

$$f_i = \phi_i p_i \qquad (7.4.12)$$
(gas mixture)

Accordingly, Eq. 7.4.11 becomes

$$\ln \phi_i = \int_0^p \left(\frac{V_i}{RT} - \frac{1}{p} \right) dp \qquad (7.4.13)$$
(gas mixture, constant T)

As p approaches zero, the integral in Eqs. 7.4.11 and 7.4.13 approaches zero, f_i approaches p_i, and ϕ_i approaches unity.

By combining Eqs. 7.4.10 and 7.4.11, we obtain

$$\mu_i = \mu_i^\circ(g) + RT \ln \frac{p_i}{p^\circ} + \int_0^p \left(V_i - \frac{RT}{p} \right) dp \qquad (7.4.14)$$
(gas mixture, constant T)

which is the analogue for a gas mixture of Eq. 5.11.1 for a pure gas. Section 5.11 describes the procedure needed to obtain formulas for various molar quantities of a pure gas from Eq. 5.11.1. By following a similar procedure with Eq. 7.4.14, we obtain the formulas for partial molar quantities of a constituent of a gas mixture shown in the first column of Table 7.1 on the next page. These formulas are obtained with the help of Eqs. 7.3.4, 7.3.5, 7.3.8, and 7.3.10.

The equation of state of a gas mixture can be written as the virial equation

$$p(V/n) = RT \left[1 + \frac{B}{(V/n)} + \frac{C}{(V/n)^2} + \cdots \right] \qquad (7.4.15)$$

This equation is the same as Eq. 2.1.3 for a pure gas except that the molar volume V_m is replaced by the *mean molar volume* V/n and the virial coefficients $B, C, \ldots$ depend on composition as well as temperature.

At low to moderate pressures, the simple equation of state

$$V = \frac{nRT}{p} + nB \qquad (7.4.16)$$

describes a gas mixture to a sufficiently high degree of accuracy (see Eq. 2.1.7 on page 15). If a gas mixture of constant composition obeys this equation of state, we can derive the mean molar quantities G/n, S/n, H/n, U/n, and C_p/n in the same way as described for the corresponding molar quantities of a pure gas in Sec. 5.11; for instance, the formula of Eq. 5.11.11 on page 133 becomes

$$U/n = \frac{\sum_i n_i U_i^\circ(g)}{n} - pT \frac{dB}{dT} \qquad (7.4.17)$$

Section 7.4 Gas Mixtures

Table 7.1 Partial molar quantities of constituent i of a gas mixture

General relation	Equation of state[a] $V = nRT/p + nB$
$\mu_i = \mu_i^\circ(g) + RT \ln \dfrac{p_i}{p^\circ} + \displaystyle\int_0^p \left(V_i - \dfrac{RT}{p}\right) dp$	$\mu_i = \mu_i^\circ(g) + RT \ln \dfrac{p_i}{p^\circ} + B_i' p$
$S_i = S_i^\circ(g) - R \ln \dfrac{p_i}{p^\circ} - \displaystyle\int_0^p \left[\left(\dfrac{\partial V_i}{\partial T}\right)_p - \dfrac{R}{p}\right] dp$	$S_i = S_i^\circ(g) - R \ln \dfrac{p_i}{p^\circ} - p \dfrac{dB_i'}{dT}$
$H_i = H_i^\circ(g) + \displaystyle\int_0^p \left[V_i - T\left(\dfrac{\partial V_i}{\partial T}\right)_p\right] dp$	$H_i = H_i^\circ(g) + p\left(B_i' - T \dfrac{dB_i'}{dT}\right)$
$U_i = U_i^\circ(g) + \displaystyle\int_0^p \left[V_i - T\left(\dfrac{\partial V_i}{\partial T}\right)_p\right] dp + RT - pV_i$	$U_i = U_i^\circ(g) - pT \dfrac{dB_i'}{dT}$
$C_{p,i} = C_{p,i}^\circ(g) - \displaystyle\int_0^p T\left(\dfrac{\partial^2 V_i}{\partial T^2}\right)_p dp$	$C_{p,i} = C_{p,i}^\circ(g) - pT \dfrac{d^2 B_i'}{dT^2}$

[a] B and B_i' are defined by Eqs. 7.4.19 and 7.4.21.

Equation 7.4.17 is needed in Prob. 8.7(m).

From statistical mechanical theory, the dependence of the second virial coefficient B of a binary gas mixture on the mole fraction composition is given by

$$B = x_A^2 B_{AA} + 2x_A x_B B_{AB} + x_B^2 B_{BB} \tag{7.4.18}$$
(binary gas mixture)

where B_{AA} and B_{BB} are the second virial coefficients of pure A and B and B_{AB} is a mixed second virial coefficient. B_{AA}, B_{BB}, and B_{AB} are functions of T only. For a mixture of any number of gases, the composition dependence of B is given by

$$B = \sum_i \sum_j x_i x_j B_{ij} \tag{7.4.19}$$
(gas mixture, $B_{kl} = B_{lk}$)

If a gas mixture obeys the equation of state of Eq. 7.4.16, the partial molar volume of constituent i is given by

$$V_i = \left(\frac{\partial V}{\partial n_i}\right)_{T,p,n_{j \neq i}} = \frac{RT}{p} + B_i' \tag{7.4.20}$$

where the quantity B_i' is defined by

$$B_i' \equiv \left[\frac{\partial (nB)}{\partial n_i}\right]_{T,p,n_{j \neq i}} \tag{7.4.21}$$

To relate B'_i to the pure gas and mixed second virial coefficients, we write

$$nB = n \sum_k \sum_l x_k x_l B_{kl} = \frac{\sum_k \sum_l n_k n_l B_{kl}}{\sum_m n_m}$$

and carry out the differentiation of Eq. 7.4.21. The result is

$$B'_i = 2 \sum_j x_j B_{ij} - B \tag{7.4.22}$$

For the constituents of a binary mixture, this becomes

$$B'_A = B_{AA} + (-B_{AA} + 2B_{AB} - B_{BB})x_B^2$$
$$B'_B = B_{BB} + (-B_{AA} + 2B_{AB} - B_{BB})x_A^2 \tag{7.4.23}$$
(binary gas mixture)

If we substitute the expression of Eq. 7.4.20 for V_i in Eq. 7.4.13, we obtain a relation between the fugacity coefficient of constituent i and the function B'_i:

$$\ln \phi_i = \frac{B'_i p}{RT} \tag{7.4.24}$$

The second column of Table 7.1 gives formulas for various partial molar quantities of constituent i in terms of B'_i and its temperature derivative. The formulas are the same as Eqs. 5.11.8–5.11.12 for molar quantities of a *pure* gas, with B'_i replacing the second virial coefficient B.

7.5 LIQUID AND SOLID MIXTURES OF NONELECTROLYTES

Homogeneous liquid and solid mixtures are condensed phases of variable composition. Most of the discussion of condensed-phase mixtures in this section focuses on liquids. The same principles, however, apply to homogeneous solid mixtures, often called solid solutions. These solids include most metal alloys, many gemstones, and doped semiconductors.

7.5.1 Raoult's law

Suppose we equilibrate a liquid mixture of the volatile substances A and B with a gas phase, as shown in Fig. 7.4(a) on the next page. The gas includes an inert constituent C, of negligible solubility in the liquid, whose partial pressure we can adjust to keep the total pressure constant as we vary the liquid composition. (We might, for instance, allow the liquid to equilibrate with air at atmospheric pressure.) The observed partial pressure of A in the gas phase of this system is p_A. In a second system, shown in Fig. 7.4(b), we equilibrate pure liquid A with a gas phase that also includes constituent C. The observed partial pressure of A in the second system, which is the vapor pressure of the liquid, is p_A^*.[8]

---

[8] Section 9.7.1 shows that p_A^* is practically the same as the saturation pressure of liquid A; that is, the pressure at which the *pure* liquid and gas phases of A are in equilibrium.

Section 7.5 Liquid and Solid Mixtures of Nonelectrolytes

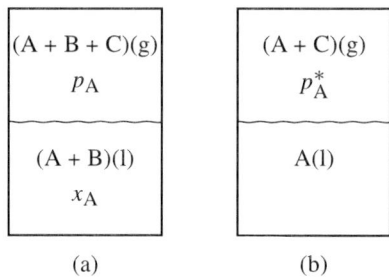

Figure 7.4 Equilibrated liquid and gas phases. (a) A binary liquid mixture in equilibrium with a gas mixture. (b) Pure liquid A in equilibrium with a gas mixture at the same temperature and pressure as (a).

Both systems are at the same temperature and same total pressure. If, as we vary the mole fraction x_A of A in the liquid mixture while keeping T and p constant, we find the partial pressure obeys the relation $p_A = x_A p_A^*$, we say that constituent A obeys **Raoult's law**.

For any constituent i of a liquid mixture, Raoult's law is

$$p_i = x_i p_i^* \tag{7.5.1}$$

(Raoult's law for partial pressure)

where x_i is the mole fraction of i in the liquid and p_i^* is the partial pressure in a gas phase equilibrated with pure liquid i at the same T and p. This relation might hold at all liquid compositions, or over only a limited range.

In an ideal gas mixture, the fugacity of each substance equals its partial pressure. If substance i obeys Raoult's law and the equilibrated gas phase is an ideal gas, then not just the partial pressure but also the fugacity is proportional to the mole fraction in the liquid:

$$f_i = x_i f_i^* \tag{7.5.2}$$

(Raoult's law for fugacity)

We can understand the meaning of the quantity f_i^* in this equation by noting that if x_i equals 1 (the liquid is pure i), then the fugacity of i in the gas mixture is f_i^*. Thus, f_i^* is the fugacity of i in a gas mixture equilibrated with pure liquid i at the temperature and pressure of interest; f_i^* is a function of both T and p.

The original form of Raoult's law was expressed with partial pressures in the form given by Eq. 7.5.1. The relation $f_i = x_i f_i^*$ (Eq. 7.5.2) expressed with fugacities is often a more useful form. The two forms, of course, are identical if the gas phase is an ideal gas mixture. When it is necessary to make a distinction between these two forms, this book refers specifically to Raoult's law for partial pressure or Raoult's law for fugacity.

We saw in Sec. 7.3.1 that if substance i can transfer between a liquid and a gas phase, its chemical potential μ_i must be the same in both equilibrated phases. Its chemical potential in the gas phase is $\mu_i(g) = \mu_i^\circ(g) + RT \ln f_i/p^\circ$ (Eq. 7.4.10). By substituting f_i from Eq.

7.5.2 in this expression and rearranging, we obtain

$$\mu_i = \left[\mu_i^\circ(g) + RT \ln \frac{f_i^*}{p^\circ}\right] + RT \ln x_i$$

The expression enclosed in brackets is independent of the mixture composition. We replace the expression by a quantity μ_i^*, a function of T and p, and write:

$$\mu_i = \mu_i^* + RT \ln x_i \qquad (7.5.3)$$
$$\text{(Raoult's law for fugacity)}$$

By setting x_i equal to 1, we see that μ_i^* is the chemical potential of pure liquid i at the temperature and pressure of the mixture. Equation 7.5.3 is valid for any constituent whose fugacity obeys Raoult's law and can be used as an alternative statement of Raoult's law for fugacity.

7.5.2 Ideal mixtures

We define an *ideal liquid mixture* as a liquid mixture in which *each* constituent obeys Raoult's law for fugacity (Eq. 7.5.2 or 7.5.3) at a given temperature and pressure and over a given range of composition. Equation 7.5.2 applies only to a volatile constituent, whereas Eq. 7.5.3 applies regardless of whether the constituent is volatile.

Raoult's "law" is a misnomer. In reality, few liquid mixtures are found to approximate the behavior of an ideal liquid mixture. To do so, the constituents must have similar molecular size and structure, and the pure liquids must be miscible in all proportions. Benzene and toluene, for instance, satisfy these requirements, and liquid mixtures of benzene and toluene are found to obey Raoult's law quite closely. In contrast, water and methanol, although miscible in all proportions, form liquid mixtures that deviate considerably from Raoult's law. The most commonly encountered situation for mixtures of organic liquids is that each constituent deviates from Raoult's law behavior by having a *higher* fugacity than predicted by Eq. 7.5.2—a *positive* deviation from Raoult's law.

Similar statements apply to ideal *solid* mixtures. In addition, Eq. 7.5.3 describes the chemical potential of each constituent of an ideal *gas* mixture, as the following derivation shows.

In an ideal gas mixture at a given T and p, the chemical potential of substance i is given by Eq. 7.4.4:

$$\mu_i = \mu_i^\circ(g) + RT \ln \frac{p_i}{p^\circ} = \mu_i^\circ(g) + RT \ln \frac{x_i p}{p^\circ}$$

The chemical potential of the pure ideal gas ($x_i = 1$) is

$$\mu_i^* = \mu_i^\circ(g) + RT \ln \frac{p}{p^\circ}$$

By eliminating $\mu_i^\circ(g)$ between these equations and rearranging, we obtain Eq. 7.5.3.

Thus, an **ideal mixture**, whether a liquid, solid, or gas, is a mixture in which the logarithm of the chemical potential μ_i of each constituent i, at a given T and p, is a linear function of the mole fraction x_i:

$$\mu_i = \mu_i^* + RT \ln x_i \tag{7.5.4}$$
(ideal mixture)

If we are able to increase x_i to 1 at constant T and p while the phase remains an ideal mixture, then μ_i^* is the chemical potential of pure i in the same physical state as the mixture. Otherwise μ_i^* represents the chemical potential of pure i in an unstable phase or hypothetical reference state.

7.5.3 Partial molar quantities in ideal mixtures

Using Eq. 7.5.4 for the chemical potential of a constituent of an ideal mixture, we now find expressions for partial molar quantities of the constituent. These expressions find their greatest use for ideal liquid and solid mixtures.

For the partial molar entropy of substance i, we have $S_i = -(\partial \mu_i / \partial T)_{p,n}$ (from Eq. 7.3.5) or, for the ideal mixture,

$$S_i = -\left(\frac{\partial \mu_i^*}{\partial T}\right)_p - R \ln x_i = S_i^* - R \ln x_i \tag{7.5.5}$$
(ideal mixture)

Since $\ln x_i$ is negative, the partial molar entropy of a constituent of an ideal mixture is greater than the molar entropy of the pure substance at the same T and p.

For the partial molar enthalpy, we have $H_i = \mu_i + TS_i$ (from Eq. 7.3.4). Using the expressions for μ_i and S_i gives us

$$H_i = \mu_i^* + TS_i^* = H_i^* \tag{7.5.6}$$
(ideal mixture)

Thus, H_i in an ideal mixture is independent of the mixture composition and is equal to the molar enthalpy of pure i at the same T and p as the mixture. In the case of an ideal *gas* mixture, H_i is also independent of p since the molar enthalpy of an ideal gas depends only on T.

The partial molar volume is given by $V_i = (\partial \mu_i / \partial p)_{T,n}$ (Eq. 7.3.6), so we have

$$V_i = \left(\frac{\partial \mu_i^*}{\partial p}\right)_T = V_i^* \tag{7.5.7}$$
(ideal mixture)

Finally, from Eqs. 7.3.8 and 7.3.10 and the prior expressions for H_i and V_i, we obtain

$$U_i = H_i^* - pV_i^* = U_i^* \tag{7.5.8}$$
(ideal mixture)

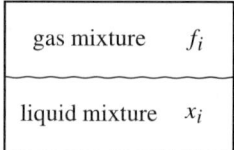

Figure 7.5 Equilibrated liquid and gas mixtures. Substance i is present in both phases.

and
$$C_{p,i} = (\partial H_i^*/\partial T)_{p,n} = C_{p,i}^* \qquad (7.5.9)$$
(ideal mixture)

Note that neither V_i, U_i, nor $C_{p,i}$ varies with the composition of the ideal mixture.

7.5.4 Henry's law

Consider the system shown in Fig. 7.5, in which a gas mixture is equilibrated with a liquid mixture. Transfer equilibrium exists for substance i, a constituent of both mixtures. The system is similar to that shown in Fig. 7.4(a). We can vary the mole fraction x_i of i in the liquid and measure the fugacity f_i of i in the gas phase.

If we allow x_i to approach zero while keeping the relative amounts of the other liquid constituents constant, and provided substance i has the same molecular form in both phases (and is not, for instance, an electrolyte), we find experimentally that the fugacity f_i becomes proportional to x_i:
$$f_i \to K_{x,i} x_i \quad \text{as} \quad x_i \to 0$$

This behavior is called **Henry's law**. $K_{x,i}$ is a **Henry's law constant** whose value depends on the temperature and the total pressure, and also on the mole fractions of the constituents other than i in the liquid mixture. If the liquid happens to be an ideal liquid mixture, Eq. 7.5.2 shows that f_i is proportional to x_i at *all* values of x_i, and $K_{x,i}$ is then equal to f_i^*, the fugacity when the gas phase is equilibrated with pure liquid i.

If we treat the liquid mixture as a solution and i is a volatile nonelectrolyte solute B, Henry's law behavior is found in the limit of infinite dilution:
$$f_B \to K_{x,B} x_B \quad \text{as} \quad x_A \to 1$$

An example of this behavior is shown in Fig. 7.6(a) on the next page. The limiting slope of the plot of f_B versus x_B is finite, not zero or infinite. (The fugacity of a volatile *electrolyte*, such as HCl dissolved in water, displays a much different behavior as we see in Sec. 7.8.)

Since the mole fraction, concentration, and molality of a solute become proportional to one another in the limit of infinite dilution (Eq. 7.1.10), in a very dilute solution the fugacity is proportional to all three of these composition variables. This leads to three versions of

Section 7.5 Liquid and Solid Mixtures of Nonelectrolytes

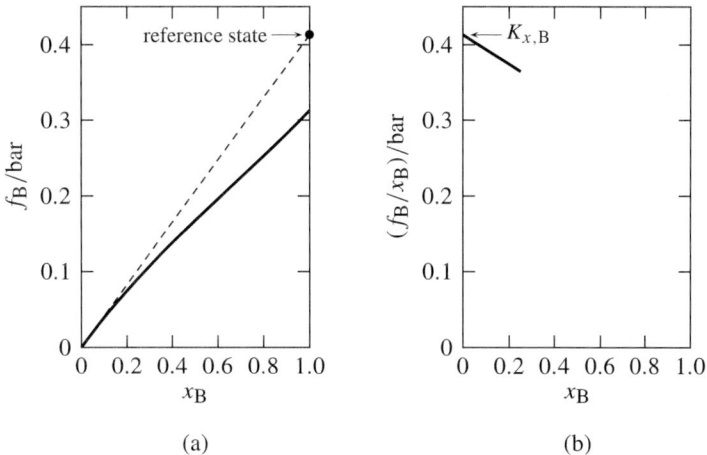

Figure 7.6 Liquid solutions of 2,3-dimethylbutane (B) in cyclooctane at 298.15 K and 1 bar. (a) Fugacity of B in an equilibrated gas phase as a function of solution composition. The dashed line, tangent to the curve at $x_B = 0$, is Henry's law behavior and its slope is $K_{x,B}$. (b) Fugacity divided by mole fraction as a function of composition; the limiting value at $x_B = 0$ is the Henry's law constant $K_{x,B}$.

Henry's law:

$$\text{mole fraction basis} \quad f_B = K_{x,B}\, x_B \quad (7.5.10)$$
(nonelectrolyte solute at infinite dilution)

$$\text{concentration basis} \quad f_B = K_{c,B}\, c_B \quad (7.5.11)$$
(nonelectrolyte solute at infinite dilution)

$$\text{molality basis} \quad f_B = K_{m,B}\, m_B \quad (7.5.12)$$
(nonelectrolyte solute at infinite dilution)

where $K_{x,B}$, $K_{c,B}$, and $K_{m,B}$ are Henry's law constants defined by[9]

$$\text{mole fraction basis} \quad K_{x,B} \equiv \lim_{x_A \to 1} \frac{f_B}{x_B} \quad (7.5.13)$$

$$\text{concentration basis} \quad K_{c,B} \equiv \lim_{x_A \to 1} \frac{f_B}{c_B} \quad (7.5.14)$$

$$\text{molality basis} \quad K_{m,B} \equiv \lim_{x_A \to 1} \frac{f_B}{m_B} \quad (7.5.15)$$

---

[9]The second edition of the IUPAC Green Book (1993) recommends using the same symbol, $k_{H,B}$, for all three of these Henry's law constants. Here, however, we use the three symbols recommended in the first edition of the IUPAC Green Book (1988) to make it easy to distinguish among them.

Note that the Henry's law constants are not dimensionless, and are functions of T and p. To evaluate one of these constants, we can plot the fugacity of B divided by a composition variable as a function of the composition variable and extrapolate to infinite dilution. The evaluation of $K_{x,B}$ by this procedure is illustrated in Fig. 7.6(b).

We can derive a general relation between the values of any two of these Henry's law constants with the use of Eqs. 7.1.10 and 7.5.10–7.5.12. For instance, the equalities $V^*_{m,A} c_B = M_A m_B$ (from Eq. 7.1.10) and $K_{c,B} c_B = K_{m,B} m_B$ (from Eqs. 7.5.11 and 7.5.12) give us

$$K_{c,B} = \frac{V^*_{m,A}}{M_A} K_{m,B}$$

7.5.5 The ideal-dilute solution

An **ideal-dilute solution** is a solution whose composition is in a dilute range that allows each solute to obey Henry's law.[10] On the microscopic level, the requirement is that solute molecules be sufficiently separated to make solute–solute interactions negligible.

Within the composition range that a solution behaves as an ideal-dilute solution, then, the fugacity of Solute B in a gas phase equilibrated with the solution varies with composition according to Eqs. 7.5.10–7.5.12. The chemical potential of B in the gas phase, which is equal to that of B in the liquid, is related to the fugacity by $\mu_B = \mu_B^\circ(g) + RT \ln(f_B/p^\circ)$ (Eq. 7.4.10). Substituting $f_B = K_{x,B} x_B$ (Eq. 7.5.10) into this equation, we obtain

$$\mu_B = \mu_B^\circ(g) + RT \ln \frac{K_{x,B} x_B}{p^\circ}$$

$$= \left[\mu_B^\circ(g) + RT \ln \frac{K_{x,B}}{p^\circ} \right] + RT \ln x_B$$

where the composition variable x_B is segregated in the last term on the right side. The expression in brackets is a function of T and p, but not of composition, and represents the chemical potential of B in a hypothetical solute reference state.

We use the symbol $\mu_{x,B}^{\text{ref}}$ for the chemical potential of the solute in this reference state. The x in the subscript reminds us that the reference state is based on mole fraction. The equation then becomes

$$\mu_B(T, p) = \mu_{x,B}^{\text{ref}}(T, p) + RT \ln x_B \qquad (7.5.16)$$
(ideal-dilute solution
of a nonelectrolyte)

Here the notation emphasizes the fact that μ_B and $\mu_{x,B}^{\text{ref}}$ are functions of T and p.

Equation 7.5.16, derived using fugacity, is valid even if the solute has such low volatility that its fugacity in an equilibrated gas phase is too low to measure. In principle, no solute is completely nonvolatile, and there is always a finite solute fugacity in the gas phase even if immeasurably small.

---

[10]Note that an ideal-dilute solution is not necessarily an ideal mixture.

It is worthwhile to describe in detail the reference state to which $\mu_{x,\text{B}}^{\text{ref}}$ refers. The general concept is also applicable to other solute reference states and solute standard states we encounter presently.

Imagine that the solution continues to exhibit the ideal-dilute behavior specified by Eqs. 7.5.10 and 7.5.16 as x_B increases beyond the ideal-dilute range at constant T and p. That is, we treat the solute fugacity as if it continues to obey Henry's law and increases as a linear function of x_B, although in reality it does not (unless we happen to be dealing with an ideal mixture). The reference state is the state of this hypothetical solution reached in the extrapolation at $x_\text{B} = 1$. An example is indicated by the filled circle in Fig. 7.6(a).

Equation 7.5.16 shows that, as x_B approaches 1 in the hypothetical solution, μ_B approaches $\mu_{x,\text{B}}^{\text{ref}}$. Thus, $\mu_{x,\text{B}}^{\text{ref}}(T, p)$ is the chemical potential of B in a fictitious reference state in which the mole fraction of B is unity and B behaves as in an ideal-dilute solution at temperature T and pressure p. This state is sometimes called the *ideal-dilute solution of unit solute mole fraction*.

By similar steps, combining Henry's law based on concentration or molality (Eqs. 7.5.11 and 7.5.12) with the relation $\mu_\text{B} = \mu_\text{B}^\circ(\text{g}) + RT \ln(f_\text{B}/p^\circ)$, we obtain for the solute chemical potential in the ideal-dilute range the equations

$$\mu_\text{B} = \mu_\text{B}^\circ(\text{g}) + RT \ln\left(\frac{K_{c,\text{B}}\, c_\text{B}}{p^\circ} \cdot \frac{c^\circ}{c^\circ}\right) = \left[\mu_\text{B}^\circ(\text{g}) + RT \ln \frac{K_{c,\text{B}}\, c^\circ}{p^\circ}\right] + RT \ln \frac{c_\text{B}}{c^\circ}$$

$$\mu_\text{B} = \mu_\text{B}^\circ(\text{g}) + RT \ln\left(\frac{K_{m,\text{B}}\, m_\text{B}}{p^\circ} \cdot \frac{m^\circ}{m^\circ}\right) = \left[\mu_\text{B}^\circ(\text{g}) + RT \ln \frac{K_{m,\text{B}}\, m^\circ}{p^\circ}\right] + RT \ln \frac{m_\text{B}}{m^\circ}$$

Note how in each equation we multiply and divide the argument of a logarithm by a constant, c° or m°, to allow the arguments of the resulting logarithms to be dimensionless. These constants are called *standard compositions* with the following values:

standard concentration $c^\circ = 1\ \text{mol dm}^{-3}$ (equal to one mole per liter, or one molar)

standard molality $m^\circ = 1\ \text{mol kg}^{-1}$ (equal to one molal)

Again in each of these equations, we replace the expression in brackets, which depends on T and p but not on composition, with the chemical potential of a solute reference state:

$$\mu_\text{B}(T, p) = \mu_{c,\text{B}}^{\text{ref}}(T, p) + RT \ln \frac{c_\text{B}}{c^\circ} \qquad (7.5.17)$$
(ideal-dilute solution
of a nonelectrolyte)

$$\mu_\text{B}(T, p) = \mu_{m,\text{B}}^{\text{ref}}(T, p) + RT \ln \frac{m_\text{B}}{m^\circ} \qquad (7.5.18)$$
(ideal-dilute solution
of a nonelectrolyte)

The quantities $\mu_{c,\text{B}}^{\text{ref}}$ and $\mu_{m,\text{B}}^{\text{ref}}$ are the chemical potentials of the solute in hypothetical reference states that are solutions of standard concentration and standard molality, respectively, behaving as ideal-dilute solutions at the temperature and pressure of interest.

For consistency with these equations, we can rewrite Eq. 7.5.16 in the form

$$\mu_B(T, p) = \mu_{x,B}^{\text{ref}}(T, p) + RT \ln \frac{x_B}{x^\circ}$$

with the definition

standard mole fraction $x^\circ = 1$.

7.5.6 Solvent behavior in the ideal-dilute solution

We now use the Gibbs–Duhem equation to derive the behavior of the solvent in an ideal-dilute solution of a nonelectrolyte. The Gibbs–Duhem equation applies to changes in a mixture at constant T and p and has the general form $\sum_i x_i \text{d}X_i = 0$ (Eq. 7.2.11), where X_i is any partial molar quantity. With μ_i as the partial molar quantity, the equation becomes

$$\sum_i x_i \, \text{d}\mu_i = 0 \qquad (7.5.19)$$
$$\text{(constant } T \text{ and } p\text{)}$$

We use subscript A for the solvent, rewrite Eq. 7.5.19 as

$$x_A \, \text{d}\mu_A + \sum_{i \neq A} x_i \, \text{d}\mu_i = 0$$

and rearrange to

$$\text{d}\mu_A = -\frac{1}{x_A} \sum_{i \neq A} x_i \, \text{d}\mu_i \qquad (7.5.20)$$
$$\text{(constant } T \text{ and } p\text{)}$$

This equation shows how changes in the solute chemical potentials, due to a composition change at constant T and p, affect the chemical potential of the solvent.

In an ideal-dilute solution, the chemical potential of solute i is given by $\mu_i = \mu_{x,i}^{\text{ref}} + RT \ln x_i$ (Eq. 7.5.16), and the differential of μ_i at constant T and p is

$$\text{d}\mu_i = RT \, \text{d} \ln x_i = RT \, \text{d}x_i / x_i$$

(Here we have used the fact that $\mu_{x,i}^{\text{ref}}$ is constant at constant T and p.) When we substitute this expression for $\text{d}\mu_i$ in Eq. 7.5.20, we obtain

$$\text{d}\mu_A = -\frac{RT}{x_A} \sum_{i \neq A} \text{d}x_i$$

Now since the sum of all mole fractions is 1, we have the equation $\sum_{i \neq A} x_i = 1 - x_A$ whose differential is $\sum_{i \neq A} \text{d}x_i = -\text{d}x_A$. Making this substitution in the preceding displayed equation gives us

$$\text{d}\mu_A = \frac{RT}{x_A} \text{d}x_A = RT \, \text{d} \ln x_A$$

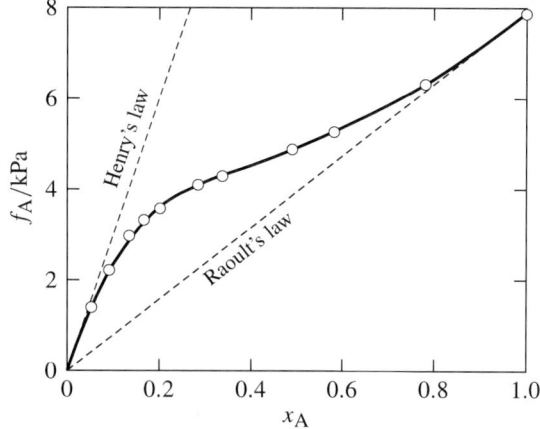

Figure 7.7 Fugacity of ethanol in a gas phase equilibrated with a binary liquid mixture of ethanol (A) and H_2O at 25 °C and 1 bar (Dobson, H. J. E. *J. Chem. Soc.* **1925**, *127*, 2866–2873). The dashed lines show Henry's law behavior and Raoult's law behavior.

Consider a process in which we add small amounts of the solute or solutes to the solvent at constant T and p, causing both μ_A and x_A to decrease. We assume the solution formed in this process is in the ideal-dilute solution range and integrate the preceding equation over the path:

$$\int_{\mu_A^*}^{\mu_A} d\mu_A = RT \int_{x_A=1}^{x_A} d\ln x_A$$

The result is $\mu_A - \mu_A^* = RT \ln x_A$ or

$$\mu_A = \mu_A^* + RT \ln x_A \tag{7.5.21}$$
(ideal-dilute solution of a nonelectrolyte)

which is Raoult's law for fugacity given by Eq. 7.5.3.

Thus, in an ideal-dilute solution of nonelectrolytes, *each solute obeys Henry's law and the solvent obeys Raoult's law.*

An equivalent statement is that a nonelectrolyte in a mixture approaches Henry's law behavior as its mole fraction approaches zero and approaches Raoult's law behavior as its mole fraction approaches unity. This is illustrated by the behavior of ethanol in its mixtures with water shown in Fig. 7.7. The ethanol exhibits positive deviations from Raoult's law and negative deviations from Henry's law.

7.5.7 Partial molar quantities in an ideal-dilute solution

Let us first consider the *solvent* of a solution that is dilute enough to be in the ideal-dilute range. In this range, the solvent fugacity obeys Raoult's law, and the partial molar quantities

Table 7.2 Partial molar quantities of solvent and nonelectrolyte solute in an ideal-dilute solution

Solvent	Solute
$\mu_A = \mu_A^* + RT \ln x_A$	$\mu_B = \mu_{x,B}^{\text{ref}} + RT \ln x_B$
	$ = \mu_{c,B}^{\text{ref}} + RT \ln(c_B/c^\circ)$
	$ = \mu_{m,B}^{\text{ref}} + RT \ln(m_B/m^\circ)$
$S_A = S_A^* - R \ln x_A$	$S_B = S_{x,B}^{\text{ref}} - R \ln x_B$
	$ = S_{c,B}^{\text{ref}} - R \ln(c_B/c^\circ)$
	$ = S_{m,B}^{\text{ref}} - R \ln(m_B/m^\circ)$
$H_A = H_A^*$	$H_B = H_B^\infty$
$V_A = V_A^*$	$V_B = V_B^\infty$
$U_A = U_A^*$	$U_B = U_B^\infty$
$C_{p,A} = C_{p,A}^*$	$C_{p,B} = C_{p,B}^\infty$

of the solvent are the same as those in an ideal mixture. Formulas for these quantities were given in Eqs. 7.5.4–7.5.9 and are collected in the first column of Table 7.2. The formulas show that the chemical potential and partial molar entropy of the solvent, at constant T and p, vary with the solution composition and, in the limit of infinite dilution ($x_A \to 1$), approach the values for the pure solvent. The partial molar enthalpy, volume, internal energy, and heat capacity, on the other hand, are constant in the ideal-dilute region and are equal to the corresponding molar values for the pure solvent.

Next we consider a *solute*, B, of a binary ideal-dilute solution. The solute obeys Henry's law, and its chemical potential is given by $\mu_B = \mu_{x,B}^{\text{ref}} + RT \ln x_B$ (Eq. 7.5.16), where $\mu_{x,B}^{\text{ref}}$ is a function of T and p, but not of composition. μ_B varies with the composition and goes to $-\infty$ as the solution becomes infinitely dilute ($x_B \to 0$).

For the partial molar entropy of the solute, we use $S_B = -(\partial \mu_B / \partial T)_{p,n}$ (Eq. 7.3.5) and obtain

$$S_B = -\left(\frac{\partial \mu_{x,B}^{\text{ref}}}{\partial T}\right)_p - R \ln x_B$$

The partial derivative $-(\partial \mu_{x,B}^{\text{ref}} / \partial T)_p$ represents the partial molar entropy $S_{x,B}^{\text{ref}}$ of B in the fictitious reference state of unit solute mole fraction. Thus, we write the expression in the form

$$S_B = S_{x,B}^{\text{ref}} - R \ln x_B \qquad (7.5.22)$$

(ideal-dilute solution of a nonelectrolyte)

This equation shows that the partial molar entropy varies with composition and goes to $+\infty$ in the limit of infinite dilution. From the expressions of Eqs. 7.5.17 and 7.5.18, we can derive similar expressions for S_B in terms of the solute reference states on a concentration or molality basis.

From the relation $H_B = \mu_B + T S_B$ (Eq. 7.3.4), combined with Eqs. 7.5.16 and 7.5.22, we have

$$H_B = \mu_{x,B}^{\text{ref}} + T S_{x,B}^{\text{ref}} = H_{x,B}^{\text{ref}}$$

showing that the partial molar enthalpy of the solute is constant throughout the ideal-dilute solution range. Therefore, we can write

$$H_B = H_B^{\infty}$$

where H_B^{∞} is the partial molar enthalpy at infinite dilution. By similar reasoning, using Eqs. 7.3.6–7.3.10, we find the partial molar volume, internal energy, and heat capacity of the solute are constant in the ideal-dilute range and equal to the values at infinite dilution. The expressions are listed in the second column of Table 7.2 on the preceding page.

7.6 ACTIVITY COEFFICIENTS

An *activity coefficient* is a kind of adjustment factor that relates the actual behavior of a species to ideal behavior.

7.6.1 Ideal mixtures

To understand the concept of an activity coefficient, first consider the following expressions for chemical potentials in ideal mixtures and ideal-dilute solutions of nonelectrolytes. We use the notation μ^{id} for these chemical potentials.

Constituent of an ideal gas mixture (Eq. 7.4.4) $\qquad \mu_i^{\text{id}} = \mu_i^{\circ}(g) + RT \ln \dfrac{p_i}{p^{\circ}}$

Constituent of an ideal liquid or solid mixture (Eq. 7.5.4) $\qquad \mu_i^{\text{id}} = \mu_i^{*} + RT \ln x_i$

Solvent of an ideal-dilute solution (Eq. 7.5.21) $\qquad \mu_A^{\text{id}} = \mu_A^{*} + RT \ln x_A$

Solute, ideal-dilute solution, mole fraction basis (Eq. 7.5.16) $\qquad \mu_B^{\text{id}} = \mu_{x,B}^{\text{ref}} + RT \ln x_B$

Solute, ideal-dilute solution, concentration basis (Eq. 7.5.17) $\qquad \mu_B^{\text{id}} = \mu_{c,B}^{\text{ref}} + RT \ln \dfrac{c_B}{c^{\circ}}$

Solute, ideal-dilute solution, molality basis (Eq. 7.5.18) $\qquad \mu_B^{\text{id}} = \mu_{m,B}^{\text{ref}} + RT \ln \dfrac{m_B}{m^{\circ}}$

Note that each equation has the general form

$$\mu_i^{\text{id}} = \mu_i^{\text{ref}} + RT \ln \left(\frac{\text{composition variable}}{\text{standard composition}} \right)$$

(Here μ_i^{ref} for a gas is the standard chemical potential, and for a mixture constituent or a solvent is the chemical potential of the pure substance.)

7.6.2 Real mixtures

If a mixture is *not* ideal, we can write expressions for the chemical potential having a form similar to the one just given for μ_i^{id}, with each composition variable now multiplied by an **activity coefficient**:

$$\mu_i = \mu_i^{\text{ref}} + RT \ln \left[(\text{activity coefficient of } i) \times \left(\frac{\text{composition variable}}{\text{standard composition}} \right) \right] \quad (7.6.1)$$

The activity coefficient of a species is a dimensionless quantity whose value depends on temperature, pressure, and mixture composition, and also on our choice of the reference state for the species. Under conditions in which the mixture behaves ideally, the activity coefficient is 1; otherwise, the activity coefficient has the value that gives the correct chemical potential.

We shall use various symbols for activity coefficients depending on the nature of the reference state. For nonelectrolytes, the expressions having the form above are as follows:

Constituent of a gas mixture	$\mu_i = \mu_i^\circ(g) + RT \ln \left(\phi_i \dfrac{p_i}{p^\circ} \right)$	(7.6.2)
Constituent of a liquid or solid mixture	$\mu_i = \mu_i^* + RT \ln (\gamma_i x_i)$	(7.6.3)
Solvent of a solution	$\mu_A = \mu_A^* + RT \ln (\gamma_A x_A)$	(7.6.4)
Solute of a solution, mole fraction basis	$\mu_B = \mu_{x,B}^{\text{ref}} + RT \ln (\gamma_{x,B}\, x_B)$	(7.6.5)
Solute of a solution, concentration basis	$\mu_B = \mu_{c,B}^{\text{ref}} + RT \ln \left(\gamma_{c,B} \dfrac{c_B}{c^\circ} \right)$	(7.6.6)
Solute of a solution, molality basis	$\mu_B = \mu_{m,B}^{\text{ref}} + RT \ln \left(\gamma_{m,B} \dfrac{m_B}{m^\circ} \right)$	(7.6.7)

Equation 7.6.2 uses ϕ_i as the activity coefficient of substance i in a gas mixture. This is the same as the *fugacity coefficient* defined by $\phi_i = f_i/p_i$ (Eq. 7.4.12), as we can see by comparing Eqs. 7.4.10 and 7.6.2.

Equation 7.6.3 refers to a constituent of a liquid or solid mixture of substances that mix in all proportions. Equation 7.6.4 refers to the solvent of a solution. The reference states of these substances are the pure liquid or solid at the temperature and pressure of the mixture. For the activity coefficient of such a substance, we use the symbol γ_i or γ_A.[11]

---

[11] The IUPAC Green Book recommends the symbol f_i for the activity coefficient when the reference state is the pure liquid or solid. This book uses γ_i instead to avoid confusion with f_i, the usual symbol for fugacity.

Section 7.6 Activity Coefficients

The symbols for the activity coefficients of a nonelectrolyte solute, $\gamma_{x,B}$, $\gamma_{c,B}$, and $\gamma_{m,B}$, indicate the choice of the solute reference state (Eqs. 7.6.5–7.6.7). While three expressions for μ_B are shown, they must all give the same *value* of μ_B. Recall that the chemical potential of a species is the partial molar Gibbs energy—the rate at which G increases with the amount of the species added at constant T and p. The chemical potential depends on the temperature and pressure of the mixture, and also on its composition. Although we cannot precisely measure or calculate an absolute value of a chemical potential, in principle it does have a definite value for given conditions, and this value is independent of our choice of a reference state.

A change in the choice of a reference state of a species does, however, usually affect the value of the activity coefficient.

Looking at the right sides of Eqs. 7.6.2–7.6.7, we can think of $\phi_i p_i$ (the fugacity) as an *effective* partial pressure; of $\gamma_i x_i$, $\gamma_A x_A$, and $\gamma_{x,B} x_B$ as *effective* mole fractions; of $\gamma_{c,B} c_B$ as an *effective* concentration; and of $\gamma_{m,B} m_B$ as an *effective* molality. That is, these are the values of the composition variables that would give the same chemical potential in an ideal mixture as the actual chemical potential in the real mixture. These effective composition variables are an alternative way to express the escaping tendency of a substance from a phase; they are related exponentially to the chemical potential, which is also a measure of escaping tendency (Sec. 7.3).

The change at constant temperature that causes the behavior of a mixture constituent to approach ideal behavior must cause the activity coefficient to approach unity:

$$\phi_i \to 1 \quad \text{as} \quad p \to 0 \tag{7.6.8}$$
(gas, constant T)

$$\gamma_i \to 1 \quad \text{as} \quad x_i \to 1 \tag{7.6.9}$$
(constant T and p)

$$\gamma_A \to 1 \quad \text{as} \quad x_A \to 1 \tag{7.6.10}$$
(solvent, constant T and p)

$$\gamma_{x,B} \to 1 \quad \text{as} \quad x_A \to 1 \tag{7.6.11}$$
(solute, constant T and p)

$$\gamma_{c,B} \to 1 \quad \text{as} \quad x_A \to 1 \tag{7.6.12}$$
(solute, constant T and p)

$$\gamma_{m,B} \to 1 \quad \text{as} \quad x_A \to 1 \tag{7.6.13}$$
(solute, constant T and p)

7.6.3 Nonideal dilute solutions

How would we expect the activity coefficient of a nonelectrolyte solute to behave in a dilute solution as the solute mole fraction increases beyond the ideal-dilute solution range? We go through the following argument based on molecular properties at constant T and p.

We focus our attention on a single solute molecule. This molecule has interactions with nearby solute molecules. Each interaction depends on the intermolecular distance and causes a change in the internal energy compared to the interaction with solvent at the same distance.[12] The number of solute molecules in a volume element at a given distance from the solute molecule we are focusing on is proportional to the local solute concentration. If the solution is dilute and the interactions weak, we expect the local solute concentration to be proportional to the macroscopic solute mole fraction. Thus, the partial molar quantities U_B and V_B of the solute should be approximately linear functions of x_B in a dilute solution.

We assume the dependence of S_B on x_B is approximately the same as in an ideal mixture, as predicted by statistical mechanics if the solvent and solute molecules have similar sizes and shapes. Then, since the solute chemical potential is given by $\mu_B = U_B + pV_B - TS_B$ (from Eqs. 7.3.4 and 7.3.8), we expect its deviation from the ideal-dilute behavior given by Eq. 7.5.16 on page 196 can be described by adding a term proportional to x_B:

$$\mu_B = \mu_{x,B}^{\text{ref}} + RT \ln x_B + k_x x_B$$

where k_x is a positive or negative constant related to solute-solute interactions.

If we equate this expression for μ_B with the one that defines the activity coefficient, $\mu_B = \mu_{x,B}^{\text{ref}} + RT \ln(\gamma_{x,B} x_B)$ (Eq. 7.6.5) and solve for the activity coefficient, we obtain[13]

$$\gamma_{x,B} = \exp(k_x x_B / RT)$$

An expansion of the exponential in powers of x_B converts this to $\gamma_{x,B} = 1 + (k_x/RT)x_B + \cdots$. Thus, we predict $\gamma_{x,B}$ is a linear function of x_B at low x_B. An ideal-dilute solution, then, is one in which x_B is much smaller than RT/k_x. An ideal mixture (Eq. 7.5.4) requires the interaction constant k_x to be zero.

By similar reasoning, we reach analogous conclusions for solute activity coefficients on a concentration or molality basis. For instance, at low m_B, the chemical potential of B should be approximately $\mu_{m,B}^{\text{ref}} + RT \ln(m_B/m°) + k_m m_B$, where k_m is a constant; then the activity coefficient at low m_B is given by

$$\gamma_{m,B} = \exp(k_m m_B / RT) = 1 + (k_m/RT)m_B = \cdots \tag{7.6.14}$$

The prediction that a solute activity coefficient in a dilute solution is a linear function of the composition variable is borne out by experiment, as illustrated later in Figs. 7.9 and 7.10. This prediction applies only to a nonelectrolyte solute; for an electrolyte, the slope of activity coefficient versus molality approaches $-\infty$ at low molality (see page 222).

---

[12] In Sec. 8.2.5, it is shown that roughly speaking the internal energy change is negative if the average of the attractive forces between two solute molecules and two solvent molecules is greater than the attractive force between a solute molecule and a solvent molecule at the same distance and positive for the opposite situation.

[13] This is essentially the result of the McMillan–Mayer solution theory from statistical mechanics.

7.6.4 Activity coefficients from gas fugacities

If we equilibrate a gas phase with a liquid mixture containing a nonelectrolyte constituent that is sufficiently volatile to give a measurable partial pressure in the gas, we have a convenient way to evaluate the activity coefficient of the constituent in the liquid. This is the experimental arrangement shown schematically in Fig. 7.5 on page 194. For instance, we can equilibrate the liquid mixture with air at a controlled total pressure. If the air has negligible solubility in the liquid, it will have little effect on the activity coefficient. At a reasonably low pressure, the volatile constituent's fugacity will be approximately equal to its partial pressure, equal in turn to the product of the easily measured quantities mole fraction and total pressure.[14] We know the chemical potential of the constituent must be the same in both phases.

In the liquid mixture, the chemical potential of substance i is given by $\mu_i = \mu_i^* + RT \ln(\gamma_i x_i)$ using a pure-liquid reference state or a similar relation involving one of the solute reference states. The chemical potential in the gas phase is $\mu_i = \mu_i^\circ(g) + RT \ln(f_i/p^\circ)$. Equating these expressions for μ_i and solving for γ_i, we obtain

$$\gamma_i = \exp\left[\frac{\mu_i^\circ(g) - \mu_i^*}{RT}\right] \times \frac{f_i}{x_i p^\circ}$$

The quantities enclosed in the brackets are not functions of the composition, and p° is a constant, so we can write

$$\gamma_i = C \frac{f_i}{x_i}$$

where C is a constant at a given T and p. For a solvent or solute, we obtain the analogous relations $\gamma_A = Cf_A/x_A$, $\gamma_{x,B} = Cf_B/x_B$, $\gamma_{c,B} = Cf_B/(c_B/c^\circ)$, and $\gamma_{m,B} = Cf_B/(m_B/m^\circ)$ (the constant C has a different value in each case).

How do we determine the proportionality constant C? When i is in its reference state in the liquid, the activity coefficient is 1 and the composition variable equals the standard composition. Each of the relations involving C can then be arranged to

$$C = \frac{1}{f_i^{\text{ref}}}$$

where f_i^{ref} is the fugacity of i in a gas phase equilibrated with a liquid phase with i in its reference state. Depending on the choice of reference state, we calculate the activity coefficient as follows.

For a *constituent of a liquid mixture* with a pure-liquid reference state, f_i^{ref} is f_i^*, the fugacity of i in a gas phase equilibrated with pure liquid i at the temperature and pressure of the mixture. Therefore, the activity coefficient is given by

$$\gamma_i = \frac{f_i}{x_i f_i^*} \qquad (7.6.15)$$

---

[14] For a precise evaluation of the fugacity, we can use the methods of Sec. 7.4.4.

The *solvent of a solution* also has a pure-liquid reference state, and so its activity coefficient is

$$\gamma_A = \frac{f_A}{x_A f_A^*} \qquad (7.6.16)$$

For a solute with a reference state based on *mole fraction*, we obtain f_B^{ref} by extrapolating the Henry's law behavior $f_B = K_{x,B} x_B$, observed in the ideal-dilute solution, to $x_B = 1$. $K_{x,B}$ is the Henry's law constant of B on a mole fraction basis at the temperature and pressure of the solution. Therefore, the activity coefficient is given by

$$\gamma_{x,B} = \frac{f_B}{K_{x,B} x_B} \qquad (7.6.17)$$

Similarly, for a solute reference state based on *concentration*, f_B^{ref} is the extrapolation of $f_B = K_{c,B} c_B$ to $c_B = c°$. This gives

$$\gamma_{c,B} = \frac{f_B}{K_{c,B} c_B} \qquad (7.6.18)$$

Finally, for a solute standard state based on *molality*, we have

$$\gamma_{m,B} = \frac{f_B}{K_{m,B} m_B} \qquad (7.6.19)$$

These relations all equate the activity coefficient of a component of a liquid mixture to the ratio of the actual fugacity in the gas phase and the fugacity in an ideal mixture (or ideal-dilute solution) of the same composition. So the activity coefficient is a correction factor by which the fugacity calculated for an ideal mixture can be converted to the actual fugacity. Here are examples of the evaluation of γ_A, $\gamma_{x,B}$, and $\gamma_{m,B}$.

Ethanol and water at 25 °C mix in all proportions, so we can use pure-liquid reference states for both components. A plot of ethanol fugacity versus mole fraction at fixed T and p, shown earlier in Fig. 7.7, is repeated in Fig. 7.8(a) on the next page. Ethanol is component A. In the figure, the filled circle is the pure-liquid reference state at $x_A = 1$, where f_A is equal to f_A^*. The open circles at $x_A = 0.4$ indicate the actual fugacity f_A in a gas phase equilibrated with a liquid mixture of this composition, and the fugacity $x_A f_A^*$, which the ethanol would have if the mixture were ideal. The ratio of these two quantities is the activity coefficient γ_A.

In the figure, the ethanol fugacity in the gas phase equilibrated with pure ethanol is $f_A^* = 7.89$ kPa (0.0789 bar). The ethanol fugacity in the gas equilibrated with an ethanol–water mixture of composition $x_A = 0.4$ is $f_A = 4.52$ kPa. The ethanol fugacity for an ideal mixture of this composition would be, according to Raoult's law for fugacity, $x_A f_A^* = 0.4 \times 4.54$ kPa $= 3.16$ kPa. Therefore, the activity coefficient is $\gamma_A = 4.52$ kPa$/3.16$ kPa $= 1.43$. This value is shown as an open circle in Fig. 7.8(b).

Figure 7.8(b) shows how γ_A varies with composition. Note how γ_A approaches 1 as x_A approaches 1, as it must according to Eq. 7.6.9.

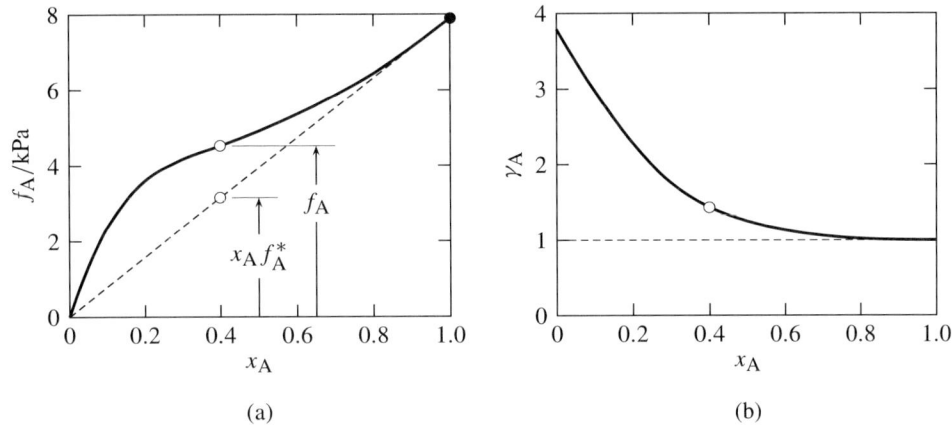

Figure 7.8 Liquid mixtures of ethanol (A) and H_2O at 25 °C and 1 bar. (a) Ethanol fugacity as a function of mixture composition. The dashed line is Raoult's law behavior, and the filled circle is the reference state. (b) Ethanol activity coefficient as a function of mixture composition.

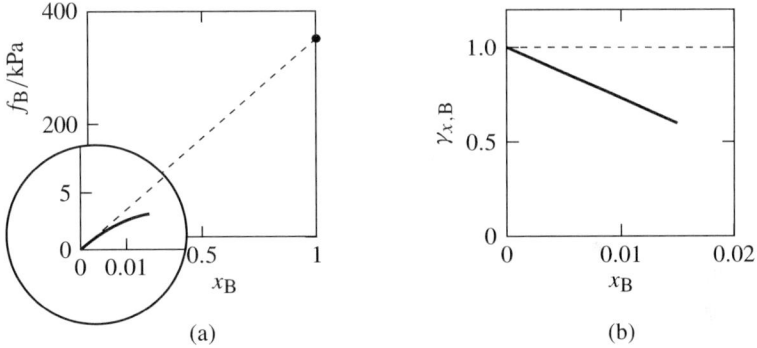

Figure 7.9 Behavior of 1-butanol (B) in dilute aqueous mixtures at 50.08 °C and 1 bar (Fischer, K.; Gmehling, J. *J. Chem. Eng. Data* **1994**, *39*, 309–315). (a) 1-butanol fugacity in an equilibrated gas phase as a function of mole fraction in the liquid phase, measured up to the solubility limit at $x_B = 0.015$. The dilute region is shown in a magnified view. The dashed line is Henry's law behavior on a mole fraction basis. The filled circle indicates the solute reference state based on mole fraction. (b) Activity coefficient as a function of mole fraction.

Water and 1-butanol are two liquids that do not mix in all proportions; that is, 1-butanol has limited solubility in water. Figures 7.9(a) and 7.10(a) show the fugacity of 1-butanol plotted as functions of both mole fraction and molality. The figures demonstrate how, treating 1-butanol as a solute, we find a solute reference state by a linear extrapolation of the fugacity to a standard composition. The fugacity f_B is quite different for the two reference

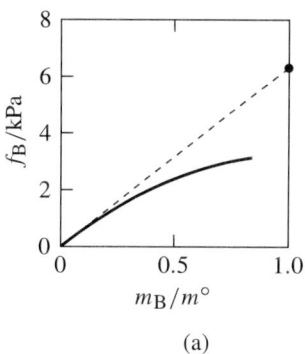 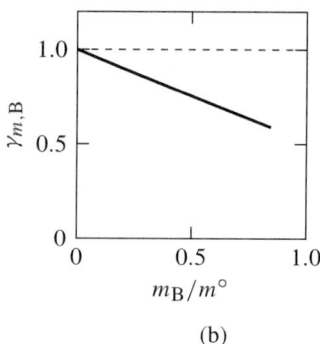

Figure 7.10 Dilute aqueous solutions of 1-butanol (same experimental data as in Fig. 7.9). (a) 1-butanol fugacity in an equilibrated gas phase as a function of molality in the liquid phase, measured up to the solubility limit at $m_B = 0.85\,\text{mol}\,\text{kg}^{-1}$. The dashed line is Henry's law behavior on a molality basis. The filled circle indicates the solute reference state on this basis. (b) Activity coefficient as a function of molality.

states. At the reference state indicated by a filled circle in Fig. 7.9(a), f_B equals the Henry's law constant $K_{x,B}$; at the reference state in Fig. 7.10(a), f_B equals $K_{m,B}m°$. Note how the activity coefficients plotted in Figs. 7.9(b) and 7.10(b) approach 1 at infinite dilution (in agreement with Eqs. 7.6.11 and 7.6.13) and how they vary linearly with x_B or m_B in the dilute solution as predicted by the theoretical argument of Sec. 7.6.3.

7.6.5 Activity coefficients from the Gibbs–Duhem equation

How can we measure the activity coefficient of a liquid mixture constituent that has low volatility? In the case of a binary mixture, if the other constituent is volatile enough for fugacity measurements over a range of liquid composition, we may use the Gibbs–Duhem equation.

First we consider a binary mixture of two liquids that mix in all proportions. We assume only one of these constituents, A, is appreciably volatile. By measuring the fugacity of A in a gas phase equilibrated with the binary mixture, we can evaluate its activity coefficient based a pure-liquid reference state with the relation $\gamma_A = f_A/x_A f_A^*$ (Eq. 7.6.16). We wish to use our fugacity measurements to determine the activity coefficient of the nonvolatile component, B, as well.

For a binary mixture, the Gibbs–Duhem equation in the form given by Eq. 7.5.19 is

$$x_A \, d\mu_A + x_B \, d\mu_B = 0$$

where $d\mu_A$ and $d\mu_B$ refer to changes at constant T and p. Now from $\mu_A = \mu_A^* + RT \ln(\gamma_A x_A)$ (Eq. 7.6.3), we obtain

$$d\mu_A = RT \, d\ln \gamma_A + RT \, d\ln x_A = RT \, d\ln \gamma_A + \frac{RT}{x_A} dx_A$$

For component B, we obtain in the same way $d\mu_B = RT\, d\ln \gamma_B + RT\, dx_B/x_B$; since dx_B is equal to $-dx_A$, this becomes

$$d\mu_B = RT\, d\ln \gamma_B - \frac{RT}{x_B} dx_A$$

Substituting these expressions for $d\mu_A$ and $d\mu_B$ in the Gibbs–Duhem equation and solving for $d\ln \gamma_B$, we obtain

$$d\ln \gamma_B = -\frac{x_A}{x_B} d\ln \gamma_A$$

Integration from $x_B = 1$, where γ_B equals 1 and $\ln \gamma_B$ equals 0, to another composition x'_B gives

$$\ln \gamma_B(x'_B) = -\int_{x_B=1}^{x_B=x'_B} \frac{x_A}{x_B} d\ln \gamma_A \qquad (7.6.20)$$
$$\text{(binary mixture, constant } T \text{ and } p)$$

Next consider a binary liquid mixture in which constituent B is neither volatile nor able to mix in all proportions with A. In this case, it is appropriate to treat B as a solute and to base its activity coefficient on a solute reference state. We could obtain an expression for $\ln \gamma_{x,B}$ similar to Eq. 7.6.20, but the integration would have to start at $x_B = 0$, where the integrand x_A/x_B would be infinite. Instead, it is convenient to use the method described in the next section.

7.6.6 Activity coefficients from osmotic coefficients

It is customary to evaluate the activity coefficient of a nonvolatile solute with a function ϕ_m called the **molal osmotic coefficient**, or osmotic coefficient on a molality basis, defined by

$$\phi_m \equiv \frac{\mu_A^* - \mu_A}{RT M_A \sum_{i \neq A} m_i} \qquad (7.6.21)$$
$$\text{(solution)}$$

The sum in the denominator on the right side is the total molality of all solute species.

The definition of ϕ_i has the following significance. In an ideal-dilute solution, the solvent chemical potential is $\mu_A = \mu_A^* + RT \ln x_A = \mu_A^* + RT \ln(1 - \sum_{i \neq A} x_i)$. In the limit of infinite dilution, the mole fraction of solute i becomes $x_i = M_A m_i$ (see Eq. 7.1.10). The expansion of the function $\ln(1-X)$ in powers of X gives the power series $\ln(1-X) = -X - X^2/2 - X^3/3 - \cdots$. Thus, in the limit of infinite dilution, the solvent chemical potential is related to solute molalities by

$$\mu_A = \mu_A^* - RT M_A \sum_{i \neq A} m_i$$

The deviation of ϕ_m from unity is a measure of the deviation of μ_A from infinite-dilution behavior, as we can see by comparing the preceding equation with a rearrangement of Eq. 7.6.21:

$$\mu_A = \mu_A^* - \phi_m RT M_A \sum_{i \neq A} m_i$$

The function ϕ_m is called the osmotic coefficient because Π/m_B for a solution is equal to ϕ_m times the limiting value of Π/m_B at infinite dilution, where Π is the osmotic pressure. This is explained in Sec. 9.4.4.

Any method that measures $\mu_A^* - \mu_A$, the lowering of the solvent chemical potential caused by the presence of the solute, allows us to evaluate ϕ_m through Eq. 7.6.21. These methods include solvent fugacity measurements described in Sec. 7.6.7 and freezing point and osmotic pressure measurements described in Sec. 9.2.

Suppose we have a solution of a nonelectrolyte solute B whose activity coefficient $\gamma_{m,B}$ we wish to evaluate as a function of m_B. For a binary solution, Eq. 7.6.21 becomes

$$\phi_m = \frac{\mu_A^* - \mu_A}{RT M_A m_B} \qquad (7.6.22)$$
(binary nonelectrolyte solution)

Solving for μ_A and taking its differential at constant T and p, we obtain

$$d\mu_A = -RT M_A \, d(\phi_m m_B) = -RT M_A (\phi_m \, dm_B + m_B \, d\phi_m)$$

From $\mu_B = \mu_{m,B}^{\text{ref}} + RT \ln(\gamma_{m,B} m_B/m°)$ (Eq. 7.6.7), we obtain

$$d\mu_B = RT \, d\ln \frac{\gamma_{m,B} m_B}{m°} = RT \left(d\ln \gamma_{m,B} + \frac{dm_B}{m_B} \right)$$

We substitute these expressions for $d\mu_A$ and $d\mu_B$ in the Gibbs–Duhem equation in the form given by Eq. 7.2.10, $n_A \, d\mu_A + n_B \, d\mu_B = 0$, make the substitution $n_A M_A = n_B/m_B$ (from Eq. 7.1.6 on page 172), and obtain after some rearrangement the relation

$$d\ln \gamma_{m,B} = d\phi_m + \frac{\phi_m - 1}{m_B} \, dm_B$$

We integrate both sides of this equation for a composition change at constant T and p from $m_B = 0$ (where $\ln x_B$ is 0 and ϕ_m is 1) to any desired value of m_B, with the result

$$\ln \gamma_{m,B}(m_B) = \phi_m(m_B) - 1 + \int_0^{m_B} \frac{\phi_m - 1}{m_B} \, dm_B \qquad (7.6.23)$$
(binary nonelectrolyte solution)

Once we know ϕ_m as a function of molality from zero up to the molality of interest, we may evaluate the solute activity coefficient $\gamma_{m,B}$ with Eq. 7.6.23. Suppose, to take a

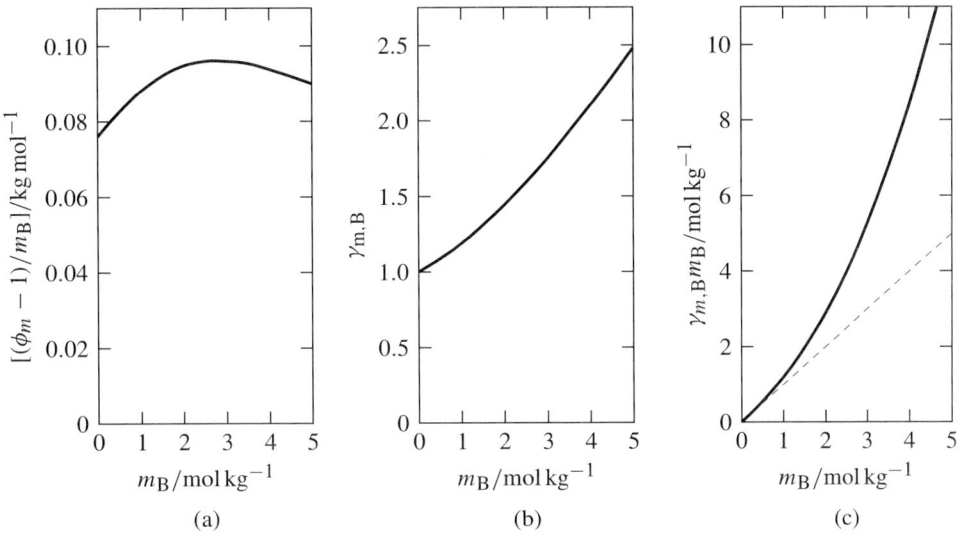

Figure 7.11 Aqueous sucrose solutions at 25 °C. (a) Integrand of the integral in Eq. 7.6.23 as a function of solution composition. (b) Solute activity coefficient on a molality basis. (c) Product of activity coefficient and molality as a function of composition. The dashed line is the extrapolation of ideal-dilute behavior.

simple example, that in a range of dilute molalities of a binary solution the solvent chemical potential is given by

$$\mu_A = \mu_A^* - RT M_A m_B (1 + a m_B)$$

where the parameter a depends only on T. Then from Eq. 7.6.22 the molal osmotic coefficient has the functional form

$$\phi_m = \frac{\mu_A^* - \mu_A}{RT M_A m_B} = 1 + a m_B$$

and Eq. 7.6.23 becomes

$$\ln \gamma_{m,B} = 2 a m_B$$

which is of the form predicted by Eq. 7.6.14 on page 204 for a solute activity coefficient in dilute solution.

When the solvent chemical potential varies in a more complicated way with molality, we can evaluate the integral on the right side of Eq. 7.6.23 by numerical integration. The integrand $(\phi_m - 1)/m_B$ is a slowly varying function of m_B and approaches a finite value as m_B approaches zero if the solute is a nonelectrolyte.

Figure 7.11(a) shows the function $(\phi_m - 1)/m_B$ for aqueous sucrose solutions over a wide range of molality at 25 °C; ϕ_m was measured by the isopiestic vapor pressure method described in the next section. Equation 7.6.23 was used to generate the dependence of the solute activity coefficient on molality shown in Fig. 7.11(b). Figure 7.11(c) shows the effective sucrose molality $\gamma_{m,B}m_B$ as a function of composition. Note how the activity coefficient becomes greater than unity beyond the ideal-dilute region, and how consequently the effective molality becomes considerably greater than the actual molality.

7.6.7 Evaluation of the osmotic coefficient

Values of the molal osmotic coefficient ϕ_m, defined by Eq. 7.6.21, over a range of molalities are needed to determine a solute activity coefficient. Making the substitutions $\mu_A = \mu_A^\circ(g) + RT \ln(f_A/p^\circ)$ and $\mu_A^* = \mu_A^\circ(g) + RT \ln(f_A^*/p^\circ)$ gives us

$$\phi_m = \frac{\ln(f_A^*/f_A)}{M_A \sum_{i \neq A} m_i}$$

which allows us to evaluate ϕ_m for a solution of any composition from the fugacity of the solvent in an equilibrated gas phase.

Several experimental methods are available for measuring solvent fugacities in gas phases equilibrated with the pure solvent and with the solution. If the solute is nonvolatile, we may pump out the air above the solution and use a manometer to measure the pressure, which is the partial pressure of the solvent. Dynamic methods involve saturating a stream of inert gas with the solvent vapor and analyzing the gas mixture to evaluate the solvent partial pressure. For instance, we could pass dry air successively through an aqueous solution and a desiccant and measure the weight gained by the desiccant. A correction for gas nonideality, such as that given by Eq. 7.4.11, may be needed to convert p_A to f_A.

One of the most useful methods for aqueous solutions of nonvolatile solutes with molalities greater than about $0.1 \, \text{mol kg}^{-1}$ is the **isopiestic vapor pressure** technique. This is a comparative method using solutions of two different solutes. Dishes, each containing water and an accurately weighed sample of one of the solutes, are placed in wells drilled in a block made of metal for good thermal equilibration. The assembly is placed in a gas-tight chamber, the air is evacuated, and the apparatus is gently rocked in a thermostat for a period up to several days. During this period, H_2O is transferred among the dishes through the vapor space until the chemical potential of the water is equal in each solution. The solutions are then said to be *isopiestic*. Finally the dishes are removed from the apparatus and weighed to establish the molality of each solution. If the H_2O fugacity is known as a function of the molality of one of the solutes, the measurements give the relation between fugacity and molality for the other solute. Some commonly used reference solutes for which data are available are sucrose, NaCl, $CaCl_2$, and H_2SO_4.

The isopiestic vapor pressure method can also be used for nonaqueous solutions.

7.7 ACTIVITIES

The **activity** a_i of Species i is defined by the relation

$$a_i = \exp\left(\frac{\mu_i - \mu_i^\circ}{RT}\right) \tag{7.7.1}$$

or

$$\mu_i = \mu_i^\circ + RT \ln a_i \tag{7.7.2}$$

where μ_i°, the standard chemical potential, is the chemical potential of Species i in a standard state.[15]

An important application of the activity concept is in the definition of equilibrium constants (Sec. 8.8.1).

The activity of a species is a dimensionless quantity whose value depends on the temperature, pressure, and composition of the mixture and on our choice of a standard state for the species. When the species is in its standard state, its chemical potential is the standard chemical potential. It is important to note from Eq. 7.7.2 that for μ_i to equal μ_i°, the logarithm of the activity must be zero and the activity in the standard state is therefore 1.

7.7.1 Standard states of mixture constituents

The standard states used for the constituents of a mixture are the same as the reference states, with the additional requirement that the pressure is the standard pressure p°. The choices of standard states are as follows.

Constituent of a gas mixture: The standard state is the hypothetical state in which the substance is a pure gas at the same temperature as the mixture and at pressure p°, and the gas behaves as an ideal gas. (This repeats the definition given in Sec. 7.4.3.)

Constituent of a liquid or solid mixture of substances that mix in all proportions or the solvent of a solution: The standard state is the pure substance in the same physical state (liquid or solid) and at the same temperature as the mixture, at pressure p°. You will recognize this as the same standard state we use for a pure liquid or solid (Sec. 5.9). If the pure liquid or solid is stable at p° and the temperature of the mixture, the standard state is real rather than hypothetical. The standard state is called a *pure-liquid*, *pure-solid*, or *solvent* standard state as appropriate.

Solute species in a solution: The standard state of solute B is based on a composition variable, usually x_B, c_B, or m_B. The standard state is the state of B in a hypothetical solution in which the temperature is the same as the solution under consideration; the pressure is p°; B behaves according to Henry's law based on the particular composition variable; and the composition variable has its standard value $x^\circ = 1$, $c^\circ = 1 \text{ mol dm}^{-3}$, or

---

[15] Some chemists define the activity by $\mu_i = \mu_i^{\text{ref}} + RT \ln a_i$. The activity defined this way is not the same as the activity used in this book unless the species has a gas standard state or the system is at the standard pressure.

$m° = 1\,\mathrm{mol\,kg^{-1}}$. (These hypothetical solute standard states are the same as the solute reference states described in Sec. 7.5.5 with the pressure fixed at the standard pressure.) If the solute species is an ion, we must specify that the hypothetical solution is electrically neutral (see Sec. 7.8.1).

For a solute of a solution, we can use subscripts to indicate the basis of the standard state, just as we did for solute activity coefficients in Sec. 7.6.2:

$$\text{mole fraction basis} \quad \mu_B = \mu°_{x,B} + RT \ln a_{x,B} \qquad (7.7.3)$$

$$\text{concentration basis} \quad \mu_B = \mu°_{c,B} + RT \ln a_{c,B} \qquad (7.7.4)$$

$$\text{molality basis} \quad \mu_B = \mu°_{m,B} + RT \ln a_{m,B} \qquad (7.7.5)$$

In Sec. 7.5.7, we found that the values of the partial molar enthalpy, volume, internal energy, and heat capacity of a solute are independent of composition in an ideal-dilute solution and are equal to the values at infinite dilution. Since a solute in any of the solute standard states behaves as in an ideal-dilute solution at the standard pressure, it has the infinite-dilution values of these partial molar quantities:

$$H°_B = H^\infty_B (p = p°) \qquad (7.7.6)$$
$$V°_B = V^\infty_B (p = p°)$$
$$U°_B = U^\infty_B (p = p°)$$
$$C°_{p,B} = C^\infty_{p,B} (p = p°)$$

Now we need to relate activities to mixture composition. If we substitute in Eq. 7.7.1 the general expression for μ_i given by Eq. 7.6.1 on page 202, we obtain

$$a_i = \exp\left(\frac{\mu_i^{\text{ref}} - \mu°_i}{RT}\right)\left[(\text{activity coefficient of } i) \times \left(\frac{\text{composition variable}}{\text{standard composition}}\right)\right] \qquad (7.7.7)$$

The exponential factor on the right side depends on our choice of a reference state and a standard state, but not on the mixture composition. We represent this factor by the symbol Γ_i introduced by Pitzer and Brewer:[16]

$$\Gamma_i \equiv \exp\left(\frac{\mu_i^{\text{ref}} - \mu°_i}{RT}\right) \qquad (7.7.8)$$

Equation 7.7.7 then becomes

$$a_i = \Gamma_i\left[(\text{activity coefficient of } i) \times \left(\frac{\text{composition variable}}{\text{standard composition}}\right)\right] \qquad (7.7.9)$$

Definitions of standard states and explicit expressions for activities are collected in Table 7.3 on the next page. For completeness, the table includes the case of an ion solute species, which is discussed in Sec. 7.8.

---

[16] Lewis, G. N.; Randall, M.; revised by Pitzer, K. S.; Brewer, L. *Thermodynamics*, 2nd ed.; McGraw-Hill: New York, 1961, p. 249.

Section 7.7 Activities

Table 7.3 Standard states and activities

Component or species	Standard state[a]	Activity[b]
Pure gas	The pure gas at pressure $p°$, behaving as an ideal gas[c]	$a = \phi \dfrac{p}{p°} = \dfrac{f}{p°}$
Pure liquid or solid	The pure liquid or solid at pressure $p°$	$a = \Gamma$
Substance i in a gas mixture	Pure i at pressure $p°$, behaving as an ideal gas[c]	$a_i = \phi_i \dfrac{p_i}{p°} = \dfrac{f_i}{p°}$
Nonelectrolyte i in a liquid or solid mixture	Pure i in the same physical state as the mixture, at pressure $p°$	$a_i = \Gamma_i \gamma_i x_i = \Gamma_i \dfrac{f_i}{f_i^*}$
Solvent A of a solution	Pure A in the same physical state as the solution, at pressure $p°$	$a_\mathrm{A} = \Gamma_\mathrm{A} \gamma_\mathrm{A} x_\mathrm{A} = \Gamma_\mathrm{A} \dfrac{f_\mathrm{A}}{f_\mathrm{A}^*}$
Nonelectrolyte solute B, mole fraction basis	B at pressure $p°$ and mole fraction 1, behavior extrapolated from infinite dilution on a mole fraction basis[c]	$a_{x,\mathrm{B}} = \Gamma_{x,\mathrm{B}} \gamma_{x,\mathrm{B}} x_\mathrm{B} = \Gamma_{x,\mathrm{B}} \dfrac{f_\mathrm{B}}{K_{x,\mathrm{B}}}$
Nonelectrolyte solute B, concentration basis	B at pressure $p°$ and concentration $c°$, behavior extrapolated from infinite dilution on a concentration basis[c]	$a_{c,\mathrm{B}} = \Gamma_{c,\mathrm{B}} \gamma_{c,\mathrm{B}} \dfrac{c_\mathrm{B}}{c°} = \Gamma_{c,\mathrm{B}} \dfrac{f_\mathrm{B}}{K_{c,\mathrm{B}} c°}$
Nonelectrolyte solute B, molality basis	B at pressure $p°$ and molality $m°$, behavior extrapolated from infinite dilution on a molality basis[c]	$a_{m,\mathrm{B}} = \Gamma_{m,\mathrm{B}} \gamma_{m,\mathrm{B}} \dfrac{m_\mathrm{B}}{m°} = \Gamma_{m,\mathrm{B}} \dfrac{f_\mathrm{B}}{K_{m,\mathrm{B}} m°}$
Ion solute species	The ion species in an electrically neutral solution at pressure $p°$ and molality $m°$, behavior extrapolated from infinite dilution on a molality basis[c]	$a_+ = \Gamma_+ \gamma_+ \dfrac{m_+}{m°} \quad a_- = \Gamma_- \gamma_- \dfrac{m_-}{m°}$
Electrolyte solute substance B	B at pressure $p°$ and molality $m°$, behavior extrapolated from infinite dilution on a molality basis[c]	$a_{m,\mathrm{B}} = \nu_+^{\nu_+} \nu_-^{\nu_-} \Gamma_{m,\mathrm{B}} \gamma_\pm^\nu \left(\dfrac{m_\mathrm{B}}{m°}\right)^\nu$

[a]In each standard state, the temperature is the same as that of the phase under consideration.
[b]For a nonelectrolyte constituent of a condensed phase mixture, f_i, f_A, and f_B refer to the fugacity in a gas phase equilibrated with the condensed phase.
[c]A hypothetical standard state.

7.7.2 The pressure factor Γ_i

What is the significance of the quantity Γ_i defined by Eq. 7.7.8? Suppose Species i is in its reference state; then μ_i equals μ_i^{ref} and, according to Eq. 7.6.1, the factor shown in brackets in Eq. 7.7.9 is unity. Equation 7.7.9 in this situation then reads $a_i = \Gamma_i$. So Γ_i is the activity of Species i in its reference state.

We know that the activity of a species in its *standard* state is always 1, so Γ_i is 1 when the reference state is the same as the standard state. The value of Γ_i for a *gas* constituent is 1 at *all* pressures because the reference and standard states of a gas are identical.

The only way the reference state of a constituent of a liquid or solid mixture can be different from the standard state is if the pressure is different from the standard pressure. In this case, Γ_i is different from 1. Since Γ_i depends on pressure, we call it a *pressure factor* for activity. It is a function of T and p, but not of the mixture composition.

We proceed as follows to evaluate the quantity $\mu_i^{\text{ref}} - \mu_i^{\circ}$ appearing in the definition of Γ_i, to derive formulas for Γ_i. When the reference state of i undergoes an isothermal pressure change from p to p°, it becomes the standard state of i. From the relation $(\partial \mu_i/\partial p)_{T,n} = V_i$ (Eq. 7.3.6), we write $\mathrm{d}\mu_i = V_i\,\mathrm{d}p$ and integrate from p° to p at constant temperature and composition:

$$\mu_i^{\text{ref}} - \mu_i^{\circ} = \int_{p^{\circ}}^{p} \mathrm{d}\mu_i = \int_{p^{\circ}}^{p} V_i\,\mathrm{d}p \qquad (7.7.10)$$

The partial molar volume V_i appropriate to the reference state is, for a pure-liquid, pure-solid, or solvent reference state, the molar volume of the pure substance, $V_{\text{m},i}^*$. For a solute, it is the partial molar volume at infinite dilution, V_{B}^{∞} (see Table 7.2 on page 200).

The expression for $\mu_i^{\text{ref}} - \mu_i^{\circ}$ in the case of a solute reference state on a *concentration* basis is not quite the same as given in Eq. 7.7.10 because c_{B} varies (slightly) with pressure at constant temperature and composition. From the relation $\mu_{\text{B}} = \mu_{c,\text{B}}^{\text{ref}} + RT\ln(c_{\text{B}}/c^{\circ})$, valid for an ideal-dilute solution (Table 7.2), we find

$$\left(\frac{\partial \mu_{\text{B}}}{\partial p}\right)_{T,n} = \left(\frac{\partial \mu_{c,\text{B}}^{\text{ref}}}{\partial p}\right)_{T} + RT\left[\frac{\partial \ln(c_{\text{B}}/c^{\circ})}{\partial p}\right]_{T,n} \qquad (7.7.11)$$

The partial derivative $(\partial \mu_{\text{B}}/\partial p)_{T,n}$ is equal to the partial molar volume V_{B} (Eq. 7.3.6), which has its infinite-dilution value V_{B}^{∞} in the solute reference state. We rewrite the second partial derivative on the right side of Eq. 7.7.11 as follows:

$$\left[\frac{\partial \ln(c_{\text{B}}/c^{\circ})}{\partial p}\right]_{T,n} = \frac{1}{c_{\text{B}}}\left(\frac{\partial c_{\text{B}}}{\partial p}\right)_{T,n} = \frac{1}{n_{\text{B}}/V}\left(\frac{\partial (n_{\text{B}}/V)}{\partial p}\right)_{T,n}$$

$$= -\frac{1}{V}\left(\frac{\partial V}{\partial p}\right)_{T,n} = \kappa_T \qquad (7.7.12)$$

and Eq. 7.7.11 becomes

$$V_{\text{B}}^{\infty} = \left(\frac{\partial \mu_{c,\text{B}}^{\text{ref}}}{\partial p}\right)_{T} + RT\kappa_T$$

Section 7.7 Activities

Table 7.4 Expressions for the dependence of Γ_i on p. The approximate expressions assume the phase is incompressible or the solute partial molar volume is independent of pressure.

Constituent i of a gas mixture, or the pure gas	$\Gamma_i = 1$
Nonelectrolyte constituent i of a liquid or solid mixture, or the pure liquid or solid	$\Gamma_i = \exp\left(\int_{p^\circ}^{p} \dfrac{V_{m,i}^*}{RT}\,dp\right) \approx \exp\left[\dfrac{V_{m,i}^*(p-p^\circ)}{RT}\right]$
Solvent A of a solution	$\Gamma_A = \exp\left(\int_{p^\circ}^{p} \dfrac{V_{m,A}^*}{RT}\,dp\right) \approx \exp\left[\dfrac{V_{m,A}^*(p-p^\circ)}{RT}\right]$
Solute substance B, mole fraction or molality basis	$\Gamma_{x,B} = \Gamma_{m,B} = \exp\left(\int_{p^\circ}^{p} \dfrac{V_B^\infty}{RT}\,dp\right) \approx \exp\left[\dfrac{V_B^\infty(p-p^\circ)}{RT}\right]$
Solute substance B, concentration basis	$\Gamma_{c,B} = \exp\left[\int_{p^\circ}^{p} \left(\dfrac{V_B^\infty}{RT} - \kappa_T\right) dp\right] \approx \exp\left[\dfrac{V_B^\infty(p-p^\circ)}{RT}\right]$
Ion solute species	$\Gamma_+ = \exp\left(\int_{p^\circ}^{p} \dfrac{V_+^\infty}{RT}\,dp\right) \quad \Gamma_- = \exp\left(\int_{p^\circ}^{p} \dfrac{V_-^\infty}{RT}\,dp\right)$

Solving for $d\mu_{c,B}^{\text{ref}}$ and integrating from p° to p at constant T, we obtain finally

$$\mu_{c,B}^{\text{ref}} - \mu_i^\circ = \int_{p^\circ}^{p} \left(V_B^\infty - RT\kappa_T\right) dp$$

We are now able to write out explicit formulas for Γ_i for each of the reference states. They are collected in Table 7.4. Formulas for ion solute species are included.

Just how different from unity is Γ_i likely to be for a component of a condensed-phase mixture? Suppose we take the values $V_i^* = 100\,\text{cm}^3\,\text{mol}^{-1}$, $T = 300\,\text{K}$, and $p^\circ = 1\,\text{bar}$ and assume the mixture is incompressible. Then Γ_i is 1.004 at 2 bar, 1.04 at 10 bar, and 1.5 at 100 bar. It is only at high pressure that Γ_i differs appreciably from 1. For this reason, it is frequently omitted as a factor in the expression for an activity.

> In principle, we can specify any convenient value for the standard pressure p°. For a chemist making measurements at high pressures, it would be convenient to specify a value of p° within the range of the experimental pressures, for example $p^\circ = 1\,\text{kbar}$, in order that each Γ_i be close to unity.

A substance has an activity coefficient and an activity in a *pure* phase, just as it does in a mixture. In the case of a pure gas, the activity coefficient is the fugacity coefficient ϕ and the activity is f/p°. For a pure liquid or solid, the activity coefficient is 1 and the activity is $\Gamma = \exp\left[\int_{p^\circ}^{p}(V_m^*/RT)\,dp\right]$.

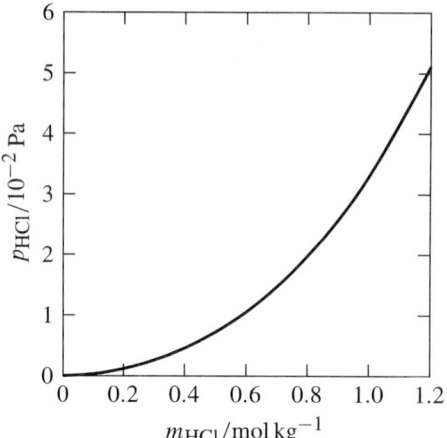

Figure 7.12 Partial pressure of HCl in a gas phase equilibrated with aqueous HCl at 25 °C and 1 bar.

7.8 ELECTROLYTE SOLUTIONS

Pure HCl (hydrogen chloride) is a gas that is very soluble in water. The curve of a plot of the partial pressure of gaseous HCl in equilibrium with aqueous HCl as a function of the solution molality (Fig. 7.12) shows that the limiting slope at infinite dilution is not finite, but zero. What is the reason for this non-Henry's law behavior? It must be that HCl is an electrolyte—it dissociates (ionizes) in the aqueous environment.

It is customary to use a *molality* basis for the reference and standard states of electrolyte solutes. This is the only basis we use in this section even when not explicitly indicated for ions. The symbol μ_+°, for instance, denotes the chemical potential of a cation in a standard state based on molality.

In dealing with a solute that is an electrolyte, we can refer to the solute (a substance) as a whole and to the individual charged species (ions) that result from dissociation. We can apply the same general definitions of chemical potential, activity coefficient, and activity to all of these species, but there are some differences in what we can actually evaluate experimentally.

7.8.1 Single-ion quantities

Let us consider a binary solution in which the solute is a strong electrolyte. The molality of the solute, m_B, is defined as the amount of solute formula unit divided by the mass of solvent. We assume the solute is completely dissociated into one kind of cation and one kind of anion with amounts denoted by n_+ and n_-, respectively.

We may in principle define the chemical potentials of the ions as follows:

$$\mu_+ \equiv \left(\frac{\partial G}{\partial n_+}\right)_{T,p,n_A,n_-} \qquad \mu_- \equiv \left(\frac{\partial G}{\partial n_-}\right)_{T,p,n_A,n_+}$$

Section 7.8 Electrolyte Solutions

That is, the cation chemical potential μ_+ is the rate at which G increases with the amount of the cation added to the solution. μ_+ is a function of T, p, m_+, and m_-. It is also a function of the *shape* of the solution phase because, unless n_+ and n_- are present in the same proportions as in the electrolyte substance, the solution has a net electric charge and an electrostatic energy that depends on the shape. Similar comments apply to μ_-.

Of course there is no experimental way to evaluate either μ_+ or μ_- because it is impossible to add cations or anions by themselves to a solution.

We can nevertheless write some theoretical relations involving μ_+ and μ_- in an electrically neutral solution, which is the only kind of solution we can deal with experimentally. Theoretically, μ_+ under these conditions is the rate at which G increases with the amount of the cation added to a solution that initially has a net negative charge, just at the point in the process at which the solution becomes electrically neutral. In the same way, μ_- in an electrically neutral solution is the rate at which G increases with the amount of added anion at the point of electrical neutrality. In the electrically neutral binary solution, μ_+ and μ_- depend only on T, p, and m_B and are independent of system shape.

For a given temperature and pressure, we can write the dependence of the chemical potentials of the ions on their molalities in the same form as for a nonelectrolyte solute (Eq. 7.6.7):

$$\mu_+ = \mu_+^{\text{ref}} + RT \ln\left(\gamma_+ \frac{m_+}{m^\circ}\right) \qquad \mu_- = \mu_-^{\text{ref}} + RT \ln\left(\gamma_- \frac{m_-}{m^\circ}\right) \qquad (7.8.1)$$

Here μ_+^{ref} and μ_-^{ref} are the chemical potentials of the ions in hypothetical solute reference states based on molality, and γ_+ and γ_- are single-ion activity coefficients on a molality basis.

The ion activity coefficients approach unity in the limit of infinite dilution:

$$\gamma_+ \to 1 \quad \text{and} \quad \gamma_- \to 1 \quad \text{as} \quad x_A \to 1 \qquad (7.8.2)$$
$$\text{(constant } T \text{ and } p\text{)}$$

That is, we assume that in an extremely dilute, electrically neutral electrolyte solution, each individual ion behaves like a nonelectrolyte solute species in an ideal-dilute solution. At a finite solute molality, the values of γ_+ and γ_- are the ones that allow Eq. 7.8.1 to give the correct values of the quantities $(\mu_+ - \mu_+^{\text{ref}})$ and $(\mu_- - \mu_-^{\text{ref}})$. Of course, we have no way to actually measure these quantities experimentally and so we cannot evaluate either γ_+ or γ_-.

We can also define single-ion activities a_+ and a_- with a relation having the form of Eq. 7.7.5 on page 214:

$$\mu_+ = \mu_+^\circ + RT \ln a_+ \qquad \mu_- = \mu_-^\circ + RT \ln a_- \qquad (7.8.3)$$

The standard chemical potentials μ_+° and μ_-° are the chemical potentials of the ions in the hypothetical solute reference states at the standard pressure. From Eqs. 7.8.1 and 7.8.3, the single-ion activities are given by

$$a_+ = \Gamma_+ \gamma_+ \frac{m_+}{m^\circ} \qquad a_- = \Gamma_- \gamma_- \frac{m_-}{m^\circ} \qquad (7.8.4)$$

where Γ_+ and Γ_- are pressure factors defined by

$$\Gamma_+ \equiv \exp\left(\frac{\mu_+^{\text{ref}} - \mu_+^\circ}{RT}\right) \qquad \Gamma_- \equiv \exp\left(\frac{\mu_-^{\text{ref}} - \mu_-^\circ}{RT}\right) \tag{7.8.5}$$

(see Eq. 7.7.8 on page 214). Like γ_+ and γ_-, the values of a_+, a_-, Γ_+, and Γ_- cannot be determined by experiment.

7.8.2 Symmetrical electrolytes

Now let us consider properties of the electrolyte solute as a whole. The simplest case is that of a binary solution in which the solute is a symmetrical strong electrolyte—a substance whose formula unit has one cation and one anion that dissociate completely. We indicate this condition by $\nu = 2$, where ν is the number of ions per formula unit. The solute with ν equal to 2 might be a 1:1 salt such as NaCl, a 2:2 salt such as $MgSO_4$, or a strong monoprotic acid such as HCl.

In this binary solution, the chemical potential of the solute as a whole is defined in the usual way as the partial molar Gibbs energy

$$\mu_B \equiv \left(\frac{\partial G}{\partial n_B}\right)_{T,p,n_A}$$

(Eq. 7.3.1) and is a function of T, p, and the solute molality m_B. Although μ_B under given conditions must in principle have a definite value, we are unable to evaluate it because we have no way to measure precisely the energy brought into the system by the solute. This energy contributes to the internal energy and thus to G. We can, however, evaluate the difference $\mu_B - \mu_{m,B}^\circ$.

We may write the additivity rule (Eq. 7.2.9) for G as either

$$G = n_A \mu_A + n_B \mu_B$$

or

$$G = n_A \mu_A + n_+ \mu_+ + n_- \mu_-$$

A comparison of these equations for the symmetrical electrolyte ($n_B = n_+ = n_-$) gives the relation

$$\mu_B = \mu_+ + \mu_- \tag{7.8.6}$$
$$(\nu = 2)$$

The solute chemical potential is thus the *sum* of the single-ion chemical potentials.

By substituting the expressions of Eq. 7.8.1 in Eq. 7.8.6 and setting the ion molalities m_+ and m_- equal to m_B, we obtain

$$\mu_B = \mu_{m,B}^{\text{ref}} + RT \ln\left[\gamma_+ \gamma_- \left(\frac{m_B}{m^\circ}\right)^2\right] \tag{7.8.7}$$
$$(\nu = 2)$$

Section 7.8 Electrolyte Solutions

where $\mu_{m,B}^{\text{ref}} = \mu_+^{\text{ref}} + \mu_-^{\text{ref}}$ is the chemical potential of the solute in a hypothetical reference state based on molality. The important feature of this relation is the appearance of the *second* power of $m_B/m°$ instead of the first power as in the case of a nonelectrolyte.

Although we cannot evaluate γ_+ or γ_- individually, we *can* evaluate the product $\gamma_+\gamma_-$. This product is the square of the **mean ionic activity coefficient** $\gamma_\pm$, defined for a symmetrical electrolyte by

$$\gamma_\pm \equiv \sqrt{\gamma_+\gamma_-} \tag{7.8.8}$$
$$(\nu = 2)$$

With this definition, Eq. 7.8.7 becomes

$$\mu_B = \mu_{m,B}^{\text{ref}} + RT \ln\left[(\gamma_\pm)^2 \left(\frac{m_B}{m°}\right)^2\right] \tag{7.8.9}$$
$$(\nu = 2)$$

Since it is possible to determine the value of $\mu_B - \mu_{m,B}^{\text{ref}}$ for a solution of known molality, $\gamma_\pm$ is a measurable quantity.

The solute activity on a molality basis, $a_{m,B}$, is defined by Eq. 7.7.5 on page 214:

$$\mu_B = \mu_{m,B}° + RT \ln a_{m,B}$$

Here $\mu_{m,B}°$ is the chemical potential of the solute in its standard state, which is the hypothetical solute reference state at the standard pressure. By equating the expressions for μ_B in this equation and Eq. 7.8.9 and solving for the activity, we obtain

$$a_{m,B} = \Gamma_{m,B} (\gamma_\pm)^2 \left(\frac{m_B}{m°}\right)^2 \tag{7.8.10}$$
$$(\nu = 2)$$

where $\Gamma_{m,B}$ is the pressure factor defined by Eq. 7.7.8 on page 214:

$$\Gamma_{m,B} = \exp\left(\frac{\mu_{m,B}^{\text{ref}} - \mu_{m,B}°}{RT}\right) \tag{7.8.11}$$

We can use the appropriate expression in Table 7.4 on page 217 to evaluate $\Gamma_{m,B}$:

$$\Gamma_{m,B} = \exp\left(\int_{p°}^{p} \frac{V_B^\infty}{RT} dp\right) \approx \exp\left[\frac{V_B^\infty(p - p°)}{RT}\right] \tag{7.8.12}$$

The value of $\Gamma_{m,B}$ is 1 at the standard pressure and close to 1 at any pressure that is not very high.

Equation 7.8.10 predicts that the activity of HCl in aqueous solutions is proportional, in the limit of infinite dilution, to the *square* of the HCl molality. In contrast, the activity of a *non*electrolyte solute is proportional to the *first* power of the molality in this limit. This predicted behavior of aqueous HCl is consistent with the data plotted in Fig. 7.12 on page 218 and is confirmed by the data for dilute HCl solutions shown in Fig. 7.13(a)

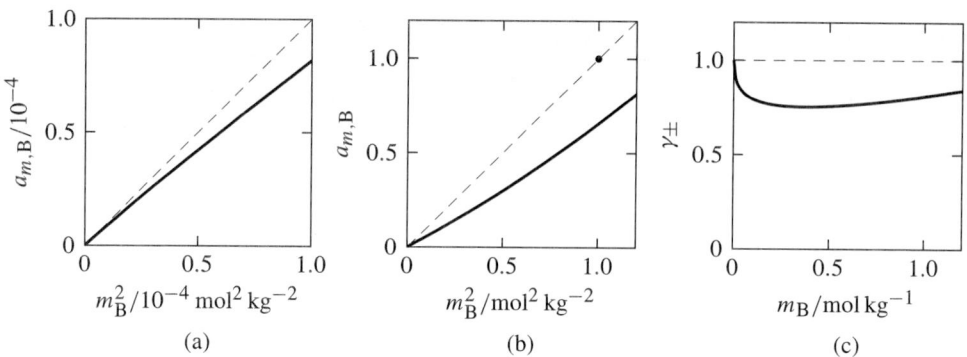

Figure 7.13 Aqueous HCl at 25 °C and 1 bar. (a) HCl activity on a molality basis as a function of molality squared. The dashed line is the extrapolation of the ideal-dilute behavior. (b) Same as (a) at a greatly reduced scale; the filled circle indicates the solute reference state. (c) Mean ionic activity coefficient of HCl as a function of molality.

222. The dashed line in Fig. 7.13(a) is the extrapolation of the ideal-dilute behavior given by $a_{m,\text{B}} = \Gamma_{m,\text{B}}(m_\text{B}/m°)^2$. The extension of this line to $m_\text{B} = m°$ establishes the hypothetical solute reference state based on molality, indicated by a filled circle in Fig. 7.13(b). (Since the pressure is 1 bar, the solute reference state is also the solute standard state.)

The curve of Fig. 7.13(c) shows how the mean ionic activity coefficient of HCl varies with molality in approximately the same range of molalities as the data shown in Fig. 7.13(b). The slope of this curve approaches $-\infty$ as the molality approaches zero, quite unlike the behavior described in Sec. 7.6.3 for the activity coefficient of a nonelectrolyte solute.

For a symmetrical strong electrolyte, $\gamma_\pm$ is the geometric average of the single-ion activity coefficients γ_+ and γ_-. We have no way of evaluating γ_+ or γ_- individually even when we know the value of $\gamma_\pm$. For instance, we cannot assume that γ_+ and γ_- are equal.

7.8.3 Electrolytes in general

The formula unit of a *non*symmetrical electrolyte solute has more than two ions. General formulas for the solute as a whole are more complicated than those for the symmetrical case treated in the preceding section, but are derived by the same reasoning.

Again we assume the solute dissociates completely into its constituent ions. We define the following symbols:

ν_+ = the number of cations per solute formula unit;

ν_- = the number of anions per solute formula unit;

ν = the sum $\nu_+ + \nu_-$.

For example, if the solute formula is $Al_2(SO_4)_3$, the values are $\nu_+ = 2$, $\nu_- = 3$, and $\nu = 5$.

In a solution of a single electrolyte solute, the ion molalities are related to the overall

Section 7.8 Electrolyte Solutions

solute molality by

$$m_+ = \nu_+ m_B \qquad m_- = \nu_- m_B \tag{7.8.13}$$

From the additivity rule for the Gibbs energy, we have

$$G = n_A \mu_A + n_B \mu_B = n_A \mu_A + \nu_+ n_B \mu_+ + \nu_- n_B \mu_-$$

giving the relation

$$\mu_B = \nu_+ \mu_+ + \nu_- \mu_- \tag{7.8.14}$$

in place of Eq. 7.8.6. By combining Eqs. 7.8.1, 7.8.13, and 7.8.14, we obtain

$$\mu_B = \mu_B^{\text{ref}} + RT \ln\left[\left(\nu_+^{\nu_+} \nu_-^{\nu_-}\right)\left(\gamma_+^{\nu_+}\right)\left(\gamma_-^{\nu_-}\right)\left(\frac{m_B}{m^\circ}\right)^\nu\right] \tag{7.8.15}$$

where $\mu_B^{\text{ref}} = \nu_+ \mu_+^{\text{ref}} + \nu_- \mu_-^{\text{ref}}$ is the chemical potential of the hypothetical reference state in which the solute has the standard molality and behaves as at infinite dilution. Equation 7.8.15 is the generalization of Eq. 7.8.7.

We define the mean ionic activity coefficient $\gamma_\pm$ in the general case by

$$\gamma_\pm^\nu = \left(\gamma_+^{\nu_+}\right)\left(\gamma_-^{\nu_-}\right) \tag{7.8.16}$$

or

$$\gamma_\pm = \left(\gamma_+^{\nu_+} \gamma_-^{\nu_-}\right)^{1/\nu} \tag{7.8.17}$$

Thus, $\gamma_\pm$ is a geometric average of γ_+ and γ_- weighted according to the numbers of the ions in the solute formula unit. With a substitution from Eq. 7.8.16, Eq. 7.8.15 becomes

$$\mu_B = \mu_B^{\text{ref}} + RT \ln\left[\left(\nu_+^{\nu_+} \nu_-^{\nu_-}\right) \gamma_\pm^\nu \left(\frac{m_B}{m^\circ}\right)^\nu\right] \tag{7.8.18}$$

Since $\mu_B - \mu_B^{\text{ref}}$ is a measurable quantity, so is $\gamma_\pm$.

The solute activity, defined by $\mu_B = \mu_{m,B}^\circ + RT \ln a_{m,B}$, is

$$a_{m,B} = \left(\nu_+^{\nu_+} \nu_-^{\nu_-}\right) \Gamma_{m,B} \gamma_\pm^\nu \left(\frac{m_B}{m^\circ}\right)^\nu \tag{7.8.19}$$

where $\Gamma_{m,B}$ is the pressure factor that we can evaluate with Eq. 7.8.12. Equation 7.8.19 is the generalization of Eq. 7.8.10. From Eqs. 7.8.5 and 7.8.11 and the relations $\mu_B^{\text{ref}} = \nu_+ \mu_+^{\text{ref}} + \nu_- \mu_-^{\text{ref}}$ and $\mu_B^\circ = \nu_+ \mu_+^\circ + \nu_- \mu_-^\circ$, we obtain the relation

$$\Gamma_{m,B} = \Gamma_+^{\nu_+} \Gamma_-^{\nu_-} \tag{7.8.20}$$

If the solution contains more than one electrolyte solute, Eq. 7.8.14 applies to each solute, and from Eqs. 7.8.1 and 7.8.16 we obtain

$$\mu_B = \mu_B^{\text{ref}} + RT \ln\left[\gamma_\pm^\nu \left(\frac{m_+}{m^\circ}\right)^{\nu_+} \left(\frac{m_-}{m^\circ}\right)^{\nu_-}\right] \tag{7.8.21}$$

This equation does not assume that the ions of solute B are present in stoichiometric amounts, as did Eq. 7.8.18. For instance, the solution might contain Na^+, K^+, Cl^-, and Br^+ ions at arbitrary molalities, subject only to the electroneutrality condition $m_{Na^+} + m_{K^+} = m_{Cl^-} + m_{Br^-}$. Then Eq. 7.8.21 relates $\mu_B - \mu_B^{ref}$ to $\gamma_\pm$, where B can be NaCl, NaBr, KCl, or KBr and m_+ and m_- are the molalities of the constituent ions of B.

As an example of how we might use Eq. 7.8.21, suppose that we have measured $\gamma_\pm$ for KCl in an aqueous solution in which KCl is the only solute and which is in equilibrium with solid KCl (a saturated solution).[17] We then obtain $\mu_B - \mu_B^{ref}$ from Eq. 7.8.21, where B is KCl. Since μ_B must be equal in the equilibrated solution and solid, $\mu_B - \mu_B^{ref}$ has the same value in a second solution at the same T and p, also in equilibrium with solid KCl, that contains additional solutes and may even have unequal molalities of K^+ and Cl^-.

Up to this point, we have assumed that the electrolyte solutes are completely dissociated into their constituent ions at all molalities. However, some solutions contain *ion pairs* (closely associated ions of opposite charge). Furthermore, in solutions of some electrolytes (often called "weak" electrolytes), an equilibrium is established between ions and electrically neutral molecules. In these kinds of solutions, the relations between solute molality and ion molalities given by Eq. 7.8.13 are no longer valid. The relation between $\gamma_\pm$ and $\mu_B - \mu_{m,B}^{ref}$ is still given by Eq. 7.8.18, but $\gamma_\pm$ does not have the physical significance of being the geometric average of the activity coefficients of the actual dissociated ions. When dissociation is not necessarily complete, $\gamma_\pm$ obtained from Eq. 7.8.18 is called the **stoichiometric activity coefficient** of the electrolyte.

7.8.4 The Debye–Hückel theory

The theory of Peter Debye and Erich Hückel (1923) provides theoretical expressions for single-ion activity coefficients and mean ionic activity coefficients in electrolyte solutions. The expressions in one form or another are very useful for extrapolation of quantities that include mean ionic activity coefficients to low solute molality or infinite dilution.

The only interactions the theory considers are the electrostatic interactions between ions. These interactions are much greater than those between uncharged molecules and die off more slowly with distance. If the positions of ions in an electrolyte solution were completely random, the net effect of electrostatic ion–ion interactions would be zero because each cation–cation or anion–anion repulsion would be balanced by a cation–anion attraction. The positions are not random, however; each cation has a surplus of anions in its immediate environment, and each anion has a surplus of neighboring cations. Each ion then has a net attractive interaction with the surrounding ion atmosphere. The result in the case of the cation at low electrolyte molality is a decrease of μ_+ compared to the same molality in the absence of ion–ion interactions, meaning that γ_+ becomes less than 1 as the electrolyte molality is increased beyond the ideal-dilute range. Similarly, γ_- also becomes less than 1.

According to the Debye–Hückel theory, the single-ion activity coefficient γ_i of ion i in

---

[17] A method by which $\gamma_\pm$ in a solution of a single electrolyte can be found from osmotic coefficients is described in Sec. 7.8.6.

a solution of one or more electrolytes is given by

$$\ln \gamma_i = -\frac{A z_i^2 \sqrt{I_m}}{1 + Ba\sqrt{I_m}} \tag{7.8.22}$$

where
z_i = the charge number of ion i (+1, −2, etc.);
I_m = the **ionic strength** of the solution on a molality basis, defined by[18]

$$I_m \equiv \tfrac{1}{2} \sum_{\text{all ions}} m_j z_j^2 \tag{7.8.23}$$

A and B are defined functions of the kind of solvent and the temperature;
a is an adjustable parameter, equal to the mean effective distance of closest approach of other ions in the solution to one of the i ions.

The definitions of the quantities A and B appearing in Eq. 7.8.22 are

$$A = \left(N_A^2 e^3/8\pi\right) (2\rho_A^*)^{1/2} (\epsilon_r \epsilon_0 RT)^{-3/2}$$
$$B = N_A e \left(2\rho_A^*\right)^{1/2} (\epsilon_r \epsilon_0 RT)^{-1/2}$$

where N_A is the Avogadro constant, e is the elementary charge (the charge of a proton), ρ_A^* and ϵ_r are the density and relative permittivity (dielectric constant) of the solvent, and ϵ_0 is the permittivity of vacuum.

When the solvent is water at 25 °C, the quantities A and B have the values

$$A = 1.1744 \, \text{kg}^{1/2}\,\text{mol}^{-1/2}$$
$$B = 3.285 \times 10^9 \, \text{m}^{-1}\,\text{kg}^{1/2}\,\text{mol}^{-1/2}$$

We obtain an expression for the mean ionic activity coefficient $\gamma_\pm$ of an electrolyte solute from Eqs. 7.8.17 and 7.8.22 with the help of the electroneutrality condition $\nu_+ z_+ = \nu_- z_-$:

$$\ln \gamma_\pm = -\frac{A |z_+ z_-| \sqrt{I_m}}{1 + Ba\sqrt{I_m}} \tag{7.8.24}$$

In this equation, z_+ and z_- are the charge numbers of the cation and anion comprising the solute to which $\gamma_\pm$ refers. Since the right side of Eq. 7.8.24 is negative at finite solute molalities and zero at infinite dilution, $\gamma_\pm$ is less than 1 at finite solute molalities and approaches 1 at infinite dilution.

Figure 7.14 on the next page shows that with the proper choice of the parameter a, the

---

[18] Lewis and Randall introduced the term *ionic strength*, defined by this equation, 2 years before the Debye–Hückel theory was published (Lewis, G. N.; Randall, M. *J. Am. Chem. Soc.* **1921**, *43*, 1112–1154). They found empirically that in dilute solutions the mean ionic activity coefficient of a given strong electrolyte is the same in all solutions having the same ionic strength.

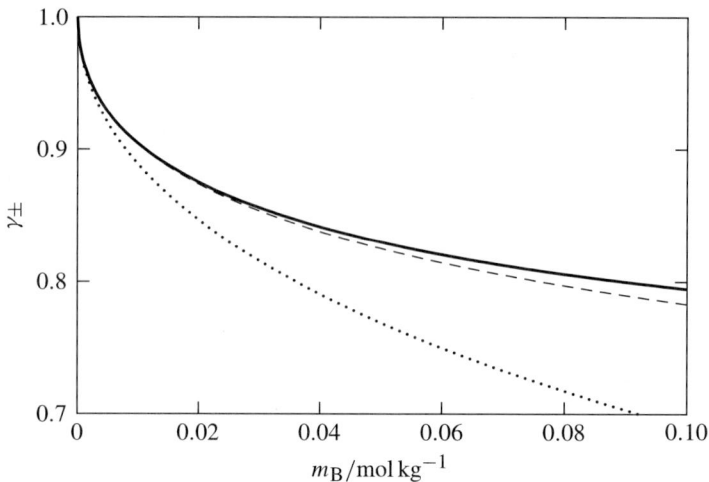

Figure 7.14 Mean ionic activity coefficient of aqueous HCl at 25 °C. Solid curve: experiment; dashed curve: Debye–Hückel theory with $a = 5 \times 10^{-10}$ m; dotted curve: Debye–Hückel limiting law.

mean ionic activity coefficient of HCl calculated from Eq. 7.8.24 (dashed curve) agrees closely with experiment (solid curve) up to 0.10 molal.

As the molalities of all solutes become small, Eq. 7.8.24 becomes

$$\ln \gamma_\pm = -A \, |z_+ z_-| \sqrt{I_m} \qquad (7.8.25)$$
(electrolyte at infinite dilution)

This form is known as the **Debye–Hückel limiting law**. Note that the limiting law contains no adjustable parameters. The dotted curve in Fig. 7.14 shows that the limiting law agrees with experiment only at quite low molality.

The ionic strength I_m is calculated from Eq. 7.8.23 with the molalities of *all* ions in the solution, not just the molality of the ion or solute whose activity coefficient we are interested in. This is because, as explained earlier, the departure of γ_i and $\gamma_\pm$ from the ideal-dilute value of 1 is caused by the interaction of each ion of interest with the ion atmosphere resulting from all other ions in the solution. If solute B is not the only electrolyte in the solution, its mean ionic activity coefficient is a function of I_m rather than m_B.

In the case of a solution containing a single 1:1 electrolyte, such as NaCl or HCl, assumed to be completely dissociated, the ionic strength is equal to the solute molality: $I_m = (1/2)[m_B \cdot (+1)^2 + m_B \cdot (-1)^2] = m_B$. For strong electrolytes with other stoichiometries, we find the relations between I_m and m_B depend on the ion charge numbers as follows:

For a 1:2 or 2:1 electrolyte, e.g., Na_2SO_4 or $CaCl_2$: $I_m = 3m_B$
For a 2:2 electrolyte, e.g., $MgSO_4$: $I_m = 4m_B$
For a 1:3 or 3:1 electrolyte, e.g., $AlCl_3$: $I_m = 6m_B$
For a 3:2 or 2:3 electrolyte, e.g., $Al_2(SO_4)_3$: $I_m = 15m_B$

Section 7.8 Electrolyte Solutions

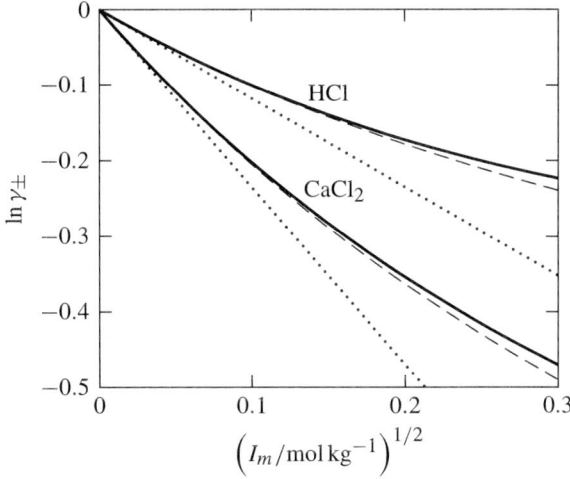

Figure 7.15 $\ln \gamma_\pm$ versus $\sqrt{I_m}$ for aqueous HCl (upper curves) and aqueous CaCl$_2$ (lower curves) at 25 °C. Solid curves: experiment; dashed curves: Debye–Hückel theory ($a = 5 \times 10^{-10}$ m for HCl, $a = 4.5 \times 10^{-10}$ m for CaCl$_2$); dotted lines: Debye–Hückel limiting law.

Figure 7.15 shows $\ln \gamma_\pm$ as a function of $\sqrt{I_m}$ for aqueous HCl and CaCl$_2$. The experimental curves have the limiting slopes predicted by the Debye–Hückel limiting law (Eq. 7.8.25), but at a low ionic strength begin to deviate significantly from the linear relations predicted by that law. The full Debye–Hückel equation (Eq. 7.8.24) fits the experimental curves over a wider range of ionic strengths.

7.8.5 Derivation of the Debye–Hückel equation

Debye and Hückel derived Eq. 7.8.22 with a combination of electrostatic theory, statistical mechanical theory, and thermodynamics. This section gives a brief outline of their derivation.

The derivation starts by focusing on one particular ion of Species i as it moves through the solution; call it the central ion. Around this central ion, the time-average spatial distribution of any ion species j is not random because of the interaction of these ions of Species j with the central ion. (Species i and j may be the same or different.) The distribution, whatever it is, must be spherically symmetric about the central ion; that is, a function only of the distance r from the center of the ion. The local concentration, c'_j, of the ions of Species j at a given value of r depends on the ion charge $z_j e$ and the electric potential ϕ at that position. The time-average electric potential in turn depends on the distribution of all ions and is symmetric about the central ion, so expressions must be found for c'_j and ϕ as functions of r that are mutually consistent.

Debye and Hückel assumed c'_j is given by the Boltzmann distribution

$$c'_j = c_j e^{-z_j e\phi/kT}$$

where $z_j e\phi$ is the electrostatic energy of an ion of Species j, and k is the Boltzmann constant ($k = R/N_A$). As r becomes large, ϕ approaches zero and c'_j approaches the macroscopic concentration c_j. As T increases, c'_j at a fixed value of r approaches c_j because of the randomizing effect of thermal energy. Debye and Hückel expanded the exponential function and retained only the first two terms: $c'_j \approx c_j(1 - z_j e\phi/kT)$. The distribution of each ion species is assumed to follow this relation. The electric potential function consistent with this distribution and with the electroneutrality of the solution as a whole is

$$\phi = (z_i e/4\pi\epsilon_r\epsilon_0 r)e^{\kappa(a-r)}/(1+\kappa a)$$

Here κ is defined by $\kappa^2 = 2N_A^2 e^2 I_c/\epsilon_r\epsilon_0 RT$, where I_c is the *ionic strength on a concentration basis* defined by $I_c = (1/2)\sum_i c_i z_i^2$.

The electric potential ϕ is assumed to be a sum of two contributions: the electric potential the central ion would cause at infinite dilution, $z_i e/4\pi\epsilon_r\epsilon_0 r$, and the electric potential due to all other ions, ϕ'. Thus, ϕ' is equal to $\phi - z_i e/4\pi\epsilon_r\epsilon_0 r$ or

$$\phi' = (z_i e/4\pi\epsilon_r\epsilon_0 r)[e^{\kappa(a-r)}/(1+\kappa a) - 1]$$

This expression for ϕ' is valid for distances down to a, the distance of closest approach of other ions. At smaller values of r, ϕ' is constant and equal to the value at $r = a$, which is $\phi'(a) = -(z_i e/4\pi\epsilon_r\epsilon_0)\kappa/(1+\kappa a)$. The interaction energy between the central ion and the surrounding ions (the ion atmosphere) is the product of the central ion charge and $\phi'(a)$.

The last step of the derivation is the calculation of the work of a hypothetical reversible process in which the surrounding ions stay in their final distribution, and the charge of the central ion gradually increases from zero to its actual value $z_i e$. Let $\alpha z_i e$ be the charge at each stage of the process, where α is a fractional advancement that changes from 0 to 1. Then the work w' due to the interaction of the central ion with its ion atmosphere is $\phi'(a)$ integrated over the charge:

$$w' = -\int_{\alpha=0}^{\alpha=1} [(\alpha z_i e/4\pi\epsilon_r\epsilon_0)\kappa/(1+\kappa a)]\,\mathrm{d}(\alpha z_i e)$$
$$= -(z_i^2 e^2/8\pi\epsilon_r\epsilon_0)\kappa/(1+\kappa a)$$

Since the infinitesimal Gibbs energy change in a reversible process is given by $\mathrm{d}G = -S\,\mathrm{d}T + V\,\mathrm{d}p + \mathrm{d}w'$ (Eq. 4.9.5), this reversible nonexpansion work at constant T and p is equal to the Gibbs energy change. The Gibbs energy change per amount of Species i is $w'N_A = -(z_i^2 e^2 N_A/8\pi\epsilon_r\epsilon_0)\kappa/(1+\kappa a)$. This quantity is $\Delta G/n_i$ for the process in which a solution of fixed composition changes from a hypothetical state lacking ion–ion interactions to the real state with ion–ion interactions present. $\Delta G/n_i$ may be equated to the difference of the chemical potentials of i in the final and initial states. If the chemical potential without ion–ion interactions is taken to be that calculated for ideal-dilute behavior on a molality basis, $\mu_{m,i}^{\text{ref}} + RT \ln(m_i/m°)$, then $-(z_i^2 e^2 N_A / 8\pi \epsilon_r \epsilon_0)\kappa/(1+\kappa a)$ is equal to $\mu_i - [\mu_{m,i}^{\text{ref}} + RT \ln(m_i/m°)] = RT \ln \gamma_{m,i}$. In a dilute solution, c_i can with little error be set equal to $\rho_A^* m_i$, and I_c to $\rho_A^* I_m$. Equation 7.8.22 follows.

7.8.6 Mean ionic activity coefficients from osmotic coefficients

Recall that $\gamma_\pm$ is the stoichiometric activity coefficient of an electrolyte or the mean ionic activity coefficient if the electrolyte dissociates completely.

Few electrolytes are sufficiently volatile, as HCl is, to permit $\gamma_\pm$ to be evaluated from partial pressure measurements of the solute. Section 11.5 describes how, in favorable cases, it is possible to evaluate $\gamma_\pm$ from the emf of a galvanic cell.

The other general procedure available for evaluating $\gamma_\pm$ requires knowledge of the molal osmotic coefficient ϕ_m as a function of molality. We may measure ϕ_m by any of the methods used for nonelectrolytes. The isopiestic method (Sec. 7.6.7) and freezing-point depression measurements (Sec. 9.4.1) are commonly used. We define the molal osmotic coefficient of a binary solution of an electrolyte by

$$\phi_m \equiv \frac{\mu_A^* - \mu_A}{RT M_A \nu m_B} \quad (7.8.26)$$
(binary electrolyte solution)

That is, for an electrolyte, the sum $\sum_{i \neq A} m_i$ appearing in the definition of ϕ_m (Eq. 7.6.21 on page 209) is taken as the sum of the ion molalities assuming complete dissociation. We now show that we can use ϕ_m defined this way to evaluate $\gamma_\pm$.

The derivation is like that described in Sec. 7.6.6 for a binary solution of a nonelectrolyte. Solving Eq. 7.8.26 for μ_A and taking the differential of μ_A at constant T and p, we obtain

$$d\mu_A = -RT M_A \nu (\phi_m \, dm_B + m_B \, d\phi_m)$$

From Eq. 7.8.18 on page 223 we obtain

$$d\mu_B = RT\nu \left(d\ln \gamma_\pm + \frac{dm_B}{m_B} \right)$$

Substitution of these expressions in the Gibbs–Duhem equation $n_A \, d\mu_A + n_B \, d\mu_B = 0$, with the substitution $n_A M_A = n_B/m_B$, yields

$$d\ln \gamma_\pm = d\phi_m + \frac{\phi_m - 1}{m_B} dm_B \quad (7.8.27)$$

Then integration from $m_B = 0$ to any desired value of m_B gives the result

$$\ln \gamma_\pm(m_B) = \phi_m(m_B) - 1 + \int_0^{m_B} \frac{\phi_m - 1}{m_B} dm_B \quad (7.8.28)$$

The right side of this equation is the same as the expression for $\ln \gamma_{m,B}$ for a nonelectrolyte (Eq. 7.6.23 on page 210).

The integrand of the integral on the right side of Eq. 7.8.28 approaches $-\infty$ as m_B approaches zero, making it difficult to evaluate the integral by numerical integration starting at $m_B = 0$. (This difficulty does not exist when the solute is a nonelectrolyte.) Instead, we can split the integral into two parts

$$\int_0^{m_B} \frac{\phi_m - 1}{m_B}\, dm_B = \int_0^{m'_B} \frac{\phi_m - 1}{m_B}\, dm_B + \int_{m'_B}^{m_B} \frac{\phi_m - 1}{m_B}\, dm_B$$

where the integration limit m'_B is a low molality at which the value of ϕ_m is available and at which $\gamma_\pm$ can be measured or else estimated from the Debye–Hückel equation. The first integral on the right side is evaluated by writing Eq. 7.8.28 in the form

$$\ln \gamma_\pm(m'_B) = \phi_m(m'_B) - 1 + \int_0^{m'_B} \frac{\phi_m - 1}{m_B}\, dm_B$$

Using the two preceding equations, Eq. 7.8.28 becomes

$$\ln \gamma_\pm(m_B) = \phi_m(m_B) - \phi_m(m'_B) + \ln \gamma_\pm(m'_B) + \int_{m'_B}^{m_B} \frac{\phi_m - 1}{m_B}\, dm_B \qquad (7.8.29)$$

The integral in this equation can easily be evaluated by numerical integration.

7.9 GRAVITATIONAL AND CENTRIFUGAL FIELDS

A tall column of a fluid in a gravitational field and a liquid-filled centrifuge cell in a rotating centrifuge rotor are both systems in which a field exerts a force on each element of mass. In the gravitational field, the force is directed downward, and in the centrifuge cell the force is directed away from the axis of rotation of the rotor. In an equilibrium state of a mixture in one of these systems, we expect intensive properties such as pressure and composition to vary along the direction of the force.

We wish to find the general conditions for the state to be an equilibrium state. As in Sec. 6.1.4, we imagine the system to be divided into a stack of many thin slab-shaped phases of fixed volume, one of which, α, serves as a reference phase. The infinitesimal change of the internal energy of phase β, with dV^β set equal to zero, is

$$dU^\beta = T^\beta\, dS^\beta + \sum_i \mu_i^\beta\, dn_i^\beta$$

The total differential of the internal energy of the whole system is

$$dU = dU^\alpha + \sum_{\beta \neq \alpha} dU^\beta$$
$$= T^\alpha\, dS^\alpha + \sum_{\beta \neq \alpha} T^\beta\, dS^\beta + \sum_i \mu_i^\alpha\, dn_i^\alpha + \sum_i \sum_{\beta \neq \alpha} \mu_i^\beta\, dn_i^\beta$$

Section 7.9 Gravitational and Centrifugal Fields

We have the relation $dS = dS^\alpha + \sum_{\beta \neq \alpha} dS^\beta$. The conditions for isolation are $dU = 0$ and $dn_i^\alpha + \sum_{\beta \neq \alpha} dn_i^\beta = 0$ (for each substance i). With appropriate substitutions for dS^α and dn_i^α, and with dU set equal to zero, the preceding displayed equation yields the following expression for dS in the isolated system:

$$dS = \sum_{\beta \neq \alpha} \frac{T^\alpha - T^\beta}{T^\alpha} dS^\beta + \sum_i \sum_{\beta \neq \alpha} \frac{\mu_i^\alpha - \mu_i^\beta}{T^\alpha} dn_i^\beta$$

At equilibrium, the coefficient multiplying each differential on the right side must be zero. We conclude that at equilibrium *the temperature and the chemical potential of each substance are uniform throughout the system*.

The next two sections give examples of this principle's application.

7.9.1 Ideal gas mixture in a gravitational field

The standard chemical potential of gaseous substance i at elevation h in a gravitational field, relative to the standard chemical potential at the reference elevation $h = 0$, is given by Eq. 6.1.5 on page 146:

$$\mu_i^\circ(h) = \mu_i^\circ(0) + M_i g h$$

Here M_i is the molar mass and g is the acceleration of free fall. If the substance is a constituent of an ideal gas mixture, its chemical potential at elevation h is related to its partial pressure at this elevation by $\mu_i(h) = \mu_i^\circ(h) + RT \ln[p_i(h)/p^\circ]$. In the equilibrium state, the chemical potential at an arbitrary elevation h is equal to the chemical potential at the reference elevation:

$$\mu_i^\circ(0) + M_i g h + RT \ln \frac{p_i(h)}{p^\circ} = \mu_i^\circ(0) + RT \ln \frac{p_i(0)}{p^\circ}$$

Solving for $p_i(h)$ gives us

$$p_i(h) = p_i(0) e^{-M_i g h / RT}$$

Thus, each constituent of an ideal gas mixture individually obeys the barometric formula given by Eq. 6.1.7 on page 146.

7.9.2 Liquid solution in a centrifugal field

In a centrifuge cell embedded in a rotor rotating at angular velocity ω (radians per unit time), the outward centrifugal force on a volume element of mass m is equal to $m\omega^2 r$, where r is the radial distance of the volume element from the axis of rotation. The potential energy of the volume element is equal to $m\phi$, where ϕ is a potential function given by

$$\phi = -\frac{\omega^2 r^2}{2} \tag{7.9.1}$$

Consider a centrifuge cell filled with a solution containing solute B. If a fixed amount of B in a solute standard state based, say, on mole fraction is moved reversibly and adiabatically from position r' to a position r'', where the potential function ϕ is different, its potential energy changes. We must add this potential energy change to the standard chemical potential at r' to obtain the standard chemical potential at the new position, just as we did in Sec. 6.1.4 for a gas in a gravitational field:

$$\mu^\circ_{x,B}(r'') = \mu^\circ_{x,B}(r') + M_B \left[\phi(r'') - \phi(r') \right] \qquad (7.9.2)$$

where M_B is the solute molar mass.

The chemical potential of solute B at position r in the cell is related to its activity by

$$\mu_B(r) = \mu^\circ_{x,B}(r) + RT \ln a_{x,B}(r)$$

In an equilibrium state, the solute has the same chemical potential at positions r' and r'':

$$\mu^\circ_{x,B}(r') + RT \ln a_{x,B}(r') = \mu^\circ_{x,B}(r'') + RT \ln a_{x,B}(r'')$$

With the use of Eq. 7.9.2, this becomes

$$\ln a_{x,B}(r'') - \ln a_{x,B}(r') = \frac{M_B}{RT} \left[\phi(r') - \phi(r'') \right]$$

The solute activity is $a_{x,B} = \Gamma_{x,B}\, \gamma_{x,B}\, x_B$ (Table 7.3). Let us assume the solution is dilute enough for us to approximate the solute activity coefficient $\gamma_{x,B}$ by 1. The pressure factor for activity is given by $\Gamma_{x,B} = \exp[V_B(p - p^\circ)/RT]$ (Table 7.4). We then have

$$\ln x''_B - \ln x'_B = \ln a_{x,B}(r'') - \ln \Gamma_{x,B}(r'') - \left[\ln a_{x,B}(r') - \ln \Gamma_{x,B}(r')\right]$$
$$= \frac{M_B}{RT} \left[\phi(r') - \phi(r'')\right] + \frac{V_B(p' - p'')}{RT} \qquad (7.9.3)$$

where x''_B and x'_B are the solute mole fractions and p'' and p' are the pressures at positions r'' and r', respectively.

From Eq. 7.9.1, we have

$$\phi(r') - \phi(r'') = \frac{\omega^2}{2} \left[(r'')^2 - (r')^2 \right]$$

The dependence of pressure on position is given by[19]

$$dp = \omega^2 r \rho\, dr$$

where ρ is the solution density. Integration from r' to r'', with the assumption that the density is independent of pressure, gives

$$p'' - p' = \frac{\omega^2 \rho}{2} \left[(r'')^2 - (r')^2 \right]$$

---

[19] This relation is like the one for a fluid in a gravitational field, $dp = -g\rho\, dh$ (Eq. 6.1.8 on page 147). The acceleration of free fall, g, is replaced with the angular acceleration in the centrifuge cell, $\omega^2 r$. The sign is changed since the centrifugal force is in the $+r$ direction and p increases with r.

Equation 7.9.3 now becomes

$$\ln x_B'' - \ln x_B' = \frac{(M_B - V_B \rho)\omega^2}{2RT}\left[(r'')^2 - (r')^2\right] \tag{7.9.4}$$

This equation shows that if the effective solute density M_B/V_B is greater than the solution density ρ, the mole fraction of B increases with increasing distance from the axis of rotation.

Equation 7.9.4 is needed for *sedimentation equilibrium*, one of the most accurate methods of determining the molar mass of a macromolecule. A dilute solution of the macromolecule is placed in the cell of an analytical ultracentrifuge, and the rotor speed is selected to produce a measurable solute concentration gradient in the cell. With a fixed value of r' and x_B'' considered to be a function of r'', Eq. 7.9.4 predicts that a plot of $\ln x_B$ versus r^2 will be linear with a slope equal to $(M_B - V_B \rho)\omega^2/2RT$. We can write this quantity as

$$\frac{(M_B - V_B \rho)\omega^2}{2RT} = \frac{M_B(1 - v_B \rho)\omega^2}{2RT}$$

where $v_B = V_B/M_B$ is the *partial specific volume* of the solute found from the variation of solution volume with mass of solute added. By this means, the molar mass M_B of the macromolecule is evaluated.

PROBLEMS

7.1 For a binary solution, find expressions for the mole fractions x_B and x_A as functions of the solute molality x_B.

7.2 The *weight fraction* (or mass fraction) w_i of constituent i of a mixture is defined by $w_i = m_i/\sum_j m_j$, where m_i and m_j are masses (*not* molalities).

   (a) Follow the method of Sec. 7.1.4 to find a formula for x_B as a function of w_B, M_A, and M_B in a binary mixture.

   (b) The densities of liquid mixtures of chloroform ($CHCl_3$) and ethanol (CH_3CH_2OH) were measured at 25 °C by G. Scatchard and C. L. Raymond (*J. Am. Chem. Soc.* **1938**, *60*, 1278–1287). The data are in Table 7.5 on the next page. Use these data to evaluate the partial molar volumes of both constituents in an equimolar mixture of chloroform and ethanol. To do this, convert the weight fractions to mole fractions and find the relation among V/n, x_B, and ρ; then prepare a plot of $V/n - x_A V_{m,A}^* - x_B V_{m,B}^*$ versus x_B as suggested in Sec. 7.2.3. You may find it helpful to carry out the calculations with a spreadsheet or simple computer routine.

7.3 Extend the derivation of Prob. 6.1, concerning a liquid droplet of radius r suspended in a gas, to the case in which the liquid and gas may both be mixtures. Show that the equilibrium conditions are $T^g = T^l$, $\mu_i^g = \mu_i^l$ (for each species i that can equilibrate between the two phases), and $p^l = p^g + 2\gamma/r$, where γ is the surface tension. (As in Prob. 6.1, the last relation is the Laplace equation.)

7.4 Consider a mixture of 4.0000×10^{-2} mol of N_2 (A) and 4.0000×10^{-2} mol of CO_2 (B) in a volume of 1.0000×10^{-3} m^3 at a temperature of 298.15 K. The second virial coefficients at this

Table 7.5 Density of chloroform–ethanol mixtures at 25 °C; w_B is the weight fraction of ethanol.

w_B	$\rho/\text{g cm}^3$	w_B	$\rho/\text{g cm}^3$
0	1.47955	0.3591	1.12584
0.0337	1.43640	0.4130	1.08789
0.0658	1.39796	0.5939	0.97374
0.0865	1.37477	0.7276	0.90360
0.1483	1.30804	0.7856	0.87544
0.1651	1.29163	0.8480	0.84774
0.2714	1.19656	1	0.78562

temperature have the values

$$B_{AA} = -4.8 \times 10^{-6} \text{ m}^3 \text{ mol}^{-1}$$
$$B_{BB} = -124.5 \times 10^{-6} \text{ m}^3 \text{ mol}^{-1}$$
$$B_{AB} = -47.5 \times 10^{-6} \text{ m}^3 \text{ mol}^{-1}$$

Compare the pressure of the real gas mixture with that predicted by the ideal gas equation. See Eqs. 7.4.15 and 7.4.18.

7.5 At 25 °C and 1 bar, the Henry's law constants of nitrogen and oxygen dissolved in water are $K_{x,N_2} = 8.68 \times 10^4$ bar and $K_{x,O_2} = 4.40 \times 10^4$ bar. The vapor pressure of water at this temperature and pressure is $p_{H_2O} = 0.032$ bar. Assume that dry air contains only N_2 and O_2 at mole fractions $y_{N_2} = 0.788$ and $y_{O_2} = 0.212$. Consider liquid–gas systems formed by equilibrating liquid water and air at 25 °C and 1.000 bar, and assume that the gas phase behaves as an ideal gas mixture. The sum of the partial pressures of N_2 and O_2 must be $(1.000 - 0.032)$ bar $= 0.968$ bar; since almost all of the N_2 and O_2 is in the predominant phase, these substances must be present in this phase in the same relative amounts as in dry air. Determine the mole fractions of N_2 and O_2 in both phases in two limiting cases:

(a) A large volume of air is equilibrated with just enough water to leave a small drop of liquid.

(b) A large volume of water is equilibrated with just enough air to leave a small bubble of gas.

7.6 Find the general relation between the Henry's law constants $K_{x,B}$ and $K_{m,B}$; then from Eqs. 7.6.17 and 7.6.19, find the relation between the values of the activity coefficients $\gamma_{x,B}$ and $\gamma_{m,B}$ for solute B in a nonideal solution.

7.7 Consider a nonideal binary gas mixture with the simple equation of state $V = nRT/p + nB$ (Eq. 7.4.16).

(a) The *rule of Lewis and Randall* states that the value of the mixed second virial coefficient B_{AB} is the average of B_{AA} and B_{BB}. Show that when this rule holds, the fugacity coefficient of A in a binary gas mixture of any composition is given by $\ln \phi_A = B_{AA}p/RT$. By comparing this expression with Eq. 5.11.7 for a pure gas, express the fugacity of A in the mixture as a function of the fugacity of pure A at the same temperature and pressure as the mixture.

(b) The rule of Lewis and Randall is not accurately obeyed when constituents A and B are chemically dissimilar. For example, at 298.15 K, the second virial coefficients of H_2O (A) and N_2 (B) are $B_{AA} = -1158\,\text{cm}^3\,\text{mol}^{-1}$ and $B_{BB} = -5\,\text{cm}^3\,\text{mol}^{-1}$, respectively, whereas the mixed second virial coefficient is $B_{AB} = -40\,\text{cm}^3\,\text{mol}^{-1}$.

When liquid water is equilibrated with nitrogen at 298.15 K and 1 bar, the partial pressure of H_2O in the gas phase is $p_A = 0.03185$ bar. Use the given values of B_{AA}, B_{BB}, and B_{AB} to calculate the fugacity of the gaseous H_2O in this binary mixture. Compare this fugacity with the fugacity calculated with the value of B_{AB} predicted by the rule of Lewis and Randall.

Table 7.6 Activity coefficient of benzene (A) in mixtures of benzene and 1-octanol at 20 °C. The reference state is the pure liquid.

x_A	γ_A	x_A	γ_A
0	2.0[a]	0.7631	1.183
0.1334	1.915	0.8474	1.101
0.2381	1.809	0.9174	1.046
0.4131	1.594	0.9782	1.005
0.5805	1.370		

[a] extrapolated

7.8 Benzene and 1-octanol are two liquids that mix in all proportions. Benzene has a measurable vapor pressure, whereas 1-octanol is practically nonvolatile. The data in Table 7.6 were obtained by the isopiestic vapor pressure method (Platford, R. F. *J. Solution Chem.* **1976**, *5*, 645–650).

(a) Use numerical integration to evaluate the integral on the right side of Eq. 7.6.20 at each of the values of x_A listed in the table, and thus find γ_B at these compositions.

(b) Draw two curves on the same graph showing the effective mole fractions $\gamma_A x_A$ and $\gamma_B x_B$ as functions of x_A. Are the deviations from ideal-mixture behavior positive or negative?

7.9 Table 7.7 on the next page lists measured values of gas-phase composition and total pressure for the binary two-phase methanol–benzene system at constant temperature and varied liquid-phase composition. x_A is the mole fraction of methanol in the liquid mixture, and y_A is the mole fraction of methanol in the equilibrated gas phase.

(a) For each of the 16 different liquid-phase compositions, tabulate the partial pressures of A and B in the equilibrated gas phase.

(b) Plot p_A and p_B versus x_A on the same graph. Notice that the behavior of the mixture is far from that of an ideal mixture. Are the deviations from Raoult's law positive or negative?

(c) Tabulate and plot the activity coefficient γ_B of the benzene as a function of x_A using a pure-liquid reference state. Assume that the fugacity f_B is equal to p_B, and ignore the effects of variable pressure.

Table 7.7 Liquid and gas compositions in the two-phase system of methanol (A) and benzene (B) at 45 °C (Toghiana, H. *Int. DATA Ser., Sel. Data Mixtures, Ser. A* **1995**, *23*, 234).

x_A	y_A	p/kPa	x_A	y_A	p/kPa
0	0	29.894	0.4201	0.5590	60.015
0.0207	0.2794	40.962	0.5420	0.5783	60.416
0.0314	0.3391	44.231	0.6164	0.5908	60.416
0.0431	0.3794	46.832	0.7259	0.6216	59.868
0.0613	0.4306	50.488	0.8171	0.6681	58.321
0.0854	0.4642	53.224	0.9033	0.7525	54.692
0.1811	0.5171	57.454	0.9497	0.8368	51.009
0.3217	0.5450	59.402	1	1	44.608

(d) Estimate the Henry's law constant $K_{x,A}$ of methanol in the benzene environment at 45 °C by the graphical method suggested in Fig. 7.6(b). Again assume that f_A and p_A are equal, and ignore the effects of variable pressure.

7.10 Consider a dilute binary nonelectrolyte solution in which the dependence of the chemical potential of solute B on composition is given by

$$\mu_B = \mu_{m,B}^{\text{ref}} + RT \ln \frac{m_B}{m^\circ} + k_m m_B$$

where $\mu_{m,B}^{\text{ref}}$ and k_m are constants at a given T and p. (The derivation of this equation is sketched in Sec. 7.6.3.) Use the Gibbs–Duhem equation in the form $d\mu_A = -(n_B/n_A)\, d\mu_B$ to obtain an expression for $\mu_A - \mu_A^*$ as a function of m_B in this solution.

7.11 By means of the isopiestic vapor pressure technique, the molal osmotic coefficients of aqueous solutions of urea at 25 °C have been measured at molalities up to the saturation limit of about 20 mol kg^{-1} (Scatchard, G.; Hamer, W. J.; Wood, S. E. *J. Am. Chem. Soc.* **1938**, *60*, 3061–3070). The experimental values are closely approximated by the function

$$\phi_m = 1.00 - \frac{0.050\, m_B/m^\circ}{1.00 + 0.179\, m_B/m^\circ}$$

where m° is 1 mol kg^{-1}. Calculate values of the solvent and solute activity coefficients γ_A and $\gamma_{m,B}$ at various molalities in the range 0–20 mol kg^{-1} and plot them versus m_B/m°. Use enough points to be able to see the shapes of the curves. What are the limiting slopes of these curves as m_B approaches zero?

7.12 The mean ionic activity coefficient of NaCl in a 0.100 molal aqueous solution at 25 °C has been evaluated with emf measurements, with the result $\ln \gamma_\pm = -0.2505$. Use this value in Eq. 7.8.29, together with the osmotic coefficient values of Table 7.8 on the next page, to evaluate $\gamma_\pm$ at each of the molalities shown in the table; then plot $\gamma_\pm$ as a function of m_B.

7.13 Assume that at sea level the atmosphere has a pressure of 1.00 bar and a composition given by $x_{N_2} = 0.788$ and $x_{O_2} = 0.212$. Find the partial pressures and mole fractions of N_2 and O_2,

Table 7.8 Molal osmotic coefficient of aqueous NaCl at 25 °C (Clarke, E. C. W.; Glew, D. N. *J. Phys. Chem. Ref. Data* **1985**, *14*, 489–610).

$m_B/\text{mol kg}^{-1}$	ϕ_m	$m_B/\text{mol kg}^{-1}$	ϕ_m
0.1	0.9325	2.0	0.9866
0.2	0.9239	3.0	1.0485
0.3	0.9212	4.0	1.1177
0.5	0.9222	5.0	1.1916
1.0	0.9373	6.0	1.2688
1.5	0.9598		

and the total pressure, at an altitude of 10.0 km, making the (drastic) approximation that the atmosphere is an ideal gas mixture in an equilibrium state at 0 °C. For g use the value of the standard acceleration of free fall listed in Appendix B.

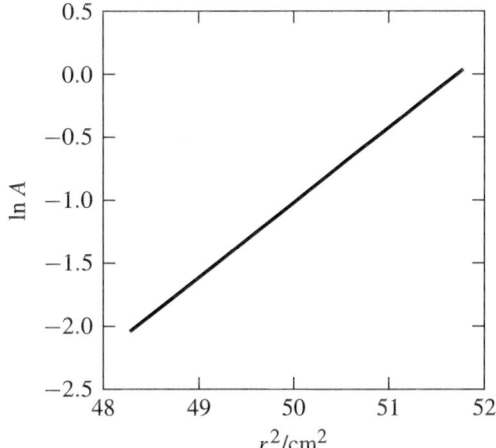

Figure 7.16 Sedimentation equilibrium of a dilute solution of the FhuA-dodecyl maltoside aggregate.

7.14 FhuA is a protein found in the outer membrane of the *Escherichia coli* bacterium. From the known amino acid sequence, its molar mass is calculated to be 78.804 kg mol^{-1}. In aqueous solution, molecules of the detergent dodecyl maltoside bind to a FhuA molecule to form an aggregate that behaves as a single solute species. Figure 7.16 shows data collected in a sedimentation equilibrium experiment with a dilute solution of the aggregate (Boulanger, P.; le Maire, M.; Bonhivers, M.; Dubois, S.; Desmadril, M.; Letellier, L. *Biochemistry* **1996**, *35*, 14216–14224). In the graph, A is the absorbance measured at a wavelength of 280 nm (a property that is a linear function of the aggregate concentration) and r is the radial distance. The experimental points fall very close to the straight line shown in the graph. The sedimentation conditions were $\omega = 838 \, \text{s}^{-1}$ and $T = 283 \, \text{K}$. The authors used the values $v_B = 0.776 \, \text{cm}^3 \, \text{g}^{-1}$ and

$\rho = 1.004 \, \text{g cm}^{-3}$.

Find the molar mass of the aggregate solute species, and use it to estimate the mass binding ratio (mass of bound detergent divided by mass of protein).

Chapter 8

REACTIONS AND OTHER CHEMICAL PROCESSES

Many processes investigated by chemists are described by balanced chemical equations that indicate the conversion of one or more reactants into one or more products. It is convenient to describe the progress of such a chemical process with a variable called the *advancement*. This chapter discusses molar reaction quantities associated with this type of process and develops the important concept of the thermodynamic equilibrium constant.

The chemical processes treated in this chapter have the following features in common: They occur in closed systems, they take place at constant temperature and pressure, and the only kind of work is expansion work. Under these conditions, the enthalpy change is equal to the heat (Eq. 3.11.4), the Gibbs energy decreases as the process proceeds spontaneously, and the equilibrium state is the one of minimum Gibbs energy (Sec. 4.9).

An equilibrium phase transition of a pure substance (Sec. 6.3) has the features just described. Other processes treated in this chapter involve mixtures and nonequilibrium states.

8.1 THE ADVANCEMENT AND MOLAR REACTION QUANTITIES

A chemical process is usually a transfer of species between phases, a rearrangement of bonds, or a combination of these changes. We can describe a chemical process by writing a chemical equation (or reaction equation). Thus, for the vaporization of water, we write

$$H_2O(l) \rightarrow H_2O(g)$$

For the dissolution of sodium chloride in water, we write

$$NaCl(s) \rightarrow Na^+(aq) + Cl^-(aq)$$

For the Haber synthesis of ammonia, the chemical equation is

$$N_2(g) + 3 H_2(g) \rightarrow 2 NH_3(g) \tag{8.1.1}$$

The essential feature of any chemical equation is that equal amounts of each element and equal net charges appear on both sides; the equation is said to be *balanced*. Thus, matter and charge are conserved, and the process may occur in a closed system. The species to the left of a single arrow are called *reactants*, the species to the right are called *products*, and the arrow indicates the *forward* direction of the process. A chemical equation is sometimes written with left and right arrows

$$N_2(g) + 3\,H_2(g) \rightleftharpoons 2\,NH_3(g)$$

to indicate that the process can occur in either direction or that it is at equilibrium.

8.1.1 An example

The ammonia synthesis of Eq. 8.1.1 serves to illustrate the concept of advancement. If the system is *open*, there are five independent variables that we may take to be T, p, and the amounts of N_2, H_2, and NH_3. We would write the total differential of the enthalpy, for instance, as

$$dH = \left(\frac{\partial H}{\partial T}\right)_{p,n} dT + \left(\frac{\partial H}{\partial p}\right)_{T,n} dp + H_{N_2}\,dn_{N_2} + H_{H_2}\,dn_{H_2} + H_{NH_3}\,dn_{NH_3} \quad (8.1.2)$$

(this is like Eq. 7.3.9). The quantities H_{N_2}, H_{H_2}, and H_{NH_3} are partial molar enthalpies. For example, H_{N_2} is defined by

$$H_{N_2} = \left(\frac{\partial H}{\partial n_{N_2}}\right)_{T,p,n_{H_2},n_{NH_3}}$$

If the system is *closed*, the amounts of the three substances can still change because of the reaction described by Eq. 8.1.1. In the closed system, the number of independent variables is reduced to three. We may choose them to be T, p, and a variable called advancement. The **advancement**, or extent of reaction, ξ, is the amount by which the reaction as described by the chemical equation has advanced in the forward direction from specified initial conditions. The quantity ξ has dimensions of amount; the usual unit for ξ is the mole.

Let the initial amounts be $n_{N_2,0}$, $n_{H_2,0}$, and $n_{NH_3,0}$. Then at any stage of the reaction process, the amounts are given by

$$n_{N_2} = n_{N_2,0} - \xi$$
$$n_{H_2} = n_{H_2,0} - 3\xi$$
$$n_{NH_3} = n_{NH_3,0} + 2\xi$$

These relations come from the stoichiometry of the reaction as expressed by the chemical equation. The relation for H_2, for example, expresses the fact that when one mole of reaction has occurred the amount of H_2 in the closed system has decreased by three moles.

Taking the differentials of the amounts, we find the infinitesimal changes are related to the change in ξ by

$$dn_{N_2} = -d\xi$$
$$dn_{H_2} = -3\,d\xi$$
$$dn_{NH_3} = 2\,d\xi$$

These relations show that in a closed system the changes in the various amounts are not independent. Making these substitutions in Eq. 8.1.2 gives

$$dH = \left(\frac{\partial H}{\partial T}\right)_{p,n} dT + \left(\frac{\partial H}{\partial p}\right)_{T,n} dp$$
$$+ \left(-H_{N_2} - 3H_{H_2} + 2H_{NH_3}\right) d\xi \qquad (8.1.3)$$
$$\text{(closed system)}$$

Since the right side of Eq. 8.1.3 has as many terms as the number of independent variables, it is an expression for the total differential of the enthalpy. The coefficient of $d\xi$ in this equation is called the **molar reaction enthalpy**, or molar enthalpy of reaction, $\Delta_r H_m$:

$$\Delta_r H_m = -H_{N_2} - 3H_{H_2} + 2H_{NH_3} \qquad (8.1.4)$$

which we identify as the partial derivative

$$\Delta_r H_m = \left(\frac{\partial H}{\partial \xi}\right)_{T,p} \qquad (8.1.5)$$

Equation 8.1.5 shows that $\Delta_r H_m$ is the rate at which the enthalpy changes with the advancement as the reaction proceeds in the forward direction at constant T and p.

> The partial molar enthalpy of a species is the rate of enthalpy increase per amount of the species added to an *open* system. To see why the particular combination of partial molar enthalpies on the right side of Eq. 8.1.4 is the rate at which enthalpy changes with advancement in the *closed* system, we can think of the following process at constant T and p: An infinitesimal amount dn of N_2 is removed from an open system, three times this amount of H_2 is removed, and twice this amount of NH_3 is added. The total enthalpy change in the open system is $dH = (-H_{N_2} - 3H_{H_2} + 2H_{NH_3})dn$. However, the net change in the system is equivalent to an advancement $d\xi = dn$ in a closed system, so $dH/d\xi$ in the closed system is equal to $(-H_{N_2} - 3H_{H_2} + 2H_{NH_3})$ in agreement with Eqs. 8.1.4 and 8.1.5.

Note that since the advancement is defined by how we write the chemical equation, the value of $\Delta_r H_m$ also depends on the chemical equation. For instance, if instead of Eq. 8.1.1 we write

$$\tfrac{1}{2}N_2(g) + \tfrac{3}{2}H_2(g) \rightarrow NH_3(g)$$

then the value of $\Delta_r H_m$ is halved.

8.1.2 Molar reaction quantities in general

Now let us generalize the relations of the preceding section for any chemical process in a closed system. Suppose the chemical equation has the form

$$a\text{A} + b\text{B} \rightarrow d\text{D} + e\text{E}$$

where A and B are reactant species, D and E are product species, and a, b, d, and e are the corresponding stoichiometric coefficients. When we rewrite the equation in the algebraic form

$$a\text{A} + b\text{B} = d\text{D} + e\text{E}$$

we can rearrange it to yield

$$0 = -a\text{A} - b\text{B} + d\text{D} + e\text{E}$$

In general, the stoichiometric relation for any chemical process is

$$0 = \sum_i \nu_i \text{A}_i$$

where ν_i is the **stoichiometric number** of Species A_i, taken as negative for a reactant and positive for a product. In the ammonia synthesis example of Sec. 8.1.1, the stoichiometric relation is $0 = -\text{N}_2 - 3\text{H}_2 + 2\text{NH}_3$; so the stoichiometric numbers are $\nu_{\text{N}_2} = -1$, $\nu_{\text{H}_2} = -3$, and $\nu_{\text{NH}_3} = +2$. In other words, each stoichiometric number is the same as the stoichiometric coefficient in the chemical equation except that the sign is negative for a reactant.

The amount of reactant or product Species i present in the closed system during the process depends on the advancement and is given by

$$n_i = n_{i,0} + \nu_i \xi \tag{8.1.6}$$
(closed system)

The infinitesimal change in the amount due to an infinitesimal change in the advancement must be

$$dn_i = \nu_i \, d\xi \tag{8.1.7}$$
(closed system)

For the independent variables of the closed system, we choose T, p, and the amounts of the species; then the total differential of any extensive property X is

$$dX = \left(\frac{\partial X}{\partial T}\right)_{p,n} dT + \left(\frac{\partial X}{\partial p}\right)_{T,n} dp + \sum_i X_i \, dn_i$$

where X_i is a partial molar quantity. Restricting the system to a closed one and substituting from Eq. 8.1.7 gives the relation

$$\mathrm{d}X = \left(\frac{\partial X}{\partial T}\right)_{p,n} \mathrm{d}T + \left(\frac{\partial X}{\partial p}\right)_{T,n} \mathrm{d}p + \Delta_\mathrm{r} X_\mathrm{m}\, \mathrm{d}\xi \qquad (8.1.8)$$
(closed system)

where the **molar reaction quantity** $\Delta_\mathrm{r} X_\mathrm{m}$ is defined by

$$\Delta_\mathrm{r} X_\mathrm{m} \equiv \sum_i \nu_i X_i \qquad (8.1.9)$$

In Eq. 8.1.8, we identify the coefficient of the last term on the right as

$$\Delta_\mathrm{r} X_\mathrm{m} = \left(\frac{\partial X}{\partial \xi}\right)_{T,p} \qquad (8.1.10)$$
(closed system)

Equation 8.1.10 shows that the molar reaction quantity $\Delta_\mathrm{r} X_\mathrm{m}$ is the rate at which the extensive property X changes with the advancement in a closed system at constant T and p. The value of $\Delta_\mathrm{r} X_\mathrm{m}$ depends on the independent variables T, p, and ξ.

Note in particular that $\Delta_\mathrm{r} X_\mathrm{m}$ is not necessarily the same as a finite change in X divided by a finite change in ξ. This latter quantity is a molar *integral* reaction quantity, which we denote by the superscript "int":

$$\Delta_\mathrm{r} X_\mathrm{m}^\mathrm{int} \equiv \frac{\Delta_\mathrm{r} X}{\Delta \xi} = \frac{X(\xi_2) - X(\xi_1)}{\xi_2 - \xi_1} \qquad (8.1.11)$$

In contrast, $\Delta_\mathrm{r} X_\mathrm{m}$ is a molar *differential* quantity.

In certain chemical processes involving more than one phase, the intensive properties of each phase remain unchanged during the course of the process. In these processes, the value of the differential quantity $\Delta_\mathrm{r} X_\mathrm{m}$ depends on T and p, but not on ξ, and is equal to the integral quantity $\Delta_\mathrm{r} X_\mathrm{m}^\mathrm{int}$.

> The use of the Δ symbol for a differential quantity can be confusing since strictly speaking this symbol should denote a finite difference between an initial and a final state. Nevertheless, notations such as $\Delta_\mathrm{r} X_\mathrm{m}$ for the rate of change of X with ξ are widely used and sanctioned by the IUPAC Green Book. Some other notations that have been suggested[1] for molar differential quantities, to distinguish them from finite changes, are $\Delta \overline{X}$, $\blacktriangle X$, and $\Delta \widetilde{X}$.

The notation for a molar reaction quantity (either differential or integral) should include a subscript following the Δ symbol to indicate the kind of chemical process. The subscript "r" denotes a reaction or process in general. The meanings of "vap," "sub," "fus," and "trs" were described in Sec. 6.3.1. Subscripts for specific kinds of reactions and processes are listed in Appendix D and are illustrated in sections to follow.

---
[1] See MacDonald, J. J. *J. Chem. Educ.* **1990**, *67*, 380–382 and references therein.

8.1.3 Standard molar reaction quantities

If a chemical process occurs while each reactant and product remains in its standard state, the molar reaction quantity $\Delta_r X_m$ is called the **standard molar reaction quantity** and is denoted by $\Delta_r X_m^\circ$. For instance, $\Delta_{vap} H_m^\circ$ is a standard molar enthalpy of vaporization (already discussed in Sec. 6.3.4), and $\Delta_r G_m^\circ$ is the standard molar Gibbs energy of a reaction.

From Eq. 8.1.9, the relation between a standard molar reaction quantity and the standard molar quantities of the reactants and products at the same temperature is

$$\Delta_r X_m^\circ \equiv \sum_i \nu_i X_i^\circ \tag{8.1.12}$$

Three comments are in order.

1. As we have seen, the standard state of a constituent of a nonideal gas mixture or of a solute in a nonideal liquid mixture is a hypothetical state, and for these mixtures the standard molar reaction quantity refers to a hypothetical process.

2. Whereas a molar reaction quantity is usually a function of T, p, and ξ, a *standard molar reaction quantity is a function only of T*. This is evident because standard-state conditions imply that each reactant and product is in a separate phase of constant defined composition and constant pressure.

3. Since a standard molar reaction quantity is independent of ξ, the standard molar differential and integral quantities are identical.

We are now ready to apply these general concepts to some specific chemical processes.

8.2 MIXING PROCESSES

A **mixing process** is a process in which pure substances mix at constant temperature and pressure. In the initial state, the system has two or more separate phases, each containing a different pure substance at the same T and p. The final state is a one-phase mixture at this T and p.

8.2.1 Mixtures in general

First let us consider changes in the Gibbs energy G. Since this is an extensive property, G in the initial state is the sum of G for each pure phase:

$$G_1 = \sum_i n_i \mu_i^*$$

where μ_i^* is the chemical potential (molar Gibbs energy) of pure substance i at the given T and p. For the final state, we use the additivity rule (Eq. 7.2.9) for a single phase:

$$G_2 = \sum_i n_i \mu_i$$

Section 8.2 Mixing Processes

where μ_i is the chemical potential of i in the mixture, a function of the composition. The overall change of G, the **Gibbs energy of mixing**, is then

$$\Delta_{\text{mix}} G = G_2 - G_1 = \sum_i n_i (\mu_i - \mu_i^*) \tag{8.2.1}$$

The **molar Gibbs energy of mixing**, $\Delta_{\text{mix}} G_{\text{m}}$, is the Gibbs energy of mixing per amount of mixture formed; that is, $\Delta_{\text{mix}} G_{\text{m}} = \Delta_{\text{mix}} G / n$, where n is the sum $\sum_i n_i$. Dividing both sides of Eq. 8.2.1 by n, we obtain

$$\Delta_{\text{mix}} G_{\text{m}} = \sum_i x_i (\mu_i - \mu_i^*) \tag{8.2.2}$$

where x_i is the mole fraction of substance i in the mixture. (For a gaseous mixture, we may write y_i in place of x_i.)

Now let us put these mixing quantities into the general framework of molar reaction quantities described in Sec. 8.1.2. For the process that forms unit total amount of mixture for unit advancement, we may write the chemical equation in the form

$$\sum_i x_i A_i^* \rightarrow \sum_i x_i A_i (\text{mixt})$$

For example, for the mixing of oxygen and nitrogen to form dry air with the composition $y_{O_2} = 0.212$, we could write

$$0.212 \, O_2(\text{pure}) + 0.788 \, N_2(\text{pure}) \rightarrow 0.212 \, O_2(\text{mixt}) + 0.788 \, N_2(\text{mixt})$$

Thus, for a mixing process, the stoichiometric number of pure substance i (a reactant) is $\nu_i = -x_i$, and the stoichiometric number of substance i in the product mixture is $\nu_i = x_i$. Then from the general expression for a molar reaction quantity, $\Delta_r X_m = \sum_i \nu_i X_i$ (Eq. 8.1.9), we obtain

$$\Delta_{\text{mix}} X_{\text{m}} = \sum_i x_i (X_i - X_i^*) \tag{8.2.3}$$

X_i is the partial molar quantity of i in the mixture, and X_i^* is the molar quantity of pure i at the same temperature and pressure as the mixture.[2]

If we let X be the Gibbs energy G, so that X_i is μ_i and X_i^* is μ_i^*, the right side of Eq. 8.2.3 becomes identical to the right side of Eq. 8.2.2. That is, the molar Gibbs energy of mixing $\Delta_{\text{mix}} G_{\text{m}}$ is a molar reaction quantity. The advancement ξ is equal to the amount n of mixture formed. Because the mixture is formed at constant composition, the differential and integral molar reaction quantities for the mixing process are equal:

$$\Delta_{\text{mix}} G_{\text{m}} = \left(\frac{\partial G}{\partial \xi} \right)_{T,p} = \frac{\Delta G}{\Delta \xi}$$

---

[2]In a phase of a single substance, a partial molar quantity of the substance is the same as the molar quantity. Thus, X_i^* and $X_{\text{m},i}^*$ are equivalent notations.

By replacing X in Eq. 8.2.3 with any extensive property, we obtain the expression for the molar mixing quantity for that property. There is no difference between the differential and integral values of any of the molar mixing quantities.

Note the similarities and differences between a molar mixing quantity and a partial molar quantity. The molar volume of mixing, $\Delta_{\text{mix}} V_{\text{m}}$, and the partial molar volume, V_i, both refer to a change at constant T and p and have dimensions of volume per amount. However, $\Delta_{\text{mix}} V_{\text{m}}$ is the volume change per amount of mixture formed from pure phases in a *closed* system, whereas V_i is the volume change per amount of i added to an *open* system.

8.2.2 Ideal mixtures

When the mixture formed is an ideal gas, liquid, or solid mixture, and the pure constituents have the same physical state as the mixture, the expressions for various molar mixing quantities are particularly simple. We denote an ideal molar mixing quantity with a superscript "id," for example, $\Delta_{\text{mix}} G_{\text{m}}^{\text{id}}$.

The chemical potential of constituent i of an ideal mixture is related to the mole fraction x_i by the relation (Eq. 7.5.4)

$$\mu_i = \mu_i^* + RT \ln x_i$$

By combining this relation with Eq. 8.2.2, we find the molar Gibbs energy of mixing to form an ideal mixture is given by

$$\Delta_{\text{mix}} G_{\text{m}}^{\text{id}} = RT \sum_i x_i \ln x_i \qquad (8.2.4)$$

Since each mole fraction is less than one and the logarithm of a fraction is negative, it follows that $\Delta_{\text{mix}} G_{\text{m}}^{\text{id}}$ is negative for every composition of the mixture. Thus, the mixing process is spontaneous, and *an ideal mixture at constant temperature and pressure cannot spontaneously separate into two phases.*

We obtain expressions for other molar mixing quantities by substituting formulas for partial molar quantities of constituents of an ideal mixture derived in Sec. 7.5.3 into Eq. 8.2.3. From $S_i = S_i^* - R \ln x_i$ (Eq. 7.5.5), we obtain

$$\Delta_{\text{mix}} S_{\text{m}}^{\text{id}} = -R \sum_i x_i \ln x_i \qquad (8.2.5)$$

This quantity is positive.

> Although the molar entropy of mixing to form an *ideal* mixture is positive, this is not true for some nonideal mixtures. McGlashan[3] cites the *negative* value $\Delta_{\text{mix}} S_{\text{m}} = -8.8 \, \text{J K}^{-1} \, \text{mol}^{-1}$ for an equimolar mixture of diethylamine and water at 322 K.

From $H_i = H_i^*$ (Eq. 7.5.6) and $U_i = U_i^*$ (Eq. 7.5.8), we have

$$\Delta_{\text{mix}} H_{\text{m}}^{\text{id}} = 0 \qquad (8.2.6)$$

---

[3] McGlashan, p. 241 (reference in Appendix I).

Section 8.2 Mixing Processes

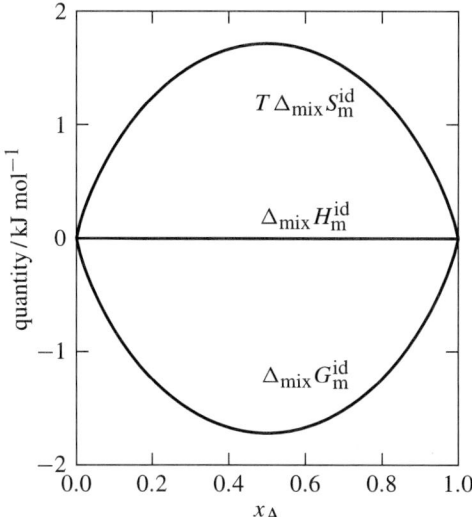

Figure 8.1 Molar mixing quantities for a binary ideal mixture at 298.15 K.

and

$$\Delta_{\text{mix}} U_m^{\text{id}} = 0 \qquad (8.2.7)$$

Thus, the mixing of liquids that form an ideal mixture is an *athermal* process, one in which no heat transfer is necessary to keep the temperature constant.

From $V_i = V_i^*$ (Eq. 7.5.7), we get

$$\Delta_{\text{mix}} V_m^{\text{id}} = 0 \qquad (8.2.8)$$

showing that the ideal molar volume of mixing is zero, and an ideal mixture has the same volume as the sum of the volumes of the pure components at the same T and p.[4]

Figure 8.1 shows how $\Delta_{\text{mix}} G_m^{\text{id}}$, $T\Delta_{\text{mix}} S_m^{\text{id}}$, and $\Delta_{\text{mix}} H_m^{\text{id}}$ depend on the composition of a binary ideal mixture. Although it is not obvious in the figure, the curves for $\Delta_{\text{mix}} G_m^{\text{id}}$ and $T\Delta_{\text{mix}} S_m^{\text{id}}$ have slopes of $+\infty$ or $-\infty$ at each end.

8.2.3 Excess quantities

For the process of mixing pure substances to form a nonideal mixture, we can refer to the value of an **excess molar mixing quantity**. This is simply the difference between the actual molar mixing quantity and the value it would have had if the mixture were ideal. Examining the dependence of excess molar mixing quantities on composition is a convenient way to characterize deviations from ideal-mixture behavior.

---

[4]From the fact mentioned on p. 173 that the volume of a mixture of water and methanol is different from the sum of the volumes of the pure liquids, we can deduce that the mixture is nonideal even though water and methanol mix in all proportions.

From Eqs. 8.2.4–8.2.8, the excess molar Gibbs energy, entropy, enthalpy, internal energy, and volume of mixing are

$$G_m^E = \Delta_{mix} G_m - RT \sum_i x_i \ln x_i$$

$$S_m^E = \Delta_{mix} S_m + R \sum_i x_i \ln x_i$$

$$H_m^E = \Delta_{mix} H_m$$

$$U_m^E = \Delta_{mix} U_m$$

$$V_m^E = \Delta_{mix} V_m$$

By substitution from Eqs. 7.6.3 and 8.2.2 in the expression for the excess molar Gibbs energy, we can relate it to the activity coefficients of the mixture constituents based on pure-liquid reference states:

$$G_m^E = RT \sum_i x_i \ln \gamma_i \tag{8.2.9}$$

It is also possible to derive the relation

$$\left[\frac{\partial (nG_m^E)}{\partial n_i} \right]_{T,p,n_{j \neq i}} = RT \ln \gamma_i \tag{8.2.10}$$

To derive Eq. 8.2.10, we consider infinitesimal changes in the mixture composition at constant T and p. From Eq. 8.2.9, we write

$$d(nG_m^E) = RT \sum_i d(n_i \ln \gamma_i) = RT \sum_i n_i \, d \ln \gamma_i + RT \sum_i (\ln \gamma_i) \, dn_i$$

From $\mu_i = \mu_i^* + RT \ln(\gamma_i x_i)$, we have $d\mu_i = RT(d \ln \gamma_i + dx_i/x_i)$. Substitution in the Gibbs–Duhem equation, $\sum_i x_i d\mu_i = 0$, gives

$$\sum_i x_i d \ln \gamma_i + \sum_i dx_i = 0$$

From $\sum_i x_i = 1$, the sum $\sum_i dx_i$ must be zero. Multiplying the preceding equation by the total amount, n, we obtain $\sum_i n_i \, d \ln \gamma_i = 0$, which turns the equation for $d(nG_m^E)$ into

$$d(nG_m^E) = RT \sum_i (\ln \gamma_i) \, dn_i$$

from which Eq. 8.2.10 follows.

8.2.4 The entropy change to form an ideal gas mixture

When pure ideal gases mix at constant T and p, the entropy change given by Eq. 8.2.5 is positive.

Consider a pure ideal-gas phase. Entropy is an extensive property, so if we divide this phase into two subsystems with a physical partition, the entropy remains unchanged. The reverse process, the removal of the partition, must also have zero entropy change. Despite the fact that the latter process allows the molecules in the two subsystems to intermingle without a change in T or p, it cannot be considered "mixing" since the entropy does not increase. The essential point is that the *same* substance is present in both of the subsystems, so there is no macroscopic change of state when the partition is removed.

From these considerations, one might conclude that the fundamental reason the entropy increases when pure ideal gases mix is that different substances become intermingled. However, this conclusion would be mistaken.

The partial molar entropy of constituent i of an ideal gas mixture is $S_i = S_i^\circ - R\ln(p_i/p^\circ)$ (Eq. 7.4.5). But the partial pressure p_i is equal to $n_i RT/V$ (Eq. 7.4.3), so at a fixed temperature and for a fixed amount of i, the partial molar entropy depends only on the *volume* and is unaffected by the presence of other substances mixed with i.

For a *pure* ideal gas, the formula for S_i gives the *molar* entropy. The entropy of the mixture is given by the additivity rule: $S = \sum_i n_i S_i$. We conclude that *the entropy of an ideal gas mixture equals the sum of the entropies of the separated pure ideal gases, each pure gas occupying the same volume and having the same temperature as in the mixture*.

We can now understand the reason the entropy change is positive when ideal gases mix at constant T and p: Each substance occupies a greater *volume* in the final state than initially. Exactly the same entropy increase would result if the volume of each of the pure ideal gases were increased without mixing. In the case of a *liquid*, the only practical way to increase the volume isothermally without vaporization is to mix the liquid with another liquid; if the resulting liquid mixture is ideal, the entropy increase is also given by Eq. 8.2.5.

8.2.5 Molecular model of a liquid mixture

We have seen that mixing two pure liquids to form an ideal liquid mixture at the same T and p causes no change of the volume and internal energy. A simple molecular model of a binary liquid mixture will elucidate the energetic molecular properties that are consistent with these macroscopic conditions. The model assumes the excess molar entropy, but not necessarily the excess molar internal energy, is zero. The model is of the type sometimes called the *quasicrystalline lattice model*, and the mixture it describes is sometimes called a *simple* mixture. Of course, a molecular model like this is outside the realm of classical thermodynamics.

The model is for Substances A and B in gas and liquid phases at a fixed temperature. Let the standard molar internal energy of pure gaseous A be $U_{m,A}^\circ(g)$. This is the molar energy in the absence of intermolecular interactions, and it depends only on the molecular constitution and the temperature. The molar internal energy of pure liquid A is lower because of the attractive intermolecular forces in the liquid phase. We assume the energy difference is equal to a sum of pairwise nearest-neighbor interactions in the liquid. Thus, the molar internal energy of pure liquid A is given by

$$U_{m,A}^* = U_{m,A}^\circ(g) + k_{AA}$$

where k_{AA} (approximately the negative of the molar internal energy of vaporization) is the interaction energy per amount of A from A–A interactions when each molecule of A is surrounded only by other molecules of A.

Similarly, the molar internal energy of pure liquid B is given by

$$U_{m,B}^* = U_{m,B}^\circ(g) + k_{BB} \tag{8.2.11}$$

where k_{BB} is for B–B interactions.

In a liquid mixture of A and B, we assume that the numbers of nearest-neighbor molecules of A and B surrounding any given molecule are in proportion to the mole fractions x_A and x_B.[5] Then the number of A–A interactions is proportional to $n_A x_A$, the number of B–B interactions is proportional to $n_B x_B$, and the number of A–B interactions is proportional to $n_A x_B + n_B x_A$. The internal energy of the liquid mixture is then

$$U(\text{mixt}) = n_A U_{m,A}^\circ(g) + n_B U_{m,B}^\circ(g)$$
$$+ n_A x_A k_{AA} + n_B x_B k_{BB} + (n_A x_B + n_B x_A) k_{AB} \tag{8.2.12}$$

where k_{AB} is the interaction energy per amount of A when each molecule of A is surrounded only by molecules of B, or the interaction per amount of B when each molecule of B is surrounded only by molecules of A.

The internal energy change for mixing amounts n_A of Liquid A and n_B of Liquid B is now

$$\Delta_{\text{mix}} U = U(\text{mixt}) - n_A U_{m,A}^* - n_B U_{m,B}^*$$
$$= n_A x_A k_{AA} + n_B x_B k_{BB} + (n_A x_B + n_B x_A) k_{AB} - n_A k_{AA} - n_B k_{BB}$$
$$= n_A (x_A - 1) k_{AA} + n_B (x_B - 1) k_{BB} + (n_A x_B + n_B x_A) k_{AB}$$

With the identities $x_A - 1 = -x_B$, $x_B - 1 = -x_A$, and $n_A x_B = n_B x_A = n_A n_B / n$ (where n is the sum $n_A + n_B$), we obtain

$$\Delta_{\text{mix}} U = \frac{n_A n_B}{n}(2k_{AB} - k_{AA} - k_{BB}) \tag{8.2.13}$$

If the internal energy change to form a mixture of any composition is to be zero, as it is for an ideal mixture, the quantity $(2k_{AB} - k_{AA} - k_{BB})$ must be zero, which means k_{AB} must equal $(k_{AA} + k_{BB})/2$. Thus, one requirement for an ideal mixture is that *an A–B interaction equals the average of an A–A interaction and a B–B interaction.*

If we write Eq. 8.2.12 in the form

$$U(\text{mixt}) = n_A U_{m,A}^\circ(g) + n_B U_{m,B}^\circ(g) + \frac{1}{n_A + n_B}(n_A^2 k_{AA} + 2 n_A n_B k_{AB} + n_B^2 k_{BB})$$

---

[5] This assumption requires the molecules of A and B to have similar sizes and shapes and to be randomly mixed in the mixture. Statistical mechanics theory shows that the molecular sizes must be approximately equal if the excess molar entropy is to be zero.

Section 8.2 Mixing Processes

we can differentiate with respect to n_B at constant n_A to evaluate the partial molar internal energy of B. The result can be rearranged to the simple form

$$U_B = U_{m,B}^* + (2k_{AB} - k_{AA} - k_{BB})(1 - x_B)^2 \qquad (8.2.14)$$

where $U_{m,B}^*$ is given by Eq. 8.2.11. This equation predicts that the value of U_B decreases with increasing x_B if k_{AB} is less negative than the average of k_{AA} and k_{BB}, increases for the opposite situation, and is equal to $U_{m,B}^*$ in an ideal liquid mixture.

The model is for what is called a *regular solution*, one with the same entropy of mixing as an ideal mixture (Eq. 8.2.5) when the volume of mixing is zero, but whose internal energy and enthalpy of mixing are not necessarily zero.

The excess molar Gibbs energy of a mixture is $G_m^E = U_m^E + pV_m^E - TS_m^E$. Using the expression of Eq. 8.2.13 and with the further assumptions that the excess molar entropy and volume are zero, this model predicts the excess molar Gibbs energy is given by

$$G_m^E = \frac{\Delta_{mix} U}{n} = x_A x_B (2k_{AB} - k_{AA} - k_{BB})$$

This is a symmetric function of x_A and x_B. It predicts, for example, that coexisting liquid layers in a binary system (Sec. 8.2.6) have the same value of x_A in one phase as the value of x_B in the other.

Measured molar excess Gibbs energies are often unsymmetric functions; to represent them, a more general function is needed. A commonly used function for a binary mixture is the **Redlich–Kister series** given by

$$G_m^E = x_A x_B \left[a + b(x_A - x_B) + c(x_A - x_B)^2 + \cdots \right]$$

where the parameters $a, b, \cdots$ depend on T and p but not on composition. This function satisfies the only requirement for the dependence of G_m^E on composition—that it equal zero when either x_A or x_B is zero.[6]

For many binary systems, the first two terms of the series can reproduce experimental G_m^E data reasonably well. This is the two-parameter Redlich–Kister series

$$G_m^E = x_A x_B [a + b(x_A - x_B)] \qquad (8.2.15)$$

in which the parameters a and b are adjusted to fit the experimental data. The activity coefficients in a mixture obeying this equation are found, from Eq. 8.2.10, to be given by

$$RT \ln \gamma_A = x_B^2 [a + (3 - 4x_B)b] \qquad RT \ln \gamma_B = x_A^2 [a + (4x_A - 3)b]$$

8.2.6 Phase separation of a liquid mixture

A binary liquid mixture in a system maintained at constant T and p can spontaneously separate into two liquid layers if any part of the curve of a plot of $\Delta_{mix} G_m$ versus x_A is concave downward. To understand this phenomenon, consider the curve shown in Fig. 8.2.

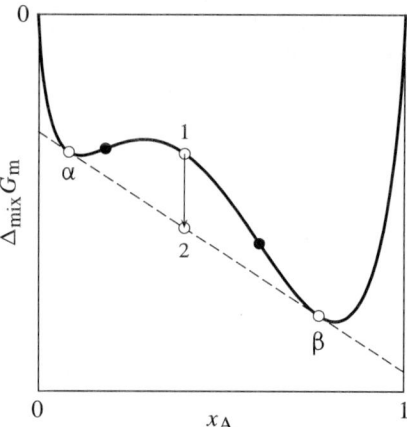

Figure 8.2 Gibbs energy of mixing versus composition in a binary liquid mixture with spontaneous phase separation. The inflection points are indicated by filled circles.

The molar Gibbs energy of mixing, which is plotted on the y (vertical) axis, can be rewritten as follows with the use of Eq. 8.2.2:

$$\Delta_{\text{mix}} G_{\text{m}} = x_{\text{A}}(\mu_{\text{A}} - \mu_{\text{A}}^*) + x_{\text{B}}(\mu_{\text{B}} - \mu_{\text{B}}^*) = \frac{n_{\text{A}}\mu_{\text{A}} + n_{\text{B}}\mu_{\text{B}}}{n} - x_{\text{A}}\mu_{\text{A}}^* - x_{\text{B}}\mu_{\text{B}}^*$$

$$= \frac{G}{n} - x_{\text{A}}\mu_{\text{A}}^* - x_{\text{B}}\mu_{\text{B}}^*$$

The plot, then, has the form $X/n - x_{\text{A}}X_{\text{A}}^* - x_{\text{B}}X_{\text{B}}^*$ versus x_{A} used in the method of intercepts (Sec. 7.2.4) to find the partial molar quantities X_{A} and X_{B}. In the present application of this method, the general extensive property X is the Gibbs energy G. By the method of intercepts, the tangent to the curve at any point on the curve intercepts the $x_{\text{A}} = 0$ axis at $y = \mu_{\text{B}} - \mu_{\text{B}}^*$ and intercepts the $x_{\text{A}} = 1$ line at $y = \mu_{\text{A}} - \mu_{\text{A}}^*$, where μ_{B} and μ_{A} are the chemical potentials in the mixture represented by the point on the curve.

For two binary liquid phases to be in transfer equilibrium, μ_{A} must be the same in both phases and μ_{B} must also be the same in both phases. The dashed line in the figure is a common tangent to the curve at the points labeled α and β. These two points are the only ones having a common tangent, and what makes the common tangent possible is the downward concavity (negative curvature) of a portion of the curve between these points. Thus, phases of compositions x_{A}^α and x_{A}^β can be in equilibrium with one another since the tangents at these points have the same intercepts; the necessary conditions $\mu_{\text{A}}^\alpha = \mu_{\text{A}}^\beta$ and $\mu_{\text{B}}^\alpha = \mu_{\text{B}}^\beta$ are satisfied.

Now consider Point 1 on the curve. A single phase of this composition, if it could exist, would spontaneously separate into the two phases of compositions x_{A}^α and x_{A}^β because the

---

[6]For a binary mixture, Eq. 8.2.9 becomes $G_{\text{m}}^{\text{E}} = RT(x_{\text{A}} \ln \gamma_{\text{A}} + x_{\text{B}} \ln \gamma_{\text{B}})$. When x_{A} is zero, γ_{B} is 1 and $\ln \gamma_{\text{B}}$ is zero. When x_{B} is zero, γ_{A} is 1 and $\ln \gamma_{\text{A}}$ is zero. Thus, G_{m}^{E} must be zero in both cases.

Section 8.2 Mixing Processes

Gibbs energy per total amount then decreases to the extent indicated by the arrow. We know that a process in which G decreases at constant T and p in a closed system, with expansion work only, is a spontaneous process (Sec. 4.9).

To show that the arrow represents the change in G/n for phase separation, we write the equation of the line through points α and β (y as a function of x_A):

$$y = y^\alpha + \left(\frac{y^\beta - y^\alpha}{x_A^\beta - x_A^\alpha}\right)(x_A - x_A^\alpha)$$

In the system either before or after phase separation occurs, x_A is the mole fraction of component A in the system as a whole. When phases α and β are present, containing amounts n^α and n^β, x_A is given by the expression

$$x_A = \frac{x_A^\alpha n^\alpha + x_A^\beta n^\beta}{n^\alpha + n^\beta}$$

By substituting this expression in the equation for the line, after some rearrangement and using $n^\alpha + n^\beta = n$, we obtain

$$y = \frac{1}{n}\left(n^\alpha y^\alpha + n^\beta y^\beta\right)$$

which equates y for a point on the line to the Gibbs energy change for mixing pure components to form an amount n^α of phase α and an amount n^β of phase β, divided by the total amount n. Thus, the difference in y between Points 1 and 2 in the figure is the decrease in molar Gibbs energy when a single phase separates into two equilibrated phases.

A mixture with any value of x_A between x_A^α and x_A^β is unstable with respect to separation into two phases of compositions x_A^α and x_A^β. Phase separation occurs only if the curve of the plot of $\Delta_{mix}G_m$ versus x_A is concave downward, which requires the curve to have at least two inflection points. However, the compositions of the two phases are not the compositions at inflection points; nor in the case of the curve shown in Fig. 8.2 are these compositions the same as those of the two local minima.

By varying the values of parameters in an expression for the excess molar Gibbs energy, we can model the onset of phase separation with a change of temperature. Figure 8.3 on the next page shows the results of using the two-parameter Redlich–Kister series (Eq. 8.2.15).

When G_m^E is zero at all mixture compositions, we have an ideal liquid mixture (or an ideal mixture in general); $\Delta_{mix}G_m$ is negative (except at the extremes $x_A = 0$ and $x_A = 1$), and the curve of the plot of $\Delta_{mix}G_m$ versus x_A is convex downward (Fig. 8.3(a), Curve 1). Thus, mixing is spontaneous, and no phase separation can occur: *The constituents of an ideal mixture mix spontaneously in all proportions.*

When the properties of the mixture are such that G_m^E is positive at each mixture composition (except at the extremes $x_A = 0$ and $x_A = 1$ where it must be zero), $\Delta_{mix}G_m$ is greater than $\Delta_{mix}G_m$ for forming a hypothetical ideal mixture of the same composition. Thus, the activity of A is greater in this nonideal mixture than in the hypothetical ideal mixture, as

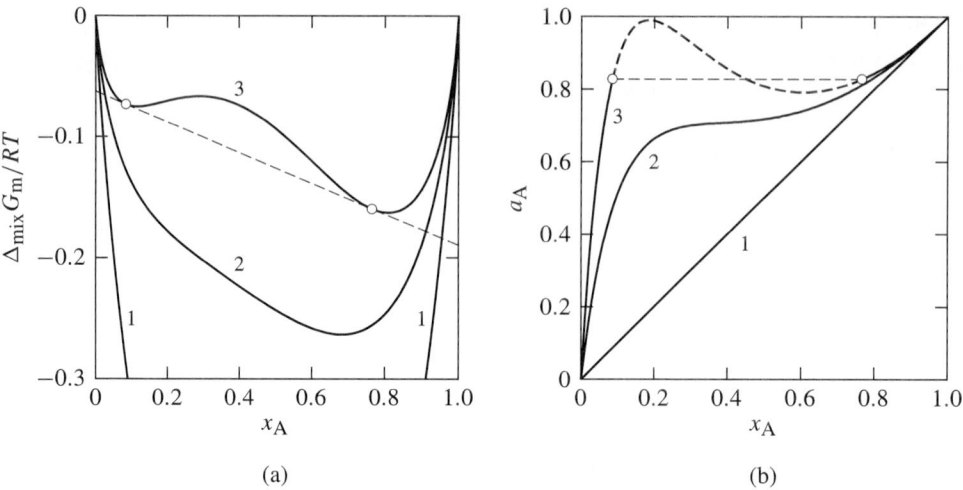

Figure 8.3 Binary liquid mixtures at 1 bar. (a) Molar Gibbs energy of mixing as a function of composition. (b) Activity of component A at $p = p°$ (using a pure-liquid standard state) as a function of composition. The curves are calculated from the two-parameter Redlich–Kister series using the following parameters. 1: $a = b = 0$ (ideal liquid mixture); 2: $a = 1.8RT, b = 0.36RT$; 3: $a = 2.4RT, b = 0.48RT$.

illustrated by Curves 2 and 3 in Fig. 8.3(b). The activity is related to the fugacity in an equilibrated gas phase by $a_A = \Gamma_A f_A/f_A^*$ (Table 7.3), so the greater activity compared to the ideal mixture corresponds to a greater fugacity, which is a positive deviation from Raoult's law.

If no portion of the $\Delta_{mix}G_m$-x_A curve of a nonideal mixture is concave downward, there is no phase separation; furthermore, a_A and f_A increase monotonically with x_A. This case is illustrated by Curve 2 in Figs. 8.3(a) and 8.3(b).

If a portion of the $\Delta_{mix}G_m$-x_A curve is concave downward, the condition needed for phase separation, then a maximum appears in the a_A-x_A curve. This case is illustrated by Curve 3, and the compositions of the coexisting phases are indicated by circles. The difference in compositions between the two circles is a *miscibility gap*. The portion of Curve 3 between these compositions in Fig. 8.3(b) is dashed to indicate it does not apply to the equilibrium system. Although the two coexisting phases have different compositions, the activity a_A must be the same in both phases because in both phases A has the same standard state and the same chemical potential.

Coexisting liquid phases are discussed further in Secs. 9.6 and 10.2.3.

8.3 REACTION ENTHALPIES

Since the reaction enthalpy $\Delta_r H$ of a reaction or other process is the enthalpy change of a process carried out at constant temperature and pressure in a closed system with no work

Section 8.3 Reaction Enthalpies

other than expansion work, it is equal under these conditions to the heat q transferred across the boundary.

8.3.1 Standard molar enthalpies of formation

The **formation reaction** of a substance is the reaction in which the substance, at a given temperature and in a given state, is formed from the constituent elements in their reference states at the same temperature. The *reference state of an element* is usually chosen to be the *standard* state of the element in the allotropic form and physical state that is stable at the given temperature and the standard pressure. For instance, at 298.15 K and 1 bar, the stable allotrope of carbon is crystalline graphite rather than diamond.

Phosphorus is an exception to the rule regarding reference states of elements. Although red phosphorus is the stable allotrope at 298.15 K, it is not well characterized. Instead, white phosphorus (crystalline P_4) at 1 bar is taken as the reference state.

At 298.15 K, the reference states of the elements are the following phases at a pressure of 1 bar. For H_2, N_2, O_2, F_2, Cl_2, and the noble gases, the reference state is the ideal gas. For Br_2 and Hg, the reference state is the liquid. For P, as just mentioned, the reference state is crystalline white phosphorus. For all other elements, the reference state is the stable crystalline allotrope.

The **standard molar enthalpy of formation** (or standard molar heat of formation), $\Delta_f H_m^\circ$, of a substance is the enthalpy change per amount of substance produced in the formation reaction of the substance in its standard state. Thus, the standard molar enthalpy of formation of gaseous methyl bromide at 298.15 K is the molar reaction enthalpy of the reaction

$$C(s, \text{graphite}, 1\text{ bar}) + \tfrac{3}{2}H_2(\text{ideal gas}, 1\text{ bar}) + \tfrac{1}{2}Br_2(l, 1\text{ bar}) \rightarrow CH_3Br(\text{ideal gas}, 1\text{ bar})$$

The value of $\Delta_f H_m^\circ$ for a given substance depends only on T. By definition, $\Delta_f H_m^\circ$ for the reference state of an element is zero.

We can use a principle called **Hess' law** to calculate the standard molar enthalpy of formation of a substance at a given temperature from standard molar reaction enthalpies at the same temperature, and to calculate a standard molar reaction enthalpy from tabulated values of standard molar enthalpies of formation. The principle is an application of the fact that enthalpy is a state function. Therefore, ΔH for a given change of the state of the system is independent of the path and is equal to the sum of ΔH values for any sequence of changes whose net result is the given change. (We may apply the same principle to a change of *any* state function.)

For example, the following combustion reactions can be carried out experimentally in a bomb calorimeter (Sec. 8.5.2), yielding the values shown of standard molar reaction enthalpies (at $T = 298.15$ K, $p^\circ = 1$ bar):

$$C(s, \text{graphite}) + O_2(g) \rightarrow CO_2(g) \qquad \Delta_r H_m^\circ = -393.51 \text{ kJ mol}^{-1}$$
$$CO(g) + \tfrac{1}{2}O_2(g) \rightarrow CO_2(g) \qquad \Delta_r H_m^\circ = -282.98 \text{ kJ mol}^{-1}$$

(Note that the first reaction, in addition to being the combustion reaction of graphite, is also the formation reaction of carbon dioxide.) The change resulting from the first reaction followed by the reverse of the second reaction is the formation reaction of carbon monoxide:

$$\text{C(s, graphite)} + \tfrac{1}{2}\text{O}_2(\text{g}) \rightarrow \text{CO(g)}$$

It would not be practical to carry out this last reaction experimentally since it would be difficult to prevent the formation of some CO_2. From Hess' law, the standard molar enthalpy of formation of CO is the sum of the standard molar enthalpies of the reactions that have the formation reaction as the net result:

$$\Delta_f H_m^\circ(\text{CO, g, 298.15 K}) = (-393.51 + 282.98) \text{ kJ mol}^{-1}$$
$$= -110.53 \text{ kJ mol}^{-1}$$

This value is one of the many standard molar enthalpies of formation to be found in compilations of thermodynamic properties of individual substances, such as the table in Appendix H. We may use the tabulated values to evaluate the standard molar reaction enthalpy $\Delta_r H_m^\circ$ of a reaction using a formula based on Hess' law. Imagine the reaction to take place in two steps: First each reactant in its standard state changes to the constituent elements in their reference states (the reverse of a formation reaction), and then these elements form the products in their standard states. The resulting formula is

$$\Delta_r H_m^\circ = \sum_i \nu_i \Delta_f H_{m,i}^\circ \qquad (8.3.1)$$

Recall that the stoichiometric number ν_i of each reactant is negative and that of each product is positive, so in Eq. 8.3.1 the standard molar reaction enthalpy is the sum of the standard molar enthalpies of formation of the products minus the sum of the standard molar enthalpies of formation of the reactants. Each term is multiplied by the appropriate stoichiometric coefficient from the chemical equation.

We may define a standard molar enthalpy of formation for a *solute in solution* to use in Eq. 8.3.1. For instance, the formation reaction of aqueous sucrose is

$$12\,\text{C(s, graphite)} + 11\,\text{H}_2(\text{g}) + \tfrac{11}{2}\text{O}_2(\text{g}) \rightarrow \text{C}_{12}\text{H}_{22}\text{O}_{11}(\text{aq})$$

and $\Delta_f H_m^\circ$ for $C_{12}H_{22}O_{11}$(aq) is the enthalpy change per amount of sucrose formed when the reactants and product are in their standard states. Note that in this formation reaction the solvent (H_2O) is *not* formed from H_2 and O_2; instead, the solute once formed combines with the amount of pure liquid water needed to form the solution.

There is no ordinary reaction that would produce an individual *ion in solution* from its element or elements without producing other species as well. We can, however, prepare a consistent set of standard molar enthalpies of formation of ions by assigning a value to a single reference ion. We can use these values for ions in Eq. 8.3.1 just like values of $\Delta_f H_m^\circ$ for substances and nonionic solutes. Aqueous hydrogen ion is the usual reference ion, to which is assigned the arbitrary value

$$\Delta_f H_m^\circ(\text{H}^+, \text{aq}) = 0 \qquad \text{(at all temperatures)}$$

Section 8.3 Reaction Enthalpies

To see how we can use this reference value, consider the reaction for the formation of aqueous HCl (hydrochloric acid):

$$\tfrac{1}{2}H_2(g) + \tfrac{1}{2}Cl_2(g) \rightarrow H^+(aq) + Cl^-(aq) \tag{8.3.2}$$

The standard molar reaction enthalpy at 298.15 K for this reaction is known, from reaction calorimetry, to be $\Delta_r H_m^\circ = -167.08 \text{ kJ mol}^{-1}$. The standard states of the gaseous H_2 and Cl_2 are, of course, the pure gases acting ideally at pressure p°, and the standard state of each of the aqueous ions is the ion at the standard molality and standard pressure, acting as if its activity coefficient on a molality basis were 1. From Eq. 8.3.1, we equate the value of $\Delta_r H_m^\circ$ to the sum of $\Delta_f H_m^\circ(H^+, aq)$ and $\Delta_f H_m^\circ(Cl^-, aq)$ (the standard molar enthalpies of formation of the H_2 and Cl_2 are zero). Since $\Delta_f H_m^\circ(H^+, aq)$ is zero, $\Delta_f H_m^\circ(Cl^-, aq)$ is $-167.08 \text{ kJ mol}^{-1}$.

Next we can combine this value of $\Delta_f H_m^\circ(Cl^-, aq)$ with the measured standard molar enthalpy of formation of aqueous sodium chloride

$$Na(s) + \tfrac{1}{2}Cl_2(g) \rightarrow Na^+(aq) + Cl^-(aq)$$

to evaluate the standard molar enthalpy of formation of aqueous sodium ion. By continuing this procedure with other reactions, we can build up a consistent set of $\Delta_f H_m^\circ$ values of various ions in aqueous solution.

8.3.2 Effect of temperature on molar reaction enthalpies

Consider a reaction occurring in a closed system at a constant temperature T_1 and at a constant pressure. The molar reaction enthalpy is $\Delta_r H_m(T_1)$. We wish to find an expression for the molar reaction enthalpy $\Delta_r H_m(T_2)$ of the same reaction at a different temperature, T_2, and at the same pressure.

For this purpose, we choose T, p, and ξ as the independent variables and write the total differential of H in the form shown by Eq. 8.1.8 on page 243:

$$dH = \left(\frac{\partial H}{\partial T}\right)_{p,\xi} dT + \left(\frac{\partial H}{\partial p}\right)_{T,\xi} dp + \Delta_r H_m \, d\xi$$

The partial derivative $(\partial H/\partial T)_{p,\xi}$ is the heat capacity at constant pressure C_p (Eq. 3.12.4) for the system at the fixed composition of each phase determined by the value of ξ. The expression for the total differential becomes

$$dH = C_p \, dT + \left(\frac{\partial H}{\partial p}\right)_{T,\xi} dp + \Delta_r H_m \, d\xi$$

from which we obtain the reciprocity relation

$$\left(\frac{\partial \Delta_r H_m}{\partial T}\right)_{p,\xi} = \left(\frac{\partial C_p}{\partial \xi}\right)_{T,p} \tag{8.3.3}$$
(closed system)

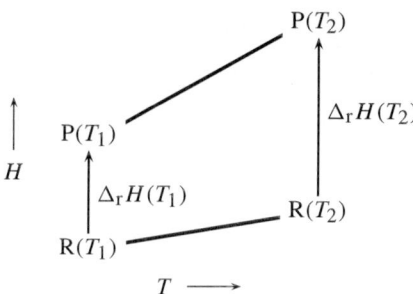

Figure 8.4 Relation between enthalpy and temperature for changes at constant pressure. R denotes reactants and P denotes products.

Now the partial derivative $(\partial C_p/\partial \xi)_{T,p}$ is the **molar heat capacity of reaction at constant pressure**, $\Delta_r C_{p,m}$ (see Eq. 8.1.10). From Eq. 8.1.9, we see it can be calculated from

$$\Delta_r C_{p,m} = \sum_i \nu_i C_{p,i}$$

where $C_{p,i}$ is the partial molar heat capacity at constant pressure of substance i. Equation 8.3.3 becomes

$$\left(\frac{\partial \Delta_r H_m}{\partial T}\right)_{p,\xi} = \Delta_r C_{p,m} \tag{8.3.4}$$

which at constant p and ξ can be arranged to

$$d\Delta_r H_m = \Delta_r C_{p,m}\, dT$$

Integration from T_1 to T_2 then gives the **Kirchhoff equation**:

$$\Delta_r H_m(T_2) = \Delta_r H_m(T_1) + \int_{T_1}^{T_2} \Delta_r C_{p,m}\, dT \tag{8.3.5}$$
$$\text{(constant } p \text{ and } \xi\text{)}$$

Figure 8.4 illustrates the principle of Eq. 8.3.5 for a reaction in which the molar integral and differential reaction enthalpies are the same. $\Delta_r C_p$ is the difference in the slopes of H versus T for products and reactants, and its integral over the temperature range T_1 to T_2 is the difference in $\Delta_r H$ for the reaction at these two temperatures. When $\Delta_r C_p$ and $\Delta_r H$ are divided by the advancement during the reaction, they become the molar quantities $\Delta_r C_{p,m}$ and $\Delta_r H_m$.

8.4 ENTHALPIES OF SOLUTION AND DILUTION

The processes of solution (dissolution) and dilution are related. The IUPAC Green Book recommends the abbreviations *sol* and *dil* for these processes.

Section 8.4 Enthalpies of Solution and Dilution

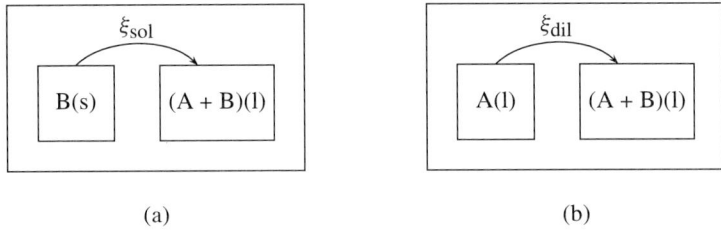

Figure 8.5 Two related processes in closed systems. (a) Solution process: Solute (assumed here to be a solid) dissolves in a solution containing a fixed amount of solvent. (b) Dilution process: Solvent mixes with a solution containing a fixed amount of solute.

An **enthalpy of solution**, $\Delta_{sol} H$, refers to the enthalpy change at constant T and p for the process of mixing pure solute B (a solid, liquid, or gas) with solvent or solution:

$$B^* \rightarrow B(sln) \tag{8.4.1}$$

(solution process, constant T and p)

An **enthalpy of dilution**, $\Delta_{dil} H$, refers to the enthalpy change at constant T and p for the process of mixing pure solvent A with solution:

$$A^*(l) \rightarrow A(sln) \tag{8.4.2}$$

(dilution process, constant T and p)

We may specify the advancement of these two kinds of processes by ξ_{sol} and ξ_{dil}, respectively. Note that both processes take place in a *closed* system that (at least initially) contains two phases. The total amounts of A and B in the system do not change, but the amount of B or A in a pure phase diminishes as the process advances and ξ_{sol} or ξ_{dil} increases (see the comparison in Fig. 8.5).

8.4.1 Enthalpy of solution

First let us consider the solution process $B^* \rightarrow B(sln)$. The **molar enthalpy of solution**, $\Delta_{sol} H_m$, is the rate of change of H with the advancement of the solution process:

$$\Delta_{sol} H_m = \left(\frac{\partial H}{\partial \xi_{sol}} \right)_{T,p} \tag{8.4.3}$$

This quantity is sometimes called the molar differential enthalpy of solution or partial molar enthalpy of solution. It is the enthalpy change per amount of solute transferred to the solution when the amount of solution is so large that the solution composition is essentially unchanged. For the solution process, the general relation $\Delta_r X_m = \sum_i \nu_i X_i$ (Eq. 8.1.9) becomes

$$\Delta_{sol} H_m = H_B - H_B^* \tag{8.4.4}$$

where H_B is the partial molar enthalpy of the solute in the solution and H_B^* is the molar enthalpy of the pure solute at the same T and p. Since both H_B and H_B^* are intensive properties, and H_B depends on the solution composition, $\Delta_{\text{sol}} H_m$ at a given T and p depends only on the solution composition and not on the amount of solution.

The molar enthalpy of solution at infinite dilution, $\Delta_{\text{sol}} H_m^\infty$, is the enthalpy change per amount of the solution process when the solute is dissolved in such a large amount of solvent that the resulting solution has the thermal properties of an infinitely dilute solution. According to Eq. 8.4.4, this quantity is given by

$$\Delta_{\text{sol}} H_m^\infty = H_B^\infty - H_B^* \tag{8.4.5}$$

A **molar integral enthalpy of solution** is the enthalpy change $\Delta_{\text{sol}} H$ at constant T and p when a finite amount n_B of solute dissolves in a specified amount of pure solvent divided by n_B. Since n_B is the advancement ξ_{sol}, the molar integral enthalpy of solution is given by

$$\Delta_{\text{sol}} H_m^{\text{int}} = \frac{\Delta_{\text{sol}} H}{\xi_{\text{sol}}} \tag{8.4.6}$$

$\Delta_{\text{sol}} H_m^{\text{int}}$ at the given T and p depends only on the composition of the solution formed. That is, if we were to double the amounts of both the solvent and solute, the solute molality would remain the same and the value of $\Delta_{\text{sol}} H_m^{\text{int}}$ would be unchanged.

Ordinarily we determine an enthalpy of solution $\Delta_{\text{sol}} H$ from the temperature change when the solution process is carried out in a constant-pressure reaction calorimeter, as is described in Sec. 8.5.1. Experimental values of $\Delta_{\text{sol}} H$ as a function of ξ_{sol} can be collected by measuring enthalpy changes during a series of successive additions of the solute to a fixed amount of solvent, resulting in a solution whose molality increases in stages. The enthalpy changes are cumulative, so the value of $\Delta_{\text{sol}} H$ after each addition is the sum of the enthalpy changes for this and the previous additions.

The relation between $\Delta_{\text{sol}} H$ and the molar integral and differential enthalpies of solution is illustrated in Fig. 8.6 on the next page with data for the solution of crystalline sodium acetate in water. The curve shows $\Delta_{\text{sol}} H$ as a function of ξ_{sol}, which is the amount of solute dissolved in a fixed amount of solvent. At any point along the curve, the ratio $\Delta_{\text{sol}} H / \xi_{\text{sol}}$ is the molar integral enthalpy of solution for the solution composition corresponding to the value of ξ_{sol} at this point. The slope of the curve at the point is the molar differential enthalpy of solution:

$$\Delta_{\text{sol}} H_m = \frac{d\Delta_{\text{sol}} H}{d\xi_{\text{sol}}} \tag{8.4.7}$$
(constant T, p, and n_A)

In the limit of infinite dilution, $\Delta_{\text{sol}} H / \xi_{\text{sol}}$ becomes equal to the slope of the curve, and both molar quantities have the same value, $\Delta_{\text{sol}} H_m^\infty$. At the standard pressure, $\Delta_{\text{sol}} H_m^\infty$ is the same as the standard molar enthalpy of solution, $\Delta_{\text{sol}} H_m^\circ$, since the standard molar enthalpy of a solute equals the molar enthalpy at $p = p^\circ$ and infinite dilution.

The solution process is equivalent to a mixing process. If the pure solute is a liquid that forms an ideal liquid mixture with the solvent, the molar enthalpy of mixing is zero

Section 8.4 Enthalpies of Solution and Dilution

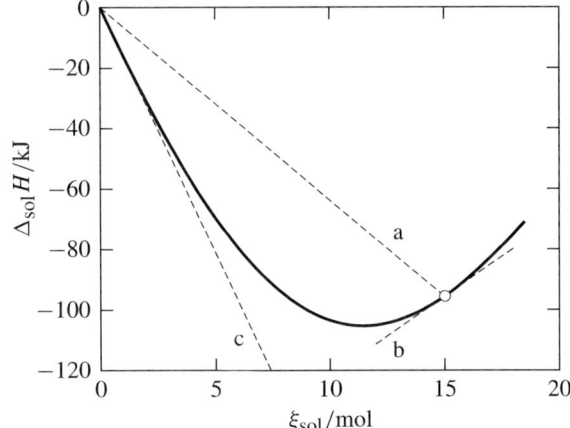

Figure 8.6 Enthalpy change for the dissolution of $NaCH_3CO_2(s)$ in one kilogram of water at 298.15 K and 1 bar in a closed system as a function of the amount of dissolved solute, ξ_{sol}. The open circle at $\xi_{sol} = 15$ mol indicates the approximate saturation limit; data to the right of this point come from supersaturated solutions. At the composition $m_B = 15$ mol kg^{-1}, the value of $\Delta_{sol}H_m^{int}$ is the slope of Line a and the value of $\Delta_{sol}H_m$ is the slope of Line b. The value of $\Delta_{sol}H_m^\infty$ is the slope of Line c.

(Eq. 8.2.6) and the enthalpy of solution is also zero. Nonelectrolyte solutes, whether solids, liquids, or gases, usually have molar differential enthalpies of solution that do not vary greatly with solution composition. This is not the case for electrolytes in aqueous solutions because strong composition-dependent electrostatic interactions exist between ions in even a fairly dilute solution.

8.4.2 Enthalpy of dilution

Next let us consider the dilution process $A^*(l) \rightarrow A(sln)$. The **molar enthalpy of dilution** (or molar differential enthalpy of dilution) is the rate of change of H with the advancement of the dilution process:

$$\Delta_{dil}H_m = \left(\frac{\partial H}{\partial \xi_{dil}}\right)_{T,p} \tag{8.4.8}$$

The general relation $\Delta_r X_m = \sum_i \nu_i X_i$ becomes

$$\Delta_{dil}H_m = H_A - H_A^* \tag{8.4.9}$$

where H_A is the partial molar enthalpy of the solvent in the solution.

In the limit of infinite dilution, H_A must approach the molar enthalpy of pure solvent, H_A^*; then Eq. 8.4.9 shows that $\Delta_{dil}H_m$ approaches zero in this limit.

8.4.3 Molar enthalpies of solute formation

Integral enthalpies of solution and dilution are conveniently expressed in terms of molar enthalpies of formation. The **molar enthalpy of formation of a solute** is the enthalpy change per amount of solute for an isothermal, isobaric process in which the solute, in the solution of a given molality m_B, is formed from its constituent elements in their reference states. We denote this quantity by $\Delta_f H_m(B, m_B)$. As explained in Sec. 8.3.1, the formation process does not include the formation of the solvent. For example, the formation process for NaOH in an aqueous solution that has 50 moles of water for each mole of NaOH ($m_B = 1.110\,\text{mol}\,\text{kg}^{-1}$) is

$$\text{Na(s)} + \tfrac{1}{2}\text{O}_2(\text{g}) + \tfrac{1}{2}\text{H}_2(\text{g}) + 50\,\text{H}_2\text{O(l)} \rightarrow \text{NaOH in 50 H}_2\text{O}$$

We may express an enthalpy of *solution* as the sum of the enthalpy changes for the following two hypothetical steps. First an amount n_B of the pure solute decomposes to the constituent elements in their reference states, and then the solution forms from these elements and pure solvent. The resulting expression in terms of molar enthalpies of formation is

$$\Delta_{\text{sol}} H = -n_B \Delta_f H_m(B^*) + n_B \Delta_f H_m(B, m_B) \tag{8.4.10}$$

where m_B is the composition of the resulting solution. Similarly, an enthalpy of *dilution* is the sum of the enthalpy changes of two steps: The initial solution (of composition m'_B) changes to pure solvent and the constituent elements of the solute in their reference states; then the final solution (of composition m''_B) is formed from these elements, the initial solvent, and additional solvent. The resulting expression, for an amount n_B of the solute, is

$$\Delta_{\text{dil}} H = -n_B \Delta_f H_m(B, m'_B) + n_B \Delta_f H_m(B, m''_B) \tag{8.4.11}$$

Equation 8.4.10 is the means by which we may evaluate the molar enthalpy of formation of a solute, at the standard pressure $p°$ and in solution of a given composition, from an experimental enthalpy of solution and the standard molar enthalpy of formation of the pure solute. Equation 8.4.11 shows how an enthalpy of dilution gives the *difference* between enthalpies of solute formation at two compositions. If we know the enthalpy of formation of the solute at one composition (such as by measuring an enthalpy of solution), we may obtain the values at other compositions from dilution measurements. Experimentally, it is sometimes more convenient to carry out the dilution process than the solution process especially when the pure solute is a gas or solid.

8.4.4 Evaluation of relative partial molar enthalpies

We can use experimental enthalpies of solution or dilution to evaluate the partial molar enthalpies of the solvent and solute relative to appropriate reference states.

The **relative partial molar enthalpy of the solvent** is defined by

$$L_A \equiv H_A - H_A^* \tag{8.4.12}$$

This is the partial molar enthalpy of the solvent, in a solution of given composition, relative to pure solvent at the same temperature and pressure.

Section 8.4 Enthalpies of Solution and Dilution

The **relative partial molar enthalpy of a solute** is defined by

$$L_B \equiv H_B - H_B^\infty \tag{8.4.13}$$

The reference state for a solute is the solute at infinite dilution.

At the standard pressure $p°$, pure solvent is in its standard state, and the partial molar enthalpy of a solute at infinite dilution is equal to its standard molar enthalpy (Eq. 7.7.6). Thus, at $p = p°$, L_A equals $H_A - H_A°$ and L_B equals $H_B - H_B°$. Values of these two quantities are useful for finding the difference $\Delta_r H_m - \Delta_r H_m°$ for a chemical process involving a solvent or solute. From Eq. 8.1.9 on page 243, this difference is given by $\sum_i \nu_i (H_i - H_i°)$.

To see how we may evaluate L_B from enthalpy of solution data, we write the identity

$$L_B = H_B - H_B^\infty = (H_B - H_B^*) - (H_B^\infty - H_B^*)$$

From Eqs. 8.4.4 and 8.4.5, this becomes

$$L_B = \Delta_{sol} H_m - \Delta_{sol} H_m^\infty$$

For the molar enthalpy of solution, we write, referring to Eq. 8.4.7,

$$\Delta_{sol} H_m = \frac{d\Delta_{sol} H}{dn_B}$$

where the enthalpy of solution $\Delta_{sol} H$ is measured for varying amounts n_B of solute dissolved in a fixed amount of solvent. Then from the definition of the molar integral enthalpy of solution, $\Delta_{sol} H_m^{int} = \Delta_{sol} H / n_B$, we obtain

$$\Delta_{sol} H_m = \frac{d(n_B \Delta_{sol} H_m^{int})}{dn_B} = \Delta_{sol} H_m^{int} + n_B \frac{d\Delta_{sol} H_m^{int}}{dn_B}$$

giving us finally for the relative partial molar enthalpy of the solute

$$L_B = \Delta_{sol} H_m^{int} + n_B \frac{d\Delta_{sol} H_m^{int}}{dn_B} - \Delta_{sol} H_m^\infty \tag{8.4.14}$$

(constant T, p, and n_A)

It is convenient to define the quantity

$$\Phi_L \equiv \Delta_{sol} H_m^{int} - \Delta_{sol} H_m^\infty \tag{8.4.15}$$

known as the *relative apparent molar enthalpy of the solute*. We can then write Eq. 8.4.14 in the compact form[7]

$$L_B = \Phi_L + n_B \frac{d\Phi_L}{dn_B} \tag{8.4.16}$$

(constant T, p, and n_A)

---

[7] Note that $\Delta_{sol} H_m^\infty$ is independent of n_B, so $d\Phi_L/dn_B$ is equal to $d\Delta_{sol} H_m^{int}/dn_B$.

Equation 8.4.16 allows us to evaluate L_B at any solution composition if we have a graph of Φ_L plotted as a function of n_B. How can we obtain the needed values of Φ_L? One possibility is to measure $\Delta_{sol}H_m^{int}$ experimentally for the formation of solutions of various molalities and extrapolate these values to $m_B = 0$ so as to evaluate $\Delta_{sol}H_m^{\infty}$. For an electrolyte solution, however, the slope of $\Delta_{sol}H_m^{int}$ versus m_B approaches $+\infty$ at $m_B = 0$, whereas the limiting slope of $\Delta_{sol}H_m^{int}$ versus $\sqrt{m_B}$ is finite and can be predicted from the Debye–Hückel limiting law (see Sec. 8.4.5). Accordingly, a satisfactory procedure for an electrolyte solution is to plot $\Delta_{sol}H_m^{int}$ as a function of $\sqrt{m_B}$ (where m_B is the molality of the solution in the final state of the solution process), perform a linear extrapolation of the experimental points to $\sqrt{m_B} = 0$, and then shift the origin of the graph to the extrapolated intercept. The result is a graph of Φ_L versus $\sqrt{m_B}$.

To relate the slope at any point of the curve on this graph to the derivative $d\Phi_L/dn_B$ of Eq. 8.4.16 (at constant n_A), we write

$$d\sqrt{m_B} = d\sqrt{n_B/M_A n_A} = \frac{dn_B}{2\sqrt{M_A n_A n_B}} = \frac{\sqrt{m_B}}{2n_B} dn_B$$

Solving for dn_B and substituting in Eq. 8.4.16, we obtain the following operational equation for evaluating L_B:

$$L_B = \Phi_L + \tfrac{1}{2}\sqrt{m_B}\,\frac{d\Phi_L}{d\sqrt{m_B}} \qquad (8.4.17)$$

(constant T and p)

We can also use the graph of Φ_L versus $\sqrt{m_B}$ to evaluate the relative partial molar enthalpy of the *solvent*. The operational equation is derived as follows. When an amount n_B of pure solute dissolves in an amount n_A of pure solvent, the enthalpy change is

$$\Delta_{sol}H = H(\text{sln}) - n_A H_A^* - n_B H_B^*$$

From the additivity rule, the enthalpy of the solution is $H(\text{sln}) = n_A H_A + n_B H_B$, so the enthalpy change is

$$\Delta_{sol}H = n_A(H_A - H_A^*) + n_B(H_B - H_B^*)$$

$\Delta_{sol}H_m^{int}$ is equal to $\Delta_{sol}H/n_B$. Combining these relations with $\Phi_L = \Delta_{sol}H_m^{int} - \Delta_{sol}H_m^{\infty}$, where $\Delta_{sol}H_m^{\infty}$ is equal to $H_B^{\infty} - H_B^*$ (Eq. 8.4.5), we obtain

$$\Phi_L = \frac{n_A(H_A - H_A^*)}{n_B} + H_B - H_B^{\infty} = \frac{H_A - H_A^*}{M_A m_B} + H_B - H_B^{\infty}$$

Then replacing $H_B - H_B^{\infty}$ with the expression in Eq. 8.4.17 and solving for $L_A = H_A - H_A^*$, we obtain the desired relation

$$L_A = -\tfrac{1}{2} M_A m_B^{3/2}\,\frac{d\Phi_L}{d\sqrt{m_B}} \qquad (8.4.18)$$

(constant T and p)

Section 8.4 Enthalpies of Solution and Dilution

Equations 8.4.17 and 8.4.18 enable us to evaluate L_B and L_A at any solution composition from a graph of Φ_L versus $\sqrt{m_B}$.

As mentioned at the end of Sec. 8.4.3, it is often more convenient to measure enthalpies of dilution than enthalpies of solution. The graph of Φ_L versus $\sqrt{m_B}$ can be obtained from either kind of measurement.

To understand how enthalpies of dilution can be used for this purpose, consider the following two possible ways of preparing a solution of composition m_B containing amounts n_A and n_B of solvent and solute. Both paths are at constant T and p in a closed system in which the solvent and solute initially are unmixed.

Path 1: The solution forms directly by dissolution of the solute in the solvent. The enthalpy change is $\Delta_{sol}H(m_B)$, where the molality of the final solution is indicated in parentheses.

Path 2: Starting with the unmixed solvent and solute, the solute dissolves in a portion of the solvent to form a solution of composition m'_B (more concentrated than m_B). The enthalpy change is $\Delta_{sol}H(m'_B)$. In a second step of this path, the remaining pure solvent mixes with the solution to dilute its composition from m'_B to m_B. The enthalpy change of the second step is $\Delta_{dil}H(m'_B \to m_B)$.

Since both paths have the same initial states and same final states, both have the same overall enthalpy change:

$$\Delta_{sol}H(m_B) = \Delta_{sol}H(m'_B) + \Delta_{dil}H(m'_B \to m_B)$$

Division by n_B converts this equation to

$$\Delta_{sol}H_m^{int}(m_B) = \Delta_{sol}H_m^{int}(m'_B) + \frac{\Delta_{dil}H(m'_B \to m_B)}{n_B}$$

Solving for $\Delta_{dil}H(m'_B \to m_B)/n_B$ and making use of the definition $\Phi_L = \Delta_{sol}H_m^{int} - \Delta_{sol}H_m^{\infty}$, we obtain the general relation

$$\frac{\Delta_{dil}H(m'_B \to m_B)}{n_B} = \Delta_{sol}H_m^{int}(m_B) - \Delta_{sol}H_m^{int}(m'_B)$$

$$= \Phi_L(m_B) - \Phi_L(m'_B) \qquad (8.4.19)$$

By taking the limit $m_B \to 0$, we find that $\Phi_L(m'_B)$ is equal to, and of opposite sign to, $\Delta_{dil}H/n_B$ for diluting a solution of molality m'_B to infinite dilution.

Suppose, in separate runs, we measure the enthalpy change for diluting a solution of fixed composition m'_B to various final compositions m_B. Then according to Eq. 8.4.19, $\Delta_{dil}H(m'_B \to m_B)/n_B$ differs by a constant from Φ_L evaluated at molality m_B. $\Phi_L(m_B)$ is zero at $m_B = 0$, so we can plot the values of $\Delta_{dil}H/n_B$ versus $\sqrt{m_B}$, extrapolate the curve to $\sqrt{m_B} = 0$, and shift the origin to the extrapolated intercept, resulting in the desired plot of Φ_L versus $\sqrt{m_B}$.

This procedure is illustrated by Fig. 8.7 on the next page in which experimental values of

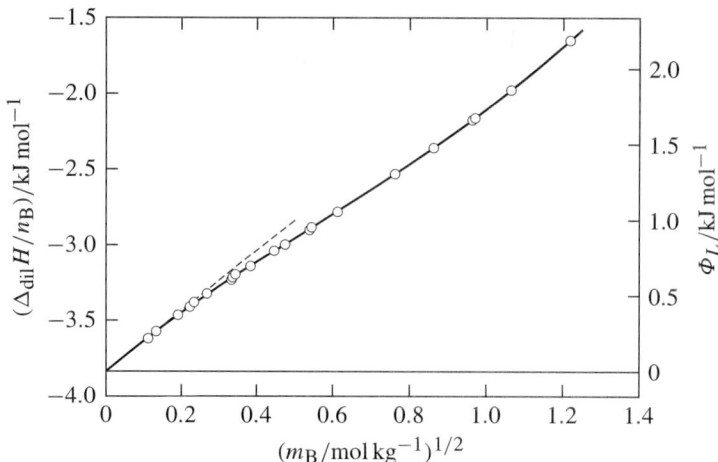

Figure 8.7 Aqueous solutions of HCl at 25.00 °C (Sturtevant, J. M. *J. Am. Chem. Soc.* **1940**, *62*, 584–587, 3265–3266). Left axis: enthalpy of dilution of 3.337 molal HCl to molality m_B, divided by the amount of solute. Circles: experimental values. Solid curve: smoothed values with initial slope calculated from the Debye–Hückel limiting law (1.99 kJ kg$^{1/2}$ mol$^{-1/2}$) as indicated by the dashed line. Right axis: relative apparent molar enthalpy of the solute.

$\Delta_{\text{dil}} H / n_B$ for dilution of a concentrated aqueous HCl solution (left axis) are plotted versus the square root of the molality m_B of the resulting diluted solution. The extrapolation to $\sqrt{m_B} = 0$ was guided by the theoretical initial slope from the Debye–Hückel limiting law (Eq. 8.4.24 on page 269). When the origin is shifted to the extrapolated intercept, the new vertical scale (right axis) gives the values of Φ_L.

Another example of a plot of Φ_L versus $\sqrt{m_B} = 0$, in this case for aqueous sodium chloride, is given in Fig. 8.8(a) on the next page. The curve for Φ_L was obtained from both solution and dilution enthalpies measured by various workers. Unlike the curve for aqueous HCl in Fig. 8.7, the slope changes sign in the molality range investigated. From this curve, the curves for L_B and L_A shown in Fig. 8.8(a) and 8.8(b) were generated with Eqs. 8.4.17 and 8.4.18.

8.4.5 Electrolyte solutions at infinite dilution

The relative partial molar enthalpy $L_B = H_B - H_B^\infty$ of an electrolyte solute is related to the temperature dependence of the mean ionic activity coefficient, as the following derivation shows.

At the standard pressure, the partial molar enthalpy of a solute at infinite dilution equals the partial molar enthalpy in the standard state, so the relative partial molar enthalpy of the solute is given by

$$L_B = H_B - H_B^\circ$$

Section 8.4 Enthalpies of Solution and Dilution

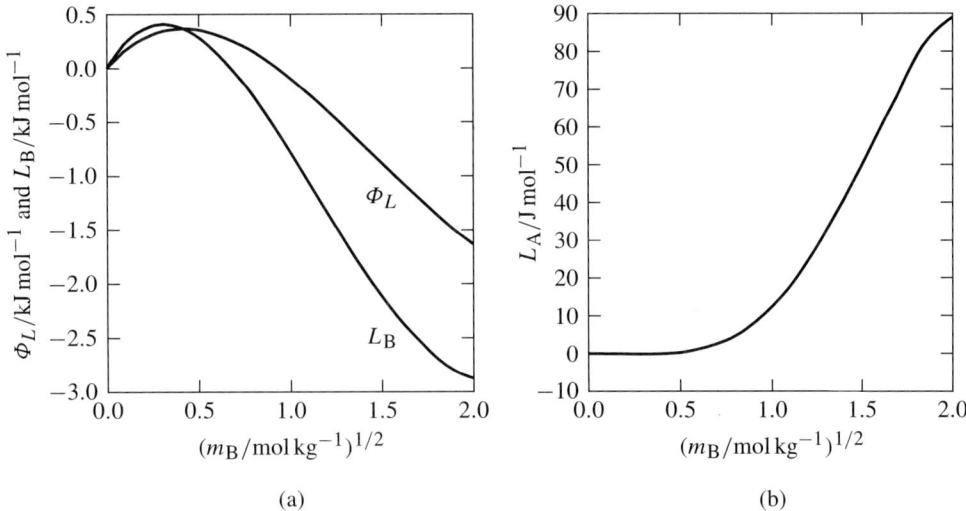

Figure 8.8 Relative thermal properties of aqueous NaCl at 25.00 °C. Data from Parker, V. B. *Thermal Properties of Aqueous Uni-univalent Electrolytes*; Nat. Stand. Data Ref. Ser., Nat. Bur. Stand. (U.S.), 1965, Table X.

From Eqs. 9.1.1 and 9.1.4 in the next chapter, we can write the relations

$$H_B = -T^2 \left[\frac{\partial(\mu_B/T)}{\partial T}\right]_{p,n} \qquad H_B^\circ = -T^2 \frac{d(\mu_{m,B}^\circ/T)}{dT}$$

Subtracting the second of these relations from the first, we obtain

$$H_B - H_B^\circ = -T^2 \left[\frac{\partial(\mu_B - \mu_{m,B}^\circ)/T}{\partial T}\right]_{p,n}$$

But the solute activity on a molality basis, $a_{m,B}$, is defined by $\mu_B - \mu_{m,B}^\circ = RT \ln a_{m,B}$. The activity of an electrolyte solute at the standard pressure, from Eq. 7.8.19, is:[8]

$$a_{m,B} = \left(\nu_+^{\nu_+} \nu_-^{\nu_-}\right) \gamma_\pm^\nu \left(\frac{m_B}{m^\circ}\right)^\nu$$

Accordingly, the relative partial molar enthalpy of the solute is related to the mean ionic activity coefficient by

$$L_B = -RT^2 \nu \left(\frac{\partial \ln \gamma_\pm}{\partial T}\right)_{p,n} \qquad (8.4.20)$$
$$(p = p^\circ)$$

---

[8] ν_+ and ν_- are the numbers of cations and anions per solute formula unit, and $\nu = \nu_+ + \nu_-$ is the total number of ions per formula unit. The pressure factor $\Gamma_{m,B}$ is equal to 1 at the standard pressure.

Suppose the solution is sufficiently dilute for the mean ionic activity coefficient to be adequately described by the Debye–Hückel limiting law, Eq. 7.8.25:

$$\ln \gamma_\pm = -A\, |z_+ z_-|\, \sqrt{I_m}$$

Here z_+ and z_- are the charge numbers of the cation and anion, I_m is the ionic strength on a molality basis, and A is a temperature-dependent physical quantity defined in Sec. 7.8.4. Then Eq. 8.4.20 becomes

$$L_\mathrm{B} = RT^2 \nu\, |z_+ z_-|\, \sqrt{I_m}\, \left(\frac{\partial A}{\partial T}\right)_{p,n} \qquad (8.4.21)$$
$$(\text{infinite dilution},\ p = p^\circ)$$

We assume the solution contains a single completely dissociated electrolyte solute. We can then write I_m, defined by $(1/2) \sum_i m_i z_i^2$, as a general function of the solute molality:

$$I_m = \tfrac{1}{2}\left(\nu_+ z_+^2 + \nu_- z_-^2\right) m_\mathrm{B}$$

With the help of the electroneutrality condition $\nu_+ z_+ = -\nu_- z_-$, the function becomes

$$I_m = \tfrac{1}{2}[-(\nu_- z_-)\nu_+ - (\nu_+ z_+) z_-] m_\mathrm{B} = \tfrac{1}{2}[-(\nu_- + \nu_+) z_+ z_-] m_\mathrm{B}$$
$$= \tfrac{1}{2} \nu\, |z_+ z_-|\, m_\mathrm{B}$$

This relation converts Eq. 8.4.21 to

$$L_\mathrm{B} = \frac{RT^2}{\sqrt{2}} \left(\frac{\partial \rho_\mathrm{A}^* A}{\partial T}\right)_{p,n} (\nu\, |z_+ z_-|)^{3/2} \sqrt{m_\mathrm{B}}$$

or

$$L_\mathrm{B} = C\left(\nu\, |z_+ z_-|\right)^{3/2} \sqrt{m_\mathrm{B}} \qquad (8.4.22)$$
$$(\text{infinite dilution},\ p = p^\circ)$$

where the coefficient C depends on T and the kind of solvent, but not on the solute molality. At 25 °C with H_2O as solvent, the coefficient is found from experimental data[9] to have the value

$$C = 1.0542 \times 10^3\ \mathrm{J\, kg^{1/2}\, mol^{-3/2}}$$

We can use the Gibbs–Duhem equation to find an expression for the relative partial molar enthalpy of the *solvent*, $L_\mathrm{A} = H_\mathrm{A} - H_\mathrm{A}^*$. The Gibbs–Duhem equation (Eq. 7.2.10 on page 180) applied to partial molar enthalpies in a binary mixture at constant T and p is $n_\mathrm{A}\, \mathrm{d}H_\mathrm{A} + n_\mathrm{B}\, \mathrm{d}H_\mathrm{B} = 0$ or

$$\mathrm{d}H_\mathrm{A} = -\frac{n_\mathrm{B}}{n_\mathrm{A}} \mathrm{d}H_\mathrm{B} = -M_\mathrm{A} m_\mathrm{B}\, \mathrm{d}H_\mathrm{B}$$

---

[9] Archer, D. G.; Wang, P. *J. Phys. Chem. Ref. Data* **1990**, *19*, 371-411.

With L_B given by Eq. 8.4.22 (for a dilute electrolyte solution at the standard pressure), the differential of H_B is given by

$$\mathrm{d}H_B = \mathrm{d}L_B = C\left(\nu\,|z_+z_-|\right)^{3/2} \mathrm{d}\sqrt{m_B} = \frac{C\left(\nu\,|z_+z_-|\right)^{3/2}}{2\sqrt{m_B}}\,\mathrm{d}m_B$$

We integrate from $m_B = 0$ to a molality m_B in the dilute range where the Debye–Hückel limiting law is assumed to hold:

$$\int_{H_A^*}^{H_A} \mathrm{d}H_A = -\frac{M_A C\left(\nu\,|z_+z_-|\right)^{3/2}}{2} \int_0^{H_B} \sqrt{m_B}\,\mathrm{d}m_B$$

with the result

$$H_A - H_A^* = L_A = -\tfrac{1}{3}M_A C \left(\nu\,|z_+z_-|\right)^{3/2} m_B^{3/2} \qquad (8.4.23)$$
$$\text{(infinite dilution, } p = p^\circ\text{)}$$

Section 8.4.4 describes how L_A and L_B can be evaluated experimentally from a plot of the quantity Φ_L versus $\sqrt{m_B}$. We can use Eq. 8.4.23 to obtain the theoretical limiting slope of this plot at infinite dilution. Substituting the infinite dilution expression for L_A into the general relation $L_A = -(1/2)M_A(m_B)^{3/2}\,\mathrm{d}\Phi_L/\mathrm{d}\sqrt{m_B}$ (Eq. 8.4.18), we obtain

$$\lim_{\sqrt{m_B}\to 0} \frac{\mathrm{d}\Phi_L}{\mathrm{d}\sqrt{m_B}} = \frac{2C}{3}\left(\nu\,|z_+z_-|\right)^{3/2} \qquad (8.4.24)$$
$$(p = p^\circ)$$

(The increase of Φ_L with increasing molality means that dilution of a very dilute solution is an exothermic process.)

From the value of C given earlier and Eqs. 8.4.22–8.4.24, the theoretical expressions for a 1:1 electrolyte at infinite dilution in H_2O at 25 °C are

$$L_A/\mathrm{J\,mol^{-1}} = -17.90\left(m_B/\mathrm{mol\,kg^{-1}}\right)^{3/2}$$

$$L_B/\mathrm{J\,mol^{-1}} = 2.982\times 10^3\left(m_B/\mathrm{mol\,kg^{-1}}\right)^{1/2}$$

$$\Phi_L/\mathrm{J\,mol^{-1}} = 1.988\times 10^3\left(m_B/\mathrm{mol\,kg^{-1}}\right)^{1/2}$$

8.5 REACTION CALORIMETRY

The usual purpose of making calorimetric measurements of a reaction or other chemical process is to evaluate the standard molar reaction enthalpy $\Delta_r H_m^\circ$.

The experimental apparatus is some kind of calorimeter, often of either the adiabatic or isothermal-jacket type described in Sec. 5.5.2 in connection with heat capacity measurements. The adiabatic type is best for reactions that take more than about 30 minutes, and the

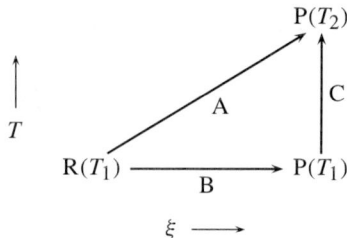

Figure 8.9 Paths for processes in a constant-pressure calorimeter (schematic). R denotes reactants and P denotes products.

isothermal-jacket type is appropriate for more rapid reactions. In either type, the chemical process takes place in a container within the calorimeter, and the thermometer is immersed either in this container or in a phase in thermal contact with it. electric heater may be included to evaluate the energy equivalent. The measured temperature change or heat flow is caused by the chemical process, instead of by electrical work as in the determination of a heat capacity.

Sections 8.5.1 and 8.5.2 describe two of the most common types of reaction calorimeters designed for the reaction process to occur either at constant pressure or at constant volume. The constant-pressure type is usually called a *reaction* calorimeter, and the constant-volume type is known as a *bomb* or *combustion* calorimeter. One important way in which these calorimeters differ from ones used for heat capacity measurements is that work is kept deliberately small so that the changes of internal energy and enthalpy are close to zero.

8.5.1 The constant-pressure reaction calorimeter

The contents of a constant-pressure calorimeter usually are open to the atmosphere, so this type of calorimeter is unsuitable for processes involving gases. It is, however, a convenient apparatus in which to study a liquid-phase chemical reaction, the dissolution of a solid or liquid solute in a liquid solvent, or the dilution of a solution with solvent.

The process is initiated in the calorimeter by allowing the reactants to mix. The temperature of the calorimeter contents is measured over a period of time starting before the initiation of the process and ending after the advancement has reached an equilibrium value. If the temperature increases, the process is **exothermic**; if it decreases, the process is **endothermic**. The heating or cooling curve (temperature as a function of time) is observed over a period of time that includes the advancement to equilibrium. For an exothermic reaction occurring in an adiabatic calorimeter, the heating curve may resemble that shown in Fig. 5.3 on page 112, and the heating curve in an isothermal-jacket calorimeter may resemble that shown in Fig. 5.4 on page 114. We designate two points on the heating or cooling curve—one at temperature T_1 before the process is initiated and one at T_2 after equilibrium has been reached. These points are indicated by open circles in Figs. 5.3 and 5.4. In Fig. 8.9, which refers to an exothermic reaction process, the change between these two points is labeled Path A.

Section 8.5 Reaction Calorimetry

The enthalpy change $\Delta_A H$ in Path A would be zero if the process were perfectly adiabatic and the only work were expansion work (see Eq. 3.11.4 on page 59), but this is rarely the case. There may be unavoidable work from stirring and from electrical temperature measurement. We can evaluate $\Delta_A H$ by one of the methods described in Sec. 5.5.2. For an adiabatic calorimeter, the appropriate expression is $\Delta_A H = \epsilon r(t_2 - t_1)$ (Eq. 5.5.6 with w_{el} set equal to zero), where ϵ is the energy equivalent of the calorimeter, r is the slope of the cooling curve when no reaction is occurring, and t_1 and t_2 are the times at temperatures T_1 and T_2. For an isothermal-jacket calorimeter, we use $\Delta_A H = -k \int_{t_1}^{t_2} (T - T_\infty) \, dt$ (Eq. 5.5.12 with w_{el} set equal to zero), where k is given by Eq. 5.5.13.

Path A is usually not isothermal. We wish to evaluate the enthalpy change of Path B of Fig. 8.9, in which the reactants change to the products at the *same* temperature. Since H is a state function, a process in which Path B is followed by Path C (in which the temperature of the products changes from T_1 to T_2 at constant pressure) has the same overall enthalpy change as Path A: $\Delta_B H + \Delta_C H = \Delta_A H$ or

$$\Delta_B H = -\Delta_C H + \Delta_A H \tag{8.5.1}$$

We do not need to actually carry out the process of Path C in the laboratory; instead we simply calculate the enthalpy change $\Delta_C H$ that would occur if we did carry it out. Since the energy equivalent is the average heat capacity, which in the absence of a reaction equals $\Delta H / \Delta T$, we may write

$$\Delta_C H = \epsilon(P)(T_2 - T_1) \tag{8.5.2}$$

where $\epsilon(P)$ denotes the energy equivalent when the calorimeter contains the products.

To measure $\epsilon(P)$, we can carry out a second experiment involving work with an electric heater included in the calorimeter, similar to the methods described in Sec. 5.5.2.

The isothermal enthalpy change divided by ξ, the advancement of the process that occurs in Path A, is the molar integral reaction enthalpy $\Delta_r H_m^{int}$. From Eqs. 8.5.1 and 8.5.2, we have

$$\Delta_r H_m^{int} = \frac{\Delta_B H}{\xi} = \frac{-\epsilon(P)(T_2 - T_1) + \Delta_A H}{\xi}$$

$$\approx -\frac{\epsilon(P)(T_2 - T_1)}{\xi} \tag{8.5.3}$$
(constant-pressure calorimeter)

Equation 8.5.3 shows that if T_2 is greater than T_1 (the process is *exothermic*), then $\Delta_r H_m^{int}$ is *negative*, reflecting the fact that after the reaction heat would have to leave the system in order for the temperature to return to its initial value. If T_2 is less than T_1 (the process is *endothermic*), $\Delta_r H_m^{int}$ is *positive*.

Most reactions cause a change in the composition of one or more phases, in which case $\Delta_r H_m^{int}$ is not the same as the molar reaction enthalpy $\Delta_r H_m$ defined by Eq. 8.1.5. Corrections, usually small, are needed to obtain the standard molar reaction enthalpy from $\Delta_r H_m^{int}$.

8.5.2 The bomb calorimeter

A bomb calorimeter typically is used to carry out the complete combustion of a solid or liquid substance in the presence of excess oxygen. The combustion reaction is initiated with electrical ignition. In addition to the main combustion reaction, there may be unavoidable side reactions, such as the formation of nitrogen oxides if N_2 is not purged from the gas phase. Sometimes auxiliary reactions are carried out deliberately to complete or moderate the main reaction.

From the measured heating curve and known properties of the calorimeter, reactants, and products, it is possible to evaluate the standard molar enthalpy of combustion, $\Delta_c H_m^\circ$, of the substance of interest at a particular temperature called the reference temperature, T_{ref}. (T_{ref} is often chosen to be 298.15 K, which is 25.00 °C.) With careful work, using temperature measurements with a resolution of 1×10^{-4} K or better and detailed corrections, the precision of $\Delta_c H_m^\circ$ can be of the order of 0.01 percent.

Bomb calorimetry is the principal means by which standard molar enthalpies of combustion of individual elements and of compounds of these elements are evaluated. From these values, using Hess' law, we can calculate the standard molar enthalpies of *formation* of the compounds as described in Sec. 8.3.1. From the formation values of only a few compounds, the standard molar enthalpies of reaction of innumerable reactions can be calculated with Hess' law (Eq. 8.3.1 on page 256).

Because of their importance, the experimental procedure and the analysis of the data it provides are now described in some detail. A comprehensive problem (Prob. 8.7) based on this material is included at the end of the chapter.

There are five main steps in the procedure of evaluating a standard molar enthalpy of combustion:

1. The combustion reaction, and any side reactions and auxiliary reactions, are carried out in the calorimeter, and the course of the resulting temperature change is observed.

2. The experimental data are used to determine the value of $\Delta_{IBP} U(T_2)$, the internal energy change of the isothermal bomb process at the final temperature of the reaction. The **isothermal bomb process** is the idealized process that would have occurred if the reaction or reactions had taken place in the calorimeter at constant temperature.

3. The internal energy change of the isothermal bomb process is corrected to yield $\Delta_{IBP} U(T_{ref})$, the value at the reference temperature of interest.

4. The standard molar internal energy of combustion, $\Delta_c U_m^\circ(T_{ref})$, is calculated. This calculation is called **reduction to standard states**.

5. The standard molar enthalpy of combustion, $\Delta_c H_m^\circ(T_{ref})$, is calculated.

The next five sections describe these steps.

Experimental

The common form of combustion bomb calorimeter shown in Fig. 8.10 on the next page consists of a thick-walled cylindrical metal vessel to contain the combustion reaction It is

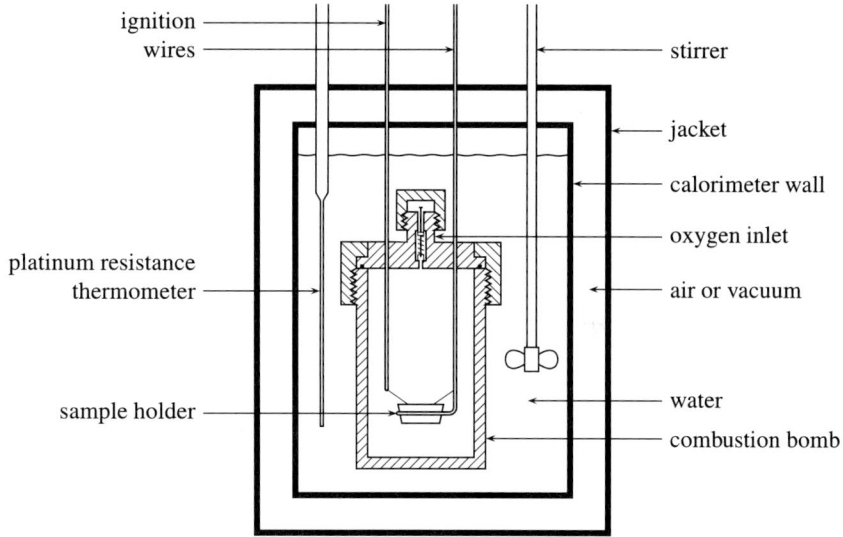

Figure 8.10 A bomb calorimeter.

called the "bomb" because it is designed to withstand high pressure. The bomb is sealed with a gas-tight screw cap. During the reaction, the sealed bomb vessel is immersed in water in the calorimeter, which is enclosed with a jacket. Conceptually, we take the *system* to be everything inside the jacket, including the calorimeter walls, water, bomb vessel, and contents of the bomb vessel.

To begin a combustion experiment, a weighed sample of the substance to be combusted is placed in a metal sample holder. The calculations are simplified if we can assume all of the sample is initially in a single phase. Thus, a volatile liquid is usually encapsulated in a bulb of thin glass (which shatters during the ignition) or confined in the sample holder by cellulose tape of known combustion properties. If one of the combustion products is H_2O, a small known mass of liquid water is placed in the bottom of the bomb vessel to saturate the gas space of the bomb vessel with H_2O. An ignition wire is placed at or near the sample, the cap is screwed onto the bomb vessel, and oxygen gas is admitted through a valve in the cap to a total pressure of about 30 bar.

To complete the setup, the sealed bomb vessel is immersed in a known mass of water in the calorimeter. A precision thermometer and a stirrer are also immersed in the water. With the stirrer turned on, the temperature is monitored until it is found to change at a slow, practically constant rate. This change is due to heat transfer through the jacket, mechanical stirring work, and electrical temperature measurement. A particular time is chosen as the initial time t_1. The measured temperature at this time is T_1, assumed to be practically uniform throughout the system. The ignition circuit is closed at or soon after time t_1 to initiate the combustion reaction in the bomb vessel. If the reaction is exothermic, the measured temperature rapidly increases over the course of several minutes as chemical

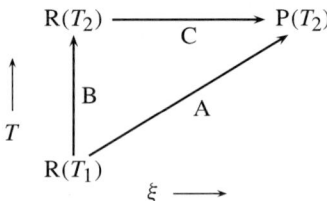

Figure 8.11 Paths for processes in a bomb calorimeter (schematic). R denotes reactants and P denotes products.

energy is converted to thermal energy. For a while the temperature in the system is far from uniform, as heat is transferred through the walls of the bomb vessel walls to the water outside. When the measured temperature is again observed to change at a slow and almost constant rate, the reaction is assumed to be complete, and the temperature is assumed once more to be uniform. A second time is now designated as the final time t_2, with final temperature T_2. For best accuracy, conditions are arranged so that T_2 is close to the desired reference temperature T_{ref}.

The isothermal bomb process

The process that actually occurs between times t_1 and t_2 is labeled Path A in Fig. 8.11. The system at uniform initial temperature T_1, with the reactants (R) in the bomb vessel, changes eventually to a uniform final temperature T_2 with products (P) in the bomb vessel. Because the jacket is not gas tight, the pressure of the water outside the bomb vessel stays constant at the atmospheric pressure p_{ext}. Inside the bomb vessel, there are changes in temperature and chemical composition at essentially constant volume, so the pressure inside the vessel is *not* constant.

Conceptually, the overall change of state that occurs in Path A could be achieved by Path B followed by Path C (Fig. 8.11). It is not necessary to actually carry out the processes of Paths B and C, but we can write simple relations for the internal energy changes of these processes.

In Path B, the system with the reactants in the bomb vessel is heated from T_1 to T_2 without chemical reaction. The initial and final states of Path B are assumed to be equilibrium states, so there may be some transfer of reactants or H_2O from one phase to another within the bomb vessel. The internal energy change in Path B is characterized by the energy equivalent ϵ of the calorimeter (page 112):

$$\Delta_B U = \epsilon (T_2 - T_1)$$

The value of ϵ is the average heat capacity dU/dT between T_1 and T_2 with reactants present in the bomb vessel.[10]

The process of Path C is the isothermal bomb process we are interested in, with reactants changing to products (and possibly some further transfer of substances between phases)

---

[10] The path in which temperature changes without reaction is best taken with reactants rather than products present because the reactants are usually more easily characterized.

within the bomb vessel at constant, uniform temperature T_2:

$$\Delta_C U = \Delta_{IBP} U(T_2)$$

From $\Delta_A U = \Delta_B U + \Delta_C U$, we obtain

$$\Delta_C = \Delta_{IBP} U(T_2) = -\epsilon(T_2 - T_1) + \Delta_A U \qquad (8.5.4)$$

We must look in detail at the possible sources of energy transfer between the system and the surroundings during the experimental Path A. These sources are (1) heat transfer, minimized but not eliminated by the jacket; (2) expansion work, $-p_{ext} \Delta V$, due to thermal expansion of the system; (3) mechanical stirring work done on the system; (4) electrical work w_{ign} done on the system by the ignition circuit; and (5) electrical work done on the system by an electrical thermometer. If none of these energy terms is significant, the system is effectively isolated from the surroundings during Path A and $\Delta_A U$ is equal to zero, giving the simple calorimetric formula

$$\Delta_{IBP} U(T_2) = -\epsilon(T_2 - T_1) \qquad (8.5.5)$$
(isolated calorimeter)

A calorimeter is never completely isolated, however, and for accurate work the energy terms just listed must be taken into account. The expansion work is usually negligible. The ignition work, w_{ign}, occurs for a short time just after time t_1, and its value is known. The effects of heat, stirring work, and temperature measurement continue throughout the course of the experiment. With these considerations, Eq. 8.5.4 becomes

$$\Delta_{IBP} U(T_2) = -\epsilon(T_2 - T_1) + w_{ign} + \Delta_A U' \qquad (8.5.6)$$

where $\Delta_A U'$ is the internal energy change in Path A due to heat, stirring, and temperature measurement. $\Delta_A U'$ can be evaluated from the energy equivalent and the observed rates of temperature change at times t_1 and t_2; the relevant relations for an isothermal jacket are Eq. 5.5.10 (with w_{el} set equal to zero) and Eq. 5.5.13.

The value of the energy equivalent ϵ needed for the combustion experiment is obtained in a separate calibration experiment. The calibration is usually carried out with the combustion of a reference substance such as benzoic acid, whose internal energy of combustion under controlled conditions is precisely known from standardization based on electrical work. If the bomb is immersed in the same mass of water in both experiments and other conditions are similar, the difference in the values of ϵ in the two experiments is equal to the known difference in the heat capacities of the initial contents (reactants, water, etc.) of the bomb vessel in the two experiments.

Correction to the reference temperature

The value of $\Delta_{IBP} U(T_2)$ evaluated from Eq. 8.5.6 refers to the internal energy change for the bomb process carried out at constant temperature T_2. We need to correct this value to refer to the desired reference temperature T_{ref}. If T_2 and T_{ref} are close in value, the correction is small and can be calculated with a simplified approximate version of the Kirchhoff equation (Eq. 8.3.5):

$$\Delta_{\text{IBP}}U(T_{\text{ref}}) = \Delta_{\text{IBP}}U(T_2) + [C_V(\text{P}) - C_V(\text{R})](T_{\text{ref}} - T_2) \tag{8.5.7}$$

where $C_V(\text{P})$ and $C_V(\text{R})$ are the heat capacities at constant volume of the contents of the bomb vessel with products and reactants, respectively, present.

Reduction to standard states

We want to obtain the value of $\Delta_c U_m^\circ(T_{\text{ref}})$, the molar internal energy change for the main combustion reaction at the reference temperature under *standard-state* conditions. Once we have this value, it is an easy matter to find the molar *enthalpy* change under standard-state conditions, our ultimate goal.

Consider a hypothetical process with the following three isothermal steps carried out at the reference temperature T_{ref}: 1. Each substance present initially in the bomb vessel changes from its standard state to its actual state at the start of the isothermal bomb process, as characterized by the initial pressure, composition, and so on. 2. The isothermal bomb process takes place, including the main combustion reaction and any side reactions and auxiliary reactions. 3. Each substance present in the final state of the isothermal bomb process changes to its standard state. The net change in this three-step process is a decrease in the amount of each reactant in its standard state and an increase in the amount of each product in its standard state. The internal energy change of Step 2 is $\Delta_{\text{IBP}}U(T_{\text{ref}})$, whose value is found from Eq. 8.5.7. The internal energy changes of Steps 1 and 3 are called **Washburn corrections**.[11]

Thus, we calculate the standard internal energy change of the main combustion reaction at temperature T_{ref} from

$$\Delta_c U^\circ(T_{\text{ref}}) = \Delta_{\text{IBP}}U(T_{\text{ref}}) + (\text{Washburn corrections}) - \sum_i \xi_i \Delta_r U_m^\circ(i) \tag{8.5.8}$$

where the sum over i is for side reactions or auxiliary reactions if present. Finally, we calculate the standard *molar* internal energy of combustion from

$$\Delta_c U_m^\circ(T_{\text{ref}}) = \frac{\Delta_c U^\circ(T_{\text{ref}})}{\xi} \tag{8.5.9}$$

where ξ is the advancement of the main combustion reaction in the bomb vessel.

Standard molar enthalpy change

The quantity $\Delta_c U_m^\circ(T_{\text{ref}})$ is the internal energy change for the main combustion reaction carried out at constant temperature T_{ref} with each reactant and product in its standard state at pressure p°. From the relations $\Delta_c H_m = \sum_i \nu_i H_i$ (Eq. 8.1.9) and $H_i = U_i + pV_i$ (from

---

[11] Washburn, E. W. *J. Res. Natl. Bur. Stand. (U.S.)* **1933**, *10*, 525.

Eq. 7.3.8), we get

$$\Delta_c H_m^\circ(T_{\text{ref}}) = \Delta_c U_m^\circ(T_{\text{ref}}) + p^\circ \sum_i \nu_i V_i^\circ$$

Molar volumes of condensed phases are much smaller than those of gases, and to a good approximation we may write

$$\Delta_c H_m^\circ(T_{\text{ref}}) = \Delta_c U_m^\circ(T_{\text{ref}}) + p^\circ \sum_i \nu_i^g V_i^\circ(g)$$

where the sum includes only gaseous reactants and products of the main combustion reaction. Since a gas in its standard state is an ideal gas with molar volume equal to RT/p°, the final relation is

$$\Delta_c H_m^\circ(T_{\text{ref}}) = \Delta_c U_m^\circ(T_{\text{ref}}) + \sum_i \nu_i^g RT_{\text{ref}} \qquad (8.5.10)$$

Washburn corrections

The Washburn corrections needed in Eq. 8.5.8 are internal energy changes for certain hypothetical physical processes occurring at the reference temperature T_{ref} and involving the substances present in the bomb vessel. In these processes, substances change from their standard states to the initial state of the isothermal bomb process and change from the final state of the isothermal bomb process to their standard states.

For example, consider the combustion of a solid or liquid compound of carbon, hydrogen, and oxygen in which the combustion products are CO_2 and H_2O and there are no side reactions or auxiliary reactions. In the initial state of the isothermal bomb process, the bomb vessel contains the pure reactant, liquid water with O_2 dissolved in it, and a gaseous mixture of O_2 and H_2O, all at a high pressure p_1. In the final state, the bomb vessel contains liquid water with O_2 and CO_2 dissolved in it and a gaseous mixture of O_2, H_2O, and CO_2, all at pressure p_2. In addition, the bomb vessel contains internal parts of constant mass such as the sample holder and ignition wires.

The Washburn corrections must use only a single standard state for each substance for Eq. 8.5.8 to correctly give the standard internal energy of combustion. In the present example we choose the following standard states: pure solid or liquid for the reactant compound, pure liquid for the H_2O, and pure ideal gases for the O_2 and CO_2, each at pressure $p^\circ = 1$ bar.

We can calculate the amount of each substance in each phase, in both the initial state and final state of the isothermal bomb process, from the following information: the internal volume of the bomb vessel; the mass of solid reactant and water initially placed in the vessel; the initial O_2 pressure; the water vapor pressure; the solubilities (estimated from Henry's law constants) of O_2 and CO_2 in the water; and the stoichiometry of the combustion reaction. Problem 8.7 on page 300 guides you through these calculations.

8.5.3 Other calorimeters

Experimenters have used great ingenuity over the years in designing calorimeters to measure reaction enthalpies and in improving their precision. In addition to the constant-pressure reaction calorimeter and bomb calorimeter described previously, three additional types are briefly mentioned.

A *phase-change calorimeter* has two coexisting phases of a pure substance in thermal contact with the reaction vessel and an adiabatic outer jacket. The two coexisting phases constitute a univariant subsystem that at constant pressure is at the fixed temperature of the phase transition. The heat released or absorbed by the reaction, instead of changing the temperature, is transferred isothermally to or from the coexisting phases and can be measured by the volume change of the phase transition. A reaction enthalpy, of course, can only be measured by this method at the temperature of the equilibrium phase transition. The well-known Bunsen ice calorimeter uses the ice–water transition at 0 °C. The solid–liquid transition of diphenyl ether has a relatively large volume change and is used for measurements at 26.9 °C. Phase-transition calorimeters are especially useful for slow reactions.

A *heat-flow calorimeter* is a variation of an isothermal-jacket calorimeter using a thermopile (Fig. 2.9) to continuously measure the temperature difference between the calorimeter and an outer jacket acting as a constant-temperature heat sink. The heat transfer takes place mostly through the thermocouple wires and to a high degree of accuracy is proportional to the temperature difference integrated over time. This is the best method for extremely slow reactions and is also used for rapid reactions.

A *flame calorimeter* is a flow system in which oxygen, fluorine, or other gaseous oxidant reacts with a gaseous fuel. The heat transfer between the flow tube and a heat sink can be measured with a thermopile, as in a heat-flow calorimeter.

8.6 ADIABATIC FLAME TEMPERATURE

With a few simple assumptions, we can estimate the temperature of a flame formed in a flowing gas mixture of oxygen or air and a fuel. We consider a moving segment of the gas mixture and treat it as a closed system in which the temperature increases as combustion takes place. We assume that the reaction occurs at constant pressure and also treat the process as adiabatic and the gas as an ideal gas mixture.

The principle of the calculation is similar to that used for a constant-pressure calorimeter as explained by the paths shown in Fig. 8.9 on page 270. The actual process corresponds to Path A, in which the combustion reaction in a segment of the gas mixture proceeds at constant pressure to extent ξ, causing the temperature to increase from T_1 to T_2. If the process is perfectly adiabatic, the enthalpy change is zero. The net enthalpy change is the same if the reaction first occurs at constant temperature T_1 and constant p (Path B), and the product mixture is then heated at constant p to temperature T_2 (Path C): $\Delta_B H + \Delta_C H = \Delta_A H$. Replacing each term in this equation by an appropriate expression, we have the relation

$$\xi \Delta_r H_m^\circ(T_1) + \int_{T_1}^{T_2} C_p(\text{P}) \, dT = 0 \tag{8.6.1}$$

where $\Delta_r H_m^\circ(T_1)$ is the standard molar reaction enthalpy at the initial temperature, and $C_p(\text{P})$ is the heat capacity at constant pressure of the product mixture.

The value of T_2 that satisfies this relation is the *maximum* flame temperature. Problem 8.9 presents an application of this calculation. Several factors cause the actual temperature in a flame to be lower: The process is never completely adiabatic, and in the high temperature of the flame there may be product dissociation and other reactions in addition to the main combustion reaction.

8.7 GIBBS ENERGY AND REACTION EQUILIBRIUM

We now want to examine the way in which the Gibbs energy varies as a chemical process advances in a closed system at constant T and p. This system has only one independent variable, for which it is convenient to choose the advancement ξ. The additivity rule (Eq. 7.2.9) for the Gibbs energy is

$$G = \sum_i n_i \mu_i \tag{8.7.1}$$

where n_i depends on ξ and μ_i may also. Thus, G is a complicated function of ξ.

8.7.1 The molar reaction Gibbs energy

Applying the general definition of a molar reaction quantity (Eq. 8.1.9) to the Gibbs energy, we obtain the definition of the **molar reaction Gibbs energy** or molar Gibbs energy of reaction, $\Delta_r G_m$:

$$\Delta_r G_m \equiv \sum_i \nu_i \mu_i \tag{8.7.2}$$

Equation 8.1.10 shows that this quantity is also given by the partial derivative

$$\Delta_r G_m = \left(\frac{\partial G}{\partial \xi}\right)_{T,p} \tag{8.7.3}$$
$$\text{(closed system)}$$

The total differential of G in a closed system with T, p, and ξ taken as the independent variables is then

$$dG = -S\,dT + V\,dp + \Delta_r G_m\,d\xi \tag{8.7.4}$$
$$\text{(closed system)}$$

The molar reaction Gibbs energy is an important quantity that we may interpret in the following alternative ways:

1. $\Delta_r G_m$ is the rate at which G changes with ξ at constant T and p (Eq. 8.7.3); it is the slope at any point on the curve of a plot of G versus ξ.

2. $\Delta_r G_m$ is the finite change ΔG divided by the finite advancement $\Delta \xi$ for reaction at constant T and p under conditions in which the chemical potential of each reactant and

```
p = 1 bar
nA = 2000 mol   pA = 0.4000 bar
nB = 2000 mol   pB = 0.4000 bar
nC = 1000 mol   pC = 0.2000 bar
```
→
```
p = 1 bar
nA = 1999 mol   pA = 0.3999 bar
nB = 1999 mol   pB = 0.3999 bar
nC = 1001 mol   pC = 0.2002 bar
```

```
p = 0.4 bar        p = 0.4 bar
nA = 1 mol    +    nB = 1 mol
```
→
```
p = 0.2 bar
nC = 1 mol
```

Figure 8.12 Two interpretations of $\Delta_r G_m$ for the reaction $A(g) + B(g) \to C(g)$ occurring in an ideal gas mixture. (Top) The reaction advances by 1 mole at constant T and p in such a large amount of the mixture that the composition change is negligible. (Bottom) Pure ideal-gas reactants form the pure ideal-gas product at the same temperature as the mixture; each substance has the same chemical potential as in the mixture because it has the same partial pressure. ΔG is the same in both processes, and $\Delta_r G_m$ is equal to $\Delta G/(1 \text{ mol})$.

product remains essentially constant. In other words, $\Delta_r G_m$ under these conditions is the same as the molar *integral* reaction Gibbs energy $\Delta_r G_m^{int}$ (see Eq. 8.1.11 on page 243). If the reaction occurs in a mixture, the amount of mixture must be large compared to $\Delta \xi$ so that the composition change is negligible.

3. According to Eq. 8.7.2, $\Delta_r G_m$ for a reaction at constant T and p in a mixture is the same as the change in G per amount of reaction when *pure* reactants having the same chemical potentials as the reactants in the mixture change to *pure* products with the same chemical potentials as the products in the mixture.

Figure 8.12 gives a schematic illustration of the second and third of these interpretations.

8.7.2 Spontaneity and equilibrium

In Sec. 4.9, we found that the spontaneous direction of a process occurring in a closed system at constant T and p, with expansion work only, is the direction of decreasing G; and that in the equilibrium state, G has its minimum value. In the case of a chemical process occurring at constant T and p, $\Delta_r G_m$ is the rate at which G changes with ξ. Thus, ξ must increase spontaneously if $\Delta_r G_m$ is negative and decrease spontaneously if $\Delta_r G_m$ is positive.[12] The condition for *reaction equilibrium* is that $\Delta_r G_m$ be zero:

$$\Delta_r G_m = \sum_i \nu_i \mu_i = 0 \qquad (8.7.5)$$
(reaction equilibrium)

---

[12] Sometimes reaction spontaneity is ascribed to the driving force of a quantity called the *affinity of reaction*. This quantity is defined as the negative of $\Delta_r G_m$. ξ increases spontaneously if the affinity is positive and decreases spontaneously if the affinity is negative; the system is at equilibrium when the affinity is zero.

Section 8.7 Gibbs Energy and Reaction Equilibrium

We can derive more general equilibrium conditions by the method of Sec. 6.1.1. Consider a system of one or more phases without internal barriers that are adiabatic or impermeable. Each phase contains one or more species, some or all of which are the reactants and products of a reaction. We single out one phase, α, as a reference phase for temperature and pressure. For each species, we designate a reference phase α' that contains that species; phase α' is not necessarily the same as phase α and may be different for each species because there may be no one phase that contains all species.

We now write a general expression for the total differential of the internal energy in the form

$$dU = T^\alpha \, dS^\alpha + \sum_{\beta \neq \alpha} T^\beta \, dS^\beta - p^\alpha \, dV^\alpha - \sum_{\beta \neq \alpha} p^\beta \, dV^\beta$$

$$+ \sum_i \left(\mu_i^{\alpha'} \, dn_i^{\alpha'} + \sum_{\beta' \neq \alpha'} \mu_i^{\beta'} \, dn_i^{\beta'} \right)$$

$$+ \sum_j \left(\mu_j^{\alpha'} \, dn_j^{\alpha'} + \sum_{\beta' \neq \alpha'} \mu_j^{\beta'} \, dn_j^{\beta'} \right) \tag{8.7.6}$$

where the sum over i is for reactant and product species of the reaction, and the sum over j is for species (if any) that are not reactants or products. The sum over β' for a species includes only those phases, other than the reference phase α', that contain the species.

We write the entropy change in the form $dS = dS^\alpha + \sum_{\beta \neq \alpha} dS^\beta$. The conditions needed for a change with zero heat and work are $dU = 0$ and $dV^\alpha + \sum_{\beta \neq \alpha} dV^\beta = 0$. The change in the amount of reactant or product Species i in the closed system is given by $dn_i^{\alpha'} + \sum_{\beta' \neq \alpha'} dn_i^{\beta'} = \nu_i \, d\xi$, and the change in the amount of a substance j that is not a reactant or product is $dn_j^{\alpha'} + \sum_{\beta' \neq \alpha'} dn_j^{\beta'} = 0$. Solving these equations for dS^α, dV^α, $dn_i^{\alpha'}$, and $dn_j^{\alpha'}$, substituting in Eq. 8.7.6, and solving for dS, we obtain finally, for changes in the isolated system

$$dS = \sum_{\beta \neq \alpha} \frac{T^\alpha - T^\beta}{T^\alpha} \, dS^\beta - \sum_{\beta \neq \alpha} \frac{p^\alpha - p^\beta}{T^\alpha} \, dV^\beta$$

$$- \frac{\sum_i \nu_i \mu_i}{T^\alpha} \, d\xi + \sum_i \sum_{\beta' \neq \alpha'} \frac{\mu_i^{\alpha'} - \mu_i^{\beta'}}{T^\alpha} \, dn_i^{\beta'} + \sum_j \sum_{\beta' \neq \alpha'} \frac{\mu_j^{\alpha'} - \mu_j^{\beta'}}{T^\alpha} \, dn_j^{\beta'}$$

The equilibrium condition is that the coefficient multiplying each differential on the right side of the equation is zero. We conclude that at equilibrium the temperature and pressure of each phase is equal to that of phase α; that each species has a chemical potential, in each phase that contains the species, equal to the chemical potential of the species in its reference phase; and that the quantity $\sum_i \nu_i \mu_i$ (which is equal to $\Delta_r G_m$) is zero.

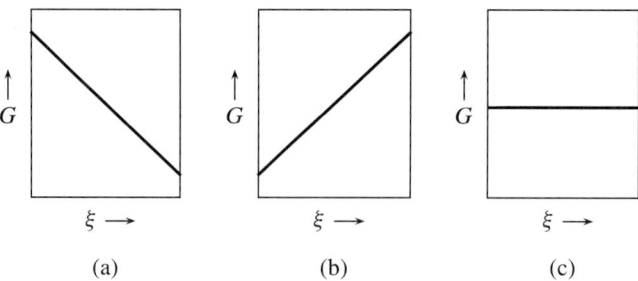

Figure 8.13 Gibbs energy versus advancement in a system of pure phases at constant T and p. G is a linear function of ξ with slope equal to $\Delta_r G_m$. (a) $\Delta_r G_m$ is negative; ξ increases spontaneously. (b) $\Delta_r G_m$ is positive; ξ decreases spontaneously. (c) $\Delta_r G_m$ is zero; the system is in reaction equilibrium at all values of ξ.

In short, *in the equilibrium state the temperature and pressure are uniform, each species has the same chemical potential in each phase containing the species, and the molar reaction Gibbs energy is zero.*

8.7.3 Pure phases

Consider a chemical process in which each reactant and product is in a separate pure phase instead of being a constituent of a mixture. For example, the decomposition of calcium carbonate, $CaCO_3(s) \rightarrow CaO(s) + CO_2(g)$, involves three pure phases if no other gas is allowed to mix with the CO_2. As this kind of reaction advances at constant T and p, the chemical potential of each substance remains constant and therefore $\Delta_r G_m$ is constant. The value of $\Delta_r G_m$ depends on T and p. If $\Delta_r G_m$ is negative, the reaction proceeds spontaneously to the right until one of the reactants is exhausted; this is a reaction that goes to completion. If $\Delta_r G_m$ is positive, the reaction proceeds spontaneously to the left until one of the products is exhausted. The reactants and products can remain in equilibrium only if T and p are such that $\Delta_r G_m$ is zero. These three cases are illustrated in Fig. 8.13.

> Note the similarity of this behavior to that of an equilibrium phase transition of a pure substance. Only one phase of a pure substance is present at equilibrium unless $\Delta_{trs} G_m$ is zero. A phase transition is a special case of a chemical process.

If, on the other hand, any of the reactants or products is in a mixture, the curve of a plot of G versus ξ (at constant T and p) turns out to exhibit a minimum with a slope of zero at the minimum point (see the example in Fig. 8.14 on the next page). ξ spontaneously changes in the direction of decreasing G until the minimum is reached, at which point $\Delta_r G_m$ (the slope of the curve) is zero and the system is in a state of reaction equilibrium.

8.7.4 Reaction in an ideal gas mixture

Let us look in detail at the source of the minimum in G for the case of a reaction occurring in an ideal gas mixture at constant T and p. For the chemical potential of each substance,

Section 8.7 Gibbs Energy and Reaction Equilibrium

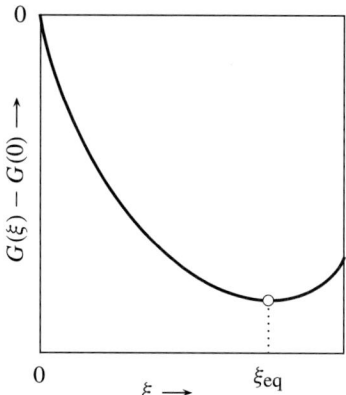

Figure 8.14 Gibbs energy as a function of advancement at constant T and p in a closed system containing a mixture. The open circle is at the minimum value of G.

we write $\mu_i = \mu_i^\circ + RT \ln(p_i/p^\circ)$ (Eq. 7.4.4), where p_i is the partial pressure of i in the mixture. Substitution in Eq. 8.7.1 gives, for the Gibbs energy at any value of ξ,

$$G(\xi) = \sum_i n_i \left(\mu_i^\circ + RT \ln \frac{p_i}{p^\circ} \right)$$

At $\xi = 0$, the amounts and partial pressures have their initial values $n_{i,0}$ and $p_{i,0}$:

$$G(0) = \sum_i n_{i,0} \left(\mu_i^\circ + RT \ln \frac{p_{i,0}}{p^\circ} \right)$$

The difference between these two expressions is

$$G(\xi) - G(0) = \sum_i (n_i - n_{i,0}) \mu_i^\circ + RT \sum_i n_i \ln \frac{p_i}{p^\circ} - RT \sum_i n_{i,0} \ln \frac{p_{i,0}}{p^\circ}$$

Converting partial pressures to mole fractions with $p_i = y_i p$ and $p_{i,0} = y_{i,0} p$ gives

$$G(\xi) - G(0) = \sum_i (n_i - n_{i,0}) \mu_i^\circ + RT \sum_i n_i \ln y_i - RT \sum_i n_{i,0} \ln y_{i,0}$$
$$+ RT \sum_i (n_i - n_{i,0}) \ln \frac{p}{p^\circ}$$

With the substitution $n_i - n_{i,0} = \nu_i \xi$ (Eq. 8.1.6) in the first and last terms on the right side, the result is

$$G(\xi) - G(0) = \xi \sum_i \nu_i \mu_i^\circ + RT \sum_i n_i \ln y_i - RT \sum_i n_{i,0} \ln y_{i,0}$$
$$+ RT \left(\sum_i \nu_i \right) \xi \ln \frac{p}{p^\circ}$$

The sum $\sum_i \nu_i \mu_i^\circ$ in the first term on the right side of the preceding equation is $\Delta_r G_m^\circ$, the standard molar reaction Gibbs energy. Making this substitution gives finally

$$G(\xi) - G(0) = \xi \Delta_r G_m^\circ + RT \sum_i n_i \ln y_i - RT \sum_i n_{i,0} \ln y_{i,0}$$

$$+ RT \left(\sum_i \nu_i \right) \xi \ln \frac{p}{p^\circ} \qquad (8.7.7)$$
(ideal gas mixture)

There are four terms on the right side of Eq. 8.7.7. The first term is the Gibbs energy change for the reaction of pure reactants to form pure products under standard-state conditions, the second is a mixing term, the third term is constant, and the last term is an adjustment of G from the standard pressure to the pressure of the gas mixture. Note that the first and last terms are proportional to the advancement and cannot cause a minimum in the curve of the plot of G versus ξ. It is the *mixing term* $RT \sum_i n_i \ln y_i$ that is responsible for the observed minimum.[13] This term is the Gibbs energy of mixing, $\Delta_{mix} G^{id} = n \Delta_{mix} G_m^{id}$ (see Eq. 8.2.4 on page 246).

Now let us consider specifically the reaction type

$$A(g) \rightarrow B(g)$$

(such as a cis-trans isomerization) in an ideal gas mixture, for which ν_A is -1 and ν_B is $+1$. We let the initial state be one of pure A: $n_{B,0} = 0$; the initial mole fractions are $y_{A,0} = 1$ and $y_{B,0} = 0$. In this reaction, the total amount $n = n_A + n_B$ is constant. Substituting these values in Eq. 8.7.7 gives[14]

$$G(\xi) - G(0) = \xi \Delta_r G_m^\circ + nRT(y_A \ln y_A + y_B \ln y_B)$$

where $\Delta_r G_m^\circ$ equals $\mu_B^\circ(g) - \mu_A^\circ(g)$. The quantity $nRT(y_A \ln y_A + y_B \ln y_B)$ is $n \Delta_{mix} G_m^{id}$, the Gibbs energy of mixing pure ideal gases A and B at constant T and p to form an ideal gas mixture with the composition y_A and y_B. Since the curve of $\Delta_{mix} G_m^{id}$ plotted against ξ has a minimum (as shown in Fig. 8.1 on page 247), $G(\xi) - G(0)$ also has a minimum.

Figure 8.15 on the next page illustrates how the position of the minimum, which is the position of reaction equilibrium, depends on the value of $\Delta_r G_m^\circ$. The more negative is $\Delta_r G_m^\circ$, the closer to the product side of the reaction is the equilibrium position. On the other hand, the more positive is $\Delta_r G_m^\circ$ the smaller is the advancement at equilibrium. These statements apply to any homogeneous reaction.

As the reaction A $\rightarrow$ B proceeds, there is no change in the total number of molecules, and therefore in an ideal gas mixture at constant temperature and pressure there is no volume change. The point of reaction equilibrium is at the minimum of G when both p and V are constant.

---

[13] This curve also causes the slope of the curve of $G(\xi) - G(0)$ versus ξ to be $-\infty$ and $+\infty$ at the left and right extremes of the curve.

[14] Note that while $\ln y_A$ approaches $-\infty$ as y_A approaches zero, the product $y_A \ln y_A$ approaches *zero* in this limit. This behavior can be proved with l'Hospital's rule (see any calculus textbook).

Section 8.7 Gibbs Energy and Reaction Equilibrium 285

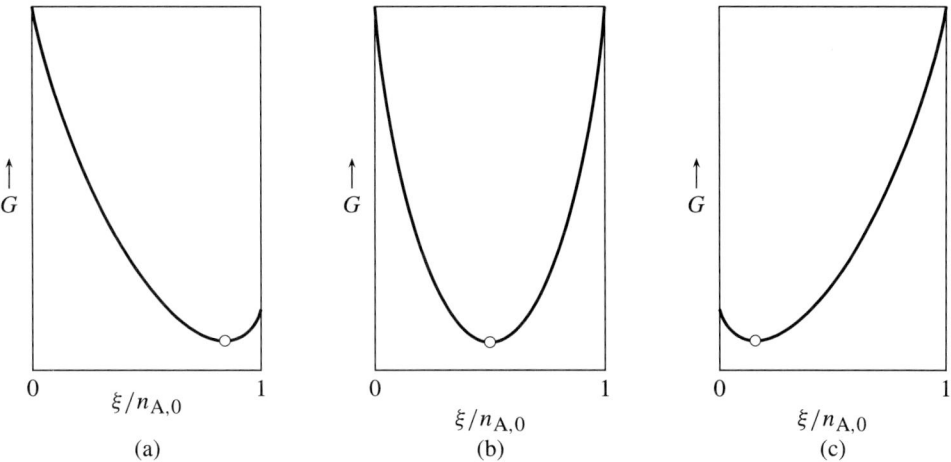

Figure 8.15 Gibbs energy as a function of the advancement of the reaction A → B in an ideal gas mixture at constant T and p. The initial amount of B is zero. The equilibrium positions are indicated by open circles. (a) $\Delta_r G_m^\circ < 0$; (b) $\Delta_r G_m^\circ = 0$; (c) $\Delta_r G_m^\circ > 0$.

The situation is different when the number of molecules changes during the reaction. Consider the reaction A → 2 B in an ideal gas mixture. As this reaction proceeds to the right at constant T, the volume increases if the pressure is held constant and the pressure increases if the volume is held constant. Figure 8.16 on the next page shows how G depends on p and V for this reaction. Movement along the horizontal dashed line in the figure corresponds to reaction at constant T and p. The minimum of G along this line is at the volume indicated by the open circle. At this volume, G has an even lower minimum at the pressure indicated by the filled circle, where the vertical dashed line is tangent to one of the contours of constant G. The condition for reaction equilibrium, however, is that $\Delta_r G_m$ be zero. This condition is satisfied along the vertical dashed line only at the position of the open circle.

This example demonstrates that for a reaction occurring at constant temperature and *volume* in which the pressure changes, the point of reaction equilibrium is not the point of minimum G. Instead, the point of reaction equilibrium in this case is at the minimum of the Helmholtz energy A (see Eq. 4.9.4 on page 91 and the discussion following it).

8.7.5 The standard molar reaction Gibbs energy

As mentioned in the preceding section, the **standard molar reaction Gibbs energy** (or standard molar Gibbs energy of reaction) of a reaction is defined by

$$\Delta_r G_m^\circ \equiv \sum_i \nu_i \mu_i^\circ \tag{8.7.8}$$

Since each standard chemical potential μ_i° is a function of T only, the value of $\Delta_r G_m^\circ$

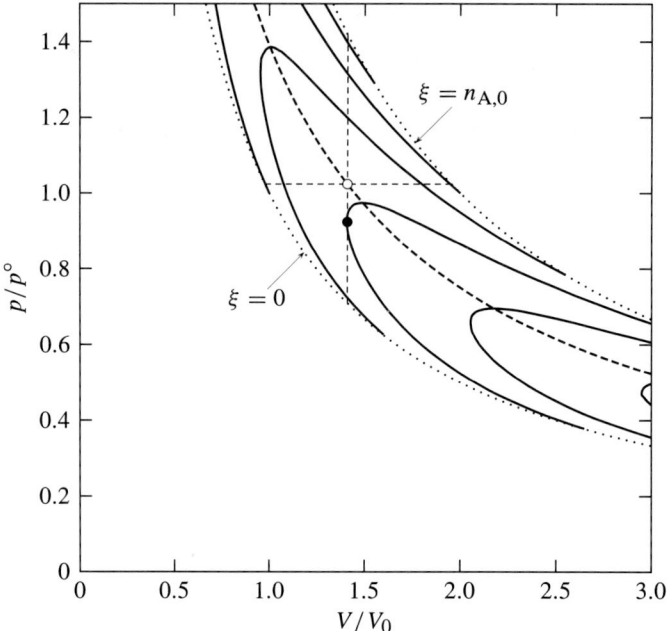

Figure 8.16 Dependence of Gibbs energy on volume and pressure, at constant temperature, in a closed system containing an ideal gas mixture of reactant and product. The reaction is A $\rightarrow$ 2B; $\Delta_r G_m^\circ$ is zero. Solid curves: contours of constant G plotted at an interval of $0.5 n_{A,0} RT$. Dashed curve: states of reaction equilibrium ($\Delta_r G_m = 0$). Dotted curves: limits of possible values of the advancement. Open circle: position of minimum G (and an equilibrium state) at the constant pressure $p = 1.02 p^\circ$. Filled circle: position of minimum G for a constant volume of $1.41 V_0$, where V_0 is the initial volume at pressure p°.

for a reaction (as defined by a chemical equation) must depend only on T and on our choice of a standard state for each reactant and product.

There are three interpretations of $\Delta_r G_m^\circ$ analogous to the interpretations of $\Delta_r G_m$ in Sec. 8.7.1:

1. $\Delta_r G_m^\circ$ is the rate at which G changes with ξ at constant T when the chemical potential of each reactant and product is equal to its standard chemical potential.

2. $\Delta_r G_m^\circ$ is the finite change ΔG divided by the finite advancement $\Delta \xi$ for reaction at constant T under conditions in which the chemical potential of each reactant and product remains essentially constant and equal to its standard chemical potential. In the case of a mixture this implies a large amount of mixture compared to $\Delta \xi$.

3. $\Delta_r G_m^\circ$ is the change in G per amount of reaction at constant T when the reactants in their standard states change to products in their standard states. Note that, unless a reactant or product is a solute, its standard state is a pure (unmixed) phase.

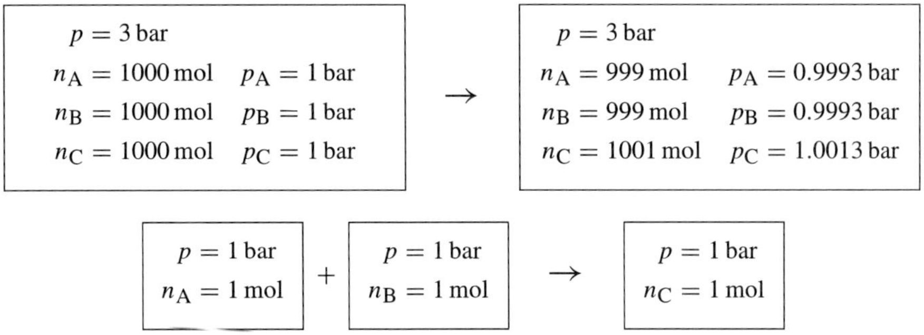

Section 8.8 The Thermodynamic Equilibrium Constant

$p = 3$ bar
$n_A = 1000$ mol $p_A = 1$ bar
$n_B = 1000$ mol $p_B = 1$ bar
$n_C = 1000$ mol $p_C = 1$ bar

$\rightarrow$

$p = 3$ bar
$n_A = 999$ mol $p_A = 0.9993$ bar
$n_B = 999$ mol $p_B = 0.9993$ bar
$n_C = 1001$ mol $p_C = 1.0013$ bar

$p = 1$ bar
$n_A = 1$ mol

$+$

$p = 1$ bar
$n_B = 1$ mol

$\rightarrow$

$p = 1$ bar
$n_C = 1$ mol

Figure 8.17 Two interpretations of $\Delta_r G_m^\circ$ for the reaction $A(g) + B(g) \rightarrow C(g)$. (Top) The reaction advances by 1 mole at constant T and p in a large amount of an ideal gas mixture in which the partial pressure of each reactant and product differs only negligibly from the standard pressure. (Bottom) The pure ideal-gas reactants in their standard states form the pure ideal-gas product in its standard state at the same temperature as the mixture. ΔG is the same in both processes, and $\Delta_r G_m^\circ$ is equal to $\Delta G/(1 \text{ mol})$.

Figure 8.17 illustrates the second and third of these interpretations for a simple gas-phase reaction.

8.8 THE THERMODYNAMIC EQUILIBRIUM CONSTANT

8.8.1 Definition

The general relation between the chemical potential μ_i and activity a_i of a species is $\mu_i = \mu_i^\circ + RT \ln a_i$ (Eq. 7.7.2). Substituting this relation in $\Delta_r G_m = \sum_i \nu_i \mu_i$ (Eq. 8.7.2) gives

$$\Delta_r G_m = \sum_i \nu_i \mu_i^\circ + RT \sum_i \nu_i \ln a_i$$

The first term on the right side is the standard molar reaction Gibbs energy $\Delta_r G_m^\circ$ (see Eq. 8.7.8). We may use properties of logarithms to write the sum in the second term as[15]

$$\sum_i \nu_i \ln a_i = \sum_i \ln \left(a_i^{\nu_i}\right) = \ln \prod_i a_i^{\nu_i}$$

The product $\prod_i a_i^{\nu_i}$ is called the **reaction quotient** or activity quotient, Q:

$$Q \equiv \prod_i a_i^{\nu_i} \qquad (8.8.1)$$

---

[15] The symbol $\prod$ stands for a continued product. If there are three species, for instance, $\prod_i a_i^{\nu_i}$ is the product $(a_1^{\nu_1})(a_2^{\nu_2})(a_3^{\nu_3})$.

Q consists of a factor for each reactant and product. Each factor is the activity raised to the power of the stoichiometric number ν_i. Since the value of ν_i is positive for a product and negative for a reactant, Q is a quotient in which the activities of the products appear in the numerator and those of the reactants appear in the denominator, with each activity raised to a power equal to the corresponding stoichiometric coefficient in the chemical equation. For instance, the reaction quotient of the ammonia synthesis reaction $N_2(g) + 3\,H_2(g) \rightarrow 2\,NH_3(g)$ is

$$Q = \frac{a_{NH_3}^2}{a_{N_2} a_{H_2}^3}$$

Q is a dimensionless quantity whose value varies with the mixture composition; that is, it changes as the reaction advances.

Thus, the molar reaction Gibbs energy under given conditions of T, p, and ξ is given by

$$\Delta_r G_m = \Delta_r G_m^\circ + RT \ln Q \tag{8.8.2}$$

The value of Q under equilibrium conditions is the **thermodynamic equilibrium constant**, K, another dimensionless quantity. The general definition of K is

$$K \equiv \prod_i (a_i)_{eq}^{\nu_i} \tag{8.8.3}$$

where the subscript "eq" indicates an equilibrium state. For the system to be in a equilibrium state, Q must equal K.

> The 1993 IUPAC Green Book gives K° or $K^\ominus$ as alternative symbols for the thermodynamic equilibrium constant, the appended superscripts denoting "standard." An IUPAC Commission on Thermodynamics has furthermore recommended[16] the name "standard equilibrium constant." Using these alternative symbols and name could cause confusion since the quantity defined by Eq. 8.8.3 does *not* refer to reactants and products in their standard states but rather to reactants and products in an *equilibrium* state.

Substituting the equilibrium conditions $\Delta_r G_m = 0$ and $Q = K$ in Eq. 8.8.2 gives an important relation between the standard molar reaction Gibbs energy for a reaction and the thermodynamic equilibrium constant:

$$\Delta_r G_m^\circ = -RT \ln K \tag{8.8.4}$$

We can solve this equation for K to obtain the equivalent relation

$$K = \exp\left(-\frac{\Delta_r G_m^\circ}{RT}\right) \tag{8.8.5}$$

---

[16] Ewing, M. B.; Lilley, T. H.; Olofsson, G. M.; Rätzsch, M. T.; Somsen, G. *Pure Appl. Chem.* **1994**, *66*, 533–552.

We have seen that the value of $\Delta_r G_m^\circ$ depends only on T and the choice of the standard states of the reactants and products. This being so, Eq. 8.8.5 shows that the value of K for a given reaction depends only on T and the choice of standard states. No other condition (such as the pressure) can affect the value of K.

Equation 8.8.5 also shows that if $\Delta_r G_m^\circ$ is positive, K is less than 1; if $\Delta_r G_m^\circ$ is negative, K is greater than 1.

The thermodynamic equilibrium constant K is a quotient of the activities of species in the equilibrium system. At typical temperatures and pressures, an activity cannot be many orders of magnitude greater than 1. For instance, a partial pressure cannot be greater than the total pressure, so at a pressure of 10 bar the activity of a gaseous constituent cannot be greater than about 10. The molarity of a solute is rarely much greater than $10 \, \text{mol dm}^{-3}$ corresponding to an activity (on a concentration basis) of about 10. Activities can however be extremely small. Consequently, in an equilibrium state of a reaction with a very *large* value of K (and a very large negative value of $\Delta_r G_m^\circ$), the activity of at least one of the *reactants* must be very small. If the reactant is a constituent of a mixture, so that its activity varies with ξ, this is a reaction that spontaneously goes practically to completion—that is, a limiting reactant is almost completely consumed.

The opposite case, a reaction with a very *small* value of K (and a very large positive value of $\Delta_r G_m^\circ$), must have a very small activity of at least one of the *products* at equilibrium. If the product is in a mixture, the reaction can only proceed spontaneously to the left. These two cases are the two extremes of the trends shown in Fig. 8.15 on page 285.

In the case of a reaction between *pure phases* at constant T and p, the activities are constant and Q does not vary with ξ. If Q is equal to K at this temperature and pressure, the system is in reaction equilibrium over the entire range of ξ; otherwise ξ spontaneously increases if Q is less than K or spontaneously decreases if Q is greater than K. This behavior of reactions between pure phases is described in Sec. 8.7.3 in relation to $\Delta_r G_m$, which depends on Q according to Eq. 8.8.2.

Equation 8.8.4 correctly relates $\Delta_r G_m^\circ$ and K only if they are both calculated with the same standard states. For instance, if we base the standard state of a particular solute species on molality in calculating $\Delta_r G_m^\circ$, the activity of that species appearing in the expression for K (Eq. 8.8.3) is also on a molality basis.

8.8.2 Reaction in a gas phase

If a reaction occurs in a gaseous mixture, the standard state of each reactant and product is the pure gas at the standard pressure p° behaving ideally (Sec. 7.4.3). In this case, each activity is given by $a_i = \phi_i p_i / p^\circ$, where ϕ_i is a fugacity coefficient (Table 7.3). From Eq. 8.8.3, we can write

$$K = \left[\prod_i (\phi_i)_{\text{eq}}^{\nu_i} \right] \left[\prod_i (p_i)_{\text{eq}}^{\nu_i} \right] \left[(p^\circ)^{-\sum_i \nu_i} \right] \quad (8.8.6)$$
(gas mixture)

which expresses the thermodynamic equilibrium constant as the product of three factors: an equilibrium fugacity-coefficient quotient, an equilibrium partial-pressure quotient, and the power of p° needed to make K dimensionless.

The equilibrium partial-pressure quotient is an **equilibrium constant on a pressure basis**, K_p:

$$K_p = \prod_i (p_i)_{\text{eq}}^{\nu_i} \qquad (8.8.7)$$
$$\text{(gas mixture)}$$

Note that K_p is not dimensionless except for a reaction in which $\sum_i \nu_i$ is equal to zero.

K_p is not a *thermodynamic* equilibrium constant because its value can vary at constant T. For instance, consider what happens as the gas pressure is increased. A fugacity coefficient ϕ_i changes from its low pressure value of 1 as p is increased and the gas mixture becomes nonideal, so K_p must also change for the product on the right side of Eq. 8.8.6 to remain constant. In other words, the reaction equilibrium *shifts* as we increase p at constant T, an effect that will be considered in more detail in Sec. 8.9.

For an example of the difference between K and K_p, consider again the ammonia synthesis $N_2(g) + 3\,H_2(g) \rightarrow 2\,NH_3(g)$. For this reaction, the expression for the thermodynamic equilibrium constant is

$$K = \left(\frac{\phi_{NH_3}^2}{\phi_{N_2}\phi_{H_2}^3}\right)_{\text{eq}} K_p (p^\circ)^2$$

where K_p is given by

$$K_p = \left(\frac{p_{NH_3}^2}{p_{N_2} p_{H_2}^3}\right)_{\text{eq}}$$

8.8.3 Reaction in solution

If any of the reactants or products are solutes in a solution, the value of K depends on which of the solute standard states described in Sec. 7.7.1 we choose. Consider the simple chemical equation

$$B(\text{sln}) \rightarrow C(\text{sln}) + D(\text{sln})$$

with thermodynamic equilibrium constant given by $K = (a_C a_D / a_B)_{\text{eq}}$. To distinguish solute standard states based on mole fraction, concentration, and molality, we can denote the thermodynamic equilibrium constant by K_x°, K_c°, or K_m°. These quantities refer to the same reaction, have values that depend only on temperature, but have different values. For a mole fraction basis the thermodynamic equilibrium constant is

$$K_x^\circ = \left(\frac{\Gamma_{x,C}\,\Gamma_{x,D}}{\Gamma_{x,B}}\right)\left(\frac{\gamma_{x,C}\,\gamma_{x,D}}{\gamma_{x,B}}\right)_{\text{eq}} K_x \qquad K_x = \left(\frac{x_C x_D}{x_B}\right)_{\text{eq}}$$

The corresponding expression for a concentration basis is

$$K_c^\circ = \left(\frac{\Gamma_{c,C}\,\Gamma_{c,D}}{\Gamma_{c,B}}\right)\left(\frac{\gamma_{c,C}\,\gamma_{c,D}}{\gamma_{c,B}}\right)_{\text{eq}} \frac{K_c}{c^\circ} \qquad K_c = \left(\frac{c_C c_D}{c_B}\right)_{\text{eq}}$$

and for a molality basis it is

$$K_m^\circ = \left(\frac{\Gamma_{m,C}\,\Gamma_{m,D}}{\Gamma_{m,B}}\right)\left(\frac{\gamma_{m,C}\,\gamma_{m,D}}{\gamma_{m,B}}\right)_{eq}\frac{K_m}{m^\circ} \qquad K_m = \left(\frac{m_C m_D}{m_B}\right)_{eq}$$

Each of these expressions is the product of a pressure-factor quotient; an equilibrium activity-coefficient quotient; an equilibrium constant on a mole fraction, concentration, or molality basis; and (in the case of concentration and molality) the power of the standard composition variable that makes the expression dimensionless.

We can derive the relations between the values of K_x°, K_c°, and K_m° in this example. At the standard pressure p°, each pressure factor equals 1. In the limit of infinite dilution, each solute activity coefficient approaches 1 and the solute composition variables approach values given by the relations in Eq. 7.1.10 on page 172: $x_B = (M_A/\rho_A^*)c_B = M_A m_B$. Making these substitutions for $p = p^\circ$ at infinite dilution in the expression for K_x° gives

$$K_x^\circ = \frac{M_A}{\rho_A^*}\left(\frac{c_C c_D}{c_B}\right)_{eq} = M_A\left(\frac{m_C m_D}{m_B}\right)_{eq}$$

and, by comparison with the expressions for K_c° and K_m° at infinite dilution:[17]

$$K_x^\circ = \frac{M_A c^\circ}{\rho_A^*} K_c^\circ = M_A m^\circ K_m^\circ$$

Since the values of K_x°, K_c°, and K_m° are not affected by pressure or dilution, these relations are valid under all conditions.

For a reaction in general, the relations among the three kinds of thermodynamic equilibrium constants are

$$K_x^\circ = \left(\frac{M_A c^\circ}{\rho_A^*}\right)^{\sum_i \nu_i} K_c^\circ = (M_A m^\circ)^{\sum_i \nu_i} K_m^\circ \qquad (8.8.8)$$

These relations can be used for reactions in which some but not all of the reactants and products are solutes; in that case, the sums over i include only the solutes.

8.8.4 Pressure-factor quotient of a reaction in solution

The pressure-factor quotient of a reaction in solution is related to the volume change of the reaction. We use Γ_r to denote the pressure-factor quotient of a reaction, a function of T and p. If we choose a molality basis for the solute standard states, the pressure-factor quotient is

$$\Gamma_r = \prod_i (\Gamma_{m,i})^{\nu_i}$$

---

[17] In these expressions, ρ_A^* should be evaluated at the standard pressure.

From Table 7.4 on page 217, we see that the pressure factor for solute i on a molality basis is

$$\Gamma_{m,i} = \exp\left(\int_{p°}^{p} \frac{V_i^\infty}{RT}\,dp\right)$$

where V_i^∞ is the partial molar volume of the solute at infinite dilution.

From these two equations, we obtain

$$\Gamma_r = \exp\left(\int_{p°}^{p} \frac{\sum_i \nu_i V_i^\infty}{RT}\,dp\right)$$

The sum $\sum_i \nu_i V_i^\infty$ is the molar reaction volume change at infinite dilution, $\Delta_r V_m^\infty$, giving finally

$$\Gamma_r = \exp\left(\int_{p°}^{p} \frac{\Delta_r V_m^\infty}{RT}\,dp\right)$$
$$\approx \exp\left[\frac{\Delta_r V_m^\infty (p - p°)}{RT}\right] \quad (8.8.9)$$

(reaction among solutes, mole fraction or molality basis)

We obtain the same expression for solute standard states based on mole fraction. The approximate form, which assumes $\Delta_r V_m^\infty$ is independent of pressure, is valid also for solute standard states based on concentration if we assume the solution is incompressible ($\kappa_T = 0$).

8.8.5 Evaluation of K

Equation 8.8.5 gives us a way to evaluate the thermodynamic equilibrium constant K of a reaction at a given temperature from the value of the standard molar reaction Gibbs energy $\Delta_r G_m°$ at that temperature. If we know the value of $\Delta_r G_m°$, we can calculate the value of K.

One method is to calculate $\Delta_r G_m°$ from values of the **standard molar Gibbs energy of formation** $\Delta_f G_m°$ of each reactant and product. These values are the standard molar reaction Gibbs energies for the formation reactions of the substances. To relate $\Delta_f G_m°$ to measurable quantities, we make the substitution $\mu_i = H_i - TS_i$ (Eq. 7.3.4) in $\Delta_r G_m = \sum_i \nu_i \mu_i$ (Eq. 8.7.2) to give $\Delta_r G_m = \sum_i \nu_i H_i - T\sum_i \nu_i S_i$ or

$$\Delta_r G_m = \Delta_r H_m - T\Delta_r S_m \quad (8.8.10)$$

When we apply this equation to a reaction with each reactant and product in its standard state, it becomes

$$\Delta_r G_m° = \Delta_r H_m° - T\Delta_r S_m° \quad (8.8.11)$$

If the reaction is the *formation* reaction of a substance, we have

$$\Delta_f G_m° = \Delta_f H_m° - T\Delta_f S_m° \quad (8.8.12)$$

Section 8.8 The Thermodynamic Equilibrium Constant

We can evaluate the standard molar Gibbs energy of formation of a substance, then, from its standard molar enthalpy and entropy of formation. Recall that standard molar enthalpies of formation may be evaluated by calorimetric methods (Sec. 8.3.1). We may evaluate the standard molar entropy of formation of a substance with a relation similar to Hess' law (Sec. 8.3.1), using the standard molar entropies of the substances involved in the formation reaction:

$$\Delta_r S_m^\circ = \sum_i \nu_i S_{m,i}^\circ \qquad (8.8.13)$$

The absolute entropy values come from heat capacity data or statistical mechanical theory by methods discussed in Sec. 4.11. Thus, it is entirely feasible to use nothing but calorimetry to evaluate an equilibrium constant, a goal sought by thermodynamicists during the first half of the 20th century.[18]

Extensive tables are available of values of $\Delta_f G_m^\circ$ for substances and ions. An abbreviated version at the single temperature 298.15 K is given in Appendix H. (Various methods by which these values can be obtained are illustrated in chaps. 9 and 11.) The tabulated values enable us to evaluate $\Delta_r G_m^\circ$ (and then K, through Eq. 8.8.4) with the expression (analogous to Hess' law)

$$\Delta_r G_m^\circ = \sum_i \nu_i \Delta_f G_{m,i}^\circ \qquad (8.8.14)$$

For *ions in aqueous solution*, the values of S_m° and $\Delta_f G_m^\circ$ found in Appendix H are based on the reference values $S_m^\circ = 0$ and $\Delta_f G_m^\circ = 0$ for H$^+$(aq) at all temperatures similar to the convention for $\Delta_f H_m^\circ$ values discussed in Sec. 8.3.1.[19] For a reaction with aqueous ions as reactants or products, these values correctly give $\Delta_r S_m^\circ$ using Eq. 8.8.13 or $\Delta_r G_m^\circ$ using Eq. 8.8.14.

The relation of Eq. 8.8.12 does not apply to an ion because we cannot write a formation reaction for a single ion; instead, the relation between $\Delta_f G_m^\circ$, $\Delta_f H_m^\circ$ and S_m° is more complicated.

Consider first a hypothetical reaction in which hydrogen ions and one or more elements form H$_2$ and a cation M^{z+} (where z_+ is the charge number of the ion):

$$z_+ \text{H}^+(\text{aq}) + \text{elements} \rightarrow (z_+/2)\text{H}_2(\text{g}) + \text{M}^{z+}(\text{aq})$$

For this reaction, using the convention that $\Delta_f H_m^\circ$, S_m°, and $\Delta_f G_m^\circ$ are zero for the aqueous H$^+$ ion and the fact that $\Delta_f H_m^\circ$ and $\Delta_f G_m^\circ$ are zero for the elements, we can write the following expressions for standard molar reaction quantities:

$$\Delta_r H_m^\circ = \Delta_f H_m^\circ(\text{M}^{z+})$$

$$\Delta_r S_m^\circ = (z_+/2) S_m^\circ(\text{H}_2) + S_m^\circ(\text{M}^{z+}) - \sum_{\text{elements}} S_{m,i}^\circ$$

---

[18] Another method, for a reaction that can be carried out reversibly in a galvanic cell, is described in Sec. 11.3.

[19] Note that the values of S_m° in Appendix H for some ions, unlike the values for substances, are *negative*; this simply means that these ions have lower standard molar entropies than H$^+$(aq).

$$\Delta_r G_m^\circ = \Delta_f G_m^\circ(M^{z+})$$

Then, from $\Delta_r G_m^\circ = \Delta_r H_m^\circ - T\Delta_r S_m^\circ$, we find

$$\Delta_f G_m^\circ(M^{z+}) = \Delta_f H_m^\circ(M^{z+}) - T\left[S_m^\circ(M^{z+}) - \sum_{\text{elements}} S_{m,i}^\circ + (z_+/2)S_m^\circ(H_2)\right]$$

For example, the standard molar Gibbs energy of the aqueous mercury(I) ion is found from

$$\Delta_f G_m^\circ(Hg_2^{2+}) = \Delta_f H_m^\circ(Hg_2^{2+}) - TS_m^\circ(Hg_2^{2+}) + 2TS_m^\circ(Hg) - \tfrac{2}{2}TS_m^\circ(H_2)$$

For an anion X^{z-} with negative charge number z_-, using the hypothetical reaction

$$|z_-/2|H_2(g) + \text{elements} \rightarrow |z_-|H^+(aq) + X^{z-}(aq)$$

we find by the same method

$$\Delta_f G_m^\circ(X^{z-}) = \Delta_f H_m^\circ(X^{z-}) - T\left[S_m^\circ(X^{z-}) - \sum_{\text{elements}} S_{m,i}^\circ - |z_-/2|S_m^\circ(H_2)\right]$$

For example, the calculation for the nitrate ion is

$$\Delta_f G_m^\circ(NO_3^-)$$
$$= \Delta_f H_m^\circ(NO_3^-) - TS_m^\circ(NO_3^-) + \tfrac{1}{2}TS_m^\circ(N_2) + \tfrac{3}{2}TS_m^\circ(O_2) + \tfrac{1}{2}TS_m^\circ(H_2)$$

8.8.6 Free-energy functions

Equation 8.8.14 is useful only if values of $\Delta_f G_m^\circ$ are available for the temperature of interest. If they are not available, it may be feasible to calculate $\Delta_r G_m^\circ$ at the desired temperature from tabulated values for each of the reactants and products of one of the following temperature-dependent functions:

$$\Phi_0(T) \equiv \frac{\mu^\circ(T) - H_m^\circ(0)}{T} \tag{8.8.15}$$

$$\Phi_{298.15}(T) \equiv \frac{\mu^\circ(T) - H_m^\circ(298.15\ \text{K})}{T} \tag{8.8.16}$$

There is no general agreement on the symbols used to represent these functions. Sometimes they are represented by symbols such as $[G^\circ - H^\circ(T_r)]/T$.

The functions are referred to by various names, including free-energy functions, Gibbs energy functions, and Giaque functions. Tables are available that tabulate values of these functions for a substance at different temperatures. The values usually vary fairly slowly with temperature, especially at high temperature, and are therefore easily interpolated. As an example, Fig. 8.18(a) on the next page shows the temperature dependence of $\Phi_{298.15}$ for gaseous ammonia.

For some substances, $\Delta_f G_m^\circ$ is a sufficiently linear function of T that it may be interpolated directly, as illustrated for ammonia in Fig. 8.18(b).

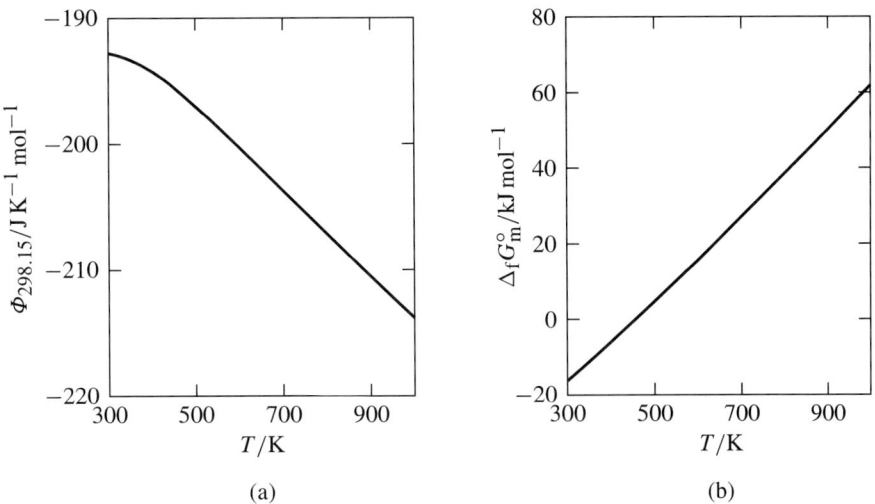

Figure 8.18 (a) Free-energy function of $NH_3(g)$ as a function of temperature. (b) Standard molar Gibbs energy of formation of $NH_3(g)$ as a function of temperature.

How may we evaluate Φ_0 or $\Phi_{298.15}$ experimentally? Substituting $\mu° = H_m° - TS_m°$ in Eq. 8.8.15 and using the third-law convention $S_m°(0) = 0$, we obtain $\Phi_0(T) = [H_m°(T) - H_m°(0)]/T - S_m°(T)$. Both terms in this expression can be evaluated with calorimetric measurements between zero kelvins and temperature T. The same substitution in Eq. 8.8.16 gives $\Phi_{298.15}(T) = [H_m°(T) - H_m°(298.15\,K)]/T - S_m°(T)$, another quantity that can be evaluated from calorimetric measurements alone.

To understand how to use values of Φ_0 or $\Phi_{298.15}$ evaluated at the temperature of interest, let us generalize the definition for substance i:

$$\Phi_{T',i}(T) \equiv \frac{\mu_i°(T) - H_i°(T')}{T}$$

Here T' is a reference temperature such as 0 K or 298.15 K. When we solve for $\mu_i°(T)$ and substitute into $\Delta_r G_m° = \sum_i \nu_i \mu_i°$ (Eq. 8.7.8), we obtain

$$\Delta_r G_m°(T) = T \sum_i \nu_i \Phi_{T',i}(T) + \sum_i \nu_i H_i°(T') \tag{8.8.17}$$

If the available function is $\Phi_{298.15}$ (interpolated to the temperature of interest), we can use Eq. 8.8.17 directly in the form

$$\Delta_r G_m°(T) = T \sum_i \nu_i \Phi_{298.15,i}(T) + \Delta_r H_m°(298.15\,K) \tag{8.8.18}$$

We may evaluate $\Delta_r H_m°(298.15\,K)$, the standard molar enthalpy of the reaction at 298.15 K, in the usual way from tabulated values of standard molar enthalpies of formation at 298.15 K (Sec. 8.3.1).

If values of Φ_0 are available, we can set T' equal to $0\,\mathrm{K}$, add and subtract the quantity $\Delta_r H_m^\circ(298.15\,\mathrm{K}) = \sum_i \nu_i H_i^\circ(298.15\,\mathrm{K})$ from the right side of Eq. 8.8.17, and rearrange the terms to obtain

$$\Delta_r G_m^\circ(T) = T\sum_i \nu_i \Phi_{0,i}(T) + \Delta_r H_m^\circ(298.15\,\mathrm{K}) - \sum_i \nu_i \left[H_i^\circ(298.15\,\mathrm{K}) - H_i^\circ(0)\right]$$

The value of $[H_i^\circ(298.15\,\mathrm{K}) - H_i^\circ(0)]$ is often available in the same table that provides values of $\Phi_{0,i}(T)$.

8.9 EFFECTS OF TEMPERATURE AND PRESSURE ON EQUILIBRIUM POSITION

When we change the temperature or pressure of a closed system having at least one of the reactants or products in a mixture, the equilibrium position given by ξ_{eq} usually changes also. To investigate these effects, we write the total differential of G with T, p, and ξ as independent variables (Eq. 8.7.4)

$$dG = -S\,dT + V\,dp + \Delta_r G_m\,d\xi$$

and obtain the reciprocity relations

$$\left(\frac{\partial \Delta_r G_m}{\partial T}\right)_{p,\xi} = -\left(\frac{\partial S}{\partial \xi}\right)_{T,p} \qquad \left(\frac{\partial \Delta_r G_m}{\partial p}\right)_{T,\xi} = \left(\frac{\partial V}{\partial \xi}\right)_{T,p}$$

We recognize the partial derivative on the right side of each of these relations as a molar reaction quantity:

$$\left(\frac{\partial \Delta_r G_m}{\partial T}\right)_{p,\xi} = -\Delta_r S_m \qquad \left(\frac{\partial \Delta_r G_m}{\partial p}\right)_{T,\xi} = \Delta_r V_m$$

We use these expressions for two of the coefficients in the total differential of $\Delta_r G_m$:

$$d\Delta_r G_m = -\Delta_r S_m\,dT + \Delta_r V_m\,dp + \left(\frac{\partial \Delta_r G_m}{\partial \xi}\right)_{T,p} d\xi \qquad (8.9.1)$$
$$\text{(closed system)}$$

Since $\Delta_r G_m$ is the partial derivative of G with respect to ξ at constant T and p, the coefficient $(\partial \Delta_r G_m/\partial \xi)_{T,p}$ is the partial *second* derivative of G with respect to ξ:

$$\left(\frac{\partial \Delta_r G_m}{\partial \xi}\right)_{T,p} = \left(\frac{\partial^2 G}{\partial \xi^2}\right)_{T,p}$$

We know that at a fixed T and p, the curve of G plotted as a function of ξ has a minimum at the position of reaction equilibrium $\xi = \xi_{eq}$. At the minimum, the slope $\Delta_r G_m$ is zero and the second derivative is positive (see Fig. 8.14 on page 283). If we change T or p, the

Section 8.9 Effects of Temperature and Pressure on Equilibrium Position

value of ξ_{eq} also changes, but $\Delta_r G_m$ at $\xi = \xi_{eq}$ remains equal to zero. By setting $\Delta_r G_m$ equal to zero in the general relation $\Delta_r G_m = \Delta_r H_m - T\Delta_r S_m$ (Eq. 8.8.10), we obtain the equation

$$\Delta_r S_m = \frac{\Delta_r H_m}{T}$$

which is valid only at reaction equilibrium where ξ equals ξ_{eq}. Making this substitution in Eq. 8.9.1 and setting $d\Delta_r G_m$ equal to zero and $d\xi$ equal to $d\xi_{eq}$, we obtain

$$0 = -\frac{\Delta_r H_m}{T}dT + \Delta_r V_m\, dp + \left(\frac{\partial^2 G}{\partial \xi^2}\right)_{T,p} d\xi_{eq} \qquad (8.9.2)$$
(closed system)

which shows how infinitesimal changes in T, p, and ξ_{eq} are related.

Now we are ready to see how ξ_{eq} is affected by changes in T or p. Solving Eq. 8.9.2 for $d\xi_{eq}$ gives

$$d\xi_{eq} = \frac{\dfrac{\Delta_r H_m}{T}dT - \Delta_r V_m\, dp}{\left(\dfrac{\partial^2 G}{\partial \xi^2}\right)_{T,p}} \qquad (8.9.3)$$
(closed system)

The right side of Eq. 8.9.3 is the expression for the total differential of ξ in a closed system that is at reaction equilibrium, with T and p as the independent variables. Thus, at constant pressure, the equilibrium shifts with temperature according to

$$\left(\frac{\partial \xi_{eq}}{\partial T}\right)_p = \frac{\Delta_r H_m}{T\left(\dfrac{\partial^2 G}{\partial \xi^2}\right)_{T,p}} \qquad (8.9.4)$$
(closed system)

and at constant temperature the equilibrium shifts with pressure according to

$$\left(\frac{\partial \xi_{eq}}{\partial p}\right)_T = -\frac{\Delta_r V_m}{\left(\dfrac{\partial^2 G}{\partial \xi^2}\right)_{T,p}} \qquad (8.9.5)$$
(closed system)

We have already seen that the second partial derivative $(\partial^2 G/\partial \xi^2)_{T,p}$ is positive. Therefore, according to Eqs. 8.9.4 and 8.9.5, $(\partial \xi_{eq}/\partial T)_p$ and $\Delta_r H_m$ have the same sign, whereas $(\partial \xi_{eq}/\partial p)_T$ and $\Delta_r V_m$ have opposite signs.

These statements express the application to temperature and pressure changes of what is known as *Le Châtelier's principle*: When a change is made to a system at equilibrium, the equilibrium shifts in the direction that tends to oppose the change. Here are two examples.

First, consider a reaction in which $\Delta_r H_m$ is negative—the reaction is exothermic. Since $(\partial \xi_{eq}/\partial T)_p$ has the same sign as $\Delta_r H_m$, an increase in temperature causes ξ_{eq} to decrease—the equilibrium shifts to the left. This is the shift that would reduce the temperature if the reaction were adiabatic.

Next consider a reaction in which $\Delta_r V_m$ is positive—the volume increases as the reaction proceeds to the right at constant T and p. $(\partial \xi_{eq}/\partial p)_T$ has the opposite sign, so if we increase the pressure by reducing the volume, the equilibrium shifts to the left. This is the shift that would reduce p, at least in a gas phase, if the reaction occurred at constant T and V.

It is easy to misuse or be misled by Le Châtelier's principle. Consider the solution process $B^*(s) \to B(sln)$ for which $(\partial \xi_{sol,eq}/\partial T)_p$, the rate of change of solubility with T, has the same sign as the molar enthalpy of solution $\Delta_{sol} H_m$. For $\Delta_{sol} H_m$, we must use the molar *differential* enthalpy of solution at saturation, which is the rate at which H changes with the amount of solute dissolved in the practically saturated solution. The sign of $\Delta_{sol} H_m$ at saturation may be different from the sign of the molar *integral* enthalpy of solution, $\Delta_{sol} H_m^{int}$.

This is the situation for the dissolution of sodium acetate shown in Fig. 8.6 on page 261. The equilibrium position (saturation) with one kilogram of water is at $\xi_{sol} \approx 15$ mol, indicated in the figure by an open circle. At this position, $\Delta_{sol} H_m$ is positive and $\Delta_{sol} H_m^{int}$ is negative. So despite the fact that the dissolution of 15 moles of sodium acetate in one kilogram of water to form a saturated solution is an exothermic process, the solubility of sodium acetate *increases* with increasing temperature contrary to what one might predict from Le Châtelier's principle.[20]

Another case in which Le Châtelier's principle is misleading is the addition of an inert gas to a gas-phase system of constant volume. Adding the inert gas increases the pressure, but has little effect on the position of a reaction equilibrium regardless of the value of $\Delta_r V_m$. This is because the inert gas affects the activities of the reactants and products only slightly (and not at all in an ideal gas mixture), so there is little effect on the value of Q. (The dependence of ξ_{eq} on p expressed by Eq. 8.9.5 does not apply to an open system.) The rigorous criterion for reaction equilibrium is always the requirement that Q must equal K or, equivalently, that $\Delta_r G_m$ must be zero.

PROBLEMS

8.1 Use values of $\Delta_f H_m^\circ$ and $\Delta_f G_m^\circ$ in Appendix H to evaluate the standard molar reaction enthalpy and the thermodynamic equilibrium constant at 298.15 K for the oxidation of nitrogen to form aqueous nitric acid:

$$\tfrac{1}{2} N_2(g) + \tfrac{5}{4} O_2(g) + \tfrac{1}{2} H_2O(l) \to H^+(aq) + NO_3^-(aq)$$

8.2 In 1982, the International Union of Pure and Applied Chemistry recommended that the value of the standard pressure p° be changed from 1 atm to 1 bar. This change affects the values of standard molar quantities of a substance calculated from experimental data.

(a) Find the changes in H_m°, S_m°, and G_m° for a gaseous substance when the standard pressure is changed isothermally from 1.01325 bar (1 atm) to exactly 1 bar. (Such a small pressure change has an entirely negligible effect on these quantities for a substance in a condensed phase.)

---

[20] Brice, L. K. *J. Chem. Educ.* **1983**, *60*, 387–389.

(b) What are the values of the corrections that need to be made to the standard molar enthalpy, entropy, and Gibbs energy of formation of $NH_3(g)$ at 298.15 K when the standard pressure is changed from 1.01325 bar to 1 bar?

8.3 From data for mercury listed in Appendix H, calculate the saturation pressure (vapor pressure) of liquid mercury at both 298.15 K and 273.15 K. You may need to make some reasonable approximations.

8.4 Given the following experimental values at $T = 298.15$ K, $p = 1$ bar:

$H^+(aq) + OH^-(aq) \rightarrow H_2O(l)$ $\qquad \Delta_r H_m^\circ = -55.82$ kJ mol^{-1}

$Na(s) + H_2O(l) \rightarrow Na^+(aq) + OH^-(aq) + \frac{1}{2}H_2(g)$ $\qquad \Delta_r H_m^\circ = -184.52$ kJ mol^{-1}

$NaOH(s) \rightarrow NaOH(aq)$ $\qquad \Delta_{sol} H_m^\infty = -44.75$ kJ mol^{-1}

NaOH in 5 H_2O $\rightarrow$ NaOH in ∞ H_2O $\qquad \Delta_{dil} H/n_B = -4.93$ kJ mol^{-1}

$NaOH(s)$ $\qquad \Delta_f H_m^\circ = -425.61$ kJ mol^{-1}

Calculate:

(a) $\Delta_f H_m^\circ$ for $Na^+(aq)$, NaOH(aq), and $OH^-(aq)$;

(b) $\Delta_f H_m$ for NaOH in 5 H_2O;

(c) $\Delta_{sol} H_m^{int}$ for the dissolution of 1 mol NaOH(s) in 5 mol H_2O.

Table 8.1 Data for Problem 8.5. The data for sodium thiosulfate are from the NBS Tables of Chemical Thermodynamic Properties (1982).

Substance	$\Delta_f H_m$/kJ mol^{-1} (298.15 K)	M/g mol^{-1}
$H_2O(l)$	−285.830	18.0153
$Na_2S_2O_3 \cdot 5H_2O(s)$	−2607.93	248.1828
$Na_2S_2O_3$ in 50 H_2O	−1135.914	
$Na_2S_2O_3$ in 100 H_2O	−1133.822	
$Na_2S_2O_3$ in 200 H_2O	−1132.236	
$Na_2S_2O_3$ in 300 H_2O	−1131.780	

8.5 Table 8.1 lists data for water, crystalline sodium thiosulfate pentahydrate, and several sodium thiosulfate solutions. Calculate ΔH to the nearest 0.01 kJ for the dissolution of 5.00 g of crystalline $Na_2S_2O_3 \cdot 5H_2O$ in 50.0 g of water at 298.15 K and 1 bar.

8.6 Consider the following experiment designed to evaluate the standard molar enthalpy of solution of gaseous HCl at 298.15 K (Gunn, S. R.; Green, L. G. *J. Chem. Eng. Data* **1963**, *8*, 180). A glass bulb containing a weighed quantity of HCl was broken in a constant-volume calorimeter bomb vessel containing a dilute aqueous solution of HCl at 298.15 K, and the temperature increase was measured. After the run, the calorimeter was calibrated with electrical work over the same temperature interval.

(a) In a run, the HCl gas dissolved not in pure water but in a dilute HCl solution. Let the molality of the initial solution be m'_B and the amount of HCl in this solution be n'_B. Let the amount of gaseous HCl added to this solution be n''_B and the molality of the final solution be m''_B. The measured enthalpy change is $\Delta_{sol} H(m'_B \to m''_B)$. The quantity of interest to be evaluated is $\Delta_{sol} H_m^\infty$, which is the molar enthalpy of solution at infinite dilution in *pure water*. Derive the relation

$$\Delta_{sol} H_m^\infty = \frac{\Delta_{sol} H(m'_B \to m''_B)}{n''_B} - L''_B - \frac{n_A(L''_A - L'_A) + n'_B(L''_B - L'_B)}{n''_B}$$

where L'_A and L'_B are the relative partial molar enthalpies of the initial solution and L''_A and L''_B are the same in the final solution.

Hint: Begin by expressing $\Delta_{sol} H(m'_B \to m''_B)$ in terms of the partial molar enthalpies of the solvent and solute before and after the dissolution of the gaseous HCl.

(b) In one run, 6.742×10^{-3} mol HCl dissolved in 374.9 g of an HCl solution of molality 2.67×10^{-4} mol kg^{-1} to form a solution of molality 0.01825 mol kg^{-1}. The enthalpy change at 298.15 K, corrected to a constant pressure of 1 bar and the ideal-gas state for the gaseous HCl, was -502.1 J.

Evaluate $\Delta_{sol} H_m^\infty$ at 298.15 K from these numerical data and the expression in Part (a). (Since this value is for gaseous HCl in its standard state, it is equal to the standard molar enthalpy of solution.) Figure 8.7 on page 266 shows that the initial and final solutions are sufficiently dilute for Φ_L to be described by Eq. 8.4.24, so values of L_A and L_B can be calculated from the equations at the end of Sec. 8.4.5.

8.7 This 16-part problem illustrates the use of experimental data from bomb calorimetry and other sources, combined with thermodynamic relations derived in this and earlier chapters, to evaluate the standard molar combustion enthalpy of a liquid hydrocarbon. The substance under investigation is *n*-hexane, and the combustion reaction in the bomb vessel is

$$C_6H_{14}(l) + \tfrac{19}{2} O_2(g) \to 6\,CO_2(g) + 7\,H_2O(l)$$

Assume that the sample is placed in a glass ampoule that shatters at ignition. Data needed for this problem are collected in Table 8.2 on the next page.

States 1 and 2 referred to in this problem are the initial and final states of the isothermal bomb process. The temperature is the reference temperature of 298.15 K.

(a) Parts (a)–(c) consist of simple calculations of some quantities needed in later parts of the problem. Begin by using the masses of C_6H_{14} and H_2O placed in the bomb vessel, and their molar masses, to calculate the amounts (moles) of C_6H_{14} and H_2O present initially in the bomb vessel. Then use the stoichiometry of the combustion reaction to find the amount of O_2 consumed and the amounts of H_2O and CO_2 present in State 2. (There is not enough information at this stage to allow you to find the amount of O_2 present, just the change.) Also find the final mass of H_2O. Assume that oxygen is present in excess and the combustion reaction goes to completion.

(b) From the molar masses and the densities of liquid C_6H_{14} and H_2O, calculate their molar volumes.

Table 8.2 Data for Problem 8.7. The values of intensive properties are for a temperature of 298.15 K and a pressure of 30 bar unless otherwise stated. Subscripts: A = H_2O, B = O_2, C = CO_2.

Properties of the bomb vessel:
internal volume .... 350.0 cm^3
mass of n-hexane placed in bomb .... 0.6741 g
mass of water placed in bomb .... 1.0016 g

Properties of liquid n-hexane:
molar mass .... $M = 86.177$ g mol^{-1}
density .... $\rho = 0.6548$ g cm^{-3}
cubic expansion coefficient .... $\alpha = 1.378 \times 10^{-3}$ K^{-1}

Properties of liquid H_2O:
molar mass .... $M = 18.0153$ g mol^{-1}
density .... $\rho = 0.9970$ g cm^{-3}
cubic expansion coefficient .... $\alpha = 2.59 \times 10^{-4}$ K^{-1}
standard molar energy of vaporization ... $\Delta_{vap}U_m^\circ = 41.53$ kJ mol^{-1}

Second virial coefficients, 298.15 K:
B_{AA} .... -1158 cm^3 mol^{-1}
B_{BB} .... -16 cm^3 mol^{-1}
dB_{BB}/dT .... 0.21 cm^3 K^{-1} mol^{-1}
B_{CC} .... -127 cm^3 mol^{-1}
dB_{CC}/dT .... 0.97 cm^3 K^{-1} mol^{-1}
B_{AB} .... -40 cm^3 mol^{-1}
B_{AC} .... -214 cm^3 mol^{-1}
B_{BC} .... -43.7 cm^3 mol^{-1}
dB_{BC}/dT .... 0.4 cm^3 K^{-1} mol^{-1}

Henry's law constants at 1 bar (solvent = H_2O):
O_2 .... $K_{m,B} = 796$ bar kg mol^{-1}
CO_2 .... $K_{m,C} = 29.7$ bar kg mol^{-1}

Partial molar volumes of solutes in water:
O_2 .... $V_B^\infty = 31$ cm^3 mol^{-1}
CO_2 .... $V_C^\infty = 33$ cm^3 mol^{-1}

Standard molar energies of solution (solvent = H_2O):
O_2 .... $\Delta_{sol}U_m^\circ = -9.7$ kJ mol^{-1}
CO_2 .... $\Delta_{sol}U_m^\circ = -17.3$ kJ mol^{-1}

(c) From the amounts present initially in the bomb vessel and the internal volume, find the volumes of liquid C_6H_{14}, liquid H_2O, and gas in State 1 and the volumes of liquid H_2O and gas in State 2. For this calculation, you can neglect the small change in the volume of liquid H_2O due to its vaporization.

(d) When the bomb vessel is charged with oxygen and before the inlet valve is closed, the pressure at 298.15 K measured on an external gauge is found to be $p_1 = 30.00$ bar. To a good approximation, the gas phase of State 1 has the equation of state of pure O_2 (since the vapor pressure of water is only 0.1 % of 30.00 bar). Assume that this equation of state is given by $V_m = RT/p + B_{BB}$ (Eq. 2.1.7 on page 15), where B_{BB} is the second virial coefficient of O_2 listed in Table 8.2. Solve for the amount of O_2 in the gas phase of State 1. The gas phase of State 2 is a mixture of O_2 and CO_2, again with a negligible partial pressure of H_2O. Assume that only small fractions of the total amounts of O_2 and CO_2 dissolve in the liquid water, and find the amount of O_2 in the gas phase of State 2 and the mole fractions of O_2 and CO_2 in this phase.

(e) You now have the information needed to find the pressure in State 2, which cannot be measured directly. For the mixture of O_2 and CO_2 in the gas phase of State 2, use Eq. 7.4.18 on page 189 to calculate the second virial coefficient. Then solve the equation of state of Eq. 7.4.16 on page 188 for the pressure. Also calculate the partial pressures of the O_2 and CO_2 in the gas mixture.

(f) Although the amounts of H_2O in the gas phases of States 1 and 2 are small, you need to know their values to take the energy of vaporization into account. In this part, you calculate the fugacities of the H_2O in the initial and final gas phases, in Part (g) you use gas equations of state to evaluate the fugacity coefficients of the H_2O (as well as of the O_2 and CO_2), and then in Part (h) you find the amounts of H_2O in the initial and final gas phases.

The pressure at which the pure liquid and gas phases of H_2O are in equilibrium at 298.15 K (the saturation pressure of water) is 0.03169 bar. Use Eq. 5.11.7 on page 132 to estimate the fugacity of $H_2O(g)$ in equilibrium with pure liquid water at this temperature and pressure. The effect of pressure on fugacity in a one-component liquid–gas system is discussed in Sec. 9.7.1; use Eq. 9.7.3 on page 339 to find the fugacity of H_2O in gas phases equilibrated with liquid water at the pressures of States 1 and 2 of the isothermal bomb process. (The mole fraction of O_2 dissolved in the liquid water is so small that you can ignore its effect on the chemical potential of the water.)

(g) Calculate the fugacity coefficients of H_2O and O_2 in the gas phase of State 1 and of H_2O, O_2, and CO_2 in the gas phase of State 2.

For State 1, in which the gas phase is practically pure O_2, you can use Eq. 5.11.7 on page 132 to calculate ϕ_{O_2}. The other calculations require Eq. 7.4.24 on page 190, with the value of B'_i found from the formulas of Eq. 7.4.22 or Eq. 7.4.23 (x_A is so small that you can set it equal to zero in these formulas).

Use the fugacity coefficient and partial pressure of O_2 to evaluate its fugacity in States 1 and 2; likewise, find the fugacity of CO_2 in State 2. [You calculated the fugacity of the H_2O in Part (f).]

(h) From the values of the fugacity and fugacity coefficient of a constituent of a gas mixture, you can calculate the partial pressure with Eq. 7.4.12 on page 188, then the mole fraction with $x_i = p_i/p$, and finally the amount with $n_i = x_i n$. Use this method to find the amounts of H_2O in the gas phases of States 1 and 2 and also calculate the amounts of H_2O in the liquid phases of both states.

(i) Next, consider the O_2 dissolved in the water of State 1 and the O_2 and CO_2 dissolved in the water of State 2. Treat the solutions of these gases as ideal dilute with the molality of solute i given by $m_i = f_i/K_{m,i}$ (Eq. 7.5.15). The values of the Henry's law constants of these gases listed in Table 8.2 are for the standard pressure of 1 bar. Use Eq. 9.7.9 on page 346 to find the appropriate values of $K_{m,i}$ at the pressures of States 1 and 2, and use these values to calculate the amounts of the dissolved gases in both states.

(j) At this point in the calculations, you know the values of all properties needed to describe the initial and final states of the isothermal bomb process and can evaluate the various Washburn corrections. These corrections are the internal energy changes, at the reference temperature of 298.15 K, of processes that connect the standard states of substances with either State 1 or State 2 of the isothermal bomb process.

First, consider the gaseous H_2O. The Washburn corrections should be based on a pure-liquid standard state for the H_2O. Section 5.11 shows that the molar internal energy of a pure gas under ideal-gas conditions (low pressure) is the same as the molar internal energy of the gas in its standard state at the same temperature. Thus, the molar internal energy change when a substance in its pure-liquid standard state changes isothermally to an ideal gas is equal to the standard molar internal energy of vaporization, $\Delta_{vap}U_m^\circ$. Using the value of $\Delta_{vap}U_m^\circ$ for H_2O given in Table 8.2, calculate ΔU for the vaporization of liquid H_2O at pressure p° to ideal gas in the amount present in the gas phase of State 1. Also calculate ΔU for the condensation of ideal gaseous H_2O in the amount present in the gas phase of State 2 to liquid at pressure p°.

(k) Next, consider the dissolved O_2 and CO_2, for which gas standard states are used. Assume that the solutions are sufficiently dilute to have infinite-dilution behavior; then the partial molar internal energy of either solute in the solution at the standard pressure $p^\circ = 1$ bar is equal to the standard partial molar internal energy based on a solute standard state (Sec. 7.7.1). Values of $\Delta_{sol}U_m^\circ$ are listed in Table 8.2. Find ΔU for the dissolution of O_2 from its gas standard state to ideal-dilute solution at pressure p° in the amount present in the aqueous phase of State 1. Find ΔU for the desolution of O_2 and of CO_2 from ideal-dilute solution at pressure p°, in the amounts present in the aqueous phase of State 2, to their gas standard states.

(l) Calculate the internal energy changes when the liquid phases of State 1 (n-hexane and aqueous solution) are compressed from p° to p_1 and the aqueous solution of State 2 is decompressed from p_2 to p°. Use the approximate formula of Eq. 5.8.9 on page 126, and treat the cubic expansion coefficient of the aqueous solutions as being the same as that of pure water.

(m) The final Washburn corrections are internal energy changes of the gas phases of States 1 and 2. H_2O has such low mole fractions in these phases that you can ignore H_2O in these calculations; that is, treat the gas phase of State 1 as pure O_2 and the gas phase of State 2 as a binary mixture of O_2 and CO_2.

One of the internal energy changes is for the compression of gaseous O_2, starting at a pressure low enough for ideal-gas behavior ($U_m = U_m^\circ$) and ending at pressure p_1 to form the gas phase present in State 1. Use Eq. 5.11.11 on page 133 to calculate $\Delta U = U(p_1) - nU_m^\circ(g)$; a value of dB/dT for pure O_2 is listed in Table 8.2.

The other internal energy change is for a process in which the gas phase of State 2 at pressure p_2 is expanded until the pressure is low enough for the gas to behave ideally, and the mixture is then separated into ideal-gas phases of pure O_2 and CO_2. The molar internal energies of the separated low-pressure O_2 and CO_2 gases are the same as the standard molar internal energies of these gases. The internal energy of unmixing ideal gases is zero (Eq. 8.2.7). Equation 7.4.17 on page 188 gives the dependence on pressure of the internal energy of a gas mixture. Calculate the value of dB/dT for the mixture of O_2 and CO_2 in State 2 (you need Eq. 7.4.18 on page 189 and the values of dB_{ij}/dT in Table 8.2) and evaluate $\Delta U = \sum_i n_i U_i^\circ(g) - U(p_2)$ for the gas expansion.

(n) Add the internal energy changes you calculated in Parts (j)–(m) to find the total internal energy change of the Washburn corrections. Note that most of the corrections occur in pairs of opposite sign and almost completely cancel one another. Which contributions are the greatest in magnitude?

(o) The internal energy change of the isothermal bomb process in the bomb vessel, corrected to the reference temperature of 298.15 K, is found to be $\Delta_{IBP} U(T_{\text{ref}}) = -32.504$ kJ. Assume that there are no side reactions or auxiliary reactions. From Eqs. 8.5.8 and 8.5.9, calculate the standard molar internal energy of combustion of n-hexane at 298.15 K.

(p) From Eq. 8.5.10, calculate the standard molar enthalpy of combustion of n-hexane at 298.15 K.

8.8 By combining the results of Prob. 8.7(p) with the values of standard molar enthalpies of formation from Appendix H, calculate the standard molar enthalpy of formation of liquid n-hexane at 298.15 K.

8.9 Consider the combustion of methane:

$$CH_4(g) + 2 O_2(g) \rightarrow CO_2(g) + 2 H_2O(g)$$

Suppose the reaction occurs in a flowing gas mixture of methane and air. Assume that the pressure is constant at 1 bar, the reactant mixture is at a temperature of 298.15 K and has stoichiometric proportions of methane and oxygen, and the reaction goes to completion with no dissociation. For the quantity of gaseous product mixture containing 1 mol CO_2, 2 mol H_2O, and the nitrogen and other substances remaining from the air, you may use the approximate formula $C_p(P) = a + bT$, where the coefficients have the values $a = 297.0$ J K^{-1} and $b = 8.520 \times 10^{-2}$ J K^{-2}. Solve Eq. 8.6.1 for T_2 to estimate (to the nearest kelvin) the maximum flame temperature.

8.10 The standard molar Gibbs energy of formation of crystalline mercury(II) oxide at 600.00 K has the value $\Delta_f G_m^\circ = -26.386$ kJ mol^{-1}. Estimate the partial pressure of O_2 in equilibrium with HgO at this temperature.

$$2 HgO(s) \rightleftharpoons 2 Hg(l) + O_2(g)$$

8.11 The combustion of hydrogen is a reaction that is known to "go to completion."

(a) Use data in Appendix H to evaluate the thermodynamic equilibrium constant at 298.15 K for the reaction

$$H_2(g) + \tfrac{1}{2} O_2(g) \rightarrow H_2O(l)$$

(b) Assume that the reaction is at equilibrium at 298.15 K in a system in which the partial pressure of O_2 is 1.0 bar. Assume ideal-gas behavior and find the equilibrium partial pressure of H_2 and the number of H_2 *molecules* in $1.0\,m^3$ of the gas phase.

(c) In the preceding part, you calculated a very small fraction for the number of H_2 molecules in $1.0\,m^3$. Statistically, this fraction can be interpreted as the fraction of a given length of time during which one molecule is present. Take the age of the universe as 1.0×10^{10} years and find the total length of time in seconds, during the age of the universe, that a H_2 molecule would exist in the equilibrium system. (This hypothetical value is a dramatic demonstration of the statement that a limiting reactant is entirely consumed during a reaction that goes to completion.)

8.12 Consider the association reaction $N_2O_4(g) \rightarrow 2\,NO_2(g)$ taking place at a constant temperature of 298.15 K and a constant pressure of 0.0500 bar. Initially (at $\xi = 0$) the system contains 1.000 mol of N_2O_4 and no NO_2. Other needed data are found in Appendix H. Assume ideal-gas behavior.

(a) For values of the advancement ξ ranging from 0 to 1 mol, at an interval of 0.1 mol or smaller, calculate $[G(\xi) - G(0)]$ to 0.01 kJ. A spreadsheet would be a convenient way to make the calculations.

(b) Plot your values of $G(\xi) - G(0)$ as a function of ξ, and draw a smooth curve through the points.

(c) On your curve indicate the estimated position of ξ_{eq}. Calculate the activities of N_2O_4 and NO_2 for this value of ξ, use them to estimate the thermodynamic equilibrium constant K, and compare your result with the value of K calculated from Eq. 8.8.5.

Table 8.3 Free-energy functions for four substances at four temperatures. The values in all but the first column are $\Phi_{298.15}/J\,K^{-1}\,mol^{-1}$.

T/K	C(s)	CO(g)	H_2(g)	H_2O(g)
800	−10.290	−208.309	−141.172	−201.329
900	−11.479	−210.631	−143.412	−204.094
1000	−12.662	−212.851	−145.537	−206.752
1100	−13.827	−214.969	−147.550	−209.303

8.13 Table 8.3 lists values of free-energy functions for four substances at several temperatures.

(a) Use this table and data in Appendix H to evaluate the thermodynamic equilibrium constant at 1000 K for the water gas reaction

$$C(s) + H_2O(g) \rightarrow CO(g) + H_2(g)$$

(b) Use an interpolation procedure to estimate the temperature at which $\Delta_r G_m^\circ$ is zero; that is, the temperature at which K is equal to 1.

Chapter 9

EQUILIBRIUM CONDITIONS IN MULTICOMPONENT SYSTEMS

This chapter applies equilibrium theory to a variety of chemical systems of more than one component. Two general approaches are illustrated: one based on the equality of the chemical potential for transfer equilibrium: $\mu_i^\alpha = \mu_i^\beta$; the other based on the thermodynamic equilibrium constant for reaction equilibrium: $K = \prod_i a_i^{\nu_i}$. The chemical potential approach can give us relations involving activity coefficients and reference states, whereas the equilibrium constant approach involves relations with activities and standard states.

9.1 EFFECTS OF TEMPERATURE

For some of the derivations in this chapter, we need an expression for the rate at which the ratio μ_i/T varies with temperature in a phase of fixed composition maintained at constant pressure. This expression leads, among other things, to an important relation between the temperature dependence of an equilibrium constant and the standard molar reaction enthalpy.

9.1.1 Variation of μ_i/T with temperature

In a phase containing Species i, either pure or in a mixture, the partial derivative of μ_i/T with respect to T at constant p and a fixed amount of each species is given by[1]

$$\left[\frac{\partial (\mu_i/T)}{\partial T}\right]_{p,n} = \frac{1}{T}\left(\frac{\partial \mu_i}{\partial T}\right)_{p,n} - \frac{1}{T^2}\mu_i$$

---

[1] This relation is obtained from the chain rule $d(uv)/dx = u\,dv/dx + v\,du/dx$ (Appendix E), where u is $1/T$, v is μ_i, and x is T.

This equality comes from a purely mathematical operation; no thermodynamics is involved. Substituting $(\partial \mu_i/\partial T)_{p,n} = -S_i$ (Eq. 7.3.5) changes the equality to

$$\left[\frac{\partial (\mu_i/T)}{\partial T}\right]_{p,n} = -\frac{1}{T}S_i - \frac{1}{T^2}\mu_i = -\frac{TS_i + \mu_i}{T^2}$$

Further substitution of $\mu_i = H_i - TS_i$ (Eq. 7.3.4) gives

$$\left[\frac{\partial (\mu_i/T)}{\partial T}\right]_{p,n} = -\frac{H_i}{T^2} \tag{9.1.1}$$

9.1.2 Variation of μ_i°/T with temperature

Next we apply Eq. 9.1.1 to Species i in its standard state. The chemical potential is then the *standard* chemical potential μ_i°. For a given standard state, μ_i° and μ_i°/T are functions only of T since the pressure is fixed at p° and the composition of the phase is determined by the definition of the standard state. When the standard state is a pure ideal gas or a pure liquid or solid, Eq. 9.1.1 becomes

$$\frac{d(\mu_i^\circ/T)}{dT} = -\frac{H_{m,i}^\circ}{T^2} \tag{9.1.2}$$
(pure-substance standard state)

The standard chemical potential of solute B of a solution requires a little more care. If we substitute the relation $\mu_B = \mu_B^\circ + RT \ln a_B$ in Eq. 9.1.1 and rearrange, we obtain

$$\frac{d(\mu_B^\circ/T)}{dT} = -\frac{H_B}{T^2} - R\left(\frac{\partial \ln a_B}{\partial T}\right)_{p,n} \tag{9.1.3}$$

The derivative on the left side of this equation is a function only of T, so we can evaluate it from the expression on the right side with any combination of composition and pressure. Let us use an ideal-dilute solution at the standard pressure. Then H_B is the standard molar enthalpy H_B° and the activity a_B is equal to either x_B, c_B/c°, or m_B/m° depending on the choice of solute standard state. When the amounts of solvent and solute are fixed, x_B and m_B/m° remain constant as T changes, so for a solute whose standard state is based on mole fraction or molality Eq. 9.1.3 becomes

$$\frac{d(\mu_{x,B}^\circ/T)}{dT} = \frac{d(\mu_{m,B}^\circ/T)}{dT} = -\frac{H_B^\circ}{T^2} \tag{9.1.4}$$

For a solute standard state based on concentration, we must consider the following behavior of the quantity $\ln(c_B/c^\circ)$:

$$\left[\frac{\partial \ln (c_B/c^\circ)}{\partial T}\right]_{p,n} = \frac{1}{c_B}\left(\frac{\partial c_B}{\partial T}\right)_{p,n} = \frac{1}{n_B/V}\left[\frac{\partial (n_B/V)}{\partial T}\right]_{p,n}$$

$$= V\left[\frac{\partial (1/V)}{\partial T}\right]_{p,n} = V\left[\frac{d(1/V)}{dV}\right]\left(\frac{\partial V}{\partial T}\right)_{p,n} = -\frac{1}{V}\left(\frac{\partial V}{\partial T}\right)_{p,n}$$

The right-most expression is the negative of the cubic expansion coefficient α of the solution (Eq. 5.2.1), which for an ideal-dilute solution is equal to the cubic expansion coefficient α_A^* of the pure solvent. For a solute with standard state based on concentration, then, Eq. 9.1.3 becomes

$$\frac{d(\mu_{c,B}^\circ/T)}{dT} = -\frac{H_B^\circ}{T^2} + R\alpha_A^* \tag{9.1.5}$$

9.1.3 Variation of ln$K$ with temperature

The thermodynamic equilibrium constant K, for a given reaction equation and a given choice of reactant and product standard states, is a function of T and *only* of T. By equating two expressions for the standard molar reaction Gibbs energy, $\Delta_r G_m^\circ = \sum_i \nu_i \mu_i^\circ$ (Eq. 8.7.8) and $\Delta_r G_m^\circ = -RT \ln K$ (Eq. 8.8.4), we obtain

$$\ln K = -\frac{1}{RT} \sum_i \nu_i \mu_i^\circ$$

The rate at which $\ln K$ varies with T is then given by

$$\frac{d \ln K}{dT} = -\frac{1}{R} \sum_i \nu_i \frac{d(\mu_i^\circ/T)}{dT} \tag{9.1.6}$$

Combining Eq. 9.1.6 with Eqs. 9.1.2, 9.1.4, or 9.1.5, and recognizing that $\sum_i \nu_i H_i^\circ$ is the standard molar reaction enthalpy $\Delta_r H_m^\circ$, we obtain the final expression for the rate at which $\ln K$ varies with T:

$$\frac{d \ln K}{dT} = \frac{\Delta_r H_m^\circ}{RT^2} - \alpha_A^* \sum_{\substack{\text{solutes,} \\ \text{conc. basis}}} \nu_i \tag{9.1.7}$$

The sum on the right side includes only solute species whose standard states are based on concentration.

To simplify the equation, we can confine the standard states of all solutes to a mole fraction or molality basis. Then Eq. 9.1.7 becomes

$$\frac{d \ln K}{dT} = \frac{\Delta_r H_m^\circ}{RT^2} \tag{9.1.8}$$
(no solute standard states based on concentration)

We can solve this for $\Delta_r H_m^\circ$ to obtain

$$\Delta_r H_m^\circ = RT^2 \frac{d \ln K}{dT} \tag{9.1.9}$$
(no solute standard states based on concentration)

Table 9.1 Comparison of the Clausius–Clapeyron and van't Hoff equations for vaporization of a liquid at a given temperature.

Clausius–Clapeyron equation	van't Hoff equation
$\Delta_{\text{vap}} H_{\text{m}} \approx -R \dfrac{\text{d}\ln(p/p^\circ)}{\text{d}(1/T)}$	$\Delta_{\text{vap}} H_{\text{m}}^\circ = -R \dfrac{\text{d}\ln K}{\text{d}(1/T)}$
Derivation assumes $V_{\text{m}}(\text{g}) \gg V_{\text{m}}(\text{l})$ and ideal-gas behavior.	An exact relation.
$\Delta_{\text{vap}} H_{\text{m}}$ is the difference of the molar enthalpies of the real gas and the liquid at the saturation pressure of the liquid.	$\Delta_{\text{vap}} H_{\text{m}}^\circ$ is the difference of the molar enthalpies of the ideal gas and the liquid at pressure p°.
p is the saturation pressure of the liquid.	K is equal to $a(\text{g})/a(\text{l}) = (\phi p/p^\circ)/\Gamma(\text{l})$, and is only approximately equal to p/p°.

which we can then convert to an equivalent form with the mathematical identity $\text{d}(1/T) = -(1/T^2)\text{d}T$:

$$\Delta_{\text{r}} H_{\text{m}}^\circ = -R \frac{\text{d}\ln K}{\text{d}(1/T)} \tag{9.1.10}$$
(no solute standard states based on concentration)

Equations 9.1.9 and 9.1.10 are two forms of the **van't Hoff equation**. They allow us to evaluate the standard molar reaction enthalpy of a reaction by a noncalorimetric method from the temperature dependence of $\ln K$. For example, we can plot $\ln K$ versus $1/T$; then according to Eq. 9.1.10, the slope of the curve at any value of $1/T$ is equal to $-\Delta_{\text{r}} H_{\text{m}}^\circ / R$ at the corresponding temperature T.

A simple way to derive this last procedure is to substitute $\Delta_{\text{r}} G_{\text{m}}^\circ = \Delta_{\text{r}} H_{\text{m}}^\circ - T \Delta_{\text{r}} S_{\text{m}}^\circ$ (Eq. 8.8.11) in $\Delta_{\text{r}} G_{\text{m}}^\circ = -RT \ln K$ (Eq. 8.8.4) and rearrange to the following equation:

$$\ln K = -\frac{\Delta_{\text{r}} H_{\text{m}}^\circ}{R} \left(\frac{1}{T}\right) + \frac{\Delta_{\text{r}} S_{\text{m}}^\circ}{R}$$

Suppose we draw a plot of $\ln K$ versus $1/T$. In a small temperature interval in which $\Delta_{\text{r}} H_{\text{m}}^\circ$ and $\Delta_{\text{r}} S_{\text{m}}^\circ$ are practically constant, the curve will appear linear. According to this equation, the curve in this interval has a slope of $-\Delta_{\text{r}} H_{\text{m}}^\circ / R$, and a straight line tangent to a point on the curve has an intercept at $1/T = 0$ equal to $\Delta_{\text{r}} S_{\text{m}}^\circ / R$.

When we apply Eq. 9.1.10 to the *vaporization process* $\text{A}(\text{l}) \to \text{A}(\text{g})$ in a one-component system, it resembles the Clausius–Clapeyron equation for the same process (Eq. 6.4.7 on page 167). These equations are not exactly equivalent, however, as the comparison in Table 9.1 shows.

9.2 SOLVENT CHEMICAL POTENTIALS FROM PHASE EQUILIBRIA

Section 7.6.6 explained how we can evaluate the activity coefficient $\gamma_{m,\text{B}}$ of a nonelectrolyte solute of a binary solution if we know the variation of the molal osmotic coefficient of the solution from infinite dilution to the molality of interest. A similar procedure for the mean ionic activity coefficient of an *electrolyte* solute was treated in Sec. 7.8.6.

The physical measurements needed to find the molal osmotic coefficient ϕ_m of a solution are directed to calculating the quantity $\mu_{\text{A}}^* - \mu_{\text{A}}$, the difference between the chemical potential of the pure solvent and the chemical potential of the solvent in the solution, both at the temperature and pressure of interest. This difference is positive because the presence of the solute reduces the solvent's chemical potential. To calculate the molal osmotic coefficient of the solution from $\mu_{\text{A}}^* - \mu_{\text{A}}$, we use Eq. 7.6.21 on page 209 for a nonelectrolyte solute or Eq. 7.8.26 on page 229 for an electrolyte solute. The sequence of steps, then, is to determine $\mu_{\text{A}}^* - \mu_{\text{A}}$ over a range of molality at constant T and p, convert these values to ϕ_m, and obtain the solute activity coefficient[2] by a suitable integration from infinite dilution to the molality of interest.

Sections 9.2.1 and 9.2.2 describe freezing-point and osmotic-pressure measurements, two much-used methods for evaluating $\mu_{\text{A}}^* - \mu_{\text{A}}$ in a binary solution at a given T and p. The isopiestic vapor-pressure method was described in Sec. 7.6.7. The freezing-point and isopiestic vapor-pressure methods are often used for electrolyte solutions, while osmotic pressure is especially useful for solutions of macromolecules.

9.2.1 Freezing-point measurements

This section explains how we can obtain a value of $\mu_{\text{A}}^* - \mu_{\text{A}}$ for a solution of a given composition and at a given T and p from the freezing point (freezing temperature) of the solution combined with additional data obtained with calorimetric measurements.

Consider a binary solution of Solvent A and Solute B. We assume that when this solution is cooled at constant pressure and composition, the solid that first appears is pure A. (Thus, for an aqueous solution, the solid would be ice.) The temperature at which solid A first appears is $T_{\text{f,A}}$, the freezing point of the solution. When the pure solvent is cooled at the same pressure, its freezing point is $T_{\text{f,A}}^*$. Both $T_{\text{f,A}}$ and $T_{\text{f,A}}^*$ are known experimentally.

At the temperature of interest, T', we wish to determine the value of $\mu_{\text{A}}^*(\text{l}, T') - \mu_{\text{A}}(\text{sln}, T')$, where $\mu_{\text{A}}^*(\text{l}, T')$ refers to pure liquid solvent and $\mu_{\text{A}}(\text{sln}, T')$ refers to the solution. Figure 9.1 explains the principle of the procedure. The figure shows curves of the function μ_{A}/T for the solvent in the pure solid and liquid phases and in the fixed-composition solution, as functions of T at constant p. Since the chemical potential of the solvent is the same in the solution and solid phases at temperature $T_{\text{f,A}}$ and is the same in the pure liquid and solid phases at temperature $T_{\text{f,A}}^*$, the curves intersect at these temperatures as shown.

Formulas for the slopes of the three curves, from Eq. 9.1.1 on page 307, are included

---

[2] A measurement of $\mu_{\text{A}}^* - \mu_{\text{A}}$ also gives us the *solvent* activity coefficient, based on the pure-solvent reference state, through the relation $\mu_{\text{A}} = \mu_{\text{A}}^* + RT \ln \gamma_{\text{A}} x_{\text{A}}$ (Eq. 7.6.4).

Section 9.2 Solvent Chemical Potentials from Phase Equilibria

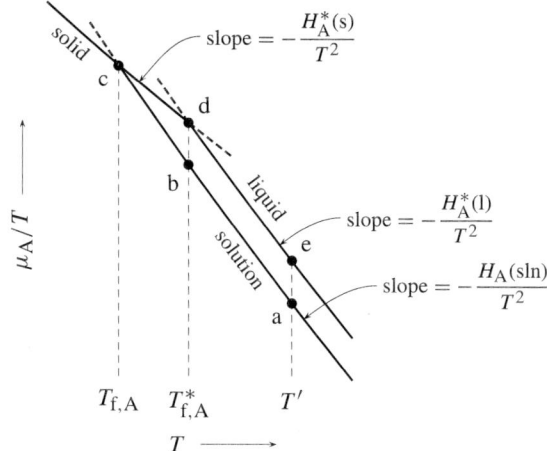

Figure 9.1 Integration path for determining $\mu_A^* - \mu_A$ at temperature T' from the freezing point of a solution (schematic). The dashed extensions of the curves represent unstable states.

in the figure. The desired value of $\mu_A^*(l, T') - \mu_A(\text{sln}, T')$ is the product of T' and the difference in μ_A/T at points e and a. To find the difference in μ_A/T at points e and a, we can integrate the slope, $d(\mu_A/T)/dT$, over T along the path abcde:

$$\frac{\mu_A^*(l, T')}{T'} - \frac{\mu_A(\text{sln}, T')}{T'} = -\int_{T'}^{T_{f,A}^*} \frac{H_A(\text{sln})}{T^2} dT - \int_{T_{f,A}^*}^{T_{f,A}} \frac{H_A(\text{sln})}{T^2} dT$$

$$- \int_{T_{f,A}}^{T_{f,A}^*} \frac{H_A^*(s)}{T^2} dT - \int_{T_{f,A}^*}^{T'} \frac{H_A^*(l)}{T^2} dT$$

By combining integrals with the same range of integration, we turn the equation into

$$\frac{\mu_A^*(l, T')}{T'} - \frac{\mu_A(\text{sln}, T')}{T'} = \int_{T_{f,A}}^{T_{f,A}^*} \frac{H_A(\text{sln}) - H_A^*(s)}{T^2} dT + \int_{T_{f,A}^*}^{T'} \frac{H_A(\text{sln}) - H_A^*(l)}{T^2} dT \quad (9.2.1)$$

For convenience of notation, we use $\Delta_{\text{sol}} H_{m,A}$ to denote the molar enthalpy difference $H_A(\text{sln}) - H_A^*(s)$. $\Delta_{\text{sol}} H_{m,A}$ is the molar differential enthalpy of solution of solid A in the solution at constant T and p.[3] The first integral on the right side of Eq. 9.2.1 requires knowledge of $\Delta_{\text{sol}} H_{m,A}$ over a temperature range, but the only temperature at which it is practical to measure this quantity calorimetrically is at the equilibrium transition temperature $T_{f,A}$. It is usually sufficient to assume $\Delta_{\text{sol}} H_{m,A}$ is a linear function of T:

$$\Delta_{\text{sol}} H_{m,A}(T) = \Delta_{\text{sol}} H_{m,A}(T_{f,A}) + \Delta_{\text{sol}} C_{p,A}(T - T_{f,A}) \quad (9.2.2)$$

---

[3] In Sec. 8.4.1, $\Delta_{\text{sol}} H_m$ was defined as the molar differential enthalpy of solution of a solute. The notation $\Delta_{\text{sol}} H_{m,A}$ indicates that it is the pure solvent, A, rather than the pure solute that enters the solution.

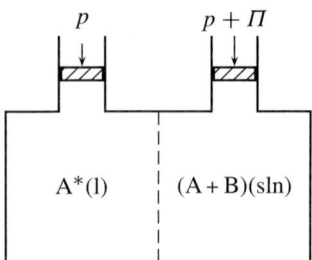

Figure 9.2 Apparatus to measure osmotic pressure. The dashed line represents a semipermeable membrane.

The partial molar heat capacity difference $\Delta_{sol}C_{p,A} = C_{p,A}(\text{sln}) - C_{p,A}(\text{s})$ is treated as a constant that can be determined from calorimetric measurements.

The quantity $H_A(\text{sln}) - H_A^*(\text{l})$ in the second integral on the right side of Eq. 9.2.1 is the molar enthalpy of dilution of the solvent in the solution, $\Delta_{dil}H_{m,A}$ (Sec. 8.4). This quantity can be measured calorimetrically at any temperature higher than $T_{f,A}^*$. Making this substitution in Eq. 9.2.1 together with that of Eq. 9.2.2, carrying out the integration of the first integral and rearranging, we obtain finally

$$\mu_A^*(l, T') - \mu_A(\text{sln}, T') = T'\left[\Delta_{sol}H_{m,A}(T_{f,A}) - T_{f,A}\Delta_{sol}C_{p,A}\right]\left(\frac{1}{T_{f,A}} - \frac{1}{T_{f,A}^*}\right)$$
$$+ T'\Delta_{sol}C_{p,A}\ln\frac{T_{f,A}^*}{T_{f,A}} + T'\int_{T_{f,A}^*}^{T'}\frac{\Delta_{dil}H_{m,A}}{T^2}\,dT$$

In principle, we could use boiling-point measurements in a similar way to evaluate $\mu_A^* - \mu_A$ for a solution containing a nonvolatile solute.

9.2.2 Osmotic-pressure measurements

We can also evaluate $\mu_A^* - \mu_A$ from the solution property called *osmotic pressure*. A simple apparatus for measuring this property is shown in Fig. 9.2. The system consists of two liquid phases separated by a semipermeable membrane. One phase is pure solvent and the other is a solution with the same solvent and at the same temperature. The semipermeable membrane is permeable to the solvent and impermeable to the solute or solutes. The pressures of the phases are controlled by external forces applied to the pistons; the presence of the membrane allows these pressures to be unequal in an equilibrium state.

When both phases have the same temperature and pressure, the chemical potential of the solvent is less in the solution than in the pure liquid because of the presence of the solute. Under these conditions, the flow of solvent through the membrane into the solution is spontaneous (the phenomenon of *osmosis*, Greek for *push*). If we use the right-hand piston to increase the pressure of the solution, the solvent chemical potential increases in this phase. The **osmotic pressure** of the solution, Π, is defined as the additional pressure the solution must have, compared with the pure solvent at the same temperature, to establish an equilibrium state with no flow of solvent in either direction through the membrane.

> No membrane is completely impermeable to a solute. All that is required for the establishment of an equilibrium state with two different pressures is that solvent transfer equilibrium be established on a time scale short compared to the period of observation, and that the amount of solute transferred in this time be negligible.

The osmotic pressure is an intensive property of a solution whose value depends on the solution's temperature, pressure, and composition. Note that in the system shown in Fig. 9.2, Π refers to the osmotic pressure of the solution at pressure p (the pressure of the solvent shown in the figure), not at pressure $p + \Pi$. In other words, the osmotic pressure of a solution at a given temperature and pressure is the *additional* pressure that would have to be exerted to establish transfer equilibrium with pure solvent at the given temperature and pressure. A solution has the property called osmotic pressure regardless of whether this additional pressure is actually present, just as a solution has a freezing point even when its actual temperature is different from the freezing point.

From the derivation in Sec. 7.3.1, we concluded that two conditions are needed for transfer equilibrium of a species that can transfer between two phases: (1) the species must have the same chemical potential in both phases, and (2) both phases must have the same temperature. We found that these conditions hold whether or not the two phases have the same pressure. For the present system, we express the first condition by the equality

$$\mu_A(p + \Pi) = \mu_A^*(p) \qquad (9.2.3)$$

where $\mu_A(p + \Pi)$ is the chemical potential of the solvent in the solution at pressure $p + \Pi$ and $\mu_A^*(p)$ is the chemical potential of the pure liquid solvent at pressure p.

For a solution of fixed temperature and composition, the dependence of μ_A on pressure is given according to Eq. 7.3.6 by

$$\left(\frac{\partial \mu_A}{\partial p}\right)_{T,n} = V_A$$

where V_A is the partial molar volume of the solvent in the solution. Rewriting this equation in the form $d\mu_A = V_A \, dp$ and integrating from p to $p + \Pi$, we obtain

$$\mu_A(p + \Pi) - \mu_A(p) = \int_p^{p+\Pi} V_A \, dp$$

Substitution from Eq. 9.2.3 changes this to

$$\mu_A^*(p) - \mu_A(p) = \int_p^{p+\Pi} V_A \, dp \qquad (9.2.4)$$
$$\text{(constant } T\text{)}$$

which is the desired expression for $\mu_A^* - \mu_A$ at a single temperature and pressure. To evaluate the integral, we need an experimental value of the osmotic pressure Π of the solution. If we assume V_A is constant in the pressure range from p to $p + \Pi$, Eq. 9.2.4 becomes simply

$$\mu_A^*(p) - \mu_A(p) = V_A \Pi \tag{9.2.5}$$

9.3 BINARY MIXTURE IN EQUILIBRIUM WITH A PURE PHASE

In this section, we consider a binary liquid mixture of Components A and B in equilibrium with either pure solid A or pure gaseous A. Our aim is to find general relations among changes of temperature, pressure, and mixture composition in the two-phase equilibrium system that we can apply to specific situations in later sections.

To avoid confusion, we assume for the first part of the derivation that the pure phase is *solid* A. Later, by changing the phase designation, we can apply the results to a system with a phase of pure *gaseous* A.

We begin by writing the total differential of $\mu_A(\text{mixt})/T$ with T, p, and x_A as the independent variables. These quantities refer to the binary liquid mixture, and we have not yet imposed a condition of equilibrium with another phase. The general expression for the total differential is

$$d[\mu_A(\text{mixt})/T] = \left[\frac{\partial \mu_A(\text{mixt})/T}{\partial T}\right]_{p,x_A} dT + \left[\frac{\partial \mu_A(\text{mixt})/T}{\partial p}\right]_{T,x_A} dp$$

$$+ \left[\frac{\partial \mu_A(\text{mixt})/T}{\partial x_A}\right]_{T,p} dx_A$$

We use Eq. 9.1.1 on page 307 to replace the first partial derivative on the right side and Eq. 7.3.6 on page 183 to replace the second:

$$d[\mu_A(\text{mixt})/T] = -\frac{H_A(\text{mixt})}{T^2} dT + \frac{V_A(\text{mixt})}{T} dp + \left[\frac{\partial \mu_A(\text{mixt})/T}{\partial x_A}\right]_{T,p} dx_A$$

Next we write the total differential of $\mu_A^*(s)/T$ for pure solid A. The independent variables are T and p; the expression is like the preceding one with the last term missing:

$$d[\mu_A^*(s)/T] = -\frac{H_A^*(s)}{T^2} dT + \frac{V_A^*(s)}{T} dp$$

If the two phases are in equilibrium, the chemical potential of A is the same in both phases. If changes occur in T, p, or x_A while the phases remain in equilibrium, the condition

$$d[\mu_A(\text{mixt})/T] = d[\mu_A^*(s)/T]$$

must be satisfied. From the three preceding equations, combining terms, we obtain

$$\frac{\Delta_{\text{sol}} H_{\text{m,A}}}{T^2} \, dT - \frac{\Delta_{\text{sol}} V_{\text{m,A}}}{T} \, dp = \left[\frac{\partial \mu_{\text{A}}(\text{mixt})/T}{\partial x_{\text{A}}}\right]_{T,p} dx_{\text{A}} \qquad (9.3.1)$$
(phases in equilibrium)

Here we have replaced $H_{\text{A}}(\text{mixt}) - H_{\text{A}}^*(\text{s})$ by $\Delta_{\text{sol}} H_{\text{m,A}}$, the molar differential enthalpy of solution of solid A in the liquid mixture. Also $V_{\text{A}}(\text{mixt}) - V_{\text{A}}^*(\text{s})$ has been replaced by $\Delta_{\text{sol}} V_{\text{m,A}}$, the molar differential volume of solution. Equation 9.3.1 is a relation between changes in the variables T, p, and x_{A}, only two of which are independent in the equilibrium system.

We can apply the equation to a system in which the pure phase is a gas instead of a solid. $\Delta_{\text{sol}} H_{\text{m,A}}$ and $\Delta_{\text{sol}} V_{\text{m,A}}$ then refer to the solution process for the gas instead of the solid.

Suppose we set dp equal to zero in Eq. 9.3.1 and solve for dT/dx_{A}. This gives us the rate at which T changes with x_{A} at constant p:

$$\left(\frac{\partial T}{\partial x_{\text{A}}}\right)_p = \frac{T^2}{\Delta_{\text{sol}} H_{\text{m,A}}} \left[\frac{\partial \mu_{\text{A}}(\text{mixt})/T}{\partial x_{\text{A}}}\right]_{T,p} \qquad (9.3.2)$$
(phases in equilibrium)

We can also set dT equal to zero in Eq. 9.3.1 and find the rate at which p changes with x_{A} at constant T:

$$\left(\frac{\partial p}{\partial x_{\text{A}}}\right)_T = -\frac{T}{\Delta_{\text{sol}} V_{\text{m,A}}} \left[\frac{\partial \mu_{\text{A}}(\text{mixt})/T}{\partial x_{\text{A}}}\right]_{T,p} \qquad (9.3.3)$$
(phases in equilibrium)

We use the two preceding relations in Secs. 9.4 and 9.5.

9.4 COLLIGATIVE PROPERTIES OF A DILUTE SOLUTION

The **colligative properties** of a solution are usually considered to be:

1. *Freezing-point depression:* the decrease in the freezing temperature of the solution, compared to pure solvent at the same pressure.

2. *Boiling-point elevation:* the increase in the boiling temperature of a solution containing nonvolatile solutes, compared to pure solvent at the same pressure.

3. *Vapor-pressure lowering:* the decrease in the vapor pressure of a solution containing nonvolatile solutes, compared to the vapor pressure of the pure solvent at the same temperature.

4. *Osmotic pressure:* the increase in the pressure of the solution that places the solvent in transfer equilibrium with pure solvent at the same temperature and pressure as the original solution.[4]

---

[4]This definition was discussed in Sec. 9.2.2.

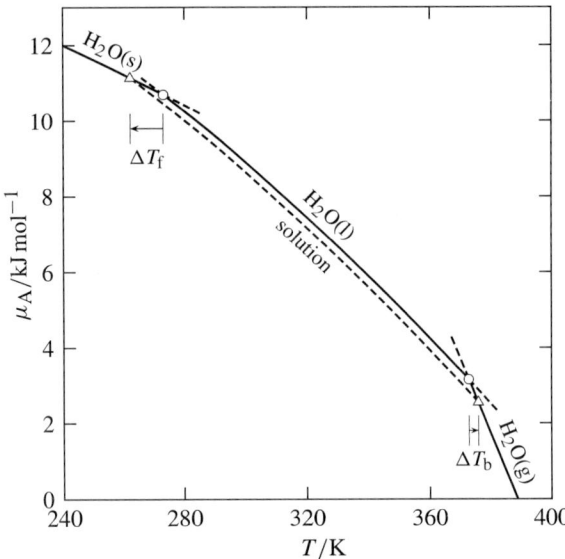

Figure 9.3 Freezing-point depression and boiling-point elevation in an aqueous solution. Curves: dependence on temperature of the chemical potential of H_2O (A) in pure phases and in an aqueous solution at 1 bar. The μ_A values have an arbitrary zero. The solution curve is calculated for an ideal-dilute solution of composition $x_A = 0.9$.

Note that all four properties are defined by an equilibrium between the liquid solution and a liquid or gas phase of the pure solvent. The properties called colligative (Latin: *tied together*) have in common a dependence on the concentration of solute particles, which affects the solvent chemical potential.

Figure 9.3 illustrates the principle of freezing-point depression and boiling-point elevation in the case of an aqueous solution. At a fixed pressure, pure liquid water is in equilibrium with ice at the freezing point and with steam at the boiling point. These are the temperatures at which H_2O has the same chemical potential in both phases at this pressure, occurring at the intersections of the chemical potential curves for the phases. The intersections are indicated by open circles in Fig. 9.3. The presence of dissolved solute in the water causes a lowering of the H_2O chemical potential at each temperature. Consequently, the curve for the chemical potential of H_2O in a solution intersects the curve for ice at a lower temperature and the curve for steam at a higher temperature (at the open triangles in Fig. 9.3). The freezing point is depressed by ΔT_f, and the boiling point (if the solute is nonvolatile) is elevated by ΔT_b.

Sections 9.4.1–9.4.4 derive theoretical relations between each of the four colligative properties and solute composition variables in the limit of infinite dilution. The expressions show that the colligative properties of a dilute binary solution depend on properties of the solvent, are proportional to the solute concentration and molality, but do not depend on the kind of solute.

Although these expressions provide no information about the activity coefficient of a solute, they are useful for estimating the solute molar mass. For example, from a measurement of any of the colligative properties of a dilute solution and the appropriate theoretical relation, we can obtain an approximate value of the solute molality m_B. (It is only approximate because, to make a measurement of reasonable precision, we cannot use an extremely dilute solution.) If we prepare the solution with a known amount n_A of solvent and a known mass of solute, we can calculate the amount of solute from $n_B = n_A M_A m_B$ (a rearrangement of Eq. 7.1.6 on page 172); then the solute molar mass is the solute mass divided by n_B.

9.4.1 Freezing-point depression

As in Sec. 9.2.1, we assume that the solid which forms when a dilute solution is cooled to its freezing point is *pure* Component A.

Equation 9.3.2 on page 315 gives the general dependence of temperature on the composition of a binary liquid mixture of A and B that is in equilibrium with pure solid A. We treat the mixture as a solution. The solvent is Component A, the solute (which can be either a nonelectrolyte or an electrolyte) is B, and the temperature is the freezing point $T_{f,A}$:

$$\left(\frac{\partial T_{f,A}}{\partial x_A}\right)_p = \frac{T_{f,A}^2}{\Delta_{sol} H_{m,A}} \left[\frac{\partial \mu_A(\text{sln})/T}{\partial x_A}\right]_{T,p} \tag{9.4.1}$$

Consider the expression on the right side of this equation in the limit of infinite dilution. In this limit, $T_{f,A}$ becomes $T_{f,A}^*$, the freezing point of the pure solvent, and $\Delta_{sol} H_{m,A}$ becomes $\Delta_{fus} H_{m,A}$, the molar enthalpy of fusion of the pure solvent.

To deal with the partial derivative on the right side of Eq. 9.4.1 in the limit of infinite dilution, we use the fact that the solvent activity coefficient γ_A approaches 1 in this limit. Then the solvent chemical potential is given by the Raoult's law relation

$$\mu_A(\text{sln}) = \mu_A^*(l) + RT \ln x_A \tag{9.4.2}$$

(solution at infinite dilution)

where $\mu_A^*(l)$ is the chemical potential of A in a pure-liquid reference state at the same T and p as the mixture.[5]

Equation 9.4.2 is valid even when the solute is an electrolyte, provided we calculate x_A from the amounts of all species present at infinite dilution. In the limit of infinite dilution any electrolyte solute is completely dissociated to its constituent ions; ion pairs and weak electrolytes are completely dissociated in this limit. Thus, for a binary solution of electrolyte B with ν ions per formula unit, we should calculate x_A from

$$x_A = \frac{n_A}{n_A + \nu n_B} \tag{9.4.3}$$

where n_B is the amount of solute formula unit. (If the solute is a nonelectrolyte, we simply set ν equal to 1 in this equation.)

---

[5] At the freezing temperature of the mixture, the reference state is an unstable supercooled liquid.

The derivation of Eq. 9.4.2 for a solution of an electrolyte is the same as described in Sec. 7.5.6 for an ideal-dilute binary solution of a nonelectrolyte. The starting point is the Gibbs–Duhem equation, $\sum_i x_i \, d\mu_i = 0$, and the relation $\mu_i = \mu_{x,i}^{\text{ref}} + RT \ln x_i$ for each solute species at infinite dilution. These equations are valid when μ_i and x_i refer to an ion solute species, just as they are for a molecular species, so x_A should be calculated from the amounts of all ion and molecular species present at infinite dilution. For a binary solution, Eq. 9.4.3 is the appropriate expression for x_A. The general expression is

$$x_A = \frac{n_A}{n_A + \sum_{i \neq A} n_i}$$

From Eq. 9.4.2, we can write

$$\left[\frac{\partial \mu_A(\text{sln})/T}{\partial x_A} \right]_{T,p} \to R \quad \text{as} \quad x_A \to 1 \quad (9.4.4)$$

In the limit of infinite dilution, then, Eq. 9.4.1 becomes

$$\lim_{x_A \to 1} \left(\frac{\partial T_{f,A}}{\partial x_A} \right)_p = \frac{R(T_{f,A}^*)^2}{\Delta_{\text{fus}} H_{m,A}} \quad (9.4.5)$$

It is customary to relate freezing-point depression to the solute concentration c_B or molality m_B. From Eq. 9.4.3, we obtain

$$1 - x_A = \frac{\nu n_B}{n_A + \nu n_B}$$

In the limit of infinite dilution, when n_B is much smaller than n_A, $1 - x_A$ approaches the value $\nu n_B / n_A$.[6] Then, using expressions in Eq. 7.1.10 on page 172, we obtain the relations[7]

$$\begin{aligned} dx_A &= -d(1 - x_A) = -\nu \, d(n_B/n_A) \\ &= -\nu V_{m,A}^* \, dc_B \\ &= -\nu M_A \, dm_B \end{aligned} \quad (9.4.6)$$

(binary solution at infinite dilution)

which transform Eq. 9.4.5 into the following:

$$\lim_{c_B \to 0} \left(\frac{\partial T_{f,A}}{\partial c_B} \right)_p = -\frac{\nu V_{m,A}^* R (T_{f,A}^*)^2}{\Delta_{\text{fus}} H_{m,A}}$$

$$\lim_{m_B \to 0} \left(\frac{\partial T_{f,A}}{\partial m_B} \right)_p = -\frac{\nu M_A R (T_{f,A}^*)^2}{\Delta_{\text{fus}} H_{m,A}} \quad (9.4.7)$$

---

[6] More rigorously, we find the derivative $d(1 - x_A)/d(n_B/n_A)$ approaches ν as n_B/n_A approaches zero.
[7] A small dependence of dc_B on dT as well as on dx_A has been ignored.

We can apply these equations to a nonelectrolyte solute by setting ν equal to 1.

As c_B or m_B approaches zero, $T_{f,A}$ approaches $T_{f,A}^*$. The freezing-point depression (a negative quantity) is $\Delta T_f = T_{f,A} - T_{f,A}^*$. In the range of molalities of a dilute solution in which $(\partial T_{f,A}/\partial m_B)_p$ is given by the expression on the right side of Eq. 9.4.7, we can write

$$\Delta T_f = -\frac{\nu M_A R (T_{f,A}^*)^2}{\Delta_{fus} H_{m,A}} m_B \qquad (9.4.8)$$

The **molal freezing-point depression constant** or cryoscopic constant, K_f, is defined for a binary solution by

$$K_f \equiv -\lim_{m_B \to 0} \frac{\Delta T_f}{\nu m_B}$$

and, from Eq. 9.4.8, has a value given by

$$K_f = \frac{M_A R (T_{f,A}^*)^2}{\Delta_{fus} H_{m,A}}$$

The value of K_f calculated from this formula depends only on the kind of solvent and the pressure (which determines $T_{f,A}^*$). For H$_2$O at 1 bar, the calculated value is $K_b = 1.860 \, \text{K kg mol}^{-1}$ (Prob. 9.4).

In the dilute binary solution, we have the relation

$$\Delta T_f = -\nu K_f m_B \qquad (9.4.9)$$
(dilute binary solution)

This relation is useful for estimating the molality of a dilute nonelectrolyte solution ($\nu = 1$) from a measurement of the freezing point. The relation is of little utility for an electrolyte solute because at any electrolyte molality that is high enough to give a measurable depression of the freezing point, the mean ionic activity coefficient deviates greatly from 1 and the relation is not accurate.

9.4.2 Boiling-point elevation

We can apply Eq. 9.3.2 to the boiling point T_b of a dilute binary solution. The pure phase of A in equilibrium with the solution is now a gas instead of a solid.[8] Following the procedure of Sec. 9.4.1, we obtain

$$\lim_{m_B \to 0} \left(\frac{\partial T_b}{\partial m_B} \right)_p = \frac{\nu M_A R (T_b^*)^2}{\Delta_{vap} H_{m,A}}$$

where $\Delta_{vap} H_{m,A}$ is the molar enthalpy of vaporization of pure solvent at its boiling point T_b^*.

---

[8] We must assume the solute is nonvolatile or has negligible partial pressure in the gas phase.

The **molal boiling-point elevation constant** or ebullioscopic constant, K_b, is defined for a binary solution by

$$K_b \equiv \lim_{m_B \to 0} \frac{\Delta T_b}{\nu m_B}$$

where $\Delta T_b = T_b - T_b^*$ is the boiling-point elevation. Accordingly, K_b has a value given by

$$K_b = \frac{M_A R (T_b^*)^2}{\Delta_{vap} H_{m,A}}$$

For the boiling point of a dilute solution, the analogy of Eq. 9.4.9 is

$$\Delta T_b = \nu K_b m_B \qquad (9.4.10)$$
$$\text{(dilute binary solution)}$$

Since K_f has a larger value than K_b (because $\Delta_{fus} H_{m,A}$ is smaller than $\Delta_{vap} H_{m,A}$), the measurement of freezing-point depression is more useful than that of boiling-point elevation for estimating the molality of a dilute solution.

9.4.3 Vapor-pressure lowering

In a binary two-phase system in which a solution of volatile solvent A and nonvolatile solute B is in equilibrium with gaseous A, the vapor pressure of the solution is equal to the system pressure p.

Equation 9.3.3 on page 315 gives the general dependence of p on x_A for a binary liquid mixture in equilibrium with pure gaseous A. $\Delta_{sol} V_{m,A}$ refers to the volume change for the dissolution of the gas in the solution. In the limit of infinite dilution, $-\Delta_{sol} V_{m,A}$ becomes the molar volume change for the vaporization of the pure solvent $\Delta_{vap} V_{m,A}$. We also apply the limiting expressions of Eqs. 9.4.4 and 9.4.6. The result is

$$\lim_{c_B \to 0} \left(\frac{\partial p}{\partial c_B} \right)_T = -\frac{\nu V_{m,A}^* RT}{\Delta_{vap} V_{m,A}} \qquad \lim_{m_B \to 0} \left(\frac{\partial p}{\partial m_B} \right)_T = -\frac{\nu M_A RT}{\Delta_{vap} V_{m,A}}$$

If we neglect the molar volume of the liquid solvent compared to that of the gas, and assume the gas is ideal, then $\Delta_{vap} V_{m,A}$ is $V_A^*(g) = RT/p_A^*$ and we obtain

$$\lim_{c_B \to 0} \left(\frac{\partial p}{\partial c_B} \right)_T \approx -\nu V_{m,A}^* p_A^* \qquad \lim_{m_B \to 0} \left(\frac{\partial p}{\partial m_B} \right)_T \approx -\nu M_A p_A^*$$

where p_A^* is the vapor pressure of the pure solvent at the temperature of the solution.

Thus, vapor-pressure lowering in the limit of infinite dilution is given by

$$\Delta p \approx -\nu V_{m,A}^* p_A^* c_B \qquad \text{or} \qquad \Delta p \approx -\nu M_A p_A^* m_B$$

We see that the lowering in this limit depends on the kind of solvent and the solution composition, but not on the kind of solute.

9.4.4 Osmotic pressure

The osmotic pressure Π is an intensive property of a solution and was defined in Sec. 9.2.2. In a dilute solution of low Π, the approximation used in Eq. 9.2.5 (that the partial molar volume V_A of the solvent is constant in the pressure range from p to $p + \Pi$) becomes valid and we can write

$$\Pi = \frac{\mu_A^*(l) - \mu_A(\text{sln})}{V_A}$$

In the limit of infinite dilution, we substitute the relation $\mu_A^*(l) - \mu_A(\text{sln}) = -RT \ln x_A$ (Eq. 9.4.2), replace V_A with the molar volume of the pure solvent V_A^*, and obtain

$$\Pi = -\frac{RT \ln x_A}{V_A^*}$$

and

$$\lim_{x_A \to 1} \left(\frac{\partial \Pi}{\partial x_A}\right)_{T,p} = -\frac{RT}{V_A^*}$$

The relations in Eq. 9.4.6 transform the preceding equation into

$$\lim_{c_B \to 0} \left(\frac{\partial \Pi}{\partial c_B}\right)_{T,p} = \nu RT \qquad (9.4.11)$$

$$\lim_{m_B \to 0} \left(\frac{\partial \Pi}{\partial m_B}\right)_{T,p} = \frac{\nu RT M_A}{V_A^*} = \nu \rho_A^* RT \qquad (9.4.12)$$

These equations show that the osmotic pressure becomes independent of the kind of solute as the solution approaches infinite dilution. The integrated forms of these equations are

$$\Pi = \nu c_B RT \qquad (9.4.13)$$
(dilute binary solution)

$$\Pi = \frac{RT M_A}{V_A^*} \nu m_B = \rho_A^* RT \nu m_B \qquad (9.4.14)$$
(dilute binary solution)

Equation 9.4.13 is **van't Hoff's equation** for osmotic pressure. If there is more than one solute, νc_B can be replaced by $\sum_{i \neq A} c_i$ and νm_B by $\sum_{i \neq A} m_i$ in these expressions.

In Sec. 7.6.6, it was stated that Π/m_B is equal to the product of ϕ_m and the limiting value of Π/m_B at infinite dilution, where $\phi_m = (\mu_A^* - \mu_A)/RT M_A \sum_{i \neq A} m_i$ is the osmotic coefficient. This relation follows directly from Eqs. 9.2.5 and 9.4.14.

9.5 SOLID–LIQUID EQUILIBRIA

A *freezing-point curve* (freezing temperature as a function of liquid composition) and a *solubility curve* (composition of a solution in equilibrium with a pure solid as a function of temperature) are complementary ways of describing the same physical situation. Thus, strange as the wording may seem, the composition of an aqueous solution at the freezing point gives the solubility of ice in the solution.

9.5.1 Freezing points of ideal binary liquid mixtures

Section 9.2.1 described the use of freezing-point measurements to determine the solvent chemical potential in a solution of arbitrary composition compared to the chemical potential of the pure solvent. The way in which freezing point varies with solution composition in the limit of infinite dilution was derived in Sec. 9.4.1. Now let us consider the freezing behavior over the entire composition range of an *ideal* liquid mixture.

The general relation between temperature and the composition of a binary liquid mixture, when the mixture is in equilibrium with pure solid A, is given by Eq. 9.3.2:

$$\left(\frac{\partial T}{\partial x_A}\right)_p = \frac{T^2}{\Delta_{sol} H_{m,A}} \left[\frac{\partial \mu_A(\text{mixt})/T}{\partial x_A}\right]_{T,p}$$

We can replace T by $T_{f,A}$ to indicate this is the temperature at which the mixture freezes to form solid A. The partial derivative on the right side of this equation is obtained from the expression for the chemical potential of Constituent A in an ideal liquid mixture

$$\mu_A(\text{mixt}) = \mu_A^*(l) + RT \ln x_A$$

giving

$$\left[\frac{\partial \mu_A(\text{mixt})/T}{\partial x_A}\right]_{T,p} = \frac{R}{x_A}$$

With this substitution, we have

$$\left(\frac{\partial T_{f,A}}{\partial x_A}\right)_p = \frac{RT_{f,A}^2}{x_A \Delta_{sol} H_{m,A}} \quad (9.5.1)$$
(ideal liquid mixture)

Figure 9.4 on the next page compares the freezing behavior of benzene predicted by this equation with experimental freezing-point data for benzene–toluene and benzene–cyclohexane mixtures. Any constituent that forms an ideal liquid mixture with benzene should give freezing points for the formation of solid benzene that fall on the curve in this figure. The agreement is good over a wide range of composition for benzene–toluene mixtures (circles), which are known to closely approximate ideal liquid mixtures. The agreement with benzene–cyclohexane mixtures (triangles), which are not ideal liquid mixtures at all compositions, is confined to the ideal-dilute region.

Section 9.5 Solid–Liquid Equilibria

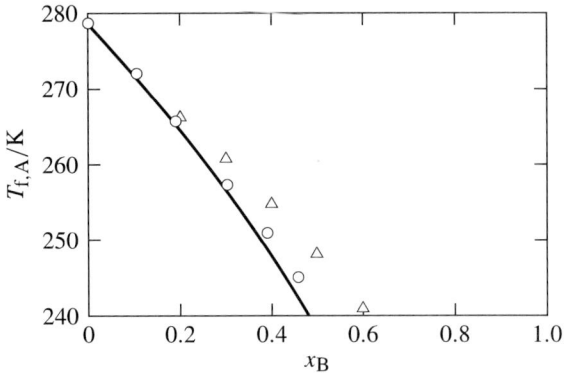

Figure 9.4 Dependence on composition of the freezing point of binary liquid mixtures with benzene as Component A. Solid curve: calculated for an ideal liquid mixture (Eq. 9.5.1), taking the temperature variation of $\Delta_{sol}H_{m,A}$ into account. Circles: B = toluene. Triangles: B = cyclohexane. (Negishi, R. *Rev. Phys. Chem. Japan* **1941**, *15*, 98-116.)

If we can assume that $\Delta_{sol}H_{m,A}$ is constant as x_A is increased to 1, we can replace it by $\Delta_{fus}H_{m,A}$, the molar enthalpy of fusion of pure solid A at its melting point. This assumption allows us to integrate Eq. 9.5.1 as follows from an arbitrary mixture composition x_A at freezing point $T_{f,A}$ to pure liquid A at freezing point $T_{f,A}^*$:

$$\int_{T_{f,A}}^{T_{f,A}^*} \frac{dT}{T^2} = \frac{R}{\Delta_{fus}H_{m,A}} \int_{x_A}^{x_A=1} d\ln x_A$$

The result, after some rearrangement, is

$$\ln x_A = \frac{\Delta_{fus}H_{m,A}}{R}\left(\frac{1}{T_{f,A}^*} - \frac{1}{T_{f,A}}\right) \qquad (9.5.2)$$

(ideal liquid mixture, $\Delta_{sol}H_{m,A} = \Delta_{fus}H_{m,A}$)

This equation was used to generate the curves shown in Fig. 9.5 on the next page. Although the shape of the freezing-point curve ($T_{f,A}$ versus x_B) shown in Fig. 9.4 is concave downward, Fig. 9.5 on the next page shows this is not always the case. When $\Delta_{fus}H_{m,A}/RT_{f,A}^*$ is less than 2, the freezing-point curve at low x_B is concave *upward*.

9.5.2 Solubility of a solid nonelectrolyte

Suppose we find that a solution of Solute B at a particular combination of temperature, pressure, and composition can exist in transfer equilibrium with pure solid B at the same temperature and pressure. This solution is said to be **saturated** with respect to the solid. We can express the **solubility** of the solid in the solvent by the value of the mole fraction, concentration, or molality of B in the saturated solution.[9] If the solution contains more than

---
[9] The IUPAC Green Book recommends the symbol s_B for solubility expressed as a concentration.

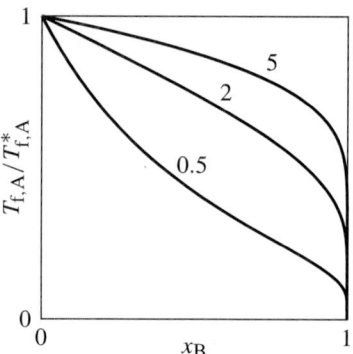

Figure 9.5 Freezing-point curves of ideal binary mixtures. The solid is Component A. Each curve is calculated from Eq. 9.5.2 and is labeled with the value of $\Delta_{\text{fus}} H_{\text{m,A}}/RT_{\text{f,A}}^*$.

one solute, we can speak of the solubility of B in the mixture of all components other than B.

We can also define solubility as the maximum value of the solute mole fraction, concentration, or molality that can exist in the solution without the possibility of spontaneous precipitation.

In this section, we consider the solubility of a *solid nonelectrolyte*. For the solution process $B(s) \rightarrow B(sln)$, the general expression for the thermodynamic equilibrium constant expression is $K = a_B(\text{sln})/a_B(s)$.[10] The activity of the pure solid is $a_B(s) = \Gamma_B(s)$. Let us use a solute standard state based on mole fraction; then the solute activity is $a_B(\text{sln}) = \Gamma_{x,B} \gamma_{x,B} x_B$. From these relations, the solubility expressed as a mole fraction is

$$x_B = \frac{\Gamma_B(s) K}{\Gamma_{x,B} \gamma_{x,B}}$$

If we measure the solubility at the standard pressure, the pressure factors $\Gamma_B(s)$ and $\Gamma_{x,B}$ are equal to unity and the solubility is given by

$$x_B = \frac{K}{\gamma_{x,B}} \quad (9.5.3)$$
$$\text{(solubility of solid B, } p = p^\circ\text{)}$$

If the pressure is not exactly equal to p° but is not very much greater, the values of the pressure factors are close to unity and it is a good approximation to use Eq. 9.5.3.

We can find the standard molar enthalpy of solution of B from the temperature dependence of the solubility. Combining Eqs. 9.1.8 and 9.5.3, we obtain

$$\Delta_{\text{sol}} H_{\text{m,B}}^\circ = RT^2 \frac{d \ln(\gamma_{x,B} x_B)}{dT} \quad (9.5.4)$$
$$(p = p^\circ)$$

---

[10]In this chapter, all states are assumed to be equilibrium states, although not indicated by the "eq" subscripts used in chap. 8.

Section 9.5 Solid–Liquid Equilibria

($\Delta_{sol}H^\circ_{m,B}$ is the same as the molar enthalpy of solution of B at infinite dilution if the pressure is p°.) If the solubility is sufficiently low for us to be able to set the solute activity coefficient equal to unity, Eq. 9.5.4 becomes

$$\Delta_{sol}H^\circ_{m,B} = RT^2 \frac{d\ln x_B}{dT} \qquad (9.5.5)$$
$$(p = p^\circ, \gamma_{x,B} = 1)$$

If the solubility x_B increases with increasing temperature, $\Delta_{sol}H^\circ_{m,B}$ must be positive and the solution process is endothermic. A decrease of solubility with increasing temperature implies an exothermic solution process. These statements refer to a solid of low solubility; see page 298 for a discussion of the general relation between the temperature dependence of solubility and the sign of the molar differential enthalpy of solution at saturation.

For a solute standard state based on *molality*, we can derive equations like Eqs. 9.5.4 and 9.5.5 with $\gamma_{x,B}$ replaced by $\gamma_{m,B}$ and x_B replaced by m_B/m°. If we use a solute standard state based on *concentration*, the expressions become slightly more complicated. The solubility is given by

$$c_B = \frac{\Gamma_B(s) \, Kc^\circ}{\Gamma_{c,B} \, \gamma_{c,B}}$$

From Eq. 9.1.7, we obtain, for a solid of low solubility, the relation

$$\Delta_{sol}H^\circ_{m,B} = RT^2 \left(\frac{d\ln(c_B/c^\circ)}{dT} + \alpha^*_A \right) \qquad (9.5.6)$$
$$(p = p^\circ, \gamma_{c,B} = 1)$$

9.5.3 Ideal solubility of a solid

The **ideal solubility** of a solid at a given temperature and pressure is the solubility calculated on the assumptions that the liquid is an ideal liquid mixture and that the molar enthalpy of solution equals the molar enthalpy of fusion of the solid. These were the assumptions used to derive Eq. 9.5.2 for the freezing-point curve of a liquid mixture. We exchange the constituent labels A and B so that B is now the constituent forming the solid phase:

$$\ln x_B = \frac{\Delta_{fus}H_{m,B}}{R} \left(\frac{1}{T^*_{f,B}} - \frac{1}{T} \right) \qquad (9.5.7)$$
(ideal solubility of solid B)

Here $T^*_{f,B}$ is the melting point of solid B.

The ideal solubility of a solid is independent of the kind of solvent and increases with increasing temperature. For solids with similar molar enthalpies of fusion, the ideal solubility is less at a given temperature the higher is the melting point. This behavior is shown in Fig. 9.6 on the next page. For the experimental solubility of a solid to agree even approximately with the ideal value, the solvent and solute must be chemically similar and the temperature must be close to the melting point of the solid (so that $\Delta_{sol}H_{m,B}$ is close in value to $\Delta_{fus}H_{m,B}$).

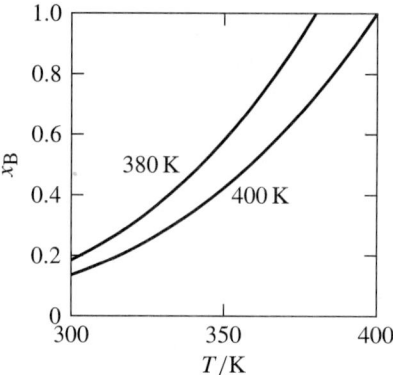

Figure 9.6 Ideal solubility of solid B as a function of T. The curves are calculated for two solids having the same molar enthalpy of fusion ($\Delta_{\text{fus}} H_{\text{m,B}} = 20\,\text{kJ}\,\text{mol}^{-1}$) and the values of $T_{\text{f,B}}^*$ indicated.

From the freezing behavior of benzene–toluene mixtures shown by the circles in Fig. 9.4, we can see that solid benzene has nearly ideal solubility in liquid toluene at temperatures not lower than about 20 K below the melting point of benzene.

9.5.4 Solid compounds

Binary liquid mixtures are known in which the solid that appears when the mixture is cooled to its freezing point is a compound containing a fixed proportion of the two components of the mixture. This kind of solid is called a **solid compound** or stoichiometric addition compound. Examples are salt hydrates (salts with fixed numbers of waters of hydration in the formula units) and certain metal alloys.

The composition of the liquid mixture in this kind of system is variable, whereas the composition of the solid compound is fixed. Suppose the components are A and B, present in the liquid mixture at mole fractions x_A and x_B, and the solid compound has the formula $A_a B_b$. We assume the compound is completely dissociated in the liquid phase; that is, that no $A_a B_b$ molecules exist in the liquid. The chemical equation for the freezing process is

$$a\text{A}(\text{mixt}) + b\text{B}(\text{mixt}) \rightarrow A_a B_b(\text{s})$$

When reaction equilibrium exists between the liquid and solid phases, the temperature is the freezing point T_f of the liquid. At equilibrium, the molar reaction Gibbs energy defined by $\Delta_r G_m = \sum_i \nu_i \mu_i$ (Eq. 8.7.2) is zero:

$$-a\mu_A - b\mu_B + \mu(\text{s}) = 0$$

Here μ_A and μ_B refer to chemical potentials in the liquid mixture, and $\mu(\text{s})$ refers to the solid compound.

Section 9.5 Solid–Liquid Equilibria

We divide both sides of the preceding equation by T and take differentials:

$$-a\,\mathrm{d}\left(\frac{\mu_A}{T}\right) - b\,\mathrm{d}\left(\frac{\mu_B}{T}\right) + \mathrm{d}\left[\frac{\mu(s)}{T}\right] = 0 \tag{9.5.8}$$

The pressure is fixed. Then μ_A/T and μ_B/T are functions of T and x_A, and $\mu(s)/T$ is a function only of T. We find expressions for the total differentials of these quantities at constant p with the help of Eq. 9.1.1 on page 307:

$$\mathrm{d}\left(\frac{\mu_A}{T}\right) = -\frac{H_A}{T^2}\,\mathrm{d}T + \frac{1}{T}\left(\frac{\partial \mu_A}{\partial x_A}\right)_{T,p}\mathrm{d}x_A$$

$$\mathrm{d}\left(\frac{\mu_B}{T}\right) = -\frac{H_B}{T^2}\,\mathrm{d}T + \frac{1}{T}\left(\frac{\partial \mu_B}{\partial x_A}\right)_{T,p}\mathrm{d}x_A$$

$$\mathrm{d}\left[\frac{\mu(s)}{T}\right] = -\frac{H_m(s)}{T^2}\,\mathrm{d}T$$

When we substitute these expressions in Eq. 9.5.8 and solve for $\mathrm{d}T/\mathrm{d}x_A$, setting T equal to T_f, we obtain

$$\frac{\mathrm{d}T_f}{\mathrm{d}x_A} = \frac{T_f}{aH_A + bH_B - H_m(s)}\left[a\left(\frac{\partial \mu_A}{\partial x_A}\right)_{T,p} + b\left(\frac{\partial \mu_B}{\partial x_A}\right)_{T,p}\right] \tag{9.5.9}$$

The quantity $aH_A + bH_B - H_m(s)$ on the right side of Eq. 9.5.9 is $\Delta_{\text{sol}}H_m$, the molar differential enthalpy of solution of the solid compound in the liquid mixture. The two partial derivatives on the right side are related through the Gibbs–Duhem equation $x_A\,\mathrm{d}\mu_A + x_B\,\mathrm{d}\mu_B = 0$ (Eq. 7.2.11 on page 180), which applies to changes at constant T and p. We rearrange the Gibbs–Duhem equation to $\mathrm{d}\mu_B = -(x_A/x_B)\,\mathrm{d}\mu_A$ and divide by $\mathrm{d}x_A$:

$$\left(\frac{\partial \mu_B}{\partial x_A}\right)_{T,p} = -\frac{x_A}{x_B}\left(\frac{\partial \mu_A}{\partial x_A}\right)_{T,p}$$

Making this substitution in Eq. 9.5.9, we obtain

$$\frac{\mathrm{d}T_f}{\mathrm{d}x_A} = \frac{x_A T_f}{\Delta_{\text{sol}}H_m}\left(\frac{a}{x_A} - \frac{b}{x_B}\right)\left(\frac{\partial \mu_A}{\partial x_A}\right)_{T,p} \tag{9.5.10}$$

Let the fixed mole fractions of A and B in the solid compound be x'_A and x'_B. Since the solid compound formula is $A_a B_b$, these mole fractions are given by $x'_A = a/(a+b)$ and $x'_B = b/(a+b)$. We can rewrite Eq. 9.5.10 as

$$\frac{\mathrm{d}T_f}{\mathrm{d}x_A} = \frac{x_A T_f(a+b)}{\Delta_{\text{sol}}H_m}\left(\frac{x'_A}{x_A} - \frac{x'_B}{x_B}\right)\left(\frac{\partial \mu_A}{\partial x_A}\right)_{T,p} \tag{9.5.11}$$

Suppose we heat a sample of the solid compound to its melting point to form a liquid mixture of the same composition, x'_A, as the solid. The molar enthalpy change of this process is the molar enthalpy of fusion of the solid compound, $\Delta_{\text{fus}}H_{\text{m}}$, a *positive* quantity. The melting process is the same process as the dissolution of the solid compound in liquid of composition x'_{A}. The value of $\Delta_{\text{sol}}H_{\text{m}}$ for the dissolution of the solid in liquid of this composition must be equal to $\Delta_{\text{fus}}H_{\text{m}}$.

We do not expect that small increases or decreases of x_{A} from the value x'_{A} will change the sign of $\Delta_{\text{sol}}H_{\text{m}}$. In the range of compositions in which $\Delta_{\text{sol}}H_{\text{m}}$ is positive, Eq. 9.5.11 shows that the slope of the freezing-point curve, T_{f} versus x_{A}, is positive when x_{A} is less than x'_{A} (and x_{B} is greater than x'_{B}), negative when x_{A} is greater than x'_{A}, and zero when x_{A} equals x'_{A}. Thus, the freezing-point curve has a maximum at the composition of the solid compound, and the slope of the curve here is zero.

If the liquid mixture is an ideal liquid mixture, in which μ_{A} obeys Raoult's law for fugacity, $\mu_{\text{A}} = \mu_{\text{A}}^* + RT \ln x_{\text{A}}$, the partial derivative $(\partial \mu_{\text{A}}/\partial x_{\text{A}})_{T,p}$ equals RT/x_{A} and Eq. 9.5.11 becomes

$$\frac{dT_{\text{f}}}{dx_{\text{A}}} = \frac{RT_{\text{f}}^2(a+b)}{\Delta_{\text{sol}}H_{\text{m}}}\left(\frac{x'_{\text{A}}}{x_{\text{A}}} - \frac{x'_{\text{B}}}{x_{\text{B}}}\right)$$

By treating $\Delta_{\text{sol}}H_{\text{m}}$ as independent of T and x_{A} and equal to $\Delta_{\text{fus}}H_{\text{m}}$, we can rearrange this equation and integrate as follows:

$$\int_{T_{\text{f}}}^{T'_{\text{f}}} \frac{dT_{\text{f}}}{T_{\text{f}}^2} = \frac{R(a+b)}{\Delta_{\text{fus}}H_{\text{m}}} \int_{x_{\text{A}}}^{x'_{\text{A}}} \left(\frac{x'_{\text{A}}}{x_{\text{A}}} - \frac{x'_{\text{B}}}{x_{\text{B}}}\right) dx_{\text{A}}$$

Here T'_{f} is the melting point of the solid compound. The result is

$$\frac{1}{T_{\text{f}}} = \frac{1}{T'_{\text{f}}} + \frac{R}{\Delta_{\text{fus}}H_{\text{m}}}\left(a \ln \frac{x'_{\text{A}}}{x_{\text{A}}} + b \ln \frac{x'_{\text{B}}}{x_{\text{B}}}\right) \quad (9.5.12)$$

(ideal liquid mixture freezing to solid compound, $\Delta_{\text{sol}}H_{\text{m}} = \Delta_{\text{fus}}H_{\text{m}}$)

Figure 9.7 on the next page shows an example of a molten metal mixture that solidifies to an alloy of fixed composition. The freezing-point curve of this system is closely approximated by Eq. 9.5.12.

9.5.5 Solubility of a solid electrolyte

Consider the equilibrium between a crystalline salt (or other kind of ionic solid) and the dissociated solvated ions:

$$M_{\nu_+}X_{\nu_-}(s) \rightleftharpoons \nu_+ M^{z+}(aq) + \nu_- X^{z-}(aq)$$

Here ν_+ and ν_- are the numbers of cations and anions in the formula unit of the salt and z_+ and z_- are the charge numbers of these ions. The thermodynamic equilibrium constant for this kind of equilibrium is called the **solubility product**, K_{s}.

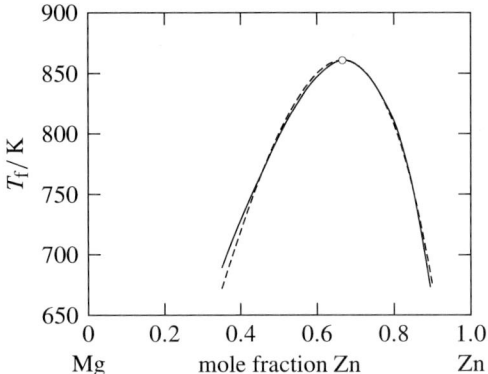

Figure 9.7 Solid curve: freezing-point curve of a liquid melt of Zn and Mg. The solid phase is the solid compound Zn_2Mg. The curve maximum (open circle) is at the compound composition $x'_{Zn} = 2/3$ and the solid compound melting point $T'_f = 861$ K. Dashed curve: calculated using Eq. 9.5.12 with $\Delta_{fus} H_m = 15.8 \text{ kJ mol}^{-1}$.

The expression for the solubility product in factors of activities is

$$K_s = \frac{(a_+^{\nu_+})(a_-^{\nu_-})}{a(s)}$$

When we replace the activities with the appropriate expressions from Table 7.3 on page 215, we obtain

$$K_s = \left[\frac{\Gamma_+^{\nu_+} \Gamma_-^{\nu_-}}{\Gamma(s)}\right] \left(\gamma_+^{\nu_+} \gamma_-^{\nu_-}\right) (m^\circ)^{-\nu} \left(m_+^{\nu_+} m_-^{\nu_-}\right)$$

where ν is the number of ions per formula unit, equal to the sum $\nu_+ + \nu_-$.

We denote the first factor enclosed by brackets on the right side, which is a reaction pressure-factor quotient, by Γ_r. Its value is exactly 1 if the system is at the standard pressure and approximately 1 unless the pressure is very high. The factor $(\gamma_+^{\nu_+} \gamma_-^{\nu_-})$ is equal to $\gamma_\pm^\nu$, where $\gamma_\pm$ is the mean ionic activity coefficient of the dissolved solute (see Eq. 7.8.16 on page 223). With these substitutions, the expression for the solubility product becomes

$$K_s = \Gamma_r \gamma_\pm^\nu (m^\circ)^{-\nu} \left(m_+^{\nu_+} m_-^{\nu_-}\right) \qquad (9.5.13)$$

If the aqueous solution is produced by allowing the salt to dissolve in pure water or in a solution of a second solute containing no ions in common with the salt, then the ion molalities in the saturated solution are $m_+ = \nu_+ m_B$ and $m_- = \nu_- m_B$. In this case, Eq. 9.5.13 becomes[11]

$$K_s = \left(\nu_+^{\nu_+} \nu_-^{\nu_-}\right) \Gamma_r \gamma_\pm^\nu \left(\frac{m_B}{m^\circ}\right)^\nu \qquad (9.5.14)$$
(no common ion)

---

[11] We could also have obtained this expression by writing $K_s = a_{m,B}/a(s)$ and replacing $a_{m,B}$ with the expression of Eq. 7.8.19 on page 223.

where m_B is the solubility of the salt expressed as a molality.

Since K_s depends only on T and Γ_r depends only on T and p, Eq. 9.5.14 shows that any change in the solution composition at constant T and p that decreases $\gamma_\pm$ must increase the solubility. For instance, the *salt effect* is the increase of the solubility of a sparingly soluble salt due to increased ionic strength when we dissolve a second salt, lacking a common ion, in the solvent.

We can use the salt effect to evaluate the solubility product by an extrapolation procedure:

$$K_s = \left(v_+^{v_+} v_-^{v_-}\right) \Gamma_r \lim_{I_m \to 0} \left(\frac{m_B}{m^\circ}\right)^v$$

If the ionic strength of the saturated salt solution is sufficiently low (i.e., the solubility is sufficiently low), it may be practical to evaluate the solubility product with Eq. 9.5.14 and an estimate of $\gamma_\pm$ from the Debye–Hückel limiting law (see Prob. 9.17). The most accurate method of measuring a solubility product is through the standard cell potential of an appropriate galvanic cell (Sec. 11.3).

Equation 9.5.13 is a general equation that applies even if the solution saturated with one salt contains a second salt with a common ion. For instance, consider the saturated solution of the sparingly soluble salt $M_{v_+}X_{v_-}$ in a solution containing the more soluble salt $M_{v'_+}Y_{v'_-}$ at molality m_C. The *cation* is the common ion in this example. The expression for the solubility product is now

$$K_s = \Gamma_r \gamma_\pm^v (m^\circ)^{-v} (v_+ m_B + v'_+ m_C)^{v_+} (v_- m_B)^{v_-} \qquad (9.5.15)$$
$$\text{(common cation)}$$

where m_B again is the solubility of the sparingly soluble salt. K_s and Γ_r are constant if T and p fixed, so any increase in m_C at constant T and p must cause a decrease in the solubility m_B; this is the *common-ion effect*.

From the measured solubility of a salt in pure solvent or in an electrolyte solution with a common cation, and the value of K_s, we can evaluate the mean ionic activity coefficient $\gamma_\pm$ through Eq. 9.5.14 or 9.5.15. (An equation similar to Eq. 9.5.15 would be used if there is a common anion.) This procedure has the disadvantage of being limited to the salt molality existing in the saturated solution.

We find the temperature dependence of K_s by applying Eq. 9.1.8:

$$\frac{d \ln K_s}{dT} = \frac{\Delta_{sol} H^\circ_{m,B}}{RT^2}$$

At the standard pressure, $\Delta_{sol} H^\circ_{m,B}$ is the same as $\Delta_{sol} H^\infty_{m,B}$, the molar enthalpy of solution at infinite dilution in the pure solvent.

9.6 LIQUID–LIQUID EQUILIBRIA

The pure liquid phases of two substances, when placed in contact with one another, may fail to mix completely. Obviously these substances do not form ideal liquid mixtures or they would mix in all proportions (Sec. 8.2.2). The thermodynamic conditions for phase separation were discussed in Sec. 8.2.6; phase separation is usually the result of positive deviations from Raoult's law. Typically, when phase separation occurs, one of the substances is polar and the other nonpolar.

We may say that two liquids are immiscible, but they are never *completely* immiscible. Even liquid mercury, when equilibrated with water, has some H_2O dissolved in it, and some mercury dissolves in the water, although the amounts may be too small to measure. It is more accurate to say that two liquids that form two liquid phases are *partially* miscible.

The relations between the compositions of the phases and the activity coefficients of the constituents depend on our choice of reference states or standard states.

9.6.1 Pure-liquid reference states

Consider the general case of two equilibrated liquid phases α and β containing two or more components. The condition for transfer equilibrium of Component i is $\mu_i^\alpha = \mu_i^\beta$. If pure component i is a liquid, we can use a pure-liquid reference state for i in both phases:

$$\mu_i^*(l) + RT \ln(\gamma_i^\alpha x_i^\alpha) = \mu_i^*(l) + RT \ln(\gamma_i^\beta x_i^\beta)$$

This leads to the following relation between the compositions and activity coefficients:

$$\frac{x_i^\alpha}{x_i^\beta} = \frac{\gamma_i^\beta}{\gamma_i^\alpha} \qquad (9.6.1)$$

(equilibrated phases, pure-liquid reference states)

If the mole fraction of Component i is close to zero in phase α and close to unity in phase β, then γ_i^β must be approximately 1 (because γ_i^β is based on a pure-liquid reference state) and γ_i^α is much greater than 1.

> Note that if we use the equilibrium constant approach for the process $A(\alpha) \rightleftharpoons A(\beta)$, with the same pure-liquid standard state for Component A in both phases, the expression for the thermodynamic equilibrium constant is
>
> $$K = \frac{a_A^\beta}{a_A^\alpha} = \frac{\Gamma_A \gamma_A^\beta x_A^\beta}{\Gamma_A \gamma_A^\alpha x_A^\alpha}$$
>
> Furthermore, $\Delta_r G_m^\circ$ for this process is zero because μ_A does not change under standard conditions. Then from the relation $\Delta_r G_m^\circ = -RT \ln K$, we see that the value of K must be 1. This leads to the same expression for x_A^α / x_A^β as in Eq. 9.6.1.

9.6.2 Solubility of a liquid

Suppose that, in a system of two equilibrated liquid phases, Component B has a low mole fraction in phase α and a mole fraction close to 1 in phase β. It is then appropriate to treat B as a *solute* in phase α and a constituent of a liquid mixture in phase β. The value of x_B^α is the *solubility* of B in liquid A.

The equilibrium when two liquid phases are present is $B(\beta) \rightleftharpoons B(\alpha)$, and the expression for the thermodynamic equilibrium constant, with the solute standard state based on mole fraction, is

$$K = \frac{a_{x,B}^{\alpha}}{a_{B}^{\beta}} = \frac{\Gamma_{x,B}^{\alpha} \gamma_{x,B}^{\alpha} x_{B}^{\alpha}}{\Gamma_{B}^{\beta} \gamma_{B}^{\beta} x_{B}^{\beta}}$$

The solubility of B in phase α is then

$$x_{B}^{\alpha} = \left(\frac{\Gamma_{B}^{\beta}}{\Gamma_{x,B}^{\alpha}} K\right) \frac{\gamma_{B}^{\beta} x_{B}^{\beta}}{\gamma_{x,B}^{\alpha}}$$

The quantity in parentheses on the right side depends only on T and p and is equal to K if the pressure is $p°$. The other quantities on the right side can vary at a given T and p if the system has more than two components.

> The expression for x_{B}^{α} shows that the solubility of B in phase α is related to the composition of phase β, the B-rich phase. The solubility is sometimes defined as the maximum quantity of a solute that can dissolve in a given quantity of solvent. One might think that the solubility of B would be different if we gradually added *pure* liquid B to the solvent. However, the result would be the same as bringing the two liquids directly into contact with one another to form two layers because the maximum amount of pure solute that can dissolve in the solvent occurs at the point of incipient phase separation and the B-rich phase that separates will not be pure B.

If there are only two components, A and B, each of which has only a limited solubility in the other, and the pressure is not very large, then the pressure factors and activity coefficients are close to unity and the solubility of B in A is given by $x_{B}^{\alpha} \approx K$.[12] In this case, the temperature dependence of the solubility is given by

$$\frac{d \ln x_{B}^{\alpha}}{dT} = \frac{d \ln K}{dT} = \frac{\Delta_{sol} H_{m,B}^{\circ}}{RT^2} \qquad (9.6.2)$$

where $\Delta_{sol} H_{m,B}^{\circ}$ is the molar enthalpy change for the transfer of *pure* liquid solute to the solution at infinite dilution, at the standard pressure. This equation can be applied to the system of water and *n*-butylbenzene, which have very small mutual solubilities. Figure 9.8 on the next page shows that the solubility of *n*-butylbenzene in water exhibits a minimum at about 12 °C, meaning that $\Delta_{sol} H_{m,B}^{\circ}$ is negative below this temperature and positive above.

We can also relate the solubility of B to its Henry's law constant $K_{x,B}^{\alpha}$. Suppose the two liquid phases are equilibrated not only with one another but also with a gas phase. Since B is equilibrated between phase α and the gas, we have $\gamma_{x,B}^{\alpha} = f_B / K_{x,B}^{\alpha} x_B^{\alpha}$ (Eq. 7.6.17). We also have $\gamma_B^{\beta} = f_B / x_B^{\beta} f_B^*$ from the equilibration of B between phase β and the gas (Eq. 7.6.15). By eliminating the fugacity f_B from these relations, we obtain the general relation

$$x_B^{\alpha} = \frac{\gamma_B^{\beta} x_B^{\beta} f_B^*}{\gamma_{x,B}^{\alpha} K_{x,B}^{\alpha}}$$

---

[12] Note how these activity coefficients and the equilibrium constant differ from those described in the preceding section. Different standard states were used there.

Section 9.6 Liquid–Liquid Equilibria

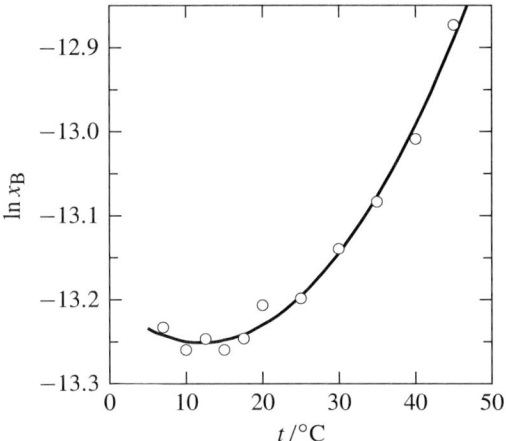

Figure 9.8 Aqueous solubility of liquid *n*-butylbenzene as a function of temperature (Owens, J. W.; Wasik, S. P.; DeVoe, H. J. *J. Chem. Eng. Data* **1986**, *31*, 47–51).

If we assume as before that the two liquid components have low mutual solubilities, so that the activity coefficients and x_B^β are close to 1, and that the gas phase behaves ideally, the solubility of B is given by $x_B^\alpha \approx p_B^*/K_{x,B}^\alpha$, where p_B^* is the vapor pressure of the pure solute.

9.6.3 Distribution of a solute between two solvents

Consider a two-component system of two equilibrated liquid phases, Component A having the highest mole fraction in phase α and Component B having the highest mole fraction in phase β. If we add a small quantity of a third component C, it will distribute itself between the two phases: $C(\beta) \rightleftharpoons C(\alpha)$. It is appropriate to treat C as a solute in *both* phases. The thermodynamic equilibrium constant, with standard states based on mole fraction, is

$$K = \frac{a_{x,C}^\alpha}{a_{x,C}^\beta} = \frac{\Gamma_{x,C}^\alpha \gamma_{x,C}^\alpha x_C^\alpha}{\Gamma_{x,C}^\beta \gamma_{x,C}^\beta x_C^\beta}$$

At a fixed T and p, the equilibrium constant and pressure factors are constant and so the ratio $K' = \gamma_{x,C}^\alpha x_C^\alpha / \gamma_{x,C}^\beta x_C^\beta$ must also be constant. From the relation $\gamma_{x,C}^\alpha = f_C / K_{x,C}^\alpha x_C^\alpha$ (Eq. 7.6.17), and the corresponding relation for phase β, we find K' is equal to the ratio $K_{x,C}^\beta / K_{x,C}^\alpha$ of Henry's law constants in the two solvents. We can vary the mole fractions x_C^α or x_C^β,[13] and when x_C is low enough in both phases for the activity coefficients to be unity the ratio x_C^α / x_C^β is constant and equal to K'. The constancy of this ratio is the **Nernst distribution law**.

---

[13] The Gibbs phase rule for a multicomponent system, to be described in Sec. 10.1, shows that a two-phase, three-component system has three degrees of freedom. This means we can independently vary the temperature, pressure, and mole fraction of C in one of the phases. With only two components, the system has only two degrees of freedom, and the compositions of both phases are determined by the temperature and pressure.

Since solute molality and concentration are proportional to mole fraction in dilute solutions, the ratios m_C^α/m_C^β and c_C^α/c_C^β also approach constant values at a given T and p. The ratio using concentrations is called the **partition coefficient** or **distribution coefficient**.

In the limit of infinite dilution of C, the two phases have the compositions that exist when only Components A and B are present. As C is added and x_C^α and x_C^β increase beyond the region of infinitely dilute solution behavior, the ratios x_B^α/x_A^α and x_B^β/x_A^β may change. Often continued addition of C increases the mutual solubilities of A and B, resulting eventually when enough C has been added in a single liquid phase containing all three components. It is easier to understand this behavior with the help of a ternary phase diagram such as Fig. 10.17 on page 381.

9.6.4 Membrane equilibria

A solvent can exist in transfer equilibrium between two solutions of different compositions, even if the solutions are miscible, if the phases are separated by a membrane that is permeable to the solvent but not to each solute. An equilibrium state in such a system is called an **osmotic membrane equilibrium** and requires in general that the two phases be at the same temperature but different pressures. We have already seen that this effect is present in an apparatus that measures osmotic pressure (Fig. 9.2 on page 312).

Consider a system in which there is transfer equilibrium of the solvent across a membrane separating phases α and β. The phases have equal solvent chemical potentials but different pressures:

$$\mu_A^\beta(p^\beta) = \mu_A^\alpha(p^\alpha)$$

The dependence of the chemical potential of a species on pressure in a phase of fixed temperature and composition is given by $(\partial \mu_i/\partial p)_{T,n} = V_i$ (Eq. 7.3.6), where V_i is the partial molar volume of the species in the phase. If we apply this relation to the solvent in phase β with composition fixed at the composition of the equilibrium system, treat the partial molar volume V_A as independent of pressure, and integrate from the pressure of phase α to that of phase β, we obtain

$$\mu_A^\beta(p^\beta) = \mu_A^\beta(p^\alpha) + V_A^\beta(p^\beta - p^\alpha) \qquad (9.6.3)$$

By equating the two expressions for $\mu_A^\beta(p^\beta)$ and rearranging, we obtain the following expression for the pressure difference:

$$p^\beta - p^\alpha = \frac{\mu_A^\alpha(p^\alpha) - \mu_A^\beta(p^\alpha)}{V_A^\beta}$$

Comparing this equation with Eq. 9.2.5 on page 314, we find that the pressure difference is related to the osmotic pressures of the two solutions, measured at pressure p^α, by

$$p^\beta - p^\alpha = \Pi^\beta - \left(\frac{V_A^\alpha}{V_A^\beta}\right)\Pi^\alpha$$

Equilibrium dialysis is a useful technique for studying the binding of a small solute species (a ligand) to an uncharged macromolecule. The macromolecule solution is placed on one side of a membrane through which it cannot pass, with a solution without the macromolecule on the other side, and the ligand is allowed to come to transfer equilibrium across the membrane. If the same solute standard state is used for the ligand in both solutions, at equilibrium the unbound ligand must have the same activity in both solutions. Measurements of the total ligand molality in the macromolecule solution and the ligand molality in the other solution, combined with estimated values of the unbound ligand activity coefficients, allow the amount of ligand bound per macromolecule to be calculated.

If the two-phase system contains certain solute *ion* species that cannot pass through the membrane while other ions can, the situation is more complicated. Usually if the membrane is impermeable to one kind of ion, an ion to which it is permeable cannot achieve transfer equilibrium across the membrane unless the phases have different electric potentials. The equilibrium state in this case is a **Donnan membrane equilibrium**, and the resulting electric potential difference across the membrane is called the **Donnan potential**. This phenomenon is related to the membrane potentials that are important in the functioning of nerve and muscle cells (although living cells are not in equilibrium states).

A Donnan potential can be measured electrically (with some uncertainty due to unknown liquid-junction potentials) by connecting a silver-silver chloride electrode (described in Sec. 11.1) to each phase through a salt bridge.

To understand the Donnan membrane equilibrium, we first need to investigate the general relation between the chemical potential of a charged species and the electric potential of the phase. The electric potential in the interior of a phase is called the *inner electric potential* or *Galvani potential*. If a particle of charge Q_{el} is brought from a place where the electric potential is zero to a phase where the electric potential (Galvani potential) is ϕ, there is an electrical contribution to the internal energy change of the phase equal to $Q_{el}\phi$. An amount n_i of ion i has a charge given by $Q_{el} = z_i F n_i$, where z_i is the charge number ($+1$, -2, etc.) of the ion, and F is the Faraday constant,[14] so an infinitesimal amount dn_i of ion i has an infinitesimal charge $dQ_{el} = z_i F \, dn_i$.

Suppose we compare two phases that have the same temperature, pressure, and composition and different electric potentials: One phase has zero electric potential and the other has a nonzero electric potential ϕ. If $U(0)$ is the internal energy of the first phase and $U(\phi)$ is the internal energy of the other phase, then the changes when an amount dn_i of ion i enters each phase are related by $dU(\phi) = dU(0) + z_i F \phi \, dn_i$. Since the Gibbs energy of a phase is defined as $G = U + pV - TS$ and the chemical potential is given by $\mu_i = (\partial G/\partial n_i)_{T,p,n_{j \neq i}}$, the chemical potentials of the ion in the two phases are related by[15]

$$\mu_i(\phi) = \mu_i(0) + z_i F \phi \tag{9.6.4}$$

Now consider an ion that can pass through the membrane separating phases α and β. We can write the condition for transfer equilibrium of this ion, the equality of the chemical

---

[14]See Sec. 11.3.

[15]Some thermodynamicists emphasize the distinction between $\mu_i(0)$ (which depends only on T, p, and the chemical composition of the phase) and $\mu_i(\phi)$ (which for an ion depends also on ϕ) by calling $\mu_i(0)$ the *chemical potential* and $\mu_i(0) + z_i F \phi$ the *electrochemical potential*.

potentials in both phases, in the form

$$\mu_i^\beta(p^\beta, \phi^\beta) = \mu_i^\alpha(p^\alpha, \phi^\alpha)$$

where the pressure and electric potential are indicated in parentheses and it is understood that the phases have the same temperature but different compositions. Substitution from Eq. 9.6.4 gives

$$\mu_i^\beta(p^\beta, 0) + z_i F \phi^\beta = \mu_i^\alpha(p^\alpha, 0) + z_i F \phi^\alpha$$

From the method used to obtain Eq. 9.6.3, we can write the effect of a pressure change on μ_i^β at constant temperature and composition as follows:

$$\mu_i^\beta(p^\beta, 0) = \mu_i^\beta(p^\alpha, 0) + V_i^\beta(p^\beta - p^\alpha)$$

Then by equating the two expressions for $\mu_i^\beta(p^\beta, 0)$ and rearranging, we obtain

$$\mu_i^\beta(p^\alpha, 0) = \mu_i^\alpha(p^\alpha, 0) + z_i F(\phi^\alpha - \phi^\beta) + V_i^\beta(p^\alpha - p^\beta)$$

The chemical potential of solute i is related to the molality m_i according to Eq. 7.6.7:

$$\mu_i = \mu_{m,i}^{\text{ref}} + RT \ln \left(\gamma_{m,i} \frac{m_i}{m^\circ} \right)$$

Here $\mu_{m,i}^{\text{ref}}$ is a molality-based reference state at the same T, p, and ϕ as the phase of interest. Therefore, the relation between μ_i in two phases that have the compositions of phase β and phase α but are both at pressure p^α and electric potential 0 is

$$\mu_i^\beta(p^\alpha, 0) = \mu_i^\alpha(p^\alpha, 0) + RT \ln \left(\frac{\gamma_{m,i}^\beta m_i^\beta}{\gamma_{m,i}^\alpha m_i^\alpha} \right)$$

The activity coefficients in this expression are for the same pressure, p^α. Equating the two expressions for $\mu_i^\beta(p^\alpha, 0)$ and rearranging, we obtain

$$RT \ln \left(\frac{\gamma_{m,i}^\beta m_i^\beta}{\gamma_{m,i}^\alpha m_i^\alpha} \right) = z_i F(\phi^\alpha - \phi^\beta) + V_i^\beta(p^\alpha - p^\beta)$$

Under typical conditions, $V_i^\beta(p^\alpha - p^\beta)$ is much smaller in magnitude than $z_i F(\phi^\alpha - \phi^\beta)$. If we neglect the second term on the right side of the previous equation and assume the activity coefficient is the same in both phases, we can rearrange the equation to

$$\phi^\alpha - \phi^\beta \approx \frac{RT}{z_i F} \ln \frac{m_i^\beta}{m_i^\alpha} \tag{9.6.5}$$

Section 9.6 Liquid–Liquid Equilibria

This approximation is valid when i is an ion that can pass through the membrane and relates the Donnan potential to the ratio of the ion molalities of the two phases in an equilibrium state.

As a specific example of a Donnan membrane equilibrium, consider a system in which a solution of a protein or other polyelectrolyte with a net negative charge, together with a counterion M^+ and a salt MX of the counterion, is equilibrated with a solution of the salt across a semipermeable membrane. The membrane is permeable to all species but the polyelectrolyte. Assume the polyelectrolyte has a negative charge of $-z$ (where z is a large positive integer) and the salt is a 1:1 electrolyte such as potassium chloride represented by the formula MX. The species in phase α are H_2O, M^+, and X^-; those in phase β are H_2O, M^+, X^-, and the polyelectrolyte P^{z-}.

Because the polyelectrolyte in this example has a negative charge, the system has more M^+ ions than X^- ions. If the net charge of each phase is to be exactly zero, then either the molality of M^+ must be greater in phase β than in phase α or the molality of X^- must be greater in phase α than in phase β, or both. To attain transfer equilibrium, some M^+ ions must transfer into phase α or some X^- ions into phase β. As this transfer occurs, phase α gains a net positive charge and phase β gains an equal net negative charge. The electric potential of phase α becomes positive relative to phase β, and according to Eq. 9.6.4 the chemical potential of M^+ increases in phase α and that of X^- decreases in phase β. The equilibrium state is reached when the chemical potentials of both M^+ and X^- are equal in the two phases. At this point, phase α has a minute excess of M^+ ions compared to X^- ions; the difference in molalities, however, is too small to be detected by chemical analysis.

Applying Eq. 9.6.5 to this system (with $z_+ = +1$ and $z_- = -1$) gives us the two approximate relations

$$\phi^\alpha - \phi^\beta \approx \frac{RT}{F} \ln \frac{m_+^\beta}{m_+^\alpha} \qquad \phi^\alpha - \phi^\beta \approx \frac{RT}{F} \ln \frac{m_-^\alpha}{m_-^\beta}$$

These approximations are consistent only if the following relation exists between the molalities of M^+ and X^-:

$$m_+^\alpha m_-^\alpha \approx m_+^\beta m_-^\beta \tag{9.6.6}$$

Although in the equilibrium state each phase has unequal amounts of positive and negative changes, the difference is so small that for all practical purposes we can equate the amounts of positive and negative charges in the phase:

$$m_+^\alpha = m_-^\alpha \qquad m_+^\beta = m_-^\beta + z m_P$$

Substituting these expressions in Eq. 9.6.6 gives the relation

$$(m_-^\alpha)^2 \approx (m_-^\beta + z m_P) m_-^\beta$$

This shows that in the equilibrium state m_-^α is greater than m_-^β. Then Eq. 9.6.6 shows that m_+^α is less than m_+^β and Eq. 9.6.5 shows that $\phi^\alpha - \phi^\beta$ is positive.

These conclusions apply to the present example in which the polyelectrolyte has a negative charge and a positive counterion, when $p^\alpha - p^\beta$ is small. A more general statement is that the counterion is more concentrated in the phase with the polyelectrolyte, the small ion with the same charge as the polyelectrolyte is more concentrated in the phase without the polyelectrolyte, and the electric potential of the phase with the polyelectrolyte relative to the other phase has the same sign as the polyelectrolyte charge.

It should be clear that the existence of a Donnan membrane equilibrium introduces complications that would make it difficult to use the pressure difference to estimate the molar mass of the polyelectrolyte by the method of Sec. 9.4 or to study the binding of a charged ligand by equilibrium dialysis.

9.7 LIQUID–GAS EQUILIBRIA

Next we consider systems in which liquid and gas phases, at least one of which is a mixture, are equilibrated.

9.7.1 Effect of liquid pressure on gas fugacity

If we vary the pressure of a pure liquid or liquid mixture, keeping temperature constant, there is a small effect on the fugacity of each volatile component in an equilibrated gas phase. We can vary the pressure of the liquid while keeping its temperature and composition constant by changing the partial pressure of an additional gaseous component that has negligible solubility in the liquid.

Constituent i has the same chemical potential μ_i in both phases. Combining the relations $(\partial \mu_i / \partial p)_{T,n} = V_i(\text{l})$ for the liquid (Eq. 7.3.6) and $\mu_i = \mu_i^\circ(\text{g}) + RT \ln(f_i/p^\circ)$ for the gas (Eq. 7.4.10), we obtain

$$\frac{\mathrm{d}\ln(f_i/p^\circ)}{\mathrm{d}p} = \frac{V_i(\text{l})}{RT} \tag{9.7.1}$$

(equilibrated liquid and gas mixtures, constant T and liquid composition)

This equation shows that an increase in pressure, at constant temperature and liquid composition, causes an increase in fugacity.

Integration of Eq. 9.7.1 between pressures p_1 and p_2 yields

$$f_i(p_2) = f_i(p_1) \exp\left[\int_{p_1}^{p_2} \frac{V_i(\text{l})}{RT} \, \mathrm{d}p\right] \tag{9.7.2}$$

(equilibrated liquid and gas mixtures, constant T and liquid composition)

The exponential factor on the right side is called the **Poynting factor**.

The integral in the Poynting factor is simplified if we treat $V_i(\text{l})$ as independent of pressure. Then we obtain the approximate relation

$$f_i(p_2) \approx f_i(p_1) \exp\left[\frac{V_i(l)(p_2 - p_1)}{RT}\right] \qquad (9.7.3)$$

(equilibrated liquid and gas mixtures, constant T and liquid composition)

For typical values of the partial molar volume $V_i(l)$, the exponential factor is close to unity unless $p_2 - p_1$ is very large. For instance, setting $V_i(l)$ equal to $100 \, \text{cm}^3 \, \text{mol}^{-1}$ and T equal to $300 \, \text{K}$, we obtain a value for $f_i(p_2)/f_i(p_1)$ of 1.008 if $p_2 - p_1$ is 2 bar, 1.04 if $p_2 - p_1$ is 10 bar, and 1.5 if $p_2 - p_1$ is 100 bar. Thus, unless the pressure change is large, we can to a good approximation neglect the effect of total pressure on fugacity. (This statement applies only to the fugacity of a substance in a gas phase that is equilibrated with a liquid phase containing the same substance. If the liquid phase is absent, the fugacity of course is approximately directly proportional to the total pressure of the gas.)

We can apply Eqs. 9.7.2 and 9.7.3 to *pure* liquid A, in which case $V_i(l)$ is the molar volume $V_A^*(l)$. The partial pressure p_A in the equilibrated gas phase is then the vapor pressure of pure liquid A. When the equilibrated gas phase is also pure A, p_A is the saturation pressure (Sec. 6.2.2). It is usually a good approximation to treat the gas phase as ideal so that p_A equals the fugacity f_A, which as we have seen depends on the total pressure. Then if the gas phase contains additional components, and provided the total pressure is not much greater than the saturation pressure, the vapor pressure is only slightly greater than the saturation pressure.

As an application of these relations, consider the effect of the size of a liquid droplet on the equilibrium vapor pressure. The calculation of Prob. 9.9(b) shows that the fugacity of H_2O in a gas phase equilibrated with liquid water in a small droplet is slightly greater than when the liquid is in a bulk phase. The smaller the radius of the droplet, the greater is the fugacity and the vapor pressure. Thus, small liquid droplets are unstable relative to the same amount of liquid in larger droplets or a bulk phase.

9.7.2 Effect of liquid composition on gas fugacities

Consider the experimental system shown in Fig. 7.4(a) on page 191. A binary liquid mixture of two volatile constituents A and B is equilibrated with a gas mixture containing A, B, and a third gaseous component C used to control the total pressure. The conditions of transfer equilibrium are the equality of μ_A in both phases and of μ_B in both phases:

$$\mu_A(l) = \mu_A^\circ(g) + RT \ln \frac{f_A}{p^\circ} \qquad \mu_B(l) = \mu_B^\circ(g) + RT \ln \frac{f_B}{p^\circ}$$

Suppose we make an infinitesimal change in the liquid composition at constant T and p. This causes infinitesimal changes in the chemical potentials and fugacities:

$$d\mu_A = RT \frac{df_A}{f_A} \qquad d\mu_B = RT \frac{df_B}{f_B}$$

By inserting these expressions in the Gibbs–Duhem equation $x_A\,d\mu_A = -x_B\,d\mu_B$ (Eq. 7.5.19), we obtain

$$\frac{x_A}{f_A}df_A = -\frac{x_B}{f_B}df_B \qquad (9.7.4)$$
(binary liquid mixture equilibrated with gas, constant T and p)

This equation is a relation between changes in gas-phase fugacities caused by a change in the liquid-phase composition. It shows that a composition change at constant T and p that *increases* the fugacity of A in the equilibrated gas must *decrease* the fugacity of B.

In any binary liquid mixture of nonelectrolytes, at compositions close to $x_B = 0$ and at constant T and p, the fugacity of B in the equilibrated gas phase is directly proportional to x_B (Henry's law):

$$f_B = kx_B$$

If we are treating B as the solute of a solution, this relation holds in the ideal-dilute region and the proportionality constant k is, of course, the Henry's law constant $K_{x,B}$ (Eq. 7.5.10). If the relation happens to hold at all compositions, it is a statement of Raoult's law for fugacity (Eq. 7.5.2), and k is equal to f_B^*. For composition changes within the composition range in which the relation holds, we can write

$$\frac{df_B}{dx_B} = k = \frac{f_B}{x_B}$$

With the substitution $dx_B = -dx_A$ and rearrangement, this becomes

$$-\frac{x_B}{f_B}df_B = dx_A$$

Combined with Eq. 9.7.4, this is $(x_A/f_A)\,df_A = dx_A$, which we can rearrange and integrate

$$\int_{f_A^*}^{f_A}\frac{df_A}{f_A} = \int_1^{x_A}\frac{dx_A}{x_A}$$

with the result $\ln(f_A/f_A^*) = \ln x_A$, or

$$f_A = x_A f_A^*$$

This relation between f_A and x_A is valid at any composition in the range in which f_B is proportional to x_B; f_A^* is the fugacity of A in the gas phase equilibrated with pure liquid A at the same T and p. The result is Raoult's law for fugacity (Eq. 7.5.2 on page 191). We can draw two conclusions:

1. In the ideal-dilute region of a binary solution, where the solute obeys Henry's law, the solvent must obey Raoult's law. (A similar result was derived in Sec. 7.5.6 for a solution with any number of solutes.)

Section 9.7 Liquid–Gas Equilibria

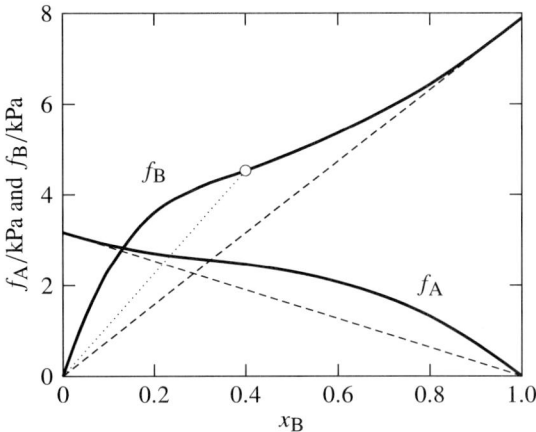

Figure 9.9 Fugacities in a gas phase equilibrated with a binary liquid mixture of H_2O (A) and ethanol (B) at 25 °C and 1 bar. The dashed lines show Raoult's law behavior.

2. If one constituent of a binary liquid mixture obeys Raoult's law at all compositions, so also must the other constituent. This is the definition of a binary *ideal* liquid mixture.

Suppose a binary liquid mixture is nonideal and Component B exhibits *positive* deviations from Raoult's law. An example is the water–ethanol system shown in Fig. 9.9. At each point on the curve of f_B versus x_B, the slope df_B/dx_B is less than the slope f_B/x_B of a line drawn from the origin to the point (see the dotted line in Fig. 9.9), except that the two slopes become equal at $x_B = 1$:

$$\frac{df_B}{dx_B} \leq \frac{f_B}{x_B}$$

As we can see from the figure, this relation must apply to any component whose fugacity curve exhibits a positive deviation from Raoult's law throughout the composition range (except in the limit $x_B \to 1$) and that has only one inflection point. With the substitution $dx_B = -dx_A$, the relation becomes

$$-\frac{df_B}{dx_A} \leq \frac{f_B}{x_B} \quad \text{or} \quad -\frac{x_B}{f_B} df_B \leq dx_A$$

Then Eq. 9.7.4 becomes

$$\frac{x_A}{f_A} df_A \leq dx_A \quad \text{or} \quad \frac{df_A}{f_A} \leq \frac{dx_A}{x_A}$$

The inequality continues to hold if we integrate in the direction of increasing x_A:

$$\int_{f_A}^{f_A^*} \frac{df_A}{f_A} \leq \int_{x_A}^{1} \frac{dx_A}{x_A} \qquad \ln \frac{f_A^*}{f_A} \leq -\ln x_A$$

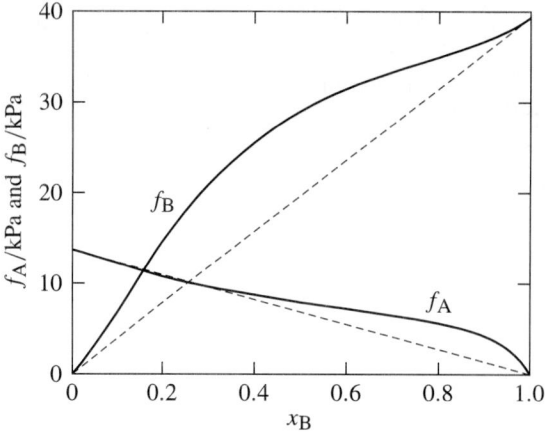

Figure 9.10 Fugacities in a gas phase equilibrated with a binary liquid mixture of chloroform (A) and ethanol (B) at 35 °C (Scatchard, G.; Raymond, C. L. *J. Am. Chem. Soc.* **1938**, *60*, 1278–1287).

Taking the exponential of both sides and solving for f_A gives

$$f_A \geq x_A f_A^*$$

Thus, *if the curve of fugacity versus mole fraction for one constituent of a binary liquid mixture exhibits only positive deviations from Raoult's law, with only one inflection point, so also must the curve of the other constituent.* In the water–ethanol system shown in Fig. 9.9, both curves have positive deviations from Raoult's law and both have a single inflection point. By the same method, we find that if the fugacity curve of one constituent has only *negative* deviations from Raoult's law with a single inflection point, the same is true of the other constituent.

Figure 9.10 shows the case of a binary mixture in which one component (B) has only positive deviations from Raoult's law but the other component (A) has both positive and negative deviations (f_A is slightly smaller than $x_A f_A^*$ for x_B less than 0.3). This unusual behavior is possible because both fugacity curves have two inflection points instead of the usual one. Other types of unusual nonideal behavior are possible.[16]

9.7.3 The Duhem–Margules equation

If we divide both sides of Eq. 9.7.4 by dx_A, we obtain

$$\frac{x_A}{f_A}\frac{df_A}{dx_A} = -\frac{x_B}{f_B}\frac{df_B}{dx_A} \qquad (9.7.5)$$

(binary liquid mixture equilibrated with gas, constant T and p)

---

[16]McGlashan, M. L. *J. Chem. Educ.* **1963**, *40*, 516–518.

This is the **Duhem–Margules equation**.

To a good approximation, by neglecting the effect of total pressure on fugacity (Sec. 9.7.1), we can apply the Duhem–Margules equation to a liquid–gas system in which the total pressure is *not* constant. Consider the case of a binary liquid–gas system in which the third component, C, is omitted, and the total pressure is the sum of p_A and p_B. Assume the gas mixture behaves ideally, so the fugacities are the same as the partial pressures. With the help of Eq. 9.7.5, we find the rate at which the total pressure changes with the liquid composition, at constant T, as follows:

$$\frac{dp}{dx_A} = \frac{d(p_A + p_B)}{dx_A} = \frac{dp_A}{dx_A}\left(1 - \frac{x_A/x_B}{p_A/p_B}\right)$$

$$= \frac{dp_A}{dx_A}\left(1 - \frac{x_A/x_B}{y_A/y_B}\right)$$

Here y_A and y_B are the mole fractions of A and B in the gas phase given by $y_A = p_A/p$ and $y_B = p_B/p$.

We can make several deductions from the preceding relation concerning a binary liquid–gas system at constant T. If the ratio y_A/y_B is greater than x_A/x_B (meaning that the mole fraction of A is greater in the gas than in the liquid), then $(x_A/x_B)/(y_A/y_B)$ is less than 1 and dp/dx_A must have the same sign as dp_A/dx_A, which is positive. Conversely, if y_A/y_B is less than x_A/x_B (the mole fraction of B is greater in the gas than in the liquid), then dp/dx_A must be negative. Thus, *the gas phase is richer than the liquid in the component whose addition to the liquid at constant temperature causes the total pressure to increase.* This statement is a version of *Konowaloff's rule*.

In some isothermal binary systems, the total pressure exhibits a maximum or minimum at a particular liquid composition. At this composition, dp/dx_A is zero but dp_A/dx_A is positive, so that x_A/x_B must equal y_A/y_B and the liquid and gas phases have identical mole fraction compositions (but different densities). The liquid with this composition is called an *azeotrope*. The behavior of systems with azeotropes is discussed in Sec. 10.2.4.

9.7.4 Gas solubility

For the solution process $B(g) \rightarrow B(sln)$, the general expression for the thermodynamic equilibrium constant is $K = a_B(sln)/a_B(g)$.

The activity of B in the gas phase is given by $a_B(g) = f_B/p°$. If the solute is a nonelectrolyte and we choose a standard state based on mole fraction, the activity in the solution is $a_B(sln) = \Gamma_{x,B}\,\gamma_{x,B}\,x_B$. The equilibrium constant is then given by

$$K_x° = \frac{\Gamma_{x,B}\,\gamma_{x,B}\,x_B}{f_B/p°}$$

and the solubility expressed as the mole fraction of solute in the solution is given by

$$x_B = \frac{K_x°\,f_B/p°}{\Gamma_{x,B}\,\gamma_{x,B}} \tag{9.7.6}$$

(nonelectrolyte gas solubility)

The notation K_x° is a reminder that the equilibrium constant uses a mole fraction basis for the solute standard state.

Any change in the solution composition that increases the value of the activity coefficient $\gamma_{x,\mathrm{B}}$ will decrease the solubility for the same gas fugacity. This solubility decrease is often what happens when a salt is dissolved in an aqueous solution and is known as the *salting-out effect* (Prob. 9.11).

Since the Henry's law constant $K_{x,\mathrm{B}}$ is related to x_B by $K_{x,\mathrm{B}} = f_\mathrm{B}/\gamma_{x,\mathrm{B}} x_\mathrm{B}$ (see Eq. 7.6.17 on page 206), it is related as follows to the equilibrium constant of the solution process:

$$K_{x,\mathrm{B}} = \frac{\Gamma_{x,\mathrm{B}}\, p^\circ}{K_x^\circ} \tag{9.7.7}$$
(nonelectrolyte solute)

Unless the pressure is much greater than p°, we can with negligible error set the pressure factor $\Gamma_{x,\mathrm{B}}$ equal to 1. $K_{x,\mathrm{B}}$ is then independent of p, and by using Eq. 9.1.9 we obtain

$$\frac{d\ln(K_{x,\mathrm{B}}/p^\circ)}{dT} = -\frac{d\ln K_x^\circ}{dT} = -\frac{\Delta_\mathrm{sol} H^\circ_{m,\mathrm{B}}}{RT^2}$$

(At the standard pressure, $\Delta_\mathrm{sol} H^\circ_{m,\mathrm{B}}$ is the same as the molar enthalpy of solution at infinite dilution.) When the gas solubility is low and the solution contains no other solutes, the activity coefficient $\gamma_{x,\mathrm{B}}$ is close to 1. If furthermore we assume ideal gas behavior, then Eq. 9.7.6 becomes $x_\mathrm{B} = K_x^\circ p_\mathrm{B}/p^\circ$ and the temperature dependence of the solubility for a fixed partial pressure of the gas is

$$\left(\frac{\partial \ln x_\mathrm{B}}{\partial T}\right)_{p_\mathrm{B}} = \frac{d\ln K_x^\circ}{dT} = \frac{\Delta_\mathrm{sol} H^\circ_{m,\mathrm{B}}}{RT^2}$$

Note the similarity of this equation to ones derived previously for the temperature dependence of the solubilities of solids (Eq. 9.5.5) and liquids (Eq. 9.6.2). An equivalent expression from the substitution $d(1/T) = -(1/T^2)\,dT$ is

$$\left[\frac{\partial \ln x_\mathrm{B}}{\partial (1/T)}\right]_{p_\mathrm{B}} = -\frac{\Delta_\mathrm{sol} H^\circ_{m,\mathrm{B}}}{R}$$

We can use this form to evaluate $\Delta_\mathrm{sol} H^\circ_{m,\mathrm{B}}$ from a plot of $\ln x_\mathrm{B}$ versus $1/T$.

Since the dissolution of a gas in a liquid is invariably an exothermic process, $\Delta_\mathrm{sol} H^\circ_{m,\mathrm{B}}$ is negative and the solubility must decrease with increasing temperature.

The solubility of a gas that dissociates in solution has a quite different dependence on partial pressure. An example is the solubility of gaseous HCl in water to form an electrolyte solution, shown in Fig. 7.12 on page 218.

The **ideal solubility** of a gas is the solubility calculated on the assumption that the solute obeys Raoult's law for partial pressure (Eq. 7.5.1 on page 191): $p_\mathrm{B} = x_\mathrm{B} p_\mathrm{B}^*$. The ideal solubility, expressed as a mole fraction, is then given as a function of partial pressure by

$$x_\mathrm{B} = \frac{p_\mathrm{B}}{p_\mathrm{B}^*} \tag{9.7.8}$$
(ideal solubility of a gas)

Here p_B^* is the vapor pressure of pure liquid solute at the same temperature and total pressure as the solution. If the solution is at a pressure that is too low for the pure solute to exist as a liquid, we can replace p_B^* with the saturation pressure of liquid B because the effect of total pressure on the vapor pressure of a liquid is negligible (Sec. 9.7.1). If the temperature is greater than the critical temperature of the solute, we can estimate a hypothetical vapor pressure by extrapolating the liquid–vapor coexistence curve beyond the critical point.

We can make several predictions for systems with a fixed value of p_B that obey Eq. 9.7.8:

1. The ideal solubility, expressed as a mole fraction, is independent of the kind of solvent.

2. The solubility expressed as a concentration, c_B, is lower the greater is the molar volume of the solvent.

3. The more volatile is the pure liquid solute at a particular temperature (i.e., the greater is p_B^*), the lower is the solubility.

4. The solubility decreases with increasing temperature since p_B^* increases.

Of course, these predictions apply only to solutions that behave approximately as ideal liquid mixtures, but even for many nonideal mixtures the predictions are found to agree with experiment.

As an example of the general validity of Prediction 1, Hildebrand and Scott[17] list the following solubilities of gaseous Cl_2 at 1.01 bar and 0 °C in several dissimilar solvents: $x_B = 0.270$ in heptane, $x_B = 0.288$ in $SiCl_4$, and $x_B = 0.298$ in CCl_4. These values are similar to one another and close to the ideal value $p_B/p_B^* = 0.273$.

9.7.5 Effect of pressure on Henry's law constants

We can find how a Henry's law constant based on mole fraction varies with the total pressure, at a fixed temperature, by substituting the expression for the solute pressure factor found in Table 7.4 into Eq. 9.7.7:

$$K_{x,B} = \frac{\Gamma_{x,B} \, p°}{K_x°} = \frac{p°}{K_x°} \exp\left(\int_{p°}^{p} \frac{V_B^\infty}{RT} dp\right)$$

Here $K_x°$ is the thermodynamic equilibrium constant for the solution process with a solute standard state based on mole fraction. Since $p°$ and $K_x°$ are independent of p, the ratio of the Henry's law constant at two pressures and the same temperature is given by

$$\frac{K_{x,B}(p_2)}{K_{x,B}(p_1)} = \frac{\exp\left(\int_{p°}^{p_2} \frac{V_B^\infty}{RT} dp\right)}{\exp\left(\int_{p°}^{p_1} \frac{V_B^\infty}{RT} dp\right)} = \exp\left(\int_{p_1}^{p_2} \frac{V_B^\infty}{RT} dp\right)$$

---

[17]Chap. XV (reference in Appendix I).

We can apply the same method to a Henry's law constant based on molality or concentration using Eqs. 7.6.18 and 7.6.19 and expressions from Table 7.4. We obtain the same expression as above for the ratio $K_{m,B}(p_2)/K_{m,B}(p_1)$. The ratio for a concentration basis is slightly different:

$$\frac{K_{c,B}(p_2)}{K_{c,B}(p_1)} = \exp\left[\int_{p_1}^{p_2}\left(\frac{V_B^\infty}{RT} - \kappa_T\right)dp\right]$$

Approximate versions of these relations, treating V_B^∞ and κ_T as independent of pressure, are

$$\frac{K_{x,B}(p_2)}{K_{x,B}(p_1)} = \frac{K_{m,B}(p_2)}{K_{m,B}(p_1)} \approx \exp\left[\frac{V_B^\infty(p_2 - p_1)}{RT}\right] \qquad (9.7.9)$$

$$\frac{K_{c,B}(p_2)}{K_{c,B}(p_1)} \approx \exp\left[\left(\frac{V_B^\infty}{RT} - \kappa_T\right)(p_2 - p_1)\right]$$

Unless the total pressure is much greater than 1 bar, the effect of pressure is small; see Prob. 9.12 for an example.

9.8 REACTION EQUILIBRIA

We can write an expression for the thermodynamic equilibrium constant of any chemical reaction as shown in Eq. 8.8.3 on page 288:

$$K = \prod_i a_i^{\nu_i}$$

Each activity in this equation is based on a particular standard state for the individual reactant or product and can be replaced with an expression listed in the last column of Table 7.3 on page 215.

The value of K for a particular reaction is a function of temperature only. A change of pressure or other conditions at constant temperature cannot affect the value of K, but can change equilibrium conditions by changing the values of pressure factors and activity coefficients.

For example, consider the following heterogeneous equilibrium that is important in the formation of limestone caverns:

$$\text{CaCO}_3(\text{cr, calcite}) + \text{CO}_2(\text{g}) + \text{H}_2\text{O}(\text{l}) \rightleftharpoons \text{Ca}^{2+}(\text{aq}) + 2\text{HCO}_3^-(\text{aq}) \qquad (9.8.1)$$

If we base the standard states on pure substances for the CaCO$_3$ and H$_2$O, on the ideal gas for the CO$_2$, and on molality for the solutes, the thermodynamic equilibrium constant is written

$$K = \frac{a_+ a_-^2}{a_{\text{CaCO}_3} a_{\text{CO}_2} a_{\text{H}_2\text{O}}} = \Gamma_r \frac{\gamma_+ \gamma_-^2 m_+ m_-^2/(m°)^3}{(f_{\text{CO}_2}/p°)\,\gamma_{\text{H}_2\text{O}}\, x_{\text{H}_2\text{O}}}$$

where the subscripts $+$ and $-$ refer to the Ca^{2+} and HCO_3^- ions. Γ_r is the pressure-factor quotient (Sec. 8.8.4) given for this reaction by[18]

$$\Gamma_r = \frac{\Gamma_+ \Gamma_-^2}{\Gamma_{CaCO_3} \Gamma_{H_2O}}$$

and is usually equal to 1 to a sufficiently good approximation. This is an example of a "mixed" equilibrium constant—one using different kinds of standard states. From the definition of the mean ionic activity coefficient (Eq. 7.8.16), we can replace the product $\gamma_+ \gamma_-^2$ by $\gamma_\pm^3$, where $\gamma_\pm$ is the mean ionic activity coefficient of aqueous $Ca(HCO_3)_2$:

$$K = \Gamma_r \frac{\gamma_\pm^3 m_+ m_-^2/(m°)^3}{(f_{CO_2}/p°) \gamma_{H_2O} x_{H_2O}}$$

With the reaction at equilibrium at a given temperature and pressure, any change in the ionic strength that alters $\gamma_\pm$ necessarily causes a compensating change in the solute molarities or CO_2 fugacity to keep the reaction in equilibrium.

Another example is the dissociation (ionization) of a weak monoprotic acid such as acetic acid

$$HA(aq) \rightleftharpoons H^+(aq) + A^-(aq)$$

for which the thermodynamic equilibrium constant (the *acid dissociation constant*) is

$$K_a = \Gamma_r \frac{\gamma_+ \gamma_- m_+ m_-}{\gamma_{m,HA} \, m_{HA} m°} = \Gamma_r \frac{\gamma_\pm^2 m_+ m_-}{\gamma_{m,HA} \, m_{HA} m°}$$

Suppose the solution is prepared from water and the acid, and H^+ from the dissociation of H_2O is negligible compared to H^+ from the acid dissociation. We may then write $m_+ = m_- = \alpha m_B$, where α is the degree of dissociation and m_B is the overall molality of the acid. The molality of the undissociated acid is then $m_{HA} = (1 - \alpha) m_B$, and the dissociation constant can be written

$$K_a = \Gamma_r \frac{\gamma_\pm^2 \alpha^2 m_B/m°}{\gamma_{m,HA}(1 - \alpha)} \tag{9.8.2}$$

From this equation, we see that a change in the ionic strength that decreases $\gamma_\pm$ at a given molality m_B must increase the degree of dissociation (Prob. 9.15).

9.9 EVALUATION OF STANDARD MOLAR QUANTITIES

Some of the most useful experimentally derived data for thermodynamic calculations are the values for various reactions and other chemical processes of standard molar reaction

---

[18]The product $\Gamma_+ \Gamma_-^2$ in the numerator is the pressure factor $\Gamma_{m,B}$ for the solute $Ca(HCO_3)_2$ (see Eq. 7.8.20 on page 223). A pressure factor does not appear for CO_2 because it is unity for a gas.

enthalpies, standard molar reaction Gibbs energies, and standard molar reaction entropies. The values of these quantities for a given reaction are related, as we know, by

$$\Delta_r G_m^\circ = \Delta_r H_m^\circ - T \Delta_r S_m^\circ$$

and a standard molar reaction entropy can be calculated from the standard molar entropies of the reactants and products using Eq. 8.8.13:

$$\Delta_r S_m^\circ = \sum_i \nu_i S_{m,i}^\circ$$

The values can be generated through a variety of experimental techniques. Reaction calorimetry can be used to evaluate $\Delta_r H_m^\circ$ for a reaction (Sec. 8.5). Calorimetric measurements of heat capacity and phase-transition enthalpies can be used to obtain the value of S_m° for a solid or liquid (Sec. 4.11.1). For a gas, spectroscopic measurements can be used to evaluate S_m° using statistical mechanics (Sec. 4.11.2). Evaluation of a thermodynamic equilibrium constant and its temperature derivative can provide values of $\Delta_r G_m^\circ$ and $\Delta_r H_m^\circ$ through the relations $\Delta_r G_m^\circ = -RT \ln K$ (Eq. 8.8.4) and $\Delta_r H_m^\circ = -R \, d\ln K/d(1/T)$ (Eq. 9.1.10). The equilibrium constant can be for any of the kinds of equilibria discussed in this chapter: vapor pressure, solubility, chemical reaction, and so on.

> A so-called *third-law method* can also be used to evaluate $\Delta_r H_m^\circ$ for a reaction. This requires the value of the equilibrium constant at the temperature of interest and the values of the free-energy function $\Phi_{298.15}(T)$ for each reactant and product at any one temperature T. $\Phi_{298.15}(T)$ is defined by Eq. 8.8.16 on page 294, and its value for a substance is found calorimetrically. The equation needed for the third-law method is found by replacing $\Delta_r G_m^\circ(T)$ in Eq. 8.8.18 with $-RT \ln K$ and rearranging to $\Delta_r H_m^\circ(298.15 \text{ K}) = -RT \ln K - T \sum_i \nu_i \Phi_{298.15,i}(T)$.

In addition to these methods, measurements of cell potentials are useful for reactions that can be carried out reversibly in galvanic cells. Section 11.3 describes how the standard cell potential and its temperature derivative allow $\Delta_r H_m^\circ$, $\Delta_r G_m^\circ$, and $\Delta_r S_m^\circ$ to be evaluated for such a reaction.

An efficient way of tabulating the results of experimental measurements, and of using them to generate the values of standard molar reaction quantities for reactions not investigated directly, is in the form of standard molar enthalpies and Gibbs energies of *formation*. The relations between the reaction and formation quantities (Sec. 8.3.1) are

$$\Delta_r H_m^\circ = \sum_i \nu_i \Delta_f H_{m,i}^\circ \qquad \Delta_r G_m^\circ = \sum_i \nu_i \Delta_f G_{m,i}^\circ$$

and for ions the conventions used are

$$\Delta_f H_m^\circ(\text{H}^+, \text{aq}) = 0 \qquad \Delta_f G_m^\circ(\text{H}^+, \text{aq}) = 0 \qquad S_m^\circ(\text{H}^+, \text{aq}) = 0$$

An abbreviated set of values of $\Delta_f H_m^\circ$, S_m°, and $\Delta_f G_m^\circ$, for the single temperature of 298.15 K (25.00 °C), is printed in Appendix H.[19]

---

[19] Most of the values in this table come from a project of the Committee on Data for Science and Technology

Examples of the evaluation of standard molar reaction quantities and standard molar formation quantities from measurements by various experimental techniques are found in Probs. 9.16–9.18, 11.2, and 11.3.

PROBLEMS

9.1 Consider the heterogeneous equilibrium $CaCO_3(s) \rightleftharpoons CaO(s) + CO_2(g)$. Table 9.2 lists pressures measured over a range of temperatures for this system.

Table 9.2 Pressure of an equilibrium system containing $CaCO_3(s)$, $CaO(s)$, and $CO_2(g)$ (Symth, F. H.; Adams, L. H. *J. Am. Chem. Soc.* **1923**, *45*, 1167–1184).

$t/°C$	$p/$Torr	$t/°C$	$p/$Torr
842.3	343.0	904.3	879.0
852.9	398.6	906.5	875.0
854.5	404.1	937.0	1350
868.9	510.9	937.0	1340

(a) What is the approximate relation between p and K?

(b) Plot these data in the form $\ln K$ versus $1/T$ or fit $\ln K$ to a linear function of $1/T$. Then evaluate the temperature at which the partial pressure of the CO_2 is 1 bar and the standard molar reaction enthalpy at this temperature.

9.2 For a homogeneous reaction among solutes in a solution, derive a rigorous relation between the standard molar reaction enthalpy and the temperature dependence of the thermodynamic equilibrium constant K_c°.

9.3 Show that the standard molar reaction entropy of a reaction can be calculated from the thermodynamic equilibrium constant and its temperature derivative, provided that no solute states are based on concentration, with the relation

$$\Delta_r S_m^\circ = R \ln K + RT \frac{d \ln K}{dT}$$

9.4 Use the data in Table 9.3 on the next page to evaluate the molal freezing-point depression constant and the molal boiling-point elevation constant for H_2O at a pressure of 1 bar.

9.5 An aqueous solution of the protein bovine serum albumin, containing 2.00×10^{-2} g of protein per cubic centimeter, has an osmotic pressure of 8.1×10^{-3} bar at $0\,°C$. Estimate the molar mass of this protein.

(CODATA) to establish a set of recommended, internally consistent values of thermodynamic properties. CODATA was established in 1966 as an interdisciplinary committee of the International Council of Scientific Unions with the aim of improving the reliability of data used in science and technology.

Table 9.3 Properties of H_2O at 1 bar

M	t_f	t_b	$\Delta_{fus}H_m$	$\Delta_{vap}H_m$
18.0153 g mol^{-1}	0.00 °C	99.61 °C	6.010 kJ mol^{-1}	40.668 kJ^{-1}

9.6 Figure 9.8 on page 333 shows a curve fitted to experimental points for the aqueous solubility of n-butylbenzene. The curve has the equation $\ln x_B = a(t/°C - b)^2 + c$, where the constants have the values $a = 3.34 \times 10^{-4}$, $b = 12.13$, and $c = -13.25$. Assume that the saturated solution behaves as an ideal-dilute solution, use the solute standard state based on mole fraction, and calculate $\Delta_{sol}H_{m,B}^\circ$ and $\Delta_{sol}S_{m,B}^\circ$ at 5.00 °C, 12.13 °C (the temperature of minimum solubility), and 25.00 °C.

9.7 In the resting state of a mammalian nerve cell at 37 °C, there is a membrane potential of 0.070 V across the cell membrane, the inside of the cell having negative polarity. The membrane is permeable to K^+ ion. Assume that the ion is in transfer equilibrium across the membrane, with equal activity coefficients on both sides, and find the ratio $m_+^{\text{inside}}/m_+^{\text{outside}}$. (The actual ratio is about 35; the resting cell is in a steady state rather than an equilibrium state.)

9.8 Consider a hypothetical system in which two aqueous solutions are separated by a membrane. Phase α is prepared by dissolving 1.00×10^{-5} mol KCl in 10.0 g water, and phase β is prepared from 1.00×10^{-5} mol KCl and 2.00×10^{-8} mol of the potassium salt of a polyelectrolyte dissolved in 10.0 g water. Each polyelectrolyte ion has a charge of -500. The membrane is permeable to the water molecules and to the K^+ and Cl^- ions but not to the polyelectrolyte. The system comes to equilibrium at 25.00 °C. Assume that a negligible amount of water is transferred across the boundary before osmotic membrane equilibrium is attained; thus, the sums $m_+^\alpha + m_+^\beta$ and $m_-^\alpha + m_-^\beta$ for the K^+ and Cl^- ions do not change as ions are transferred across the membrane. Also make the drastic approximation that both solutions behave as ideal-dilute solutions.

(a) Find the equilibrium molality of each solute species in the two phases.

(b) Describe any macroscopic transfer of ions across the membrane required to establish the equilibrium state.

(c) Calculate the Donnan potential, $\phi^\alpha - \phi^\beta$.

(d) Estimate the pressure difference across the membrane at equilibrium. (The density of liquid H_2O at 25.00 °C is 0.997 g cm^{-3}.)

(e) Check the assumption made in the derivation of Eq. 9.6.5 that $V_i^\beta(p^\alpha - p^\beta)$ is much smaller in magnitude than $z_i F(\phi^\alpha - \phi^\beta)$. Although it is not possible to measure the partial molar volume of a single ion species, it is safe to assume that neither K^+ nor Cl^- has a partial molar volume greater than 100 cm^3 mol^{-1}.

9.9 The derivation of Prob. 7.3 on page 233 shows that the pressure in a liquid droplet of radius r is greater than the pressure of the surrounding equilibrated gas by a quantity $2\gamma/r$, where γ is the surface tension.

(a) Consider a droplet of water of radius 1.00×10^{-6} m at 25 °C suspended in air of the same temperature. The surface tension of water at this temperature is $0.07197\,\mathrm{J\,m^{-2}}$. Find the pressure in the droplet if the pressure of the surrounding air is 1.00 bar.

(b) Calculate the difference between the fugacity of H_2O in the air of pressure 1.00 bar equilibrated with this water droplet and the fugacity in air equilibrated at the same temperature and pressure with a pool of liquid water having a flat surface. Liquid water at 25 °C and 1 bar has a vapor pressure of 0.032 bar and a molar volume of $1.807 \times 10^{-5}\,\mathrm{m^3\,mol^{-1}}$.

9.10 R. Crovetto (*J. Phys. Chem. Ref. Data* **1991**, *20*, 575–589) reviewed the published data for the solubility of gaseous CO_2 in water and fitted the Henry's law constant $K_{x,B}$ to a function of temperature. Her recommended values of $K_{x,B}$ at five temperatures are 1233 bar at 15.00 °C, 1433 bar at 20.00 °C, 1648 bar at 25.00 °C, 1874 bar at 30.00 °C, and 2111 bar at 35 °C.

(a) The partial pressure of CO_2 in the atmosphere is typically about 3×10^{-4} bar. Assume a fugacity of 3.0×10^{-4} bar and calculate the aqueous solubility at 25.00 °C expressed both as a mole fraction and as a molality.

(b) Find the standard molar enthalpy of solution at 25.00 °C.

(c) Dissolved carbon dioxide exists mostly in the form of CO_2 molecules, but a small fraction exists as H_2CO_3 molecules, and there is also some ionization:

$$CO_2(aq) + H_2O(l) \rightleftharpoons H^+(aq) + HCO_3^-(aq)$$

(The equilibrium constant of this reaction is often called the first ionization constant of carbonic acid.) Combine the $K_{x,B}$ data with data in Appendix H to evaluate K and $\Delta_r H_m^\circ$ for the ionization reaction at 25.00 °C. Keep in mind that the solute standard states used for the values in Appendix H are based on molality.

9.11 The solubility of gaseous O_2 at a partial pressure of 1.01 bar and a temperature of 310.2 K, expressed as a concentration, is $1.07 \times 10^{-3}\,\mathrm{mol\,dm^{-3}}$ in pure water and $4.68 \times 10^{-4}\,\mathrm{mol\,dm^{-3}}$ in a 3.0 M aqueous solution of KCl. This solubility decrease is the *salting-out effect*. Calculate the activity coefficient $\gamma_{c,B}$ of O_2 in the KCl solution.

9.12 Use Eq. 9.7.9 to find the percent change in the value of the Henry's law constant $K_{x,B}$ for aqueous CO_2 at 298.15 K when the total pressure is changed from the lowest possible value of 0.032 bar (the saturation pressure of water at this temperature) to 2.000 bar. The value of $K_{x,B}$ at 0.032 bar is 1648 bar, and the partial molar volume of $CO_2(aq)$ is $33\,\mathrm{cm^3\,mol^{-1}}$.

9.13 Liquid water and liquid benzene have very small mutual solubilities. Equilibria in the binary water–benzene system were investigated by Tucker et al. (Tucker, E. E.; Lane, E. H.; Christian, S. D. *J. Solution Chem.* **1981**, *10*, 1–20) as follows. A known amount of distilled water was admitted to an evacuated, thermostatted vessel. Part of the water vaporized to form a vapor phase. Small, precisely measured volumes of liquid benzene were then added incrementally from the sample loop of a liquid-chromatography valve. The benzene distributed itself between the liquid and gaseous phases in the vessel. After each addition, the pressure was read with a precision pressure gauge. From the known amounts of water and benzene and the total pressure, the liquid composition and the partial pressure of the benzene were calculated. The fugacity of the benzene in the vapor phase was calculated from its partial pressure and the second virial coefficient.

At a fixed temperature and for mole fractions x_B of benzene in the liquid phase up to about 3×10^{-4} (less than the solubility of benzene in water), the fugacity of the benzene in the equilibrated gas phase was found to have the following dependence on x_B:

$$\frac{f_B}{x_B} = K_{x,B} - A x_B$$

Here $K_{x,B}$ is the Henry's law constant and A is a constant related to deviations from Henry's law. At 30 °C, the measured values were $K_{x,B} = 385.5$ bar and $A = 2.24 \times 10^4$ bar.

(a) Treat benzene (B) as the solute and find its activity coefficient on a mole fraction basis, $\gamma_{x,B}$, at 30 °C in the solution of composition $x_B = 3.00 \times 10^{-4}$.

(b) The fugacity of benzene vapor in equilibrium with pure liquid benzene at 30 °C is $f_B^* = 0.1576$ bar. Estimate the mole fraction solubility of liquid benzene in water at this temperature.

(c) The calculation of $\gamma_{x,B}$ in Part (a) treated the benzene as a single solute species with deviations from infinite-dilution behavior. Tucker et al. suggested a dimerization model to explain the observed negative deviations from Henry's law. (Classical thermodynamics, of course, cannot prove such a molecular interpretation of observed macroscopic behavior.) The model assumes that there are two solute species, a monomer (M) and a dimer (D), in reaction equilibrium: $2M \rightleftharpoons D$. Let n_B be the total amount of C_6H_6 present in solution and define the mole fractions

$$x_B \equiv \frac{n_B}{n_A + n_B} \approx \frac{n_B}{n_A}$$

$$x_M \equiv \frac{n_M}{n_A + n_M + n_D} \approx \frac{n_M}{n_A} \qquad x_D \equiv \frac{n_D}{n_A + n_M + n_D} \approx \frac{n_D}{n_A}$$

where the approximations are for dilute solution. In the model, the individual monomer and dimer particles behave as solutes in an ideal-dilute solution, with activity coefficients equal to unity. The monomer is in transfer equilibrium with the gas phase: $x_M = f_B/K_{x,B}$. The equilibrium constant expression (using a mole fraction basis for the solute standard states and setting pressure factors equal to 1) is $K = x_D/x_M^2$. From the relation $n_B = n_M + 2n_D$, and because the solution is very dilute, the expression becomes

$$K = \frac{x_B - x_M}{2 x_M^2}$$

Make individual calculations of K from the values of f_B measured at $x_B = 1.00 \times 10^{-4}$, $x_B = 2.00 \times 10^{-4}$, and $x_B = 3.00 \times 10^{-4}$. Extrapolate to $x_B = 0$ to eliminate nonideal effects such as higher aggregates. Finally, find the fraction of the benzene molecules present in the dimer form at $x_B = 3.00 \times 10^{-4}$ if this model is correct.

9.14 Use data in Appendix H to evaluate the thermodynamic equilibrium constant at 298.15 K for the reaction of Eq. 9.8.1 on page 346.

9.15 For the dissociation of formic acid, $HCO_2H(aq) \rightleftharpoons H^+(aq) + HCO_2^-(aq)$, the acid dissociation constant at 298.15 K has the value $K_a = 1.77 \times 10^{-4}$.

(a) Use Eq. 9.8.2 on page 347 to find the degree of dissociation and the hydrogen ion molality in a 0.01000 molal formic acid solution. You can safely set Γ_r and $\gamma_{m,HA}$ equal to 1, and use the Debye–Hückel limiting law (Eq. 7.8.25) to calculate $\gamma_\pm$. You can do this calculation by iteration: Start with an initial estimate of the ionic strength (in this case 0), calculate $\gamma_\pm$ and α, and repeat these steps until the value of α no longer changes.

(b) Estimate the degree of dissociation of formic acid in a solution that is 0.01000 molal in both formic acid and sodium nitrate again using the Debye–Hückel limiting law for $\gamma_\pm$. Compare with the value in Part (a).

9.16 Use the following experimental information to evaluate the standard molar enthalpy of formation and the standard molar entropy of the aqueous chloride ion at 298.15 K based on the convention $\Delta_f H_m^\circ(H^+, aq) = 0$ and $S_m^\circ(H^+, aq) = 0$ (Secs. 8.3.1 and 8.8.5). (The calculated values will be close to, but not exactly the same as, those listed in Appendix H, which are based on the same data combined with data of other workers.)

The standard molar enthalpy of reaction at 298.15 K for the reaction $\tfrac{1}{2}H_2(g) + \tfrac{1}{2}Cl_2(g) \rightarrow$ HCl(g) was measured in a flow calorimeter (Rossini, F. D. *J. Res. Natl. Bur. Stand. (U.S.)* **1932**, *9*, 679–702), with the result $\Delta_r H_m^\circ = -92.312\,\text{kJ mol}^{-1}$.

The standard molar entropy of gaseous HCl at 298.15 K, calculated from spectroscopic data using statistical mechanics, is $S_m^\circ = 186.902\,\text{J K}^{-1}\,\text{mol}^{-1}$.

From five calorimetric runs in the work cited in Prob. 8.6 on page 299, the average experimental value of the standard molar enthalpy of solution of gaseous HCl at 298.15 K is $\Delta_{sol} H_{m,B}^\circ = -74.84\,\text{kJ mol}^{-1}$.

From vapor pressure measurements of concentrated aqueous HCl solutions, the value of the ratio $f_B/a_{m,B}$ for gaseous HCl in equilibrium with aqueous HCl at 298.15 K is 5.032×10^{-7} bar (Randall, M.; Young, L. E. *J. Am. Chem. Soc.* **1928**, *50*, 989–1004).

9.17 The solubility of crystalline AgCl in ultrapure water has been determined from the electrical conductivity of the saturated solution (Gledhill, J. A.; Malan, G. M. *Trans. Faraday Soc.* **1952**, *48*, 258–262). The average of five measurements at 298.15 K is $s_B = 1.337 \times 10^{-5}\,\text{mol dm}^{-3}$. The density of water at this temperature is $\rho_A^* = 0.9970\,\text{kg dm}^{-3}$.

(a) From these data and the Debye–Hückel limiting law, calculate the solubility product K_s of AgCl at 298.15 K.

(b) Evaluate the standard molar Gibbs energy of formation of aqueous Ag^+ ion at 298.15 K, using the results of Part (a) and the values $\Delta_f G_m^\circ(Cl^-, aq) = -131.22\,\text{kJ mol}^{-1}$ and $\Delta_f G_m^\circ(AgCl, s) = -109.77\,\text{kJ mol}^{-1}$ from Appendix H.

9.18 The following reaction was carried out in an adiabatic solution calorimeter:

$$AgNO_3(s) + KCl(aq, m_B = 0.101\,\text{mol kg}^{-1}) \rightarrow AgCl(s) + KNO_3(aq)$$

(Wagman, D. D.; Kilday, M. V. *J. Res. Natl. Bur. Stand. (U.S.)* **1973**, *77A*, 569–579). The reaction can be assumed to go to completion, and the amount of KCl was in slight excess, so the amount of AgCl formed was equal to the initial amount of AgNO$_3$. After correction for the enthalpies of diluting the solutes in the initial and final solutions to infinite dilution, the standard molar reaction enthalpy at 298.15 K was found to be $\Delta_r H_m^\circ = -43.042\,\text{kJ mol}^{-1}$. The same workers used solution calorimetry to obtain the molar enthalpy of solution at infinite dilution of crystalline AgNO$_3$ at 298.15 K: $\Delta_{sol} H_{m,B}^\infty = 22.727\,\text{kJ mol}^{-1}$.

(a) Show that the difference of these two values is the standard molar reaction enthalpy for the precipitation reaction

$$\text{Ag}^+(\text{aq}) + \text{Cl}^-(\text{aq}) \rightarrow \text{AgCl}(\text{s})$$

and evaluate this quantity.

(b) Evaluate the standard molar enthalpy of formation of aqueous Ag^+ ion at 298.15 K, using the results of Part (a) and the values $\Delta_f H_m^\circ(\text{Cl}^-, \text{aq}) = -167.08 \text{ kJ mol}^{-1}$ and $\Delta_f H_m^\circ(\text{AgCl}, \text{s}) = -127.01 \text{ kJ mol}^{-1}$ from Appendix H. (These values come from calculations similar to those in Probs. 9.16 and 11.3.) The calculated value will be close to, but not exactly the same as, the value listed in Appendix H, which is based on the same data combined with data of other workers.

Chapter 10

THE PHASE RULE AND PHASE DIAGRAMS FOR MULTICOMPONENT SYSTEMS

We encountered the Gibbs phase rule and phase diagrams in chap. 6 in connection with single-substance systems. As stated in that chapter, a phase diagram is a kind of two-dimensional map that shows which phase or phases are stable under a given set of conditions. The present chapter derives the full version of the Gibbs phase rule for multicomponent systems. It then discusses phase diagrams for some representative types of multicomponent systems, and shows how they are related to the phase rule and to equilibrium concepts developed in chaps. 8 and 9.

10.1 THE GIBBS PHASE RULE FOR MULTICOMPONENT SYSTEMS

In Sec. 6.1.7, the Gibbs phase rule for a pure substance was written $F = 3 - P$. In a system of more than one substance with the usual independent variables (ignoring surface effects and external fields), the phase rule may be written either in the form

$$F = 2 + C - P \qquad (10.1.1)$$

or

$$F = 2 + s - r - P \qquad (10.1.2)$$

where the symbols have the following meanings:

F = the number of degrees of freedom (or variance)

  = the maximum number of intensive variables that can be varied independently while the system remains in equilibrium;

C = the number of components

= the minimum number of substances (or fixed-composition mixtures of substances) that could be used to prepare each phase individually;

P = the number of different phases;

s = the number of different species;

r = the number of independent relations among intensive variables of individual phases other than relations needed for thermal, mechanical, and transfer equilibrium.

If we subdivide a phase, that does *not* change the number of phases P. That is, we treat noncontiguous regions of the system that have identical intensive properties as parts of the same phase.

10.1.1 Degrees of freedom

Consider a system in an equilibrium state. In this state, the system has one or more phases; each phase contains one or more species; and intensive properties such as T, p, and the mole fraction of a species in a phase have definite values. Starting with the system in this state, we can make changes that place the system in a new equilibrium state having the same kinds of phases and the same species, but different values of some of the intensive properties. The number of different independent intensive variables that we may change in this way is the **number of degrees of freedom** or **variance**, F, of the system.

Clearly, the system remains in equilibrium if we change the *amount* of a phase without changing its temperature, pressure, or composition. This, however, is the change of an extensive variable and is not counted as a degree of freedom.

The phase rule, in the form to be derived, applies to a system that continues to have complete thermal, mechanical, and transfer equilibrium as intensive variables change. This means there are no insulating walls, rigid partitions, or semipermeable or impermeable membranes separating the phases. Furthermore, every conceivable reaction among the species is either at reaction equilibrium or else frozen at a fixed advancement during the time period we observe the system.

The number of degrees of freedom is the maximum number of intensive properties of the equilibrium system we may independently vary, or fix at arbitrary values, without causing a change in the number and kinds of phases and species. We cannot, of course, change one of these properties to just any value whatever. We are able to vary the value only within a certain finite (sometimes quite narrow) range before a phase disappears or a new one appears.

The number of degrees of freedom is also the number of independent intensive variables needed to specify the equilibrium state in all necessary completeness, aside from the amount of each phase. In other words, when we specify values of F different independent intensive variables, then the values of all other intensive variables of the equilibrium state have definite values determined by the physical nature of the system.

Just as for a one-component system, we can use the terms *bivariant*, *univariant*, and *invariant* depending on the value of F (Sec. 6.1.7).

Section 10.1 The Gibbs Phase Rule for Multicomponent Systems

10.1.2 Species approach to the phase rule

This section derives an expression for the number of degrees of freedom, F, based on *species*. The following section derives one based on *components*. Both approaches yield equivalent versions of the phase rule.

Recall that a *species* is an entity, uncharged or charged, distinguished from other species by its chemical formula (Sec. 7.1.1). Thus, CO_2 and CO_3^{2-} are different species, but $CO_2(aq)$ and $CO_2(g)$ is the same species in different phases.

Consider an equilibrium system of P phases, each of which contains the same set of species. Let the number of different species be s. If we could make changes while the system remains in thermal and mechanical equilibrium, but not necessarily in transfer equilibrium, we could independently vary the temperature, pressure, and amount of each species in each phase; there would then be $2 + Ps$ independent variables.

However, transfer equilibrium requires each species to have the same chemical potential in each phase: $\mu_i^\beta = \mu_i^\alpha$, $\mu_i^\gamma = \mu_i^\alpha$, and so on. There are $P - 1$ independent relations like this for each species and a total of $s(P - 1)$ independent relations for all species. Each such independent relation introduces a constraint and reduces the number of independent variables by one. Accordingly, taking transfer equilibrium into account, the number of independent variables is $2 + Ps - s(P - 1) = 2 + s$.

We obtain the same result if a species present in one phase is totally excluded from another. For example, solvent molecules are not found in a pure solute crystal, undissociated molecules of strong acids such as HCl and HNO_3 do not exist in aqueous solution, and ions are usually absent from gas phases. For each such absent species, there is one fewer amount variable and one fewer relation for transfer equilibrium; on balance, the number of independent variables is still $2 + s$.

Next we consider the possibility that further independent relations among intensive variables exist in addition to those needed for thermal, mechanical, and transfer equilibrium.[1] If there are r of these additional relations, the total number of independent variables is reduced to $2 + s - r$. These relations may come from:

1. reaction equilibria,

2. the requirement of electroneutrality in a phase containing ions, or

3. initial conditions determined by the way the system is prepared.

In the case of a reaction equilibrium, the relation is $\Delta_r G_m = \sum_i \nu_i \mu_i = 0$ (Eq. 8.7.5) or equivalently the equation $K = \prod_i (a_i)^{\nu_i}$ for the thermodynamic equilibrium constant (Eq. 8.8.3). Thus, r is the number of independent reactions that are at equilibrium, plus 1 for each phase containing ions, plus the number of independent initial conditions. Several examples are given in Sec. 10.1.4.

There is an infinite variety of possible choices of the independent variables (both extensive and intensive) for the equilibrium system, but the total *number* of independent variables

---

[1] Relations such as $\sum_i p_i = p$ for a gas phase or $\sum_i x_i = 1$ for any phase in general have already been accounted for in the derivation by the specification of p and the amount of each species.

is fixed at $2 + s - r$. Keeping intensive properties fixed, we can always vary how much of each phase is present (e.g., its volume, mass, or amount) without destroying the equilibrium. Thus, at least P of the independent variables, one for each phase, must be extensive. It follows that the maximum number of independent *intensive* variables is the difference $(2 + s - r) - P$.

> It may be that initial conditions establish relations among the amounts of phases, as illustrated in Example 2 of Sec. 10.1.4. If present, these are relations among *extensive* variables that are not counted in r. Each such independent relation decreases the total number of independent variables without changing the number of independent *intensive* variables calculated from $(2 + s - r) - P$.

Since the maximum number of independent intensive variables is the number of degrees of freedom, our expression for F based on species is

$$F = 2 + s - r - P$$

10.1.3 Components approach to the phase rule

The second derivation of the phase rule uses the concept of **components**. There is a certain minimum number of substances (or mixtures of fixed composition) that in principle we could use to prepare each individual phase of an equilibrium state of the system. The hypothetical method by which we prepare a phase can include the addition or removal of one or more of the substances or fixed-composition mixtures, and also the conversion of some of the substances into others by means of a reaction that is at equilibrium in the actual system. This minimum number of substances is the number of components, C.

The system may contain more substances than the number of components because of the existence of reaction equilibria that produce some substances from others. When we allow a reaction to produce substances to prepare a phase, nothing must remain unused. For instance, we could not use solid $CaCO_3$ to prepare a pure phase of solid CaO by the reaction $CaCO_3(s) \to CaO(s) + CO_2(g)$ because gaseous CO_2 would be left over (unless we treat CO_2 as an additional component that we could remove).

In deriving the phase rule by the components approach, it is convenient to consider only intensive variables. Suppose we have a system of P phases in which each substance present is a component (there are no reactions) and each of the C components is present in each phase (the phases are mixtures). If we make changes to the system while it remains in thermal and mechanical equilibrium, but not necessarily in transfer equilibrium, we can independently vary the temperature and pressure, and for each phase we can independently vary the mole fraction of all but one of the substances (the value of the omitted mole fraction comes from the relation $\sum_i x_i = 1$). This is a total of $2 + P(C - 1)$ independent intensive variables.

When there also exist transfer and reaction equilibria, these variables are no longer all independent. Each substance in the system is either a component or can be formed from components by a reaction that is in equilibrium. Transfer equilibria establish $P - 1$ independent relations for each component ($\mu_i^\beta = \mu_i^\alpha$, $\mu_i^\gamma = \mu_i^\alpha$, etc.) and a total of $C(P - 1)$

Section 10.1 The Gibbs Phase Rule for Multicomponent Systems

relations for all components. Since these are relations among chemical potentials, which are intensive properties, each relation reduces the number of independent intensive variables by one. The resulting number of independent intensive variables is

$$F = [2 + P(C - 1)] - C(P - 1) = 2 + C - P$$

If an equilibrium system lacks a particular component in one phase, there is one fewer mole fraction variable and one fewer relation for transfer equilibrium. These changes cancel in the calculation of F, which is still equal to $2 + C - P$. If a phase contains a substance that is formed from components by a reaction, there is an additional mole fraction variable and also the additional relation $\sum_i \nu_i \mu_i = 0$ for the reaction; again the changes cancel.

> We may need to *remove* a component from a phase to achieve the final composition. Note that it is not necessary to consider additional relations for electroneutrality or initial conditions; they are implicit in the definitions of the components. For instance, since each component is a substance of zero electric charge, the electrical neutrality of the phase is assured.

We conclude that, regardless of the kind of system, the expression for F based on components is given by Eq. 10.1.1. Comparison of Eqs. 10.1.1 and 10.1.2 shows the number of components is related to the number of species by

$$C = s - r \qquad (10.1.3)$$

10.1.4 Examples

Five examples are discussed here to illustrate various aspects of using the phase rule.

Example 1: liquid water

For a single phase of pure water, P equals 1. If we treat the water as the single species H_2O, s is 1 and r is 0. The phase rule then predicts two degrees of freedom:

$$F = 2 + s - r - P = 2 + 1 - 0 - 1 = 2$$

Since F is the number of intensive variables that can be varied independently, we could for instance vary T and p independently, or T and ρ, or any other pair of independent intensive variables.

Next let us take into account the proton transfer reaction

$$2\,H_2O(l) \;\rightleftharpoons\; H_3O^+(aq) + OH^-(aq)$$

and consider the system to contain the three species H_2O, H_3O^+, and OH^-. Then for the species approach to the phase rule, we use $s = 3$. We can write two independent relations:

1. for reaction equilibrium, $-2\mu_{H_2O} + \mu_{H_3O^+} + \mu_{OH^-} = 0$ (or $K = a_{H_3O^+} a_{OH^-} / a_{H_2O}^2$);
2. for electroneutrality, $m_{H_3O^+} = m_{OH^-}$.

Thus, we have two relations involving intensive variables only, and we set r equal to 2. The number of degrees of freedom is given by

$$F = 2 + 3 - 2 - 1 = 2$$

which is the same value of F as before.

If we consider water to contain additional cation species (e.g., $H_5O_2^+$), each such species would add 1 to s and 1 to r, but F would remain equal to 2. Thus, no matter how complicated are the equilibria that actually exist in liquid water, the number of degrees of freedom remains 2.

Applying the components approach to water is simple. All species that may exist in pure water are formed, in whatever proportions actually exist, from the single substance H_2O. Thus, there is only one component: $C = 1$. Equation 10.1.1 gives the same result as the species approach, $F = 2$.

Example 2: carbon, oxygen, and carbon oxides

Consider a system containing solid carbon and a gaseous mixture of O_2, CO, and CO_2. There are four species and two phases. If there are no reaction equilibria, as might be the case at low temperature in the absence of a catalyst, we have $r = 0$ and $C = s - r = 4$. The four components are the four substances. The phase rule tells us the system has four degrees of freedom. We could, for instance, arbitrarily vary T, p, x_{O_2}, and x_{CO}.

Now suppose we raise the temperature or introduce an appropriate catalyst to allow the following reactions to occur:

1. $2\,C(s) + O_2(g) \rightleftharpoons 2\,CO(g)$
2. $C(s) + O_2(g) \rightleftharpoons CO_2(g)$

Equilibrium with respect to these reactions introduces two new independent relations among chemical potentials or activities. The additional reaction $2\,CO(g) + O_2(g) \rightleftharpoons 2\,CO_2(g)$ does *not* add a third independent relation because this reaction is not independent of the other two—note that the equation is obtained by subtracting the chemical equation for Reaction 1 from twice the chemical equation for Reaction 2. Following the species approach, we have $s = 4$ and $r = 2$; the number of degrees of freedom from these values is

$$F = 2 + 4 - 2 - 2 = 2$$

To calculate F by the components approach, we decide on the minimum number of substances that we could use to prepare each phase separately. (This does not refer to how we actually prepare the two-phase system, but to a hypothetical preparation of each phase by itself with any of the compositions the phase is allowed to have in the actual equilibrium system.) Assume Reactions 1 and 2 are at equilibrium. We must prepare the solid phase with carbon, and we could prepare any possible equilibrium composition of the gas phase from carbon and O_2 by using both reactions. Thus, there are two components (C and O_2) giving the same result of two degrees of freedom.

What is the significance of there being two degrees of freedom when the reactions are at equilibrium? There are two ways of viewing the situation:

1. We can arbitrarily vary the two intensive variables T and p; when we do, the mole fractions of the three substances in the gas phase change in a way determined by the reaction equilibria.

2. If we specify arbitrary values of T and p, each of the mole fractions has only one possible value that will allow the two phases and four substances to be in equilibrium.

Now to illustrate an additional complexity, suppose we prepare the system by introducing half the amount of O_2 as the amount of carbon into an evacuated container and wait for reactions 1 and 2 to come to equilibrium. This method of preparation imposes an initial condition on the system, and we must decide whether the number of degrees of freedom is affected. Equating the total amount of C to the total amount of O in the equilibrated system gives the relation

$$n_C + n_{CO} + n_{CO_2} = 2n_{O_2} + n_{CO} + 2n_{CO_2} \quad \text{or} \quad n_C = 2n_{O_2} + n_{CO_2}$$

Either equation is a relation among extensive variables of two phases. From them, we are unable to obtain any relation among *intensive* variables of the two phases. Therefore, this particular initial condition does *not* change the value of r in Eq. 10.1.2, and F remains equal to 2.

Example 3: a solid salt and saturated aqueous solution

In this example, the equilibrium system consists of crystalline $PbCl_2$ and an aqueous phase containing the three species H_2O, $Pb^{2+}(aq)$, and $Cl^-(aq)$.

Applying the components approach to this system is straightforward. The solid phase is prepared from $PbCl_2$ and the aqueous phase could be prepared by dissolving solid $PbCl_2$ in H_2O. Thus, there are two components and two phases:

$$F = 2 + 2 - 2 = 2$$

For the species approach, we note there are four species ($PbCl_2$, Pb^{2+}, and Cl^- are counted as separate species because they have different formulas). There are two independent relations among intensive variables:

1. equilibrium for the dissolution process, $-\mu_{PbCl_2} + \mu_{Pb^{2+}} + 2\mu_{Cl^-} = 0$ [or $K_s = a_{Pb^{2+}} a_{Cl^-}^2 / a_{PbCl_2}$];

2. electroneutrality of the aqueous phase, $2m_{Pb^{2+}} = m_{Cl^-}$.

The species version of the phase rule gives

$$F = 2 + 4 - 2 - 2 = 2$$

with the same result as the components approach.

Example 4: liquid water and water-saturated air

For simplicity, let "air" be a gaseous mixture of N_2 and O_2. The equilibrium system in this example has two phases: liquid water saturated with dissolved air and air saturated with gaseous H_2O. We use the symbol x_i to denote the mole fraction of i in the liquid phase and y_i to denote the mole fraction in the gas phase.

If there is no special relation among the total amounts of N_2 and O_2, there are three components and the phase rule gives

$$F = 2 + C - P = 2 + 3 - 2 = 3$$

Since there are three degrees of freedom, we could, for instance, specify arbitrary values[2] of T, p, and y_{N_2}; then the values of other intensive variables such as y_{H_2O} and x_{N_2} would have definite values.

Now suppose we prepare the system with water and dry air of a *fixed* composition. This imposes an initial condition expressed by an equation such as

$$\frac{(n^l_{N_2} + n^g_{N_2})}{(n^l_{O_2} + n^g_{O_2})} = a$$

where a is a constant equal to the mole ratio of N_2 and O_2 in the dry air. Since N_2 and O_2 are present in both phases, the equation does not lead to any relation between intensive variables; we conclude r is still zero and there are still three degrees of freedom. We can also reach this conclusion by considering the number of components: Since the mole ratio of N_2 and O_2 in the aqueous solution need not be the same as in the gas, the air does not behave like a single substance, and H_2O, N_2, and O_2 are all required to prepare each phase individually, giving $C = 3$ and $F = 3$. The fact that the compositions of both phases depend on the relative amounts of the phases is illustrated by Prob. 7.5.

Finally, let us assume that we prepare the system with air of fixed composition, as before, but consider the solubilities of N_2 and O_2 in water to be negligible. Then $n^l_{N_2}$ and $n^l_{O_2}$ are zero and the initial condition becomes $n^g_{N_2}/n^g_{O_2} = a$, or $y_{N_2} = ay_{O_2}$, which is a relation between intensive variables. In this case, the phase rule gives

$$F = 2 + s - r - P = 2 + 3 - 1 - 2 = 2$$

This reduction in the value of F is simply a consequence of our inability to detect air dissolved in the water. According to the components approach, we may prepare the liquid phase with H_2O and the gas phase with H_2O and a fixed-composition mixture that behaves as a single substance; thus, there are only two components.

Example 5: equilibrium between two solid phases and a gas phase

Consider the following reaction at equilibrium:

$$3\,CuO(s) + 2\,NH_3(g) \rightleftharpoons 3\,Cu(s) + 3\,H_2O(g) + N_2(g)$$

---

[2] Arbitrary, that is, within the limits that would allow the two phases to coexist.

There are five species and one relation (for reaction equilibrium), and thus *four* components. Since there are three phases, the number of degrees of freedom is three.

It is more difficult to use the components approach to find the number of components for this example. As components, we might choose CuO and Cu (from which we could prepare the solid phases) and also NH_3 and H_2O. Then to obtain the N_2 needed to prepare the gas phase, we could use CuO and NH_3 as reactants in the reaction and remove the other products of the reaction, Cu and H_2O. In the components approach, we are allowed to remove substances from the system provided they are components.

10.2 PHASE DIAGRAMS: BINARY SYSTEMS

10.2.1 Generalities

A binary system has two components; C equals 2, and the number of degrees of freedom is $F = 4 - P$. There must be at least one phase, so the maximum value of F is 3. Since F cannot be negative, there can be no more than four phases.

We can independently vary the temperature, pressure, and composition of the system as a whole. Instead of using these variables as the coordinates of a three-dimensional phase diagram, we usually draw a two-dimensional phase diagram that is either a temperature–composition diagram at a fixed pressure or a pressure–composition diagram at a fixed temperature. The position of the system point on one of these diagrams then corresponds to a definite temperature, pressure, and overall composition. The composition variable usually varies along the horizontal axis and can be the mole fraction, mass fraction, or mass percent of one of the components in the system as a whole, as illustrated by examples in the following sections.

If the system point falls within a *one-phase* area of the phase diagram, the composition variable refers to the composition of that one phase and is an intensive property of the phase. A binary system with a single phase has three degrees of freedom. On the phase diagram, one of the intensive variables has a fixed value so there are two other independent intensive variables. For example, on a temperature–composition phase diagram, the pressure is fixed and the temperature and composition coordinates of the system point can be changed independently within the boundaries of the one-phase area of the diagram.

A binary system with *two* phases has two degrees of freedom. Once we have selected the temperature and pressure, no degrees of freedom remain, and each phase has a fixed composition. If the system point is in a two-phase area of a phase diagram, we draw a horizontal *tie line* of constant temperature (on a temperature–composition phase diagram) or constant pressure (on a pressure–composition phase diagram). The position of the point at each end of the tie line, at the boundary of the two-phase area, gives the value of the composition variable of one of the phases and also the physical state of this phase: either the state of an adjacent one-phase area or the state of a phase of fixed composition when the boundary is a vertical line. Thus, a boundary that separates a two-phase area for phases α and β from a one-phase area for phase α is a curve that describes the composition of phase α as a function of T or p when it is in equilibrium with phase β. The curve is called a *solidus*, *liquidus*, or *vaporus* depending on whether phase α is a solid, liquid, or gas.

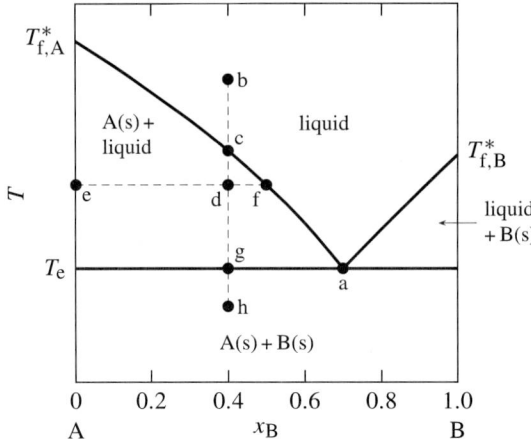

Figure 10.1 Temperature–composition phase diagram for a binary system exhibiting a eutectic point.

We can apply the lever rule described in Sec. 6.2.4 to the system point in a two-phase area. For example, if the composition variable is the mole fraction x_B, then the lever rule is $n^\alpha L^\alpha = n^\beta L^\beta$, where n^α and n^β are the total amounts in the two phases and L^α and L^β are the horizontal distances along the tie line from the system point to the ends of the tie line. The relative amounts are given by the ratio $n^\beta/n^\alpha = L^\alpha/L^\beta$. If the composition variable is a mass fraction or mass percent, amounts are replaced by masses in these equations. Thus, if we know the total amounts or masses of both components in the system as a whole, the position of the system point gives us the amounts or masses in the individual phases.

A binary system with *three* phases has only one degree of freedom and so cannot be represented by an area on the phase diagram. Instead, there is a horizontal boundary line between areas, with a special point at one position. The compositions of the three phases are given by the positions of this point and the points at the two of the line. The position of the system point on this line does not uniquely specify the relative amounts in the three phases. The examples that follow show some of the simpler kinds of phase diagrams known for binary systems.

10.2.2 Solid–liquid systems

Figure 10.1 is a temperature–composition phase diagram at a fixed pressure. The phases shown are a binary liquid mixture of A and B, pure solid A, and pure solid B. The one-phase liquid area is bounded by two curves, which we can think of either as freezing-point curves for the liquid or as solubility curves for the solids. These curves comprise the liquidus. They are drawn on the assumption that the liquid phase is an ideal liquid mixture (Eq. 9.5.2); as the mole fraction in the liquid of either component decreases from unity, the freezing point decreases. The curves meet at Point a, which is a **eutectic point**. At this point, both solid A and solid B can coexist in equilibrium with a binary liquid mixture. The composition at

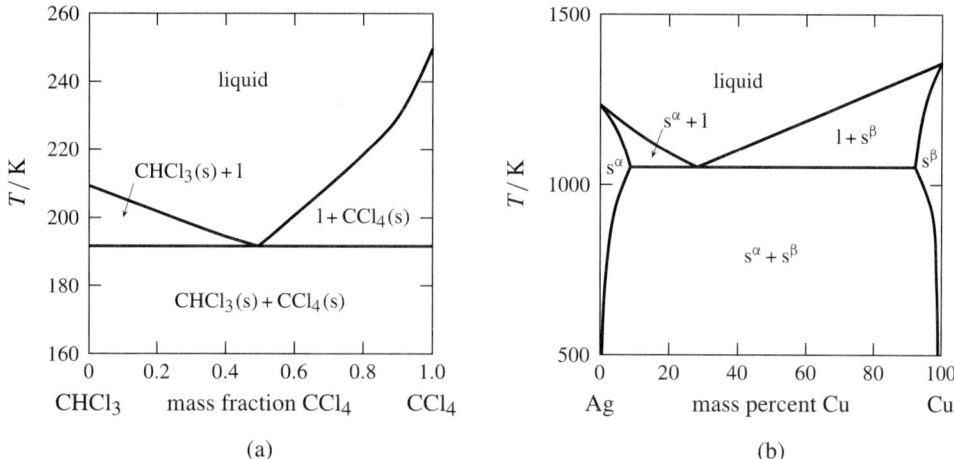

Figure 10.2 Temperature–composition phase diagrams with single eutectics: (a) two pure solids and a liquid mixture; (b) two solid solutions and a liquid mixture.

this point is the *eutectic composition*, and the temperature here (denoted T_e) is the *eutectic temperature*. T_e is the lowest temperature for the given pressure at which the liquid phase is stable ("eutectic" comes from the Greek for *easily melting*).

Suppose we combine 0.60 mol A and 0.40 mol B and adjust the temperature so as to put the system point at b. This point is in the one-phase liquid area, so the equilibrium system at this temperature has a single liquid phase. If we now place the system in thermal contact with a cold reservoir, heat is transferred out of the system and the system point moves down along the *isopleth* (path of constant overall composition) b–h. The cooling rate depends on the temperature gradient at the system boundary and the system's heat capacity.

At Point c on the isopleth, the system point reaches the boundary of the one-phase area and is about to enter the two-phase area labeled A(s) + liquid. At this point in the cooling process, the liquid is saturated with respect to solid A, and solid A is about to freeze out from the liquid. There is an abrupt decrease (break) in the cooling rate here since the freezing process involves an extra enthalpy decrease.

At the still lower temperature at Point d, the system point is within the two-phase area. The tie line through this point is line e–f. The compositions of the two phases are given by the values of x_B at the ends of the tie line: $x_B^s = 0$ for the solid and $x_B^l = 0.50$ for the liquid. From the lever rule, the ratio of the amounts in these phases is

$$\frac{n^l}{n^s} = \frac{x_B - x_B^s}{x_B^l - x_B} = \frac{0.40 - 0}{0.50 - 0.40} = 4.0$$

Since the total amount is $n^s + n^l = 1.00$ mol, the amounts of the two phases must be $n^s = 0.20$ mol and $n^l = 0.80$ mol.

When the system point reaches the eutectic temperature at Point g, cooling halts until all of the liquid freezes. Solid B freezes out as well as solid A. During this *eutectic halt*,

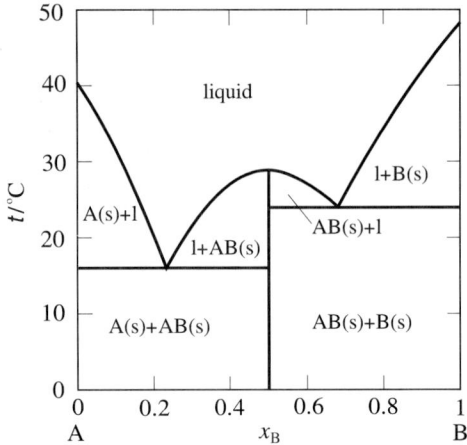

Figure 10.3 Temperature–composition phase diagram for the binary system of α-naphthylamine (A) and phenol (B) at 1 bar (Philip, J. C. *J. Chem. Soc.* **1903**, *83*, 814–834).

there are at first three phases: liquid with the eutectic composition, solid A, and solid B. As heat continues to be withdrawn from the system, the amount of liquid decreases and the amounts of the solids increase until finally only 0.60 mol of solid A and 0.40 mol of solid B are present. The temperature then begins to decrease again and the system point enters the two-phase area for solid A and solid B; tie lines in this area extend from $x_B = 0$ to $x_B = 1$.

Temperature–composition phase diagrams such as this are often mapped out experimentally by observing the cooling curve (temperature as a function of time) along isopleths of various compositions. This procedure is *thermal analysis*. A break in the slope of a cooling curve at a particular temperature indicates the system point has moved from a one-phase liquid area to a two-phase area of liquid and solid. A temperature halt indicates either the freezing point of the liquid to form a solid of the same composition or a eutectic temperature.

Figure 10.2 on the preceding page shows two temperature–composition phase diagrams with single eutectic points. The left-hand diagram is for the binary system of chloroform and carbon tetrachloride, two liquids that form nearly ideal mixtures. The solid phases are pure crystals, as in Fig. 10.1. The right-hand diagram is for the silver–copper system and involves solid phases that are solid solutions (substitutional alloys of variable composition). The area labeled s$^\alpha$ is for a solid solution that is mostly silver, and s$^\beta$ is a solid solution that is mostly copper. Tie lines in the two-phase areas do not end at a vertical line for a pure solid component as they do in the system shown in the left-hand diagram. The three phases that can coexist at the eutectic temperature of 1,052 K are the melt of the eutectic composition and the two solid solutions.

Section 9.5.4 discussed the possibility of the appearance of a *solid compound* when a binary liquid mixture is cooled. An example of this behavior is shown in Fig. 10.3, in which the solid compound contains equal amounts of the two components α-naphthylamine

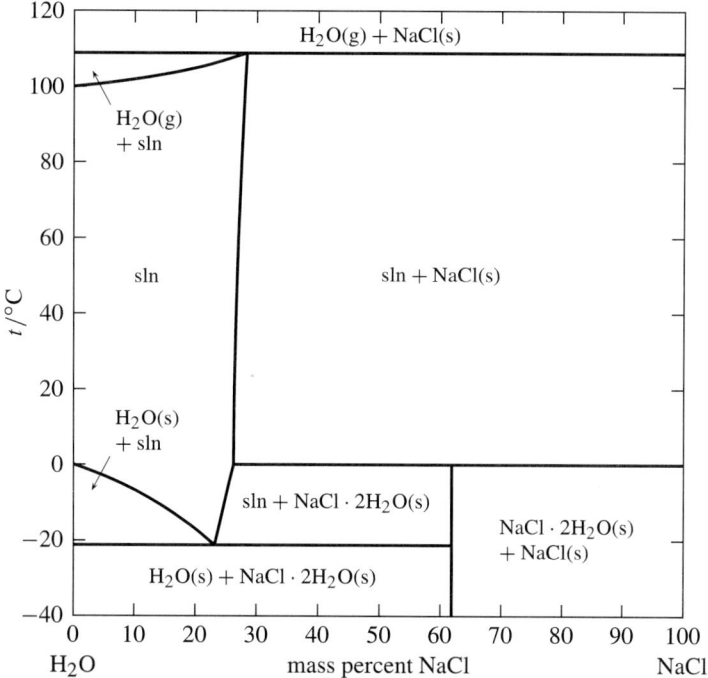

Figure 10.4 Temperature–composition phase diagram for the binary system of H_2O and NaCl at 1 bar. (*Alkali Metal and Ammonium Chlorides in Water and Heavy Water;* Cohen-Adad, R., Lorimer, J. W., Eds.; Solubility Data Series 47; Pergamon: Oxford, 1991.)

and phenol. This system has three different solid phases: pure A, pure B, and the solid compound AB. Only one or two of these solids can be present at the same time. The solid compound melts at 29 °C at the upper end of the vertical line that is drawn at the composition of the compound ($x_B = 0.5$ in this example). The solid is said to melt *congruently* to give a liquid of the same composition. The cooling curve for liquid of this composition would display a halt at the melting point.

The phase diagram shown in Fig. 10.3 has two eutectic points. It resembles two simple phase diagrams like Fig. 10.1 placed side by side. There is one important difference: The slope of the freezing-point curve (liquidus curve) is nonzero at the composition of a pure component, but is zero at the composition of a solid compound that is completely dissociated in the liquid (as theoretically derived on page 328). Thus, the curve in Fig. 10.3 has a relative maximum at $x_B = 0.5$ with a slope of zero.[3]

An example of a solid compound that does not melt congruently is shown in Fig. 10.4. The solid hydrate $NaCl \cdot 2H_2O$ is 61.9% NaCl by mass. It decomposes at 0 °C to form an

---

[3]To make an architectural analogy, the shape of the freezing-point curve at $x_B = 0.5$ is like a rounded Romanesque arch rather than a pointed Gothic arch.

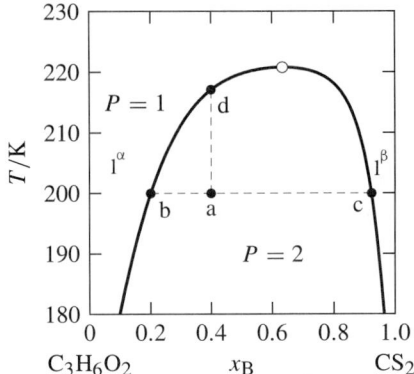

Figure 10.5 Temperature–composition phase diagram for the binary system of methyl acetate (A) and carbon disulfide (B) at 1 bar. All phases are liquids. The open circle indicates the critical point. (Data from Ferloni, P.; Spinolo, G. *Int. DATA Ser., Sel. Data Mixtures, Ser. A* **1974**, *2*, 70.)

aqueous solution of composition 26.3% NaCl by mass and a solid phase of anhydrous NaCl. These three phases can coexist at equilibrium at 0 °C. A phase transition like this, in which one solid changes into a liquid and another solid, is called *incongruent* or *peritectic* melting, and the point on the phase diagram at this temperature and the composition of the liquid is a *peritectic point*.

Figure 10.4 shows two other temperatures at which three phases can be present simultaneously: −21 °C, where the phases are ice, the solution at its eutectic point, and the solid hydrate; and 109 °C, where the phases are gaseous H_2O, a solution of composition 28.3% NaCl by mass, and solid NaCl. The positive slope of the right-hand boundary of the one-phase solution area shows that the solubilities of both the solid hydrate and the anhydrous salt increase with increasing temperature.

10.2.3 Partially miscible liquids

When two liquids that are partially miscible are combined in certain proportions, phase separation occurs (Sec. 8.2.6). Two liquid phases in equilibrium with one another are called *conjugate phases*. Obviously the two phases must have different compositions or they would be identical; the difference is called a *miscibility gap*. A binary system with two phases has two degrees of freedom, so at a given temperature and pressure each conjugate phase has a fixed composition.

The typical dependence of a miscibility gap on temperature is shown in Fig. 10.5. The miscibility gap (the difference in compositions at the left and right boundaries of the two-phase area) decreases as the temperature increases until at the *upper consolute temperature*, also called the *upper critical solution temperature*, the gap vanishes and the system becomes a single homogeneous liquid phase. The point at the maximum of the boundary curve for the two-phase area is the *consolute point* or *critical point*; at this point, the two liquid phases become identical (just as the liquid and gas phases become identical at the critical point of a pure substance).

Suppose we combine 4.0 mol of Component A (methyl acetate) and 6.0 mol of Component B (carbon disulfide) and bring the temperature to 200 K. The overall mole fraction of B is then $x_B = 0.40$. The system point, Point a, lies in the two-phase region. From Points b and c at the ends of the tie line, we find the two liquid layers have compositions $x_B^\alpha = 0.20$ and $x_B^\beta = 0.92$. Since carbon disulfide is the more dense of the two pure liquids, the bottom layer in a gravitational field is phase β, the layer that is richer in carbon disulfide. According to the lever rule (Eq. 6.2.4), the ratio of the amounts in the two phases is given by

$$\frac{n^\beta}{n^\alpha} = \frac{x_B - x_B^\alpha}{x_B^\beta - x_B} = \frac{0.40 - 0.20}{0.92 - 0.40} = 0.38$$

Combining this value with $n^\alpha + n^\beta = 10.0$ gives us $n^\alpha = 7.2$ mol and $n^\beta = 2.8$ mol.

If we gradually add more carbon disulfide to the system without changing the temperature, the system point moves to the right along the tie line. Since the ends of this tie line have fixed positions, neither phase changes composition, but the amount of phase β increases at the expense of phase α. The liquid–liquid interface moves up until, at overall composition $x_B = 0.92$, there is only one liquid phase.

Now suppose the system point is back at Point a and we raise the temperature while keeping the overall composition constant at $x_B = 0.40$. The system point moves up the isopleth a–d. Since the ratio $(x_B - x_B^\alpha)/(x_B^\beta - x_B)$ decreases, the amount of phase α increases at the expense of phase β and the liquid–liquid interface moves down toward the bottom of the container until at 217 K (Point d) there again is only one liquid phase.

10.2.4 Liquid–gas systems

Toluene and benzene form liquid mixtures that are practically ideal and closely obey Raoult's law for partial pressure. For the binary system of these components, we can use the vapor pressures of the pure liquids to generate the liquidus and vaporus curves of the pressure–composition and temperature–composition phase diagrams, as now described. The results are shown in Fig. 10.6 on the next page.

Consider a liquid mixture of Components A and B that obey Raoult's law for partial pressure (Eq. 7.5.1):

$$p_A = x_A(l) p_A^* \qquad p_B = [1 - x_A(l)] p_B^*$$

Strictly speaking, Raoult's law applies to a liquid–gas system maintained at a constant pressure by means of a third gaseous component, and p_A^* and p_B^* are the vapor pressures of the pure liquid components at this pressure and the temperature of the system. However, when a liquid phase is equilibrated with a gas phase, the partial pressure of a constituent of the liquid is practically independent of the total pressure (Sec. 9.7.1), so we apply the equations to a *binary* liquid–gas system and treat p_A^* and p_B^* as functions only of T.

When the binary system contains a liquid phase and a gas phase in equilibrium, the

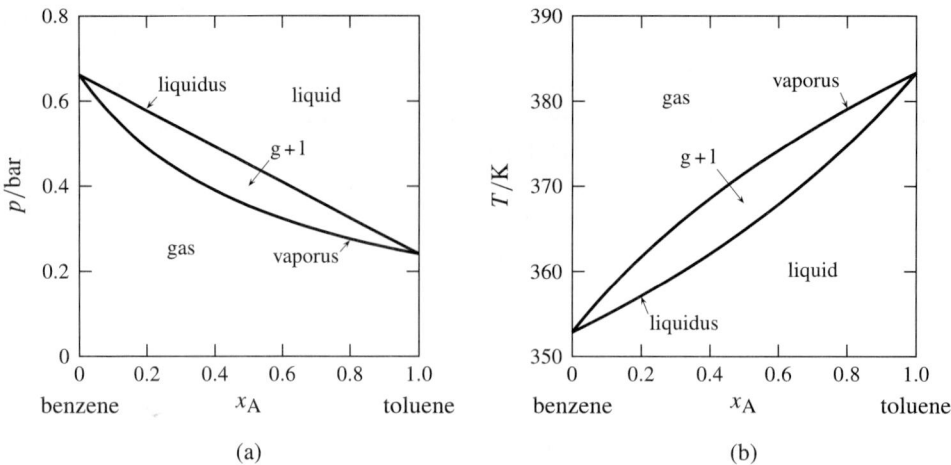

Figure 10.6 Phase diagrams for the binary system of toluene (A) and benzene (B). (a) Pressure–composition diagram at $T = 340$ K. (b) Temperature–composition diagram at $p = 1$ bar. The curves are calculated from Eqs. 10.2.2 and 10.2.3 and the saturation pressures of the pure liquids.

pressure is the sum of p_A and p_B, which from Raoult's law is given by

$$p = x_A(l)p_A^* + [1 - x_A(l)]p_B^*$$
$$= p_B^* + (p_A^* - p_B^*)x_A(l) \qquad (10.2.1)$$
$$(C = 2, \text{liquid and gas,}$$
$$\text{Raoult's law obeyed})$$

This expression shows that in the two-phase system p has a value between p_A^* and p_B^*; if T is constant, p is a linear function of the mole fraction of A in the liquid. The mole fraction composition of the gas in the two-phase system is given by[4]

$$x_A(g) = \frac{p_A}{p} = \frac{x_A(l)p_A^*}{p_B^* + (p_A^* - p_B^*)x_A(l)}$$

A binary two-phase system has two degrees of freedom. At a given T and p, each phase must have a fixed composition. We can calculate the liquid composition by rearranging Eq. 10.2.1:

$$x_A(l) = \frac{p - p_B^*}{p_A^* - p_B^*} \qquad (10.2.2)$$
$$(C = 2, \text{liquid and gas,}$$
$$\text{Raoult's law obeyed})$$

---

[4] Here we use the notation $x_A(l)$ and $x_A(g)$ for the mole fractions of Component A in the liquid and gas phases, instead of x_A and y_A. On the horizontal axis of a phase diagram, x_A is the mole fraction of A in the system as a whole.

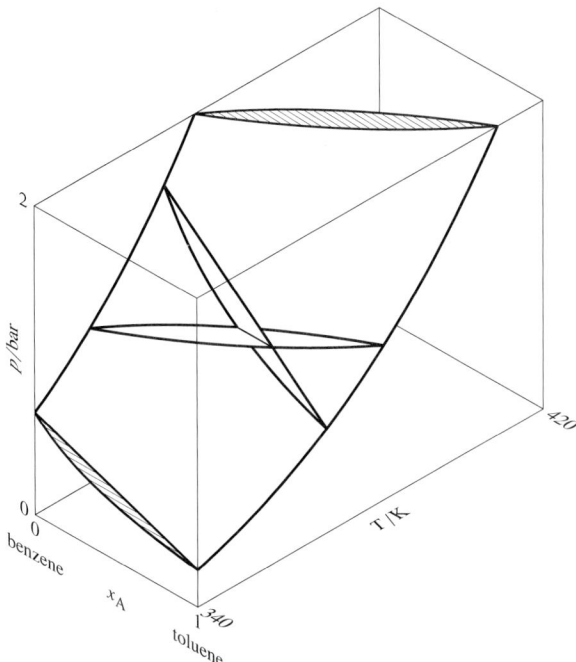

Figure 10.7 Liquidus and vaporus surfaces for the binary system of toluene (A) and benzene. Cross-sections through the two-phase region are drawn at constant temperatures of 340 K and 370 K and at constant pressures of 1 bar and 2 bar. Two of the cross-sections intersect at a tie line at $T = 370$ K and $p = 1$ bar, and the others are hatched in the direction of the tie lines.

The gas composition is then given by

$$x_A(g) = \frac{p_A}{p} = \frac{x_A(l) p_A^*}{p}$$

$$= \left(\frac{p - p_B^*}{p_A^* - p_B^*}\right) \frac{p_A^*}{p} \quad (10.2.3)$$

($C = 2$, liquid and gas, Raoult's law obeyed)

If we know p_A^* and p_B^* as functions of T, we can use Eqs. 10.2.2 and 10.2.3 to calculate the compositions for any combination of T and p at which the liquid and gas phases can coexist and thus construct a pressure–composition or temperature–composition phase diagram.

Figure 10.6(a) shows that, although p is a linear function of $x_A(l)$ when T is held constant, p is *not* a linear function of $x_A(g)$. In Fig. 10.6(b), we see that T is not a linear function of either $x_A(l)$ or $x_A(g)$ when p is held constant.

On the phase diagrams, the T-$x_A(l)$ and p-$x_A(l)$ curves are *liquidus* curves, also called *bubble-point* curves or *boiling-point* curves. The T-$x_A(g)$ and p-$x_A(g)$ curves are *vaporus*

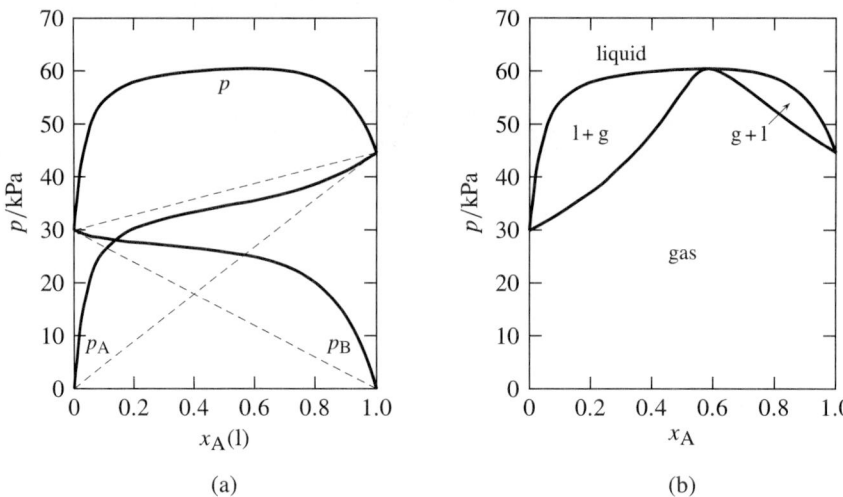

Figure 10.8 Binary system of methanol (A) and benzene at 45 °C (Toghiana, H. *Int. DATA Ser., Sel. Data Mixtures, Ser. A* **1995**, *23*, 234). (a) Partial pressures and total pressure in the gas phase equilibrated with liquid mixtures. The dashed lines indicate Raoult's law behavior. (b) Pressure–composition phase diagram at 45 °C. There is an azeotrope at $x_A = 0.59$ and $p = 60.5$ kPa.

curves, also called *dew-point* curves or *condensation* curves. These curves are actually cross-sections of liquidus and vaporus *surfaces* in a three-dimensional T–p–x_A phase diagram, as shown in Fig. 10.7 on the preceding page. In this figure, the liquidus surface is at the front and the vaporus surface is hidden behind it.

Most binary liquid mixtures do not behave ideally. The most common situation is *positive* deviations from Raoult's law.[5] Some mixtures, however, have specific A–B interactions, such as solvation or association, that prevent random mixing of the molecules of A and B, and the result is then *negative* deviations from Raoult's law. If the deviations from Raoult's law, either positive or negative, are large enough, the liquidus curve exhibits a maximum or minimum and *azeotropic* behavior results.

Figure 10.8 shows the azeotropic behavior of the binary methanol-benzene system. In Fig. 10.8(a), the experimental partial pressures of both components at a fixed temperature are plotted as a function of the liquid composition. The partial pressures of both components exhibit positive deviations from Raoult's law,[6] and the total pressure (equal to the sum of the partial pressures) has a maximum value greater than the vapor pressure of either pure component. The curve of p versus $x_A(l)$ becomes the liquidus curve of the pressure–composition phase diagram shown in Fig. 10.8(b). Points on the vaporus curve are calculated

---

[5] In the molecular theory of Sec. 8.2.5, positive deviations correspond to a less negative value of k_{AB} than the average of k_{AA} and k_{BB}.

[6] This behavior is consistent with the statement in Sec. 9.7.2—that if one constituent of a binary liquid mixture exhibits positive deviations from Raoult's law, with only one inflection point in the curve of fugacity versus mole fraction, the other constituent also has positive deviations from Raoult's law.

Section 10.2 Phase Diagrams: Binary Systems

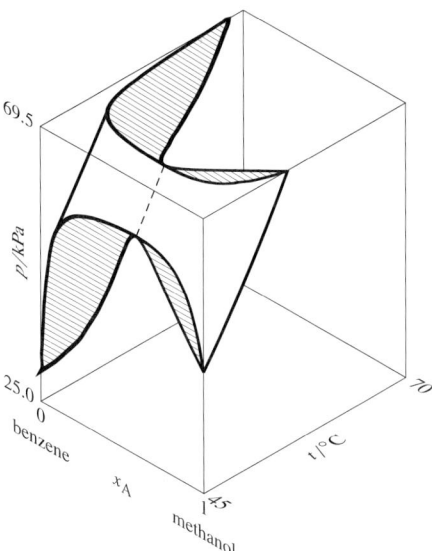

Figure 10.9 Liquidus and vaporus surfaces for the binary system of methanol (A) and benzene. Cross-sections are hatched in the direction of the tie lines. The dashed curve is the azeotrope vapor-pressure curve (Toghiana, H. *Int. DATA Ser., Sel. Data Mixtures, Ser. A* **1995**, *23*, 232–236).

from $p = p_A/x_A(g)$.

In practice, the liquidus and vaporus curves for a nonideal binary system are usually found by allowing liquid mixtures of various compositions to boil in an equilibrium still at a fixed temperature or pressure. When the liquid and gas phases have become equilibrated, samples of each are withdrawn for analysis. The partial pressures shown in Fig. 10.8(a) were calculated from the experimental gas-phase compositions with the relations $p_A = x_A(g)p$ and $p_B = p - p_A$.

If the constant-temperature liquidus curve has a maximum pressure at a liquid composition not corresponding to one of the pure components, which is the case for the methanol–benzene system, then as deduced at the end of Sec. 9.7.3 the liquid and gas phases are mixtures of identical compositions at this pressure. On the pressure–composition phase diagram, the liquidus and vaporus curves both have maxima at this pressure, and the two curves coincide here. A binary system with negative deviations from Raoult's law can have an isothermal liquidus curve with a *minimum* pressure at a particular mixture composition, in which case the liquidus and vaporus curves coincide at this minimum. The general phenomenon in which equilibrated liquid and gas mixtures have identical compositions is called *azeotropy*, and the liquid with this composition is an azeotropic mixture or **azeotrope** (Greek: *boils unchanged*). An azeotropic mixture vaporizes as if it were a pure substance, undergoing an equilibrium phase transition to a gas of the same composition.

If the liquidus and vaporus curves exhibit a *maximum* on a pressure–composition phase

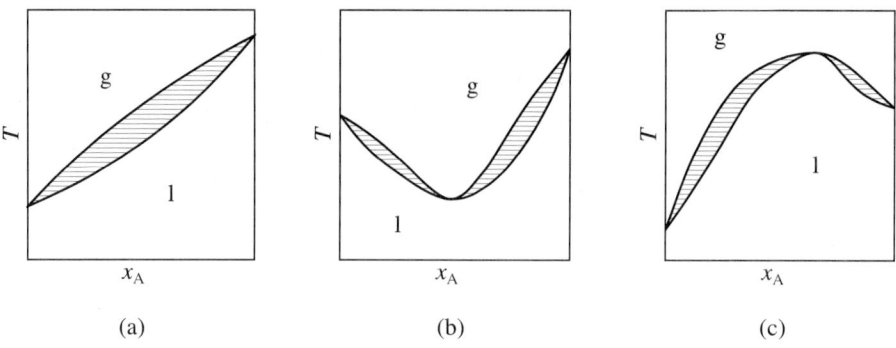

Figure 10.10 Temperature–composition phase diagrams of binary systems exhibiting (a) no azeotropy, (b) a minimum-boiling azeotrope, and (c) a maximum-boiling azeotrope. Only the one-phase areas are labeled; two-phase areas are hatched in the direction of the tie lines.

diagram, then they exhibit a *minimum* on a temperature–composition phase diagram. This relation is explained for the methanol–benzene system by the three-dimensional liquidus and vaporus surfaces drawn in Fig. 10.9 on the preceding page. In this diagram, the vaporus surface is hidden behind the liquidus surface. The hatched cross-section at the front of the figure is the same as the pressure–composition diagram of Fig. 10.8(b), and the hatched cross-section at the top of the figure is a temperature–composition phase diagram in which the system exhibits a *minimum-boiling azeotrope*.

A binary system containing an azeotropic mixture in equilibrium with its vapor has two species, two phases, and one relation among intensive variables: $x_A(\text{liq}) = x_A(\text{g})$. The number of degrees of freedom is then $F = 2 + s - r - P = 2 + 2 - 1 - 2 = 1$; the system is univariant. At a given temperature, the azeotrope can exist at only one pressure and have only one composition. As T changes, so do p and x_A along an *azeotrope vapor-pressure curve* as illustrated by the dashed curve in Fig. 10.9.

Figure 10.10 summarizes the general appearance of some relatively simple temperature–composition phase diagrams of binary systems. If the system does not form an azeotrope (*zeotropic* behavior), the equilibrated gas phase is richer in one component than the liquid phase at all liquid compositions, and the liquid mixture can be separated into its two components by fractional distillation. The gas in equilibrium with an azeotropic mixture, however, is not enriched in either component. Fractional distillation of a system with an azeotrope leads to separation into one pure component and the azeotropic mixture.

More complicated behavior is shown in the phase diagrams of Fig. 10.11 on the next page. These are binary systems with partially miscible liquids in which the boiling point is reached before an upper consolute temperature can be observed.

10.2.5 Solid–gas systems

As an example of a two-component system in which solid and gas phases can be in equilibrium, consider copper(II) sulfate and water. The anhydrous salt and its hydrates (solid com-

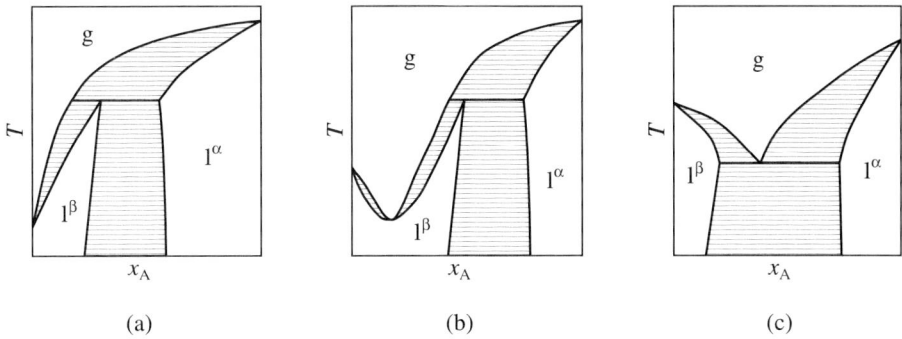

Figure 10.11 Temperature–composition phase diagrams of binary systems with partially miscible liquids exhibiting (a) the ability to be separated into pure components by fractional distillation, (b) a minimum-boiling azeotrope, and (c) boiling at a lower temperature than the boiling point of either component. Only the one-phase areas are labeled; two-phase areas are hatched in the direction of the tie lines.

pounds) form the series of solids $CuSO_4$, $CuSO_4 \cdot H_2O$, $CuSO_4 \cdot 3H_2O$, and $CuSO_4 \cdot 5H_2O$. In the pressure–composition phase diagram shown in Fig. 10.12 on the next page, these formulas are abbreviated A, AB, AB$_3$, and AB$_5$. The following dissociation (dehydration) equilibria are possible:

$$CuSO_4 \cdot H_2O(s) \rightleftharpoons CuSO_4(s) + H_2O(g)$$
$$\tfrac{1}{2}CuSO_4 \cdot 3H_2O(s) \rightleftharpoons \tfrac{1}{2}CuSO_4 \cdot H_2O(s) + H_2O(g)$$
$$\tfrac{1}{2}CuSO_4 \cdot 5H_2O(s) \rightleftharpoons \tfrac{1}{2}CuSO_4 \cdot 3H_2O(s) + H_2O(g)$$

The equations are written with coefficients that make the coefficient of $H_2O(g)$ unity. When one of these equilibria is established in the system, there are two components and three phases; the phase rule then tells us the system is univariant and the pressure has only one possible value at a given temperature. This pressure is called the *dissociation pressure* of the higher hydrate (the hydrate that dissociates in the chemical equation).

The dissociation pressures of the three hydrates are indicated by horizontal lines in Fig. 10.12. For instance, the dissociation pressure of $CuSO_4 \cdot 5H_2O$ is 1.04×10^{-2} bar. At the pressure of each horizontal line, the equilibrium system can have one, two, or three phases, with compositions given by the intersections of the line with other lines. A fourth three-phase equilibrium is shown at $p = 3.09 \times 10^{-2}$ bar; this is the equilibrium between solid $CuSO_4 \cdot 5H_2O$, the saturated aqueous solution of this hydrate, and water vapor.

Consider the thermodynamic equilibrium constant of one of the dissociation reactions. At the low pressures shown in the phase diagram, the activities of the solids are practically unity and the fugacity of the water vapor is practically the same as the pressure, so the equilibrium constant is almost exactly equal to $p_d/p°$, where p_d is the dissociation pressure of the higher hydrate in the reaction. Thus, a hydrate cannot exist in equilibrium with water vapor at a pressure below the dissociation pressure of the hydrate because dissociation would

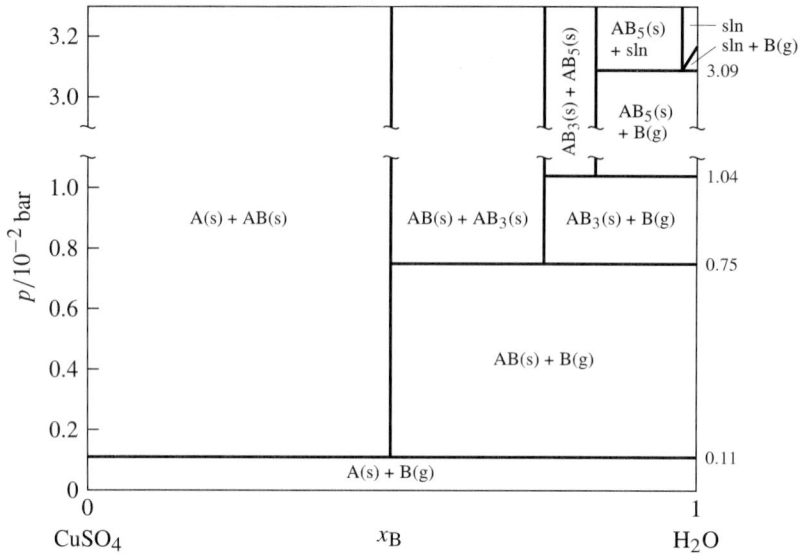

Figure 10.12 Pressure–composition phase diagram for the binary system of CuSO$_4$ (A) and H$_2$O (B) at 25 °C.

be spontaneous under these conditions. Conversely, the salt formed by the dissociation of a hydrate cannot exist in equilibrium with water vapor at a pressure above the dissociation pressure because hydration would be spontaneous.

> If the system contains dry air as an additional gaseous component and one of the dissociation equilibria is established, the partial pressure p_{H_2O} of H$_2$O is equal (approximately) to the dissociation pressure p_d of the higher hydrate. The prior statements regarding dissociation and hydration now depend on the value of p_{H_2O}. If a hydrate is placed in air in which p_{H_2O} is less than p_d, dehydration is spontaneous; this phenomenon is called **efflorescence** (Latin: *blossoming*). If p_{H_2O} is greater than the vapor pressure of the saturated solution of the highest hydrate that can form in the system, the anhydrous salt and any of its hydrates will spontaneously absorb water and form the saturated solution; this is **deliquescence** (Latin: *becoming fluid*).

If the two-component equilibrium system contains only two phases, it is bivariant corresponding to one of the areas in Fig. 10.12. Here the pressure can be varied at a fixed temperature. In the case of areas labeled with two *solid* phases, the pressure has to be applied to the solids by a fluid (other than H$_2$O) that is not considered part of the system.

10.2.6 Systems at high pressure

Binary phase diagrams begin to look different when the pressure is greater than the critical pressure of either of the pure components. Many different types of behavior have been observed in this region. One common type, that found in the binary system of heptane and

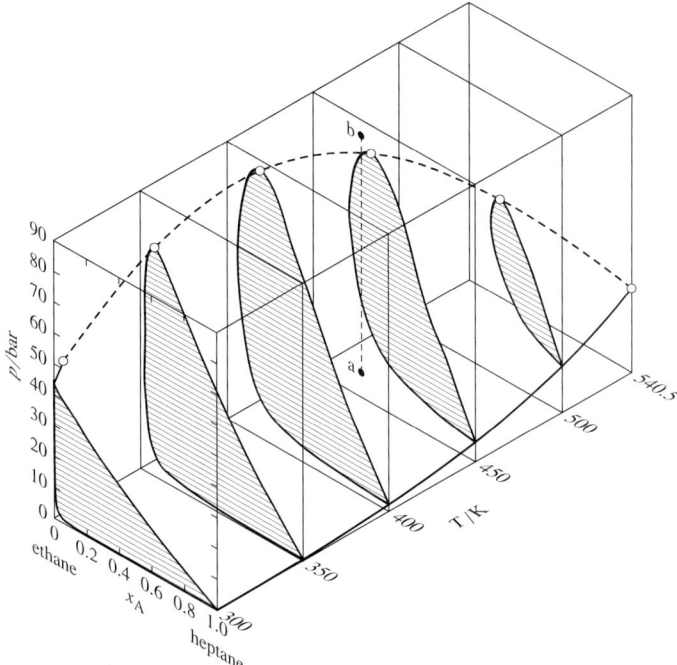

Figure 10.13 Pressure–temperature–composition behavior in the binary heptane–ethane system (Kay, W. B. *Ind. Eng. Chem.* **1938**, *30*, 459–465). The open circles are critical points; the dashed curve is the critical curve. The dashed line a–b illustrates retrograde condensation.

ethane, is shown in Fig. 10.13. This figure shows sections of a three-dimensional phase diagram at five temperatures. Each section is a pressure–composition phase diagram at constant T. The two-phase areas are hatched in the direction of the tie lines; at the left end of each tie line (at low x_A) is a vaporus curve, and at the right end is a liquidus curve. The vapor pressure curve of pure ethane ($x_A = 0$) ends at the critical point of ethane at 305.4 K; between this point and the critical point of heptane at 540.5 K, there is a continuous *critical curve*, which is the locus of critical points at which gas and liquid mixtures equilibrated at different temperatures become identical in composition and density.

Consider what happens when the system point is at Point a in Fig. 10.13 and the pressure is then increased by isothermal compression along Line a–b. The system point moves from the area for a single gas phase into the two-phase gas–liquid area and then out again. This curious phenomenon, condensation followed by vaporization, is called *retrograde condensation* (backward condensation). Under some conditions, an isobaric increase of T can result in vaporization followed by condensation; this is *retrograde vaporization*.

A different type of high-pressure behavior, that found in the xenon–helium system, is shown in Fig. 10.14 on the next page. Here the critical curve begins at the critical point of the less volatile component (xenon) and continues to *higher* temperatures and pressures

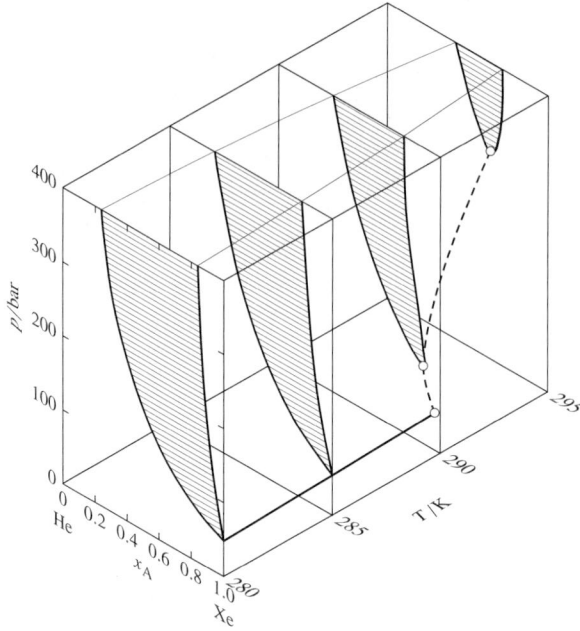

Figure 10.14 Pressure–temperature–composition behavior in the binary xenon–helium system (de Swann Arons, J.; Diepen, G. A. M. *J. Chem. Phys.* **1966**, *44*, 2322–2330). The open circles are critical points; the dashed curve is the critical curve.

than the critical temperature and pressure of either pure component. The two-phase region at pressures above this critical curve is sometimes said to represent *gas–gas equilibrium* or *gas–gas immiscibility* because we would not usually consider a liquid to exist beyond the critical points of the pure components. Of course, the coexisting phases in this two-phase region are not gases in the ordinary sense of being tenuous fluids, but are instead high-pressure fluids of liquidlike densities. If we call both phases gases, then we have to say that pure gaseous substances do not necessarily mix spontaneously in all proportions at high pressure as we assume they do at ordinary pressures.

If the pressure of any system is increased isothermally, eventually solid phases will appear; these are not shown in Fig. 10.13 or Fig. 10.14.

10.3 PHASE DIAGRAMS: TERNARY SYSTEMS

A ternary system is one with three components. We can independently vary the temperature, pressure, and two independent composition variables for the system as a whole. A two-dimensional phase diagram for a ternary system is usually drawn for conditions of constant T and p.

Although we could use the two dimensions of the diagram to express the mole fractions of two of the components in Cartesian coordinates, there are advantages in using instead the

Section 10.3   Phase Diagrams: Ternary Systems

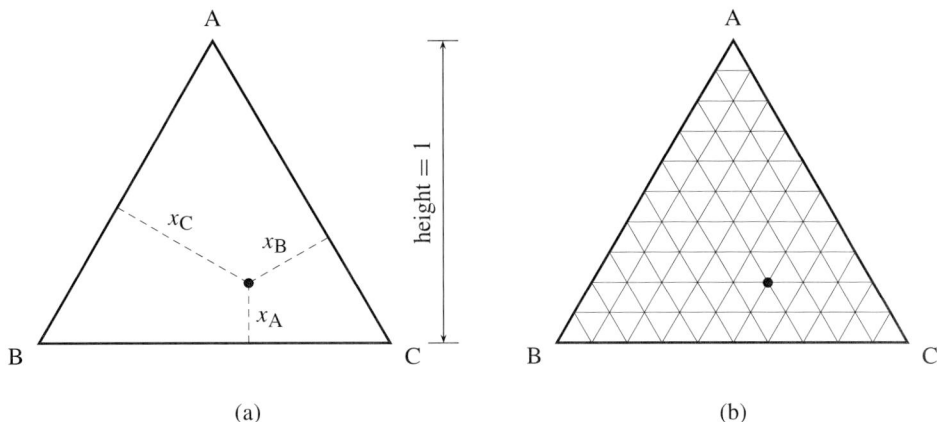

Figure 10.15   Representing the composition of a ternary system using an equilateral triangle.

triangular coordinates shown in Fig. 10.15. Each vertex of the equilateral triangle represents one of the pure components A, B, or C. A point on the side of the triangle opposite a vertex represents a binary mixture of the other two components, and a point within the triangle represents a mixture of all three components. To determine the mole fraction of a component at any point, we measure the distance to the point from the side of the triangle that is opposite the vertex for the pure component, then express this distance as a fraction of the height of the triangle. The concept is shown in Fig. 10.15(a).

To aid in reading the position of a point or in placing a point for a given composition, equally spaced lines can be drawn within the triangle parallel to each side as shown in Fig. 10.15(b). One of these lines, being at a constant distance from a side of the triangle, represents a constant mole fraction of one component. In the figure, the lines divide the distance from each side to the opposite vertex into 10 equal parts; thus, adjacent parallel lines represent a difference of 0.1 in the mole fraction of a component, starting with 0 at the side of the triangle and ending with 1 at the vertex. Using the lines, we see the filled circle in the figure represents the composition $x_A = 0.20$, $x_B = 0.30$, and $x_C = 0.50$.

The sum of x_A, x_B, and x_C must be 1. The method of representing compositions works because the sum of the lines drawn from any point within an equilateral triangle to the three sides, perpendicular to the sides, equals the height of the triangle. The proof of this fact is shown in Fig. 10.16 on the next page.

Two useful properties of this way of representing a ternary composition are as follows:

1. Points on a line parallel to a side of the triangle represent systems in which one of the mole fractions remains constant.

2. Points on a line passing through a vertex represent systems in which the ratio of two of the mole fractions remains constant.

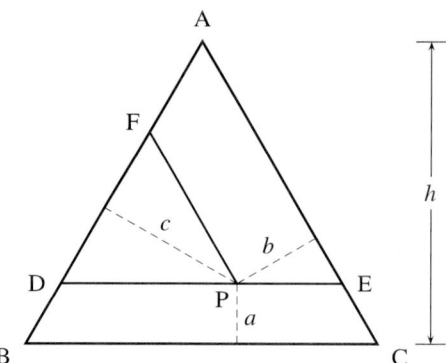

Figure 10.16 Proof that the sum of the lengths a, b, and c is equal to the height h of the large equilateral triangle ABC. ADE and FDP are two smaller equilateral triangles. The height of triangle ADE is equal to $h - a$. The height of triangle FDP is equal to the height of triangle ADE minus length b, and is also equal to length c: $h - a - b = c$. Therefore, $a + b + c = h$.

10.3.1 Three liquids

Figure 10.17 on the next page is the ternary phase diagram of liquid phases of a system of ethanol, benzene, and water at a constant temperature and pressure. When the system point is in the area labeled $P = 1$, there is a single liquid phase whose composition is described by the position of the point. The one-phase area extends to the side of the triangle representing binary mixtures of ethanol and benzene and to the side representing binary mixtures of ethanol and water. In other words, ethanol and benzene mix in all proportions and so do ethanol and water.

When the overall composition is such that the system point falls in the area labeled $P = 2$, two liquid phases are present. The compositions of these phases are given by the positions of the ends of a tie line through the system point. Four representative tie lines are included in the diagram, and these must be determined experimentally. In the limit of zero mole fraction of ethanol, the tie line falls along the horizontal side of the triangle and displays a miscibility gap for the binary system of benzene and water. (The conjugate phases are very nearly pure benzene and pure water).

The *plait point* shown in the figure is a kind of critical point. As the system point approaches the plait point from within the two-phase area, the length of the tie line through the system point approaches zero, the miscibility gap disappears, and the compositions of the two conjugate liquid phases become identical. At any other position of the system point within the two-phase area, the relative amounts of the two phases can be determined from the lever rule (Sec. 6.2.4).[7]

Suppose we have the binary system of benzene and water represented by Point a. Two layers are present: one is wet benzene and the other is water containing a small mole fraction

---

[7] The lever rule works, according to the general derivation in Sec. 6.2.4, because the ratio n_A/n, which is equal to x_A, varies linearly with the position of the system point along a tie line on the triangular phase diagram.

Problems

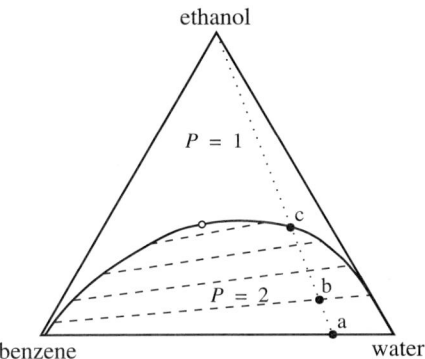

Figure 10.17 Ternary phase diagram for ethanol, benzene, and water at 30 °C and 1 bar (Brandani, V.; Chianese, A.; Rossi, M. *J. Chem. Eng. Data* **1985**, *30*, 27–29). The dashed lines are tie lines; the open circle indicates the plait point.

of benzene. If we gradually stir ethanol into this system, the system point moves along the dotted line from Point a toward the vertex for pure ethanol (but can never quite reach the vertex). At Point b, there are still two layers, and we can consider the ethanol to have distributed itself between two solvents (Sec. 9.6.3). From the positions of the ends of the tie line, we see that the mole fraction of ethanol is greater in the water-rich layer. As we continue to add ethanol, the amount of the water-rich phase increases and the amount of the benzene-rich phase decreases until at Point c the benzene-rich layer completely disappears. The added ethanol has increased the mutual solubilities of benzene and water and resulted in a single liquid phase.

10.3.2 Two solids and a solvent

The phase diagram in Fig. 10.18 on the next page is for a ternary system of water and two salts with an ion in common. There is a one-phase area labeled "sln" (for solution), 2 two-phase areas in which the phases are solid salt and saturated solution, and a triangular three-phase area. Some representative tie lines are drawn in the two-phase areas.

A system of three components and three phases has two degrees of freedom; at fixed values of T and p, each phase must have a fixed composition. The fixed compositions of the phases that are present when the system point falls in the three-phase area are the compositions at the three vertices of the inner triangle; the phases are solid NaCl, solid KCl, and a solution that is saturated with respect to both salts ($x_{NaCl} = 0.20$, $x_{KCl} = 0.11$).

From the position of the curved boundary that separates the one-phase solution area from the two-phase area for solution and solid KCl, we can see that adding NaCl to the saturated solution of KCl decreases the mole fraction of KCl in the saturated solution. Although it is not obvious in the phase diagram, adding KCl to a saturated solution of NaCl decreases the mole fraction of NaCl. This decrease in solubility when a common ion is added is the *common-ion effect* mentioned in Sec. 9.5.5.

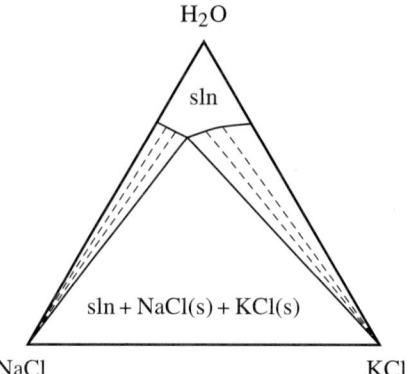

Figure 10.18 Ternary phase diagram for NaCl, KCl, and water at 25 °C and 1 bar. The dashed lines are tie lines in the two-phase areas.

PROBLEMS

10.1 Consider a one-phase system that is a gaseous mixture of N_2, H_2, and NH_3. For each of the following cases, find the number of degrees of freedom and give an example of the independent intensive variables that could be used to specify the equilibrium state, apart from the total amount of gas.

  (a) There is no reaction.

  (b) The reaction $N_2(g) + 3\,H_2(g) \rightleftharpoons 2\,NH_3(g)$ is at equilibrium.

  (c) The reaction is at equilibrium and the system is prepared from NH_3 only.

10.2 How many components has a mixture of water and deuterium oxide in which the equilibrium $H_2O + D_2O \rightleftharpoons 2\,HDO$ exists?

10.3 Consider a system containing only $NH_4Cl(s)$, $NH_3(g)$, and $HCl(g)$. Assume that equilibrium exists for the reaction $NH_4Cl(s) \rightleftharpoons NH_3(g) + HCl(g)$.

  (a) Suppose you prepare the system by placing solid NH_4Cl in an evacuated flask and heating to 400 K. Use the phase rule to decide whether you could vary the pressure while both phases remained in equilibrium at 400 K.

  (b) According to the phase rule, if the system is not prepared as described in Part (a) could you vary the pressure while both phases remain in equilibrium at 400 K?

  (c) Rationalize your conclusions for these two cases on the basis of the thermodynamic equilibrium constant. Assume that the gas phase is an ideal gas mixture and use the approximate expression $K = p_{NH_3} p_{HCl}/(p^\circ)^2$.

10.4 Consider the lime-kiln process $CaCO_3(s) \rightleftharpoons CaO(s) + CO_2(g)$. Find the number of intensive variables that can be varied independently in the following situations:

  (a) The system is prepared by placing calcium carbonate, calcium oxide, and carbon dioxide in a container.

(b) The system is prepared from calcium carbonate only.

(c) The temperature is fixed at 1,000 K.

10.5 What are the values of C and F in systems consisting of solid AgCl in equilibrium with an aqueous phase containing H_2O, Ag^+, Cl^-, Na^+, and NO_3^- prepared in the following ways? Give examples of intensive variables that could be varied independently.

(a) The system is prepared by equilibrating excess solid AgCl with an aqueous solution of $NaNO_3$.

(b) The system is prepared by mixing aqueous solutions of $AgNO_3$ and $NaCl$; some solid AgCl forms by precipitation.

10.6 How many degrees of freedom has a system consisting of solid NaCl in equilibrium with an aqueous phase containing H_2O, $Na^+(aq)$, $Cl^-(aq)$, $H^+(aq)$, and $OH^-(aq)$? Would it be possible to independently vary T, p, and m_{OH^-}? If so, explain how you could do this.

10.7 Consult the phase diagram shown in Fig. 10.4 on page 367. Suppose the system contains 36.0 g (2.00 mol) H_2O and 58.4 g (1.00 mol) NaCl at 25 °C and 1 bar.

(a) Describe the phases present in the equilibrium system and their masses.

(b) Describe the changes that occur at constant pressure if the system is placed in thermal contact with a heat reservoir at $-30\,°C$.

(c) Describe the changes that occur if the temperature is raised from 25 °C to 120 °C at constant pressure.

(d) Describe the system after 200 g H_2O is added at 25 °C.

Table 10.1 Aqueous solubilities of sodium sulfate decahydrate and anhydrous sodium sulfate

$Na_2SO_4 \cdot 10H_2O$		Na_2SO_4	
$t/°C$	x_B	$t/°C$	x_B
10	0.011	33	0.059
15	0.016	40	0.058
20	0.024	50	0.056
25	0.034		
30	0.048		

10.8 Use the following information to draw a temperature–composition phase diagram for the binary system of H_2O (A) and Na_2SO_4 (B) at $p = 1$ bar, confining t to the range -20 to $50\,°C$ and x_B to the range 0–0.1. The solid decahydrate, $Na_2SO_4 \cdot 10H_2O$, is stable below 32.4 °C. The anhydrous salt, Na_2SO_4, is stable above this temperature. There is a peritectic point for these two solids and the solution at $x_B = 0.059$ and $t = 32.4\,°C$. There is a eutectic point for ice, $Na_2SO_4 \cdot 10H_2O$, and the solution at $x_B = 0.006$ and $t = -1.3\,°C$. Table 10.1 gives the temperature dependence of the solubilities of the ionic solids.

Table 10.2 Breaks and halts in binary water–ethylene glycol mixtures.

x_B	Break (°C)	Halt (°C)	x_B	Break (°C)	Halt (°C)
0.100	−10.5	−51.0	0.500	−53.3	−63.3
0.200	−30.2	−51.0	0.600	(none)	−40.5
0.290	(none)	−51.0	0.650	−44.0	−49.5
0.300	−50.0	−51.0	0.700	(none)	−49.5
0.333	(none)	−49.2	0.800	−30.0	−49.5
0.400	−54.5	−63.3	0.900	−20.0	−49.5
0.450	(none)	−63.3			

10.9 Ethylene glycol (permanent antifreeze) forms some solid hydrates at low temperatures. The normal freezing points of water (A) and ethylene glycol (B) are 0.0 °C and −13.2 °C, respectively. Thermal analysis of mixtures of these two components at 1 atm gives the temperatures listed in Table 10.2 for breaks and halts.

(a) Use these data to construct a T-x phase diagram for this system. Label each area and indicate the probable formulas of any hydrates.

(b) Estimate the *lowest* mole fraction of ethylene glycol in an antifreeze mixture that will protect a car's radiator (i.e., prevent solid from forming) at temperatures down to −40 °C.

10.10 Figure 10.19 is a temperature–composition phase diagram for the binary system of water (A)

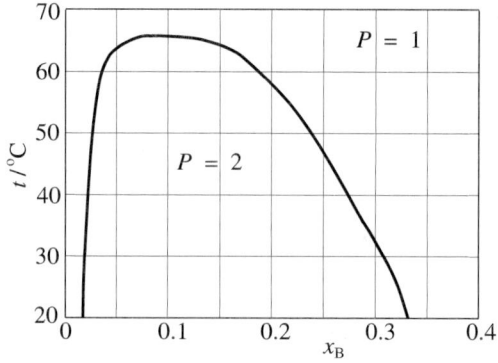

Figure 10.19 Temperature–composition phase diagram for the binary system of water (A) and phenol (B) at 1 bar. Only liquid phases are present. (See Prob. 10.10.)

and phenol (B) at 1 bar. These liquids are partially miscible below 67 °C. Phenol is more dense than water, so the layer with the higher mole fraction x_B of phenol is the bottom layer. Suppose you place 4.0 mol of H_2O and 1.0 mol of phenol in a beaker at 30 °C and gently stir to allow the layers to equilibrate.

(a) What are the compositions of the top and bottom layers?

(b) Find the amount of each component in the bottom layer.

(c) As you gradually stir more phenol into the beaker, maintaining the temperature at 30 °C, what changes occur in the volumes and compositions of the two layers? If one layer eventually disappears, what additional amount of phenol is needed to cause this to happen?

Table 10.3 Saturation vapor pressures of propane (A) and n-butane (B).

$t/°C$	p_A^*/bar	p_B^*/bar
−10.0	3.360	0.678
−20.0	2.380	0.441
−30.0	1.633	0.275

10.11 The standard boiling point of propane is −41.8 °C and that of n-butane is −0.2 °C. Table 10.3 lists vapor pressure data for the pure liquids. Assume that the liquid mixtures obey Raoult's law.

(a) Calculate the compositions, $x_A(\text{l})$, of the liquid mixtures with boiling points of −10.0 °C, −20.0 °C, and −30.0 °C at a pressure of 1 bar.

(b) Calculate the compositions, $x_A(\text{g})$, of the equilibrium vapor at these three temperatures.

(c) Plot the temperature–composition phase diagram at $p = 1$ bar using these data, and label the areas appropriately.

(d) Suppose a system containing 10.0 mol propane and 10.0 mol n-butane is brought to a pressure of 1 bar and a temperature of −25 °C. From your phase diagram, estimate the compositions and amounts of both phases.

Table 10.4 Liquid and gas compositions in the two-phase system of 2-propanol (A) and benzene at 45 °C (Brown, I.; Fock, W.; Smith, F. *Aust. J. Chem.* **1956**, *9*, 364–372).

$x_A(\text{l})$	$x_A(\text{g})$	p/kPa	$x_A(\text{l})$	$x_A(\text{g})$	p/kPa
0	0	29.89	0.5504	0.3692	35.32
0.0472	0.1467	33.66	0.6198	0.3951	34.58
0.0980	0.2066	35.21	0.7096	0.4378	33.02
0.2047	0.2663	36.27	0.8073	0.5107	30.28
0.2960	0.2953	36.45	0.9120	0.6658	25.24
0.3862	0.3211	36.29	0.9655	0.8252	21.30
0.4753	0.3463	35.93	1.0000	1.0000	18.14

10.12 Use the data tabulated in Table 10.4 to draw a pressure–composition phase diagram for the 2-propanol–benzene system at 45 °C. Label the axes and each area.

Table 10.5 Liquid and gas compositions in the two-phase system of acetone (A) and chloroform at 35.2 °C.

$x_A(l)$	$x_A(g)$	p/kPa	$x_A(l)$	$x_A(g)$	p/kPa
0	0	39.08	0.634	0.727	36.29
0.083	0.046	37.34	0.703	0.806	38.09
0.200	0.143	34.92	0.815	0.896	40.97
0.337	0.317	33.22	0.877	0.936	42.62
0.413	0.437	33.12	0.941	0.972	44.32
0.486	0.534	33.70	1.000	1.000	45.93
0.577	0.662	35.09			

10.13 Use the data tabulated in Table 10.5 to draw a pressure–composition phase diagram for the acetone–chloroform system at 35.2 °C. Label the axes and each area.

Chapter 11

GALVANIC CELLS

An *electrochemical cell* is a system in which a cell reaction is accompanied by electrical work. A **galvanic cell**, or voltaic cell, is an electrochemical cell that, when isolated, has an electric potential difference between its terminals. The cell is said to be a *seat of electromotive force*.

The reaction in a galvanic cell differs in a fundamental way from the same reaction occurring in a reaction vessel. In a reaction vessel, the reactant or reactants are in a single phase or in phases in contact with one another, and the reaction is able to advance spontaneously in one direction or the other in the isolated system until reaction equilibrium is reached. A galvanic cell, in contrast, is arranged with the reactants physically separated so that ideally the reaction cannot occur unless an electric current passes through the cell. When the cell is isolated from its surroundings, the reaction is constrained not to occur. The isolated cell can therefore in principle be in an equilibrium state that has thermal, mechanical, and transfer equilibrium but not reaction equilibrium.

Measurements of the electric potential of a galvanic cell are capable of yielding precise values of thermodynamic equilibrium constants and molar reaction quantities for certain kinds of reactions, and of mean ionic activity coefficients in electrolyte solutions.

11.1 CELL REACTIONS AND CELL DIAGRAMS

A galvanic cell contains at least one phase that is an ionic conductor—an electrolyte solution or fused salt—through which ions but not electrons can move. If there is more than one ionic conductor phase, they are in contact at liquid junctions. The ionic conductor (or series of ionic conductors) is in contact with two metal *electrodes* that are electron conductors. Finally, each electrode is attached to one of the terminals of the cell or the electrode is a terminal. The sequence of physical elements of a galvanic cell is thus,

terminal–electrode–ionic conductor–electrode–terminal

Both terminals must have identical chemical compositions for the electric potential difference between them to be measurable.

To describe a galvanic cell, it is conventional to distinguish the *left* and *right* terminals and electrodes. In this way, we establish a left–right association with the reactants and products of the half-cell reactions. The **cell reaction** is the reaction that occurs when

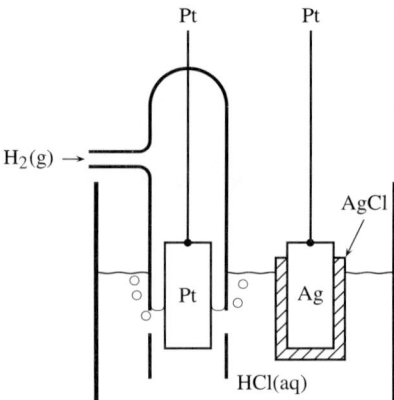

Figure 11.1 A galvanic cell without liquid junction.

electrons leave the system at the left terminal and enter the system at the right terminal. Thus, the cell reaction is the sum of two half-cell reactions, one for oxidation at the left electrode and the other for reduction at the right electrode.

Consider, for example, the galvanic cell pictured in Fig. 11.1. This consists of two electrodes or half-cells:[1] a hydrogen electrode at the left and a silver-silver chloride electrode at the right. The hydrogen electrode is a strip of platinum in contact with hydrogen gas and aqueous hydrochloric acid, which in this cell is the ionic conductor. The silver-silver chloride electrode is a silver strip or wire coated with solid silver chloride immersed in the same electrolyte. Both terminals are platinum.

The cell shown in Fig. 11.1 is compactly described by a *cell diagram* as follows:

$$\text{Pt} \mid \text{H}_2(\text{g}) \mid \text{HCl}(\text{aq}) \mid \text{AgCl}(\text{s}) \mid \text{Ag} \mid \text{Pt} \qquad (11.1.1)$$

The cell diagram includes the formulas of the reactants and products of the two half-cell reactions. Each single vertical bar represents a phase boundary. We can write the half-cell reactions as follows:

left half-cell reaction (oxidation) $\text{H}_2(\text{g}) \rightarrow 2\,\text{H}^+(\text{aq}) + 2\,\text{e}^-$

right half-cell reaction (reduction) $\text{AgCl}(\text{s}) + \text{e}^- \rightarrow \text{Ag}(\text{s}) + \text{Cl}^-(\text{aq})$

(e^- is an electron). The sum of the half-cell reactions is the cell reaction

$$\text{H}_2(\text{g}) + 2\,\text{AgCl}(\text{s}) \rightarrow 2\,\text{Ag}(\text{s}) + 2\,\text{H}^+(\text{aq}) + 2\,\text{Cl}^-(\text{aq}) \qquad (11.1.2)$$

(The stoichiometric coefficients of the right half-cell reaction have been multiplied by 2 to allow the electrons to cancel in the sum.)

It is entirely arbitrary whether we show a particular electrode at the left or the right (although often there is a preference to place the positive terminal at the right). If we exchange the two electrodes, then we must reverse the equation for the cell reaction.

---

[1] The term *electrode* can refer to either a half-cell or the electron conductor in the half-cell.

Section 11.2   The Cell Potential

The galvanic cell of Fig. 11.1, with only one electrolyte solution with essentially the same composition at both electrodes, is an example of a *cell without liquid junction* or *cell without transference*. The cell diagram

$$\text{Cu} \,|\, \text{Zn} \,|\, \text{Zn}^{2+}(\text{aq}) \,\vdots\, \text{Cu}^{2+}(\text{aq}) \,|\, \text{Cu}$$

describes a *cell with transference*, a galvanic cell with two electrolyte phases separated by a liquid junction represented by the dashed vertical bar. This is the Daniell cell drawn in Fig. 3.11 on page 54. If the liquid-junction potential is assumed to be eliminated, the liquid junction is shown by double-dashed vertical bars:

$$\text{Cu} \,|\, \text{Zn} \,|\, \text{Zn}^{2+}(\text{aq}) \,\|\, \text{Cu}^{2+}(\text{aq}) \,|\, \text{Cu} \tag{11.1.3}$$

For the cells of both cell diagrams, the cell reaction is

$$\text{Zn} + \text{Cu}^{2+}(\text{aq}) \rightarrow \text{Zn}^{2+}(\text{aq}) + \text{Cu} \tag{11.1.4}$$

11.2   THE CELL POTENTIAL

The **electric potential** ϕ at a point in space is defined as the change in the potential energy of a charge when it is brought from infinity to the point divided by the electric charge. We are concerned with the electric potential within a phase—the inner electric potential, or Galvani potential. We can measure the difference of this electric potential in the terminals provided they have the same chemical composition. If the terminals were of different metals, at least one of them would have an unknown metal-metal junction potential in its connection to the measuring circuit.

The **cell potential** of a galvanic cell is the electric potential difference between terminals of the same metal and is defined by

$$\Delta\phi \equiv \phi_R - \phi_L$$

where the subscripts R and L refer to the right and left terminals. The **emf** (electromotive force), E, of the cell is the cell potential measured under conditions of zero current when the cell is in an equilibrium state.

For example, Cell 11.1.3 with typical molalities of the zinc and copper(II) ions has its positive terminal at the right. $\Delta\phi$ and E are therefore positive. If we connect the two terminals by an external wire, electrons will flow through the wire from the left terminal to the right terminal and the cell reaction of Eq. 11.1.4 will occur spontaneously in the forward direction.

If, however, we draw the cell diagram the other way around,

$$\text{Cu} \,|\, \text{Cu}^{2+}(\text{aq}) \,\|\, \text{Zn}^{2+}(\text{aq}) \,|\, \text{Zn} \,|\, \text{Cu}$$

then the positive terminal is the left terminal, $\Delta\phi$ and E are negative, and electrons will flow through an external wire from the right terminal to the left terminal. Since the cell reaction

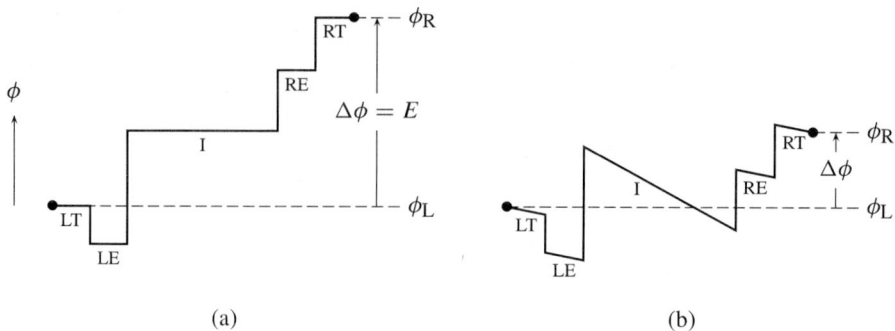

Figure 11.2 Profile of the Galvani potential in a galvanic cell (schematic). LT and RT are the left and right terminals, LE and RE are the left and right electrodes, and I denotes an ionic conductor such as an electrolyte solution. (a) Cell with zero current. (b) Cell with nonzero current.

shows oxidation at the left half-cell and reduction at the right, we must now write it as the reverse of Reaction 11.1.4

$$\mathrm{Cu} + \mathrm{Zn}^{2+}(\mathrm{aq}) \rightarrow \mathrm{Cu}^{2+}(\mathrm{aq}) + \mathrm{Zn}$$

which is the reverse of the spontaneous reaction.

What is the source of a cell potential? Within the cell, there is a jump in the Galvani potential at each junction between phases. We can think of this jump as due to two effects: a migration of charged species across the interface and orientation of polar molecules at the interface. The concept of potential jumps is shown schematically in Fig. 11.2(a) for a cell in an open circuit. We can measure the cell potential $\Delta\phi$, but not the individual jumps in ϕ at the phase junctions. The value of $\Delta\phi$ under these zero-current conditions is the cell emf E.

Figure 11.2(b) shows what happens if the terminals are short-circuited with an external resistor: Current flows through the cell, there are still jumps at the phase junctions, and the internal resistance of the conductors causes $\Delta\phi$ to be reduced in magnitude compared to E.

Some galvanic cells require two electrolyte solutions to keep solute reactants apart that would otherwise react directly. In these cells, a liquid junction is needed, and this may introduce an additional jump in ϕ called a *liquid-junction potential*. The liquid junction may be at a porous barrier or may include a bridging solution in a salt bridge. A commonly used kind of salt bridge is a tube filled with gel made from agar and concentrated aqueous KCl; this type of liquid junction is believed to reduce the liquid-junction potential to several millivolts or less. However the liquid junction is constructed, the liquid-junction potential is caused by diffusion of mobile ions. Unlike the potential jump at the interface of a solid, it involves a continuous irreversible process.

Figure 11.3 on the next page shows how we can use a potentiometer to determine the zero-current cell potential. Consider Fig. 11.3(a). Above the cell is shown an external circuit with a battery that allows an electric current to flow through a slidewire resistor. The cell's negative terminal is connected to the negative terminal of the battery. Since the cell is

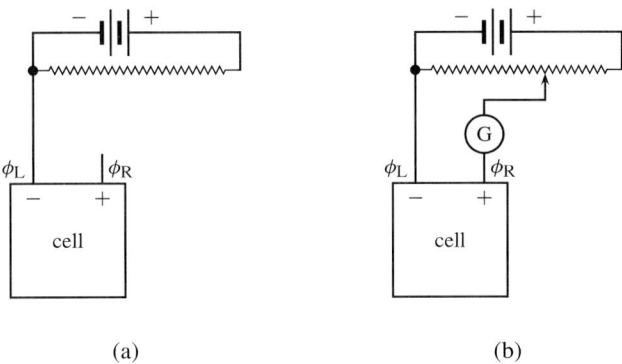

Figure 11.3 Potentiometer to measure the emf of a galvanic cell. (a) Cell with zero current. (b) Cell included in potentiometer circuit; G is a galvanometer.

not part of the circuit, no current flows through the cell, and $\phi_R - \phi_L$ is the zero-current cell potential, or emf of the cell. The left end of the slidewire has the same electric potential as the left terminal of the cell. In the setup shown, the electric potential increases as a linear function of distance to the right along the slidewire.

At some position along the slidewire, the electric potential is equal to ϕ_R. We can determine this position by connecting the right terminal of the cell to a slidewire contact as shown in Fig. 11.3(b). When we place the contact at this particular position along the slidewire, there is no electric potential gradient in the connecting wire, and we detect a condition of zero current in this wire with the galvanometer. It is now an easy matter to evaluate $\phi_R - \phi_L$ from the zero-current position of the contact; this value is still equal to the cell emf. When we keep the slidewire contact in this position, no current flows through the cell; but if we displace the contact from this position in either direction along the slidewire, current will flow in one direction or the other through the cell.

In practice, it is more convenient to measure the zero-current cell potential with a high-impedance volt meter (a volt meter that draws negligible current) instead of with a potentiometer circuit. The purpose of describing the potentiometer is to allow us to develop the thermodynamic theory in the next section.

11.3 MOLAR REACTION QUANTITIES

The **charge number**, ν_e, of a cell reaction is the amount of electrons transferred through the external circuit per unit advancement of the reaction. The charge number is a dimensionless quantity. For instance, for Reaction 11.1.2, the value of ν_e is 2, as we can see from the half-cell reactions.

The *system* is the galvanic cell. During a period that the cell reaction advances by a finite amount $\Delta\xi$, an amount of electrons $\nu_e \Delta\xi$ leaves the system at the left terminal and an equal amount enters at the right terminal. The **Faraday constant** F is a physical constant defined as the charge per amount of protons, and is equal to the product of the elementary charge (the charge of a proton) and the Avogadro constant: $F = eN_A$. Its value to five significant figures is $F = 96,485\,\text{C}\,\text{mol}^{-1}$. Thus, the charge entering the right terminal during advancement $\Delta\xi$ is $Q_{el} = -\nu_e F \Delta\xi$ (the minus sign is present because the charge of an electron is negative). For an infinitesimal advancement $d\xi$, the infinitesimal charge entering the right terminal is

$$dQ_{el} = -\nu_e F\,d\xi$$

Usually we need to consider only expansion work and electrical work. The general formula for electrical work, w_{el}, when the system is part of an electrical circuit is given by Eq. 3.8.1 on page 53: $dw_{el} = \Delta\phi\,dQ_{el}$, so the electrical work is

$$dw_{el} = -\nu_e F \Delta\phi\,d\xi$$

To relate $\Delta\phi$ to thermodynamic properties of the system, we must assume that the cell is in an equilibrium state when no current flows through it—that is, when $\Delta\phi$ is equal to E. We must furthermore assume that the advancement of the cell reaction in either direction, when we move the slidewire contact of the potentiometer shown in Fig. 11.3(b) in either direction from the zero-current position, approaches a reversible process as we make the displacement smaller and smaller.

According to Eq. 4.9.5 on page 91, the change dG of the Gibbs energy of a closed system during a reversible process at constant T and p is equal to the nonexpansion work dw'.[2] For the cell, dw' is the same as dw_{el}. According to our assumptions, $\Delta\phi$ equals the cell potential E when the cell reaction is carried out reversibly. Making these substitutions in the equation for dw_{el} gives us

$$dG = -\nu_e F E\,d\xi \qquad (11.3.1)$$
$$\text{(constant } T \text{ and } p\text{)}$$

> No real galvanic cell, even under conditions of zero current, is strictly speaking in an equilibrium state, and so the cell reaction cannot be truly reversible. For instance, in Cell 11.1.1, the electrolyte solution is saturated with both H_2 and $AgCl$, and the direct reaction $H_2 + 2\,Ag^+ \to 2\,Ag + 2\,H^+$ occurs irreversibly and continuously in the solution. The direct reaction is usually so slow that we can assume the measured emf would be the same in a hypothetical reversible cell in which the direct reaction is constrained to have zero rate. Diffusion of ions through a liquid junction is another source of irreversibility, and only if this diffusion is very slow can we treat a cell with a liquid junction as a reversible cell.

The molar reaction Gibbs energy $\Delta_r G_m$ is equal to the derivative $dG/d\xi$ at constant T and p (Eq. 8.7.3). Dividing both sides of Eq. 11.3.1 by $d\xi$, we obtain

$$\Delta_r G_m = -\nu_e F E \qquad (11.3.2)$$

---

[2] Although strictly speaking a galvanic cell is not a closed system, since electrons pass through the boundary, the number of electrons in the system is constant and Eq. 4.9.5 remains valid.

Section 11.3 Molar Reaction Quantities

This relation shows that a measured value of E for given conditions of a galvanic cell enables us to evaluate $\Delta_r G_m$ under these conditions.

We must now inquire about the meaning of the molar reaction Gibbs energy considering that different parts of the cell have different electric potentials. An argument in Sec. 9.6.4 showed that the chemical potential of ion i with charge number z_i, in a phase of electric potential ϕ, is related to the chemical potential in a phase of the same chemical composition and zero electric potential by

$$\mu_i(\phi) = \mu_i(0) + z_i F \phi \tag{11.3.3}$$

The molar reaction Gibbs energy is equal to $\sum_i \nu_i \mu_i$ (Eq. 8.7.2), where the chemical potential μ_i of an ion depends, according to Eq. 11.3.3, on the electric potential of the phase the ion is in. If all ions that are reactants or products of the cell reaction are present in the *same* electrolyte phase, or in two phases of equal electric potential, then we have from Eq. 11.3.3

$$\Delta_r G_m = \sum_i \nu_i \mu_i(\phi) = \sum_i \nu_i \mu_i(0) + F\phi \sum_i \nu_i z_i$$

Since charge is conserved in the cell reaction, the relation $\sum_i \nu_i z_i = 0$ holds and $\Delta_r G_m$ has the same value as it would if each reactant and product were in a phase of zero electric potential. The value of $\Delta_r G_m$ in this case is equal to the value for the same reaction taking place in the ordinary way in a reaction vessel.

If, however, some ions of the cell reaction are in one electrolyte of the cell and some in another, and the two phases have different electric potentials, the value of $\Delta_r G_m$ obtained from Eq. 11.3.2 depends on the value of the liquid-junction potential and is useless for further thermodynamic analysis unless we assume the liquid-junction potential is negligible or eliminated.

Consider a hypothetical cell in which each reactant and product of the cell reaction is in its standard state and any liquid-junction potential is negligible. We call it a *hypothetical cell* because, unless each reactant and product is a pure solid or liquid, hypothetical states are involved. The emf of this cell is called the **standard emf** or the **standard potential** of the cell reaction, $E°$. An experimental procedure for evaluating $E°$ is described in Sec. 11.5.

In this hypothetical cell, $\Delta_r G_m$ becomes the *standard* molar reaction Gibbs energy:

$$\Delta_r G_m° = -\nu_e F E° \tag{11.3.4}$$

Then since $\Delta_r G_m°$ is equal to $-RT \ln K$ (Eq. 8.8.4), we obtain the relation

$$\ln K = \frac{\nu_e F}{RT} E° \tag{11.3.5}$$

Equation 11.3.5 allows us to evaluate the thermodynamic equilibrium constant K of the cell reaction by a noncalorimetric method. Consider for example the cell

$$\text{Ag} \,|\, \text{Ag}^+(\text{aq}) \,\|\, \text{Cl}^-(\text{aq}) \,|\, \text{AgCl(s)} \,|\, \text{Ag}$$

The half-cell reactions are

$$\text{Ag(s)} \rightarrow \text{Ag}^+(\text{aq}) + \text{e}^-$$
$$\text{AgCl(s)} + \text{e}^- \rightarrow \text{Ag(s)} + \text{Cl}^-(\text{aq})$$

and the cell reaction is

$$\text{AgCl(s)} \rightarrow \text{Ag}^+(\text{aq}) + \text{Cl}^-(\text{aq})$$

The equilibrium constant of this reaction the solubility product K_s of silver chloride (Sec. 9.5.5). At 298.15 K, the standard cell emf is found to be $E° = -0.5770$ V. We can use this value in Eq. 11.3.5 to evaluate K_s at 298.15 K (Prob. 11.4).

Equation 11.3.5 also allows us to evaluate the standard molar reaction enthalpy by substitution in Eq. 9.1.9 on page 308:

$$\Delta_r H_m° = \nu_e F \left(T \frac{dE°}{dT} - E° \right) \qquad (11.3.6)$$
(no solute standard states based on concentration)

Finally, by combining Eq. 11.3.6 with $\Delta_r G_m° = \Delta_r H_m° - T \Delta_r S_m°$, we obtain an expression for the standard molar reaction entropy:

$$\Delta_r S_m° = \nu_e F \frac{dE°}{dT} \qquad (11.3.7)$$
(no solute standard states based on concentration)

11.4 THE NERNST EQUATION

The standard potential $E°$ of a cell reaction is the emf of the hypothetical galvanic cell in which each reactant and product is in its standard state. The value of $E°$ for a given cell reaction with given choices of standard states is a function only of temperature. The measured emf E of an actual cell, however, depends on the activities of the reactants and products as well as on temperature.

To derive the relation between E and activities, we substitute Eqs. 11.3.2 and 11.3.4 into the relation $\Delta_r G_m = \Delta_r G_m° + RT \ln Q$ (Eq. 8.8.2) and solve for E:

$$E = E° - \frac{RT}{\nu_e F} \ln Q \qquad (11.4.1)$$

Equation 11.4.1 is the **Nernst equation**. Here Q is the reaction quotient for the cell reaction defined by $Q = \prod_i a_i^{\nu_i}$ (Eq. 8.8.1).

If each reactant and product of the cell reaction are in their standard states, then each activity is 1 and $\ln Q$ is zero. The Nernst equation shows that the emf in this case is equal to the standard emf, which is what we expect. A decrease in product activities or an increase in reactant activities decreases the value of $\ln Q$ and increases the emf, as we would expect since E should be greater the more spontaneous is the forward cell reaction.

Section 11.4 The Nernst Equation

If the cell reaction comes to reaction equilibrium, as it will if we short-circuit the terminals with an external wire, the value of Q becomes equal to the thermodynamic equilibrium constant K, and the Nernst equation becomes $E = E° - (RT/\nu_e F)\ln K$. The term $(RT/\nu_e F)\ln K$ is equal to $E°$ (Eq. 11.3.5), so the emf becomes zero—the cell is "dead."

At $T = 298.15$ K $(25.00\,°\text{C})$, the value of RT/F is 0.02569 V, and we can write the Nernst equation in the compact form

$$E = E° - \frac{0.02569\text{ V}}{\nu_e}\ln Q \qquad (11.4.2)$$
$$(T = 298.15\text{ K})$$

We now illustrate the application of the Nernst equation with cell reaction 11.1.2 for Cell 11.1.1: $H_2(g) + 2\,\text{AgCl}(s) \rightarrow 2\,\text{Ag}(s) + 2\,H^+(aq) + 2\,Cl^-(aq)$. The expression for the reaction quotient is

$$Q = \frac{a_{\text{Ag}}^2 a_+^2 a_-^2}{a_{H_2} a_{\text{AgCl}}^2}$$

We may usually with negligible error set the pressure factors of the solids and solutes equal to unity. The activities of the solids are then 1, the solute activities are $a_+ = \gamma_+ m_+/m°$ and $a_- = \gamma_- m_-/m°$, and the hydrogen activity is $a_{H_2} = f_{H_2}/p°$. The ion molalities m_+ and m_- are equal to the HCl molality m_B. The expression for Q becomes

$$Q = \frac{\gamma_+^2 \gamma_-^2 (m_B/m°)^4}{f_{H_2}/p°} = \frac{\gamma_\pm^4 (m_B/m°)^4}{f_{H_2}/p°}$$

and the Nernst equation for this cell is

$$E = E° - \frac{RT}{2F}\ln\frac{\gamma_\pm^4 (m_B/m°)^4}{f_{H_2}/p°}$$
$$= E° - \frac{2RT}{F}\ln\gamma_\pm - \frac{2RT}{F}\ln\frac{m_B}{m°} + \frac{RT}{2F}\ln\frac{f_{H_2}}{p°} \qquad (11.4.3)$$

By measuring E for a cell with known values of m_B and f_{H_2}, and with a derived value of $E°$, we can use this equation to find the mean ionic activity coefficient $\gamma_\pm$ of the HCl solute.

We can always multiply the stoichiometric coefficients of the chemical equation of a cell reaction by an arbitrary factor. How does this affect the Nernst equation? Suppose we decide to multiply the coefficients of Eq. 11.1.2 by one half:

$$\tfrac{1}{2}H_2(g) + \text{AgCl}(s) \rightarrow H^+(aq) + Cl^-(aq) + \text{Ag}(s)$$

With this changed way of writing the reaction, the value of ν_e is changed from 2 to 1 and the Nernst equation becomes

$$E = E° - \frac{RT}{F}\ln\frac{\gamma_\pm^2 (m_B/m°)^2}{(f_{H_2}/p°)^{1/2}}$$

which yields the same value of E for given cell conditions as does Eq. 11.4.3. Of course, this must be the case since physically the cell is the same no matter how we write its cell reaction, and measurable physical quantities such as E are unaffected. However, calculated molar reaction quantities such as $\Delta_r G_m$ and $\Delta_r G_m^\circ$ *do* depend on how we write the cell reaction because they are changes of quantities per extent of reaction.

11.5 EVALUATION OF THE STANDARD POTENTIAL

The standard potential E° of the cell reaction of a galvanic cell is a quantity whose value, as we have seen, is useful for several purposes in thermodynamics. The value of E° for a particular cell reaction depends only on temperature. To evaluate it, we can extrapolate an appropriate function to infinite dilution where ionic activity coefficients are unity.

To see how this procedure works, consider again Cell 11.1.1 for which the cell potential depends on the molality m_B of the HCl solute according to Eq. 11.4.3. We can rearrange the equation to

$$E^\circ = E + \frac{2RT}{F} \ln \gamma_\pm + \frac{2RT}{F} \ln \frac{m_B}{m^\circ} - \frac{RT}{2F} \ln \frac{f_{H_2}}{p^\circ}$$

For given conditions of the cell, we can measure all quantities on the right side of this equation, except the term $(2RT/F) \ln \gamma_\pm$ containing the mean ionic activity coefficient of the electrolyte. We do not know the exact value of $\ln \gamma_\pm$ for any given molality until we have evaluated E°. We do know that, as m_B approaches zero, $\gamma_\pm$ approaches unity and $\ln \gamma_\pm$ must approach zero. The Debye–Hückel formula of Eq. 7.8.24 on page 225 is a theoretical expression for $\ln \gamma_\pm$, which more closely approximates the actual value the lower is the ionic strength. Accordingly, we define the quantity

$$E' \equiv E + \frac{2RT}{F} \left(-\frac{A\sqrt{m_B}}{1 + Ba\sqrt{m_B}} \right) + \frac{2RT}{F} \ln \frac{m_B}{m^\circ} - \frac{RT}{2F} \ln \frac{f_{H_2}}{p^\circ} \quad (11.5.1)$$

The expression in parentheses is the Debye–Hückel formula for $\ln \gamma_\pm$ (with I_m replaced by m_B), in which A and B have known values at any temperature (Sec. 7.8.4) and a is an ion-size parameter for which we can choose a reasonable value. At a given temperature, we can evaluate E' experimentally as a function of m_B. The value of E' differs from E° by contributions to $(2RT/F) \ln \gamma_\pm$ not accounted for by the Debye–Hückel formula. Since these contributions approach zero in the limit of infinite dilution, the extrapolation of measured values of E' to $m_B = 0$ gives a precise value of E°.

Figure 11.4 on the next page shows this extrapolation using a set of data at 298.15 K available in the literature. The extrapolated value indicated by the filled circle is $E^\circ = 0.2222$ V, and the uncertainty is on the order of only 0.1 mV.

Once we have established the value of E°, we may use the cell potential measured for a solution of any molality to evaluate the mean ionic activity coefficient in that solution by applying the Nernst equation, Eq. 11.4.3. This is how the experimental curve for HCl in Fig. 7.14 on page 226 was obtained.

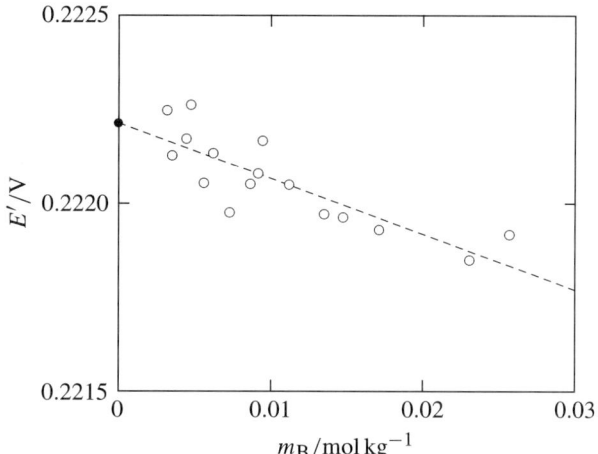

Figure 11.4 E' (defined by Eq. 11.5.1) as a function of HCl molality for the cell of Fig. 11.1 at 298.15 K. The dashed line is a least-squares fit to a linear relation. (Data of Harned, H. S.; Ehlers, R. W. *J. Am. Chem. Soc.* **1932**, *54*, 1350–1357; f_{H_2} was set equal to p_{H_2}, and the parameter a was set equal to 4.3×10^{-10} m.)

11.6 STANDARD ELECTRODE POTENTIALS

Section 11.5 explained how a series of measurements of the emf of a galvanic cell at different electrolyte molalities enables us to evaluate the standard emf, $E°$, which is the standard potential of the cell reaction. It is not necessary to carry out this involved experimental procedure for each individual cell reaction that may be of interest. Instead, we can calculate $E°$ from the values of the *standard electrode potentials* of the two electrodes of the cell of interest. Standard electrode potentials use a *standard hydrogen electrode* as a reference electrode.

A **standard hydrogen electrode** is a hydrogen electrode, such as the electrode shown at the left in Fig. 11.1, in which the $H_2(g)$ and $H^+(aq)$ species are in their standard states. Since these are *hypothetical* gas and solute standard states, the standard hydrogen electrode is a hypothetical electrode—not one that we can actually construct.

The **standard electrode potential** of an electrode (i.e., of a half-cell) is defined as the standard emf of a cell in which the electrode at the left is a hydrogen electrode and the electrode at the right is the electrode in question. In other words, the standard electrode potential is the emf of a hypothetical galvanic cell that has a standard hydrogen electrode at the left and the electrode of interest at the right, with each reactant and product of the cell reaction present in its standard state.

For example, because Cell 11.1.1 has a hydrogen electrode at the left and a silver-silver chloride electrode at the right, the standard emf of this cell is equal to the standard electrode potential of the silver-silver chloride electrode. A standard electrode potential is also called a standard *reduction* potential because the half-cell reaction for the electrode at the right is a *reduction*. The same symbol, $E°$, is used for a standard electrode potential as for a standard cell emf, which can be confusing.

Another example of a cell with a hydrogen electrode at the left is

$$\text{Pt} | \text{H}_2(\text{g}) | \text{H}^+(\text{aq}) \,\|\, \text{Zn}^{2+}(\text{aq}) | \text{Zn} | \text{Pt}$$

The standard emf of this cell at 298.15 K is found to be -0.762 V, so the standard electrode potential of the zinc electrode is -0.762 V.

Since a cell with hydrogen electrodes at both the left and right has a standard emf of zero, the standard electrode potential of the hydrogen electrode is *zero*.

The standard electrode potential of a given kind of electrode is a function only of temperature (except that of the hydrogen electrode, which is zero at all temperatures). As mentioned at the beginning of this section, we can calculate the standard emf of any cell from the standard electrode potentials of the two electrodes. To see how this works, consider a cell that combines a zinc electrode and a silver-silver chloride electrode:

$$\text{Ag} | \text{Zn} | \text{Zn}^{2+}(\text{aq}) \,\|\, \text{Cl}^-(\text{aq}) | \text{AgCl}(\text{s}) | \text{Ag} \tag{11.6.1}$$

The cell reaction and standard molar reaction Gibbs energy are

$$\text{Zn} + 2\,\text{AgCl} \rightarrow \text{Zn}^{2+} + 2\,\text{Ag} + 2\,\text{Cl}^- \qquad \Delta_r G_m^\circ = -\nu_e F E^\circ \tag{11.6.2}$$

The standard electrode potential of the zinc electrode is the standard emf of the cell

$$\text{Pt} | \text{H}_2(\text{g}) | \text{H}^+(\text{aq}) \,\|\, \text{Zn}^{2+}(\text{aq}) | \text{Zn} | \text{Pt} \qquad (E°)' = -0.762 \text{ V} \tag{11.6.3}$$

and the standard electrode potential of the silver-silver chloride electrode is the standard emf of the cell

$$\text{Pt} | \text{H}_2(\text{g}) | \text{H}^+(\text{aq}), \text{Cl}^-(\text{aq}) | \text{AgCl}(\text{s}) | \text{Ag} | \text{Pt} \qquad (E°)'' = 0.2222 \text{ V} \tag{11.6.4}$$

(The standard emf values are for 298.15 K.) The cell reactions of Cells 11.6.3 and 11.6.4, and their standard molar reaction Gibbs energies, are

$$\text{H}_2 + \text{Zn}^{2+} \rightarrow 2\,\text{H}^+ + \text{Zn} \qquad (\Delta_r G_m^\circ)' = -\nu_e F (E°)' \tag{11.6.5}$$

$$\text{H}_2 + 2\,\text{AgCl} \rightarrow 2\,\text{Ag} + 2\,\text{H}^+ + 2\,\text{Cl}^- \qquad (\Delta_r G_m^\circ)'' = -\nu_e F (E°)'' \tag{11.6.6}$$

All three cell reactions are written so as to have the same value of the charge number, $\nu_e = 2$. Since Reaction 11.6.2 is equal to Reaction 11.6.6 minus Reaction 11.6.5, a given advancement of Reaction 11.6.2 causes the same changes as the same advancement of Reaction 11.6.6 combined with the same advancement in the reverse direction of Reaction 11.6.5. Therefore, the standard molar reaction Gibbs energy of Reaction 11.6.2 is equal to the difference of the standard molar reaction Gibbs energies of Reactions 11.6.6 and 11.6.5:

$$\Delta_r G_m^\circ = (\Delta_r G_m^\circ)'' - (\Delta_r G_m^\circ)'$$

Section 11.6 Standard Electrode Potentials

We substitute from Eq. 11.3.4:

$$-\nu_e F E^\circ = -\nu_e F (E^\circ)'' + \nu_e F (E^\circ)'$$

This gives us the relation $E^\circ = (E^\circ)'' - (E^\circ)'$. Since $(E^\circ)''$ and $(E^\circ)'$ are equal to the standard electrode potentials of the right and left electrodes, respectively, of Cell 11.6.1, we can write the relation in the form

$$E^\circ = E_R^\circ - E_L^\circ \tag{11.6.7}$$

Equation 11.6.7 is a relation applicable to any galvanic cell: E° is the standard emf of the cell, and E_R° and E_L° are the standard electrode potentials of the right and left electrodes. For Cell 11.6.1 at 298.15 K, the calculation is

$$E^\circ = 0.2222 \text{ V} - (-0.762 \text{ V}) = 0.984 \text{ V}$$

The fact that E° is positive tells us that, under standard conditions, the right terminal of this cell is positive.

It should be apparent that we can use Eq. 11.6.7 to calculate a standard electrode potential from another standard electrode potential and the standard emf of a cell that contains both electrodes. Neither electrode has to be a hydrogen electrode.

Note that if we multiply the stoichiometric coefficients of a cell reaction or half-cell reaction by an arbitrary factor, we should *not* multiply E°, E_R°, or E_L° by this factor. The values of standard cell potentials and standard electrode potentials do not depend on how we write a reaction equation.

The derivation leading to Eq. 11.6.7 shows that if a cell reaction is the sum or difference of two other cell reactions, the standard potential of the cell reaction is the sum or difference of the standard potentials of the other cell reactions. This additive property does not always work for standard *electrode* potentials, however. Instead, the fundamental principle is the additive property of standard molar reaction Gibbs energies of cell reactions. For example, consider the following galvanic cells which include a hydrogen electrode and aqueous iron(II) and iron(III) ions (the standard emf values are for 298.15 K):

(1) Pt | H_2(g) | H^+(aq) ‖ Fe^{3+}(aq), Fe^{2+}(aq) | Pt $E^\circ = 0.77$ V

(2) Pt | H_2(g) | H^+(aq) ‖ Fe^{2+}(aq) | Fe | Pt $E^\circ = -0.44$ V

(3) Pt | H_2(g) | H^+(aq) ‖ Fe^{3+}(aq) | Fe | Pt

The reactions of the half-cells shown at the right of these cells are

(1) $Fe^{3+} + e^- \rightarrow Fe^{2+}$ $\nu_e = 1$

(2) $Fe^{2+} + 2e^- \rightarrow Fe$ $\nu_e = 2$

(3) $Fe^{3+} + 3e^- \rightarrow Fe$ $\nu_e = 3$

Because the galvanic cells each have a hydrogen electrode at the left, their standard emfs are the standard electrode potentials of the half-cells with the reactions written above. How can we find the standard electrode potential of half-cell Reaction (3) from the values of $E°$ for half-cell Reactions (1) and (2)? We reason as follows. The standard molar reaction Gibbs energy of the cell reaction of each cell is given by $\Delta_r G_m° = -\nu_e F E°$ (Eq. 11.3.4). Since the half-cell reactions are written with coefficients that make the sum of Reactions (1) and (2) equal to Reaction (3), the sum of $\Delta_r G_m°$ for the cell reactions of Cells (1) and (2) is equal to $\Delta_r G_m°$ for the cell reaction of Cell (3):

$$-F(0.77 \text{ V}) - 2F(-0.44 \text{ V}) = -3FE°$$

Here $E°$ is the standard electrode potential for half-cell Reaction (3). The value of this standard electrode potential at 298.15 K is therefore

$$E° = \tfrac{1}{3}(0.77 \text{ V}) + \tfrac{2}{3}(-0.44 \text{ V}) = -0.04 \text{ V}$$

PROBLEMS

11.1 Before 1982, the standard pressure was usually taken as 1 atm. For the cell shown in Fig. 11.1, what correction is needed, for a value of $E°$ obtained at 25 °C and using the older convention, to change the value to one corresponding to a standard pressure of 1 bar? Equation 11.3.4 can be used for this calculation.

11.2 Careful measurements of the emf of the cell $\text{Pt} \,|\, \text{H}_2(\text{g}) \,|\, \text{HCl(aq)} \,|\, \text{AgCl(s)} \,|\, \text{Ag} \,|\, \text{Pt}$ (Bates, R. G.; Bower, V. E. *J. Res. Natl. Bur. Stand. (U.S.)* **1954**, *53*, 283–290) yielded, at 298.15 K and using a standard pressure of 1 bar, the values $E° = 0.22217$ V and $dE°/dT = -6.462 \times 10^{-4}$ V K^{-1}. (The requested calculated values are close to, but not exactly the same as, the values listed in Appendix H, which are based on the same data combined with data of other workers.)

(a) Evaluate $\Delta_r G_m°$, $\Delta_r S_m°$, and $\Delta_r H_m°$ at 298.15 K for the reaction

$$\tfrac{1}{2}\text{H}_2(\text{g}) + \text{AgCl(s)} \rightarrow \text{H}^+(\text{aq}) + \text{Cl}^-(\text{aq}) + \text{Ag(s)}$$

(b) Problem 9.16 showed how the standard molar enthalpy of formation of the aqueous chloride ion may be evaluated based on the convention $\Delta_f H_m°(\text{H}^+, \text{aq}) = 0$. If this value is combined with the value of $\Delta_r H_m°$ obtained in Part (a) of the present problem, the standard molar enthalpy of formation of crystalline silver chloride may be evaluated. Carry out this calculation for $T = 298.15$ K using the value $\Delta_f H_m°(\text{Cl}^+, \text{aq}) = -167.08$ kJ mol^{-1} from Appendix H.

(c) By a similar procedure, evaluate the standard molar entropy, the standard molar entropy of formation, and the standard molar Gibbs energy of formation of crystalline silver chloride at 298.15 K. You need the following standard molar entropies evaluated from spectroscopic and calorimetric data:
$S_m°(\text{H}_2, \text{g}) = 130.68$ J K^{-1} mol^{-1} $S_m°(\text{Cl}_2, \text{g}) = 223.08$ J K^{-1} mol^{-1}
$S_m°(\text{Cl}^-, \text{aq}) = 56.60$ J K^{-1} mol^{-1} $S_m°(\text{Ag}, \text{s}) = 42.55$ J K^{-1} mol^{-1}

11.3 The standard emf was determined over a range of temperatures for the cell

$$\text{Pt} \,|\, \text{Ag} \,|\, \text{AgCl(s)} \,|\, \text{HCl(aq)} \,|\, \text{Cl}_2(\text{g}) \,|\, \text{Pt}$$

(Faita, G.; Longhi, P.; Mussini, T. *J. Electrochem. Soc.* **1967**, *114*, 340–343). At $T = 298.15\,\text{K}$, the standard emf was found to be $E° = 1.13579\,\text{V}$, and its temperature derivative was found to be $dE°/dT = -5.9863 \times 10^{-4}\,\text{V K}^{-1}$.

(a) Write the cell reaction for this cell.

(b) Use the data to evaluate the standard molar enthalpy of formation and the standard molar Gibbs energy of formation of crystalline silver chloride at 298.15 K. (Note that this calculation provides values of quantities also calculated in Prob. 11.2 using independent data.)

11.4 Use data in Sec. 11.3 to evaluate the solubility product of silver chloride at 298.15 K.

11.5 The emf of the galvanic cell

$$\text{Pt} \mid \text{H}_2(\text{g}, f = 1\,\text{bar}) \mid \text{HCl}(\text{aq}, 0.500\,\text{mol kg}^{-1}) \mid \text{Cl}_2(\text{g}, f = 1\,\text{bar}) \mid \text{Pt}$$

is found to be $E = 1.410\,\text{V}$ at $25.00\,°\text{C}$. The standard emf is $E° = 1.360\,\text{V}$.

(a) Find the cell reaction and calculate its thermodynamic equilibrium constant at $25.00\,°\text{C}$.

(b) Use the cell measurement to calculate the mean ionic activity coefficient of aqueous HCl at $25.00\,°\text{C}$ and a molality of $0.500\,\text{mol kg}^{-1}$.

11.6 Consider the following galvanic cell, which combines a hydrogen electrode and a calomel electrode:

$$\text{Pt} \mid \text{H}_2(\text{g}) \mid \text{HCl}(\text{aq}) \mid \text{Hg}_2\text{Cl}_2(\text{s}) \mid \text{Hg}(\text{l}) \mid \text{Pt}$$

(a) Write the cell reaction.

(b) At 298.15 K, the standard emf of this cell is $E° = 0.2680\,\text{V}$. Using the value of $\Delta_\text{f}G°_\text{m}$ for the aqueous chloride ion in Appendix H, calculate the standard molar Gibbs energy of formation of crystalline mercury(I) chloride (calomel) at 298.15 K.

(c) Calculate the solubility product of mercury(I) chloride at 298.15 K. The dissolution reaction is $\text{Hg}_2\text{Cl}_2(\text{s}) \rightleftharpoons \text{Hg}_2^{2+}(\text{aq}) + 2\,\text{Cl}^-(\text{aq})$. Take values for the standard molar Gibbs energies of formation of the aqueous ions from Appendix H.

11.7 The standard electrode potentials of the half-cells with the following reactions have been measured at 298.15 K, with the results shown:

$$\text{Cu}^{2+}(\text{aq}) + 2\,\text{e}^- \rightarrow \text{Cu}(\text{s}) \qquad E° = 0.34\,\text{V}$$
$$\text{Cu}^+(\text{aq}) + \text{e}^- \rightarrow \text{Cu}(\text{s}) \qquad E° = 0.52\,\text{V}$$

Calculate the standard emf at 298.15 K of the galvanic cell

$$\text{Pt} \mid \text{H}_2(\text{g}) \mid \text{H}^+(\text{aq}) \mid\mid \text{Cu}^{2+}(\text{aq}), \text{Cu}^+(\text{aq}) \mid \text{Pt}$$

Appendix A

DEFINITIONS OF THE SI BASE UNITS

The official definitions of the base units given in the IUPAC Green Book are as follows. The paragraphs in smaller type have been added.

The **meter** is the length of path traveled by light in vacuum during a time interval of 1/299 792 458 of a second.

The **kilogram** is the unit of mass; it is equal to the mass of the international prototype of the kilogram.

> The international prototype is a platinum-iridium cylinder stored in a vault of the International Bureau of Weights and Measures in Sèvres near Paris, France.

The **second** is the duration of 9 192 631 770 periods of the radiation corresponding to the transition between the two hyperfine levels of the ground state of the cesium-133 atom.

The **kelvin**, unit of thermodynamic temperature, is the fraction 1/273.16 of the thermodynamic temperature of the triple point of water.

The **mole** is the amount of substance of a system which contains as many elementary entities as there are atoms in 0.012 kilogram of carbon-12. When the mole is used, the elementary entities must be specified and may be atoms, molecules, ions, electrons, other particles, or specified groups of such particles.

> The Avogadro constant N_A is the number of elementary entities per mole. Its value to four significant digits is found to be $N_A = 6.022 \times 10^{23}$ mol^{-1}.

The **ampere** is that constant current which, if maintained in two straight parallel conductors of infinite length, of negligible circular cross-section, and placed 1 meter apart in vacuum, would produce between these conductors a force equal to 2×10^{-7} newton per meter of length.

The **candela** is the luminous intensity, in a given direction, of a source that emits monochromatic radiation of frequency 540×10^{12} hertz and that has a radiant intensity in that direction of (1/683) watt per steradian.

Appendix B

PHYSICAL CONSTANTS

The following table lists values of fundamental physical constants used in thermodynamic calculations. The values, except for those marked "exact," are adjusted values recommended in 1998 by the Committee on Data for Science and Technology (CODATA).[1] The number in parentheses at the end of a value is the standard deviation uncertainty in the right-most digits of the value.

Physical quantity	Symbol	Value in SI units
Avogadro constant	N_A	$6.022\,141\,99(47) \times 10^{23}\,\text{mol}^{-1}$
elementary charge	e	$1.602\,176\,462(63) \times 10^{-19}\,\text{C}$
Faraday constant	F	$9.648\,534\,15(39) \times 10^{4}\,\text{C}\,\text{mol}^{-1}$
gas constant	R	$8.314\,472(15)\,\text{J}\,\text{K}^{-1}\,\text{mol}^{-1}$
permeability of vacuum	μ_0	$4\pi \times 10^{-7}\,\text{N}\,\text{A}^{-2}$ (exact)
permittivity of vacuum	ϵ_0	$(10^7\,\text{C}^2\,\text{m}\,\text{J}^{-1}\,\text{s}^{-2})/4\pi c_0^2$ (exact) $= 8.854\,187\,816\ldots \times 10^{-12}\,\text{C}^2\,\text{J}^{-1}\,\text{m}^{-1}$
speed of light in vacuum	c_0	$2.997\,924\,58 \times 10^{8}\,\text{m}\,\text{s}^{-1}$ (exact)
standard acceleration of free fall	g_n	$9.806\,65\,\text{m}\,\text{s}^{-2}$ (exact)

---

[1] P. J. Mohr and B. N. Taylor, "The 1998 CODATA Recommended Values of the Fundamental Physical Constants, Web Version 3.0," available at http://physics.nist.gov/constants.

Appendix C

SYMBOLS FOR PHYSICAL QUANTITIES

This appendix lists the symbols for physical quantities (variables) used in this book. The symbols are those recommended in the IUPAC Green Book except for quantities followed by an asterisk (*).

Symbol	Physical quantity	SI unit
Greek letters		
alpha		
α	degree of reaction, dissociation, etc.	(dimensionless)
	cubic expansion coefficient	K^{-1}
gamma		
γ	surface tension	$N\,m^{-1}$, $J\,m^{-2}$
γ_i	activity coefficient of species i, pure liquid or solid standard state*	(dimensionless)
$\gamma_{m,i}$	activity coefficient of species i, molality basis	(dimensionless)
$\gamma_{c,i}$	activity coefficient of species i, concentration basis	(dimensionless)
$\gamma_{x,i}$	activity coefficient of species i, mole fraction basis	(dimensionless)
$\gamma_{\pm}$	mean ionic activity coefficient	(dimensionless)
Γ	pressure factor (activity of a reference state)*	(dimensionless)
epsilon		
ϵ	efficiency of a heat engine	(dimensionless)
	energy equivalent of a calorimeter*	$J\,K^{-1}$
theta		
θ	angle of rotation	(dimensionless)
kappa		
κ	reciprocal radius of ionic atmosphere	m^{-1}

(continued from previous page)

Symbol	Physical quantity	SI unit
κ_T	isothermal compressibility	Pa^{-1}
mu		
μ	chemical potential	$J\,mol^{-1}$
μ_{JT}	Joule–Thomson coefficient	$K\,Pa^{-1}$
nu		
ν	number of ions per formula unit	(dimensionless)
	stoichiometric number	(dimensionless)
ν_e	charge number of electrochemical cell reaction	(dimensionless)
ν_+	number of cations per formula unit	(dimensionless)
ν_-	number of anions per formula unit	(dimensionless)
xi		
ξ	advancement (extent of reaction)	mol
pi		
Π	osmotic pressure	Pa
rho		
ρ	density	$kg\,m^{-3}$
tau		
τ	torque*	J
phi		
ϕ	fugacity coefficient	(dimensionless)
	electric potential	V
$\Delta\phi$	electric potential difference of a galvanic cell	V
ϕ_m	osmotic coefficient, molality basis	(dimensionless)
Φ	free-energy function*	$J\,K^{-1}\,mol^{-1}$
Φ_L	relative apparent molar enthalpy of solute*	$J\,mol^{-1}$
omega		
ω	angular velocity	s^{-1}

Roman letters

A	Helmholtz energy	J
A_s	surface area	m^2
a	activity	(dimensionless)
B	second virial coefficient	$m^3\,mol^{-1}$
C	number of components*	(dimensionless)
C_p	heat capacity at constant pressure	$J\,K^{-1}$
C_V	heat capacity at constant volume	$J\,K^{-1}$
c	concentration	$mol\,m^{-3}$
E	energy	J
	emf of galvanic cell	V

(continued from previous page)

Symbol	Physical quantity	SI unit
$\boldsymbol{E}$	electric field strength	$V\,m^{-1}$
$E°$	standard emf	V
	standard electrode potential	V
E_k	kinetic energy	J
E_p	potential energy	J
F	force	N
	number of degrees of freedom*	(dimensionless)
f	fugacity	Pa
g	acceleration of free fall	$m\,s^{-2}$
G	Gibbs energy	J
h	height	m
H	enthalpy	J
$\boldsymbol{H}$	magnetic field strength	$A\,m^{-1}$
I	electric current	A
I_m	ionic strength, molality basis	$mol\,kg^{-1}$
I_c	ionic strength, concentration basis	$mol\,m^{-3}$
K	thermodynamic equilibrium constant	(dimensionless)
K_a	acid dissociation constant	(dimensionless)
$K_c°$	thermodynamic equilibrium constant, concentration basis for solute*	(dimensionless)
$K_m°$	thermodynamic equilibrium constant, molality basis for solutes*	(dimensionless)
$K_x°$	thermodynamic equilibrium constant, mole fraction basis for solutes*	(dimensionless)
K_c	equilibrium constant, concentration basis	$(mol\,m^{-3})^{\sum \nu}$
K_m	equilibrium constant, molality basis	$(mol\,kg^{-1})^{\sum \nu}$
K_p	equilibrium constant, pressure basis	$Pa^{\sum \nu}$
K_s	solubility product	(dimensionless)
$K_{c,i}$	Henry's law constant of species i, concentration basis*	$Pa\,m^3\,mol^{-1}$
$K_{m,i}$	Henry's law constant of species i, molality basis*	$Pa\,kg\,mol^{-1}$
$K_{x,i}$	Henry's law constant of species i, mole fraction basis*	Pa
l	length	m
L	relative partial molar enthalpy*	$J\,mol^{-1}$
M	molar mass	$kg\,mol^{-1}$
$\boldsymbol{M}$	magnetization	$A\,m^{-1}$
M_r	relative molecular mass (molecular weight)	(dimensionless)

(continued from previous page)

Symbol	Physical quantity	SI unit
m	mass	kg
m_i	molality of species i	mol kg^{-1}
N	number of entities (molecules, atoms, ions, formula units, etc.)	(dimensionless)
n	amount of substance	mol
P	number of phases*	(dimensionless)
p	pressure	Pa
	partial pressure	Pa
$\boldsymbol{P}$	dielectric polarization	C m^{-2}
Q	reaction quotient	(dimensionless)
Q_{el}	electric charge*	C
q	heat	J
R_{el}	electric resistance*	Ω
S	entropy	J K^{-1}
s	solubility	mol m^{-3}
	number of species*	(dimensionless)
T	thermodynamic temperature	K
t	time	s
	Celsius temperature	°C
U	internal energy	J
V	volume	m^3
v	velocity, speed	m s^{-1}
w	work	J
w_{el}	electrical work*	J
w'	nonexpansion work*	J
x	mole fraction	(dimensionless)
	Cartesian space coordinate	m
y	mole fraction in gaseous mixture	(dimensionless)
	Cartesian space coordinate	m
Z	compression factor (compressibility factor)	(dimensionless)
z	charge number of an ion	(dimensionless)
	Cartesian space coordinate	m

Appendix D

MISCELLANEOUS ABBREVIATIONS AND SYMBOLS

D.1 PHYSICAL STATES

These abbreviations for physical states (states of aggregation) may be appended in parentheses to chemical formulas or used as superscripts to symbols for physical quantities. All but "mixt" are recommended in the IUPAC Green Book.

g	gas or vapor
l	liquid
s	solid
cr	crystalline
mixt	mixture
sln	solution
aq	aqueous solution
aq, ∞	aqueous solution at infinite dilution

D.2 SUBSCRIPTS FOR CHEMICAL PROCESSES

These abbreviations are used as subscripts to the Δ symbol (e.g., $\Delta_{vap} H$). They are recommended in the IUPAC Green Book.

vap	vaporization, evaporation (l $\to$ g)		ads	adsorption
sub	sublimation (s $\to$ g)		dpl	displacement
fus	melting, fusion (s $\to$ l)		imm	immersion
trs	transition between two phases		r	reaction in general
mix	mixing of fluids		at	atomization
sol	solution of a solute in solvent		c	combustion reaction
dil	dilution		f	formation reaction

D.3 SUPERSCRIPTS

These abbreviations and symbols are used as superscripts to symbols for physical quantities. All but "int" and "ref" are recommended in the IUPAC Green Book.

°	standard
*	pure substance
∞	infinite dilution
id	ideal
int	integral
E	excess quantity
ref	reference state

Appendix E

CALCULUS REVIEW

E.1 DERIVATIVES

Let f be a function of the variable x, and let Δf be the change in f when x changes by Δx. Then the **derivative** df/dx is the ratio $\Delta f/\Delta x$ in the limit as Δx approaches zero. The derivative is also the slope of the curve of f plotted as a function of x.

If f is a function of the independent variables x, y, and z, the **partial derivative** $(\partial f/\partial x)_{y,z}$ is the derivative df/dx, with y and z held constant. The notation $\partial f/\partial x$ denotes the derivative df/dx with all independent variables except x held constant.

The following is a short list of formulas likely to be needed. In these formulas, u and v are arbitrary functions of x and a is a constant.

$$\frac{d(u^a)}{dx} = au^{a-1}\frac{du}{dx}$$

$$\frac{d(uv)}{dx} = u\frac{dv}{dx} + v\frac{du}{dx}$$

$$\frac{d(u/v)}{dx} = \left(\frac{1}{v^2}\right)\left(v\frac{du}{dx} - u\frac{dv}{dx}\right)$$

$$\frac{d(\ln x)}{dx} = \frac{1}{x}$$

$$\frac{d(e^{ax})}{dx} = ae^{ax}$$

$$\frac{df(u)}{dx} = \frac{df(u)}{du} \cdot \frac{du}{dx}$$

E.2 INTEGRALS

Let f be a function of the variable x. Imagine the range of x between the limits x' and x'' to be divided into many small increments of size Δx_i ($i = 1, 2, \ldots$). Let f_i be the value of f when x is in the middle of the range of the ith increment. Then the **integral**

$$\int_{x'}^{x''} f\, dx$$

Section E.2 Integrals

is the sum $\sum_i f_i \Delta x_i$ in the limit as each Δx_i approaches zero and the number of terms in the sum approaches infinity. The integral is also the area under the curve of f plotted as a function of x, from $x = x'$ to $x = x''$. The function f is the **integrand**, which is integrated over the integration variable x.

This book uses the following integrals:

$$\int_{x'}^{x''} dx = x'' - x'$$

$$\int_{x'}^{x''} \frac{dx}{x} = \ln\left|\frac{x''}{x'}\right|$$

$$\int_{x'}^{x''} x^a \, dx = \frac{1}{a+1}\left[(x'')^{a+1} - (x')^{a+1}\right] \qquad (a \text{ is a constant other than } -1)$$

$$\int_{x'}^{x''} \frac{dx}{ax+b} = \frac{1}{a}\ln\left|\frac{ax''+b}{ax'+b}\right| \qquad (a \text{ is a constant})$$

Here are examples of the use of the expression for the third integral with a set equal to 1 and to -2:

$$\int_{x'}^{x''} x \, dx = \frac{1}{2}\left[(x'')^2 - (x')^2\right]$$

$$\int_{x'}^{x''} \frac{dx}{x^2} = -\left(\frac{1}{x''} - \frac{1}{x'}\right)$$

Appendix F

MATHEMATICAL PROPERTIES OF STATE FUNCTIONS

The **differential** df of a state function f is an infinitesimal change of f. Since the value of a state function by definition depends only on the state of the system, the integral $\int df$ between an initial and final state equals the change in f, and this change is independent of the path. A differential with this property is called an *exact* differential.

A state function f treated as a dependent variable is a function of a certain number of independent variables that are state functions. The **total differential** of f is df expressed in terms of the differentials of the independent variables and has the form

$$df = \left(\frac{\partial f}{\partial x}\right) dx + \left(\frac{\partial f}{\partial y}\right) dy + \left(\frac{\partial f}{\partial z}\right) dz + \ldots$$

There are as many terms as there are independent variables.

Figure F.1 on the next page explains this expression in the case of a function of two independent variables, $f = f(x, y)$. For small changes dx and dy of the independent variables, the surface representing $f(x, y)$ looks like a plane and is shaded in the figure.

We may write the total differential of a function f of two variables x and y in the form

$$df = M\,dx + N\,dy$$

where M is the partial derivative $(\partial f/\partial x)_y$ and N is $(\partial f/\partial y)_x$. Both M and N are, in general, functions of x and y and may be differentiated to give mixed second partial derivatives:

$$\left(\frac{\partial M}{\partial y}\right)_x = \frac{\partial^2 f}{\partial y\,\partial x} \qquad \left(\frac{\partial N}{\partial x}\right)_y = \frac{\partial^2 f}{\partial x\,\partial y}$$

$\partial^2 f/\partial y\,\partial x$, for instance, is the partial derivative with respect to y of the partial derivative of f with respect to x. It is a theorem of calculus that if f is single valued and has continuous derivatives, and df is an exact differential, the order of differentiation in a mixed derivative

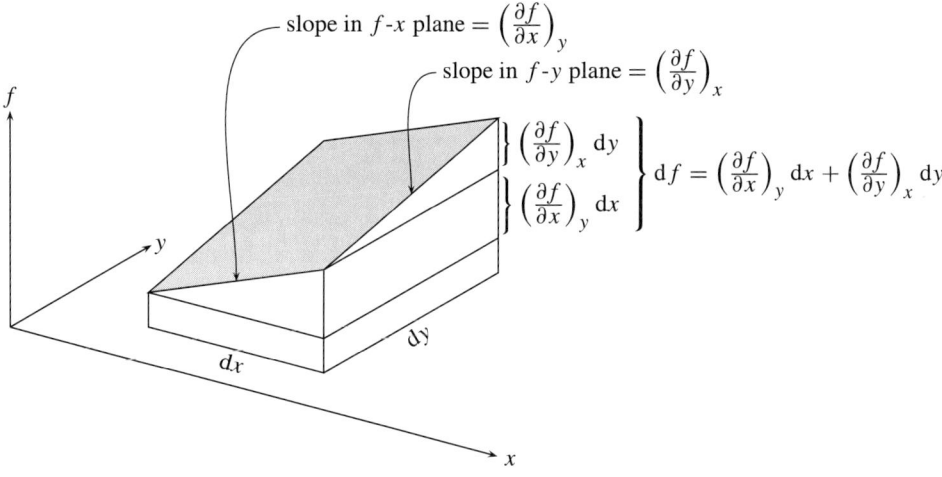

Figure F.1

is immaterial; that is, $\partial^2 f/\partial y \partial x$ and $\partial^2 f/\partial x \partial y$ are equal:

$$\left(\frac{\partial M}{\partial y}\right)_x = \left(\frac{\partial N}{\partial x}\right)_y$$

This relation is used with total differentials of various state functions of systems of two independent variables to derive so-called *Maxwell relations* (Section 5.3).

The general relation that applies to a function of any number of independent variables is

$$\left(\frac{\partial M}{\partial y}\right) = \left(\frac{\partial N}{\partial x}\right)$$

where x and y are *any* two of the independent variables, M is $\partial f/\partial x$, N is $\partial f/\partial y$, and each partial derivative has all independent variables held constant except the variable shown in the denominator. This general relation is the Euler reciprocity relation, or **reciprocity relation** for short. A necessary and sufficient condition for df to be an exact differential is that the reciprocity relation is satisfied for each pair of independent variables.

Appendix G

FORCES, ENERGY, AND WORK

To provide a detailed understanding of the meaning of energy and work in macroscopic systems, this appendix applies fundamental principles of classical mechanics to a collection of material particles representing a closed system and its surroundings. Although classical mechanics cannot duplicate all features of a chemical system (for one thing, quantum properties of atoms are ignored), this treatment suggests how we should calculate the work that appears in the first law of thermodynamics.

If you do not wish to wade through the following rather involved derivation, you may skip to the final result expressed by Eq. G.17 on page 420 and read the comments following that equation.

A *material particle* is a body that has mass and is so small that it behaves as a point, with no rotational energy. We assume each particle has a constant mass, ignoring relativistic effects that are important only when the particle moves at a speed close to the speed of light.

Consider a collection of an arbitrary number of material particles that have no interactions with any particles outside the collection. Later we consider some of the particles to constitute a thermodynamic *system* and the others to be the surroundings. The particles and their interactions will then be a model to help us analyze how to evaluate thermodynamic work.

Classical mechanics is based on the idea that one material particle acts on another by means of a *force*. Let the vector $\boldsymbol{F}_{ij}$ denote the force exerted on Particle i by Particle j.[1] The *total* force $\boldsymbol{F}_i$ acting on Particle i is the vector sum of the individual forces exerted on it by the other particles:[2]

$$\boldsymbol{F}_i = \sum_{j \neq i} \boldsymbol{F}_{ij} \qquad \text{(G.1)}$$

(The term in which j equals i has to be omitted because a particle does not act on itself.) From Newton's second law of motion, the total force $\boldsymbol{F}_i$ acting on Particle i is equal to the

---

[1] This and several other footnotes are included in case you are not familiar with vector notation. The quantity $\boldsymbol{F}_{ij}$ is printed in boldface to indicate it is a *vector* having both magnitude and direction.

[2] The rule for adding vectors, as in the summation shown here, is that the sum is a vector whose component in each direction of a Cartesian coordinate system is the sum of the components in that direction of the vectors being added.

product of its mass m_i and its acceleration:

$$F_i = m_i \frac{\mathrm{d}v_i}{\mathrm{d}t} \tag{G.2}$$

Here v_i is the particle's velocity in an inertial reference frame and t is time.

A nonzero total force causes Particle i to accelerate and its velocity and position to change. The *work*[3] done by the total force acting on the particle as it moves from an initial position at time t_1 to a final position at time t_2 is defined by the integral[4] $W_i = \int F_i \cdot \mathrm{d}r_i$, where r_i is the position vector of the particle—a vector from the origin of the coordinate system to the position of the particle. By substituting the expression for F_i (Eq. G.2) in the integrand and performing the integration, we obtain

$$\begin{aligned} W_i &= \int_{(r_i)_1}^{(r_i)_2} m_i \frac{\mathrm{d}v_i}{\mathrm{d}t} \cdot \mathrm{d}r_i = \int_{(v_i)_1}^{(v_i)_2} m_i \frac{\mathrm{d}r_i}{\mathrm{d}t} \cdot \mathrm{d}v_i = \int_{(v_i)_1}^{(v_i)_2} m_i v_i \cdot \mathrm{d}v_i \\ &= \left(\tfrac{1}{2} m_i v_i^2\right)_2 - \left(\tfrac{1}{2} m_i v_i^2\right)_1 \end{aligned} \tag{G.3}$$

The first integral has been changed to the second by interchanging $\mathrm{d}v_i$ and $\mathrm{d}r_i$. Integration over distance then changes to integration over velocity as shown.

The quantity $(1/2)m_i v_i^2$ is called the *kinetic energy* of Particle i. Kinetic energy depends only on the magnitude of the velocity (i.e., on the speed) and not on the particle's position. Equation G.3 shows that the work done by the total force acting on a particle during a time interval is equal to the change in the kinetic energy of the particle.

The *total* work W_{tot} done by all forces acting on all particles is the sum of W_i for all particles:

$$W_{\mathrm{tot}} = \sum_i W_i = \sum_i \left(\tfrac{1}{2} m_i v_i^2\right)_2 - \sum_i \left(\tfrac{1}{2} m_i v_i^2\right)_1 \tag{G.4}$$

Thus, the total work is equal to the change in the total kinetic energy.

Next we look in detail at the forces. The force that Particle j exerts on Particle i depends on the nature of the two particles and on the distance between them. For instance, Newton's law of universal gravitation gives the magnitude of a *gravitational* force as $Gm_i m_j / r_{ij}^2$, where G is the gravitational constant and r_{ij} is the interparticle distance. Coulomb's law gives the magnitude of an *electrical* force between charged particles as $Q_i Q_j / (4\pi \epsilon_0 r_{ij}^2)$, where Q_i and Q_j are the charges and ϵ_0 is the permittivity of vacuum. These two kinds of forces obey Newton's third law of action and reaction: The forces exerted by two particles on one another are equal in magnitude and opposite in direction and are directed along the line joining the two particles. (This would not be true of a *magnetic* force between moving particles.)

---

[3] The work W_i defined here is not the same as the thermodynamic work appearing in the first law.

[4] The dot between the vectors in the integrand indicates a dot product (scalar product), which is a *nonvector* quantity. The general definition of the dot product of two vectors, u and v, in terms of their components is $u \cdot v = u_x v_x + u_y v_y + u_z v_z$.

We shall assume the force F_{ij} exerted on Particle i by Particle j has a magnitude that depends only on the distance r_{ij} between the particles, and that it is directed along the line between i and j (as is true of gravitational and electrical forces). Then we can define a *potential function* Φ_{ij} for this force with the integral

$$\Phi_{ij} = \int_{r_{ij}}^{\infty} (F_{ij} \cdot e_{ij}) \, dr_{ij}$$

where e_{ij} is a unit vector pointing from j to i: $e_{ij} = (r_i - r_j)/r_{ij}$. Φ_{ij} is a function only of the distance r_{ij}. If the force is attractive (i.e., F_{ij} points from i to j), then Φ_{ij} increases with increasing r_{ij}; if the force is repulsive, Φ_{ij} decreases with increasing r_{ij}.

From Eq. G.1, we can obtain a second expression for W_{tot}:

$$W_{\text{tot}} = \sum_i W_i = \sum_i \int F_i \cdot dr_i = \sum_i \sum_{j \neq i} \int F_{ij} \cdot dr_i \tag{G.5}$$

We can rewrite the double sum of integrals on the right side of Eq. G.5 as a sum over pairs of particles, the terms for the pair i and j being

$$\int F_{ij} \cdot dr_i + \int F_{ji} \cdot dr_j = \int F_{ij} \cdot d(r_i - r_j) = \int_{(r_{ij})_1}^{(r_{ij})_2} (F_{ij} \cdot e_{ij}) \, dr_{ij}$$

$$= \int_{(r_{ij})_1}^{\infty} (F_{ij} \cdot e_{ij}) \, dr_{ij} + \int_{\infty}^{(r_{ij})_2} (F_{ij} \cdot e_{ij}) \, dr_{ij}$$

$$= -\left[(\Phi_{ij})_2 - (\Phi_{ij})_1 \right] \tag{G.6}$$

Here we have made use of the relations $F_{ji} = -F_{ij}$ (from Newton's third law) and $(r_i - r_j) = e_{ij} r_{ij}$. Equation G.5 becomes

$$W_{\text{tot}} = -\left[\sum_i \sum_{j>i} (\Phi_{ij})_2 - \sum_i \sum_{j>i} (\Phi_{ij})_1 \right] \tag{G.7}$$

By equating the two expressions for W_{tot} (Eqs. G.4 and G.7) and rearranging, we obtain

$$\sum_i \left(\tfrac{1}{2} m_i v_i^2 \right)_2 + \sum_i \sum_{j>i} (\Phi_{ij})_2 = \sum_i \left(\tfrac{1}{2} m_i v_i^2 \right)_1 + \sum_i \sum_{j>i} (\Phi_{ij})_1$$

This equation shows that the quantity

$$E_{\text{total}} = \sum_i \tfrac{1}{2} m_i v_i^2 + \sum_i \sum_{j>i} \Phi_{ij} \tag{G.8}$$

is constant over time as the particles move in response to the forces acting on them. The first term on the right side of Eq. G.8 is the total kinetic energy of the particles. The second term is the pairwise sum of particle–particle potential functions; this term is called the *potential energy* of the particles. Note that the kinetic energy depends only on particle speeds and the potential energy depends only on particle positions.

Now we are ready to divide the particles into two groups: those in the system and those in the surroundings. In the expressions to follow, we use the following convention: Indices i and j refer to particles in the system; Indices k and l refer to particles in the surroundings. With this change in notation, Eq. G.8 becomes

$$E_{\text{total}} = \sum_i \tfrac{1}{2} m_i v_i^2 + \sum_i \sum_{j>i} \Phi_{ij}$$
$$+ \sum_i \sum_k \Phi_{ik} + \sum_k \tfrac{1}{2} m_k v_k^2 + \sum_k \sum_{l>k} \Phi_{kl} \tag{G.9}$$

The surroundings may create a conservative force field (an "external" field) for a particle in the system. A conservative force field is associated with (1) a force that depends only on the particle's position, and (2) a potential energy that has only one value for each position. In general, an external field, such as a gravitational or electric field, is the result of forces exerted by stationary particles in the surroundings, and the net force depends only on the position of the particle on which it is exerted. We can replace the sum of appropriate terms in $\sum_i \sum_k \Phi_{ik}$ appearing in Eq. G.9 with $\sum_i \Phi_i^{\text{ext}}$, where Φ_i^{ext} is the potential energy of Particle i in a time-independent external field, a function only of the position of this particle. Using the index k' for those particles in the surroundings that are *not* the source of an external field, we rewrite Eq. G.9 in the form

$$E_{\text{total}} = \left[\sum_i \tfrac{1}{2} m_i v_i^2 + \sum_i \sum_{j>i} \Phi_{ij} + \sum_i \Phi_i^{\text{ext}} \right]$$
$$+ \left[\sum_i \sum_{k'} \Phi_{ik'} \right] + \left[\sum_k \tfrac{1}{2} m_k v_k^2 + \sum_k \sum_{l>k} \Phi_{kl} \right] \tag{G.10}$$

The terms on the right side of Eq. G.10 are shown grouped with brackets into three quantities. The first quantity depends only on the speeds and positions of the particles in the system, and thus represents the energy of the system, E_{sys}:

$$E_{\text{sys}} = \sum_i \tfrac{1}{2} m_i v_i^2 + \sum_i \sum_{j>i} \Phi_{ij} + \sum_i \Phi_i^{\text{ext}} \tag{G.11}$$

The last bracketed quantity on the right side of Eq. G.10 depends only on the speeds and positions of the particles in the surroundings, and so this quantity is the energy of the surroundings, E_{surr}. Thus, we can write

$$E_{\text{tot}} = E_{\text{sys}} + \sum_i \sum_{k'} \Phi_{ik'} + E_{\text{surr}} \tag{G.12}$$

The quantity $\sum_i \sum_{k'} \Phi_{ik'}$ represents potential energy shared by both the system and surroundings on account of forces other than those associated with a time-independent external field acting across the system boundary. These particular forces presently turn out to be the crucial ones for evaluating thermodynamic work.

A complication arises when we use a laboratory-fixed coordinate system for our measurements. The acceleration appearing in Newton's second law, Eq. G.2, is supposed to be measured relative to an *inertial* reference frame, such as one fixed or moving at a constant velocity with respect to local stars. A particle that is stationary relative to a laboratory-fixed frame on earth is actually accelerating with respect to an inertial frame because the earth circles the sun and spins about its axis. The total force exerted on a particle that we would calculate from Newton's second law is different in a laboratory-fixed frame than in an inertial frame because the accelerations are different in the two frames. Specifically, if $\boldsymbol{F}$ is the total force on a particle in the inertial frame and $\boldsymbol{v}$ is the particle's velocity in the laboratory-fixed frame, then the effective total force in the laboratory-fixed frame calculated from $\boldsymbol{F}_{\text{eff}} = m\,d\boldsymbol{v}/dt$ is given by[5]

$$\boldsymbol{F}_{\text{eff}} = \boldsymbol{F} + \boldsymbol{F}_{\text{centr}} + \boldsymbol{F}_{\text{Cor}}$$

where the correction terms are a centrifugal force $\boldsymbol{F}_{\text{centr}}$ and a Coriolis force $\boldsymbol{F}_{\text{Cor}}$. The earth's movement around the sun makes only a minor contribution to the correction terms, and $\boldsymbol{F}_{\text{Cor}}$ (which occurs only if the particle is moving in the laboratory-fixed frame) is usually so small that it can be neglected. This leaves as the only significant correction the contribution made to $\boldsymbol{F}_{\text{centr}}$ by the earth's spin about its axis. In the case of the gravitational force on a particle of mass m, $\boldsymbol{F}_{\text{centr}}$ is directed perpendicular to the earth's axis and its magnitude is $m\omega^2 d$, where ω is the earth's angular velocity and d is the distance of the particle from the earth's axis. The resulting correction to the gravitational force depends on latitude and is at most (at the equator) only about 0.3% of the gravitational force measured in an inertial reference frame. Not only is the correction small, but it is completely taken into account in the laboratory-fixed frame by calculating the effective gravitational force from $\boldsymbol{F}_{\text{grav}} = -mg\boldsymbol{e}_z$, where g is the acceleration of free fall and $\boldsymbol{e}_z$ is a unit vector in the $+z$ (upward) direction. The value of g is an experimental quantity that includes the effect of $\boldsymbol{F}_{\text{centr}}$, and thus depends on latitude as well as elevation above the earth's surface. Since $\boldsymbol{F}_{\text{grav}}$ depends only on position (if $\boldsymbol{F}_{\text{Cor}}$ is neglected), we can treat gravity as an conservative force field in the laboratory-fixed frame. (In contrast, if the reference frame is fixed in a satellite vehicle orbiting the earth, then the effective gravitational force is zero.)

In thermodynamics, we want to be able to evaluate the *change* in E_{sys}. To obtain a useful relation for this purpose, we carry out a procedure like that leading to Eqs. G.4 and G.7. For each of the particles in the *system* (but no others), we integrate the dot product of the total force $\boldsymbol{F}_i$ and the displacement in a time interval and then add the integrals to obtain a quantity W_{sys}. The resulting expression is exactly like Eq. G.4 except that only particles in the *system* are included:

$$W_{\text{sys}} = \sum_i W_i = \sum_i \left(\tfrac{1}{2} m_i v_i^2\right)_2 - \sum_i \left(\tfrac{1}{2} m_i v_i^2\right)_1 \tag{G.13}$$

Next we write the total force on Particle i in the form

$$\boldsymbol{F}_i = \sum_{j \neq i} \boldsymbol{F}_{ij} + \boldsymbol{F}_i^{\text{ext}} + \boldsymbol{F}_i^{\text{surr}}$$

---

[5] Goldstein, H.; *Classical Mechanics*; Addison-Wesley: Reading, Mass., 1950.

where F_{ij} is the force exerted on Particle i by Particle j, both particles being in the system; F_i^{ext} is the force on Particle i from a time-independent external field; and $F_i^{\text{surr}} = \sum_{k'} F_{ik'}$ is the force exerted on Particle i by particles in the surroundings that are not the source of a time-independent external field. Using this expression, we obtain a second expression for W_{sys}:

$$W_{\text{sys}} = \sum_i \int F_i \cdot dr_i$$
$$= \sum_i \sum_{j \neq i} \int F_{ij} \cdot dr_i + \sum_i \int F_i^{\text{ext}} \cdot dr_i + \sum_i \int F_i^{\text{surr}} \cdot dr_i \qquad (G.14)$$

By the same reasoning that led to Eq. G.7, the sum of integrals in the first term on the right side of Eq. G.14 may be written

$$\sum_i \sum_{j \neq i} \int F_{ij} \cdot dr_i = -\left[\sum_i \sum_{j>i} (\Phi_{ij})_2 - \sum_i \sum_{j>i} (\Phi_{ij})_1 \right]$$

Each integral $\int F_i^{\text{ext}} \cdot dr_i$ in the second term on the right side of Eq. G.14 is the negative of the change in the potential energy of Particle i in a time-independent external field. To see this, we write $F_i^{\text{ext}} = \sum_{k''} F_{ik''}$ (where the index k'' refers to the particles in the surroundings that are the source of the external field); the integral becomes

$$\int F_i^{\text{ext}} \cdot dr_i = \sum_{k''} \int F_{ik''} \cdot dr_i$$

To each integral $\int F_{ik''} \cdot dr_i$ we add another integral, $\int F_{k''i} \cdot dr_{k''}$, which is zero because $r_{k''}$ has a fixed position; then using Eq. G.6, we obtain

$$\int F_i^{\text{ext}} \cdot r_i = -\left[\sum_{k''} (\Phi_{ik''})_2 - \sum_{k''} (\Phi_{ik''})_2 \right] = -\left[(\Phi_i^{\text{ext}})_2 - (\Phi_i^{\text{ext}})_1 \right]$$

Equation G.14, with these substitutions, becomes

$$W_{\text{sys}} = -\left[\sum_i \sum_{j>i} (\Phi_{ij})_2 - \sum_i \sum_{j>i} (\Phi_{ij})_1 \right] - \left[\sum_i (\Phi_i^{\text{ext}})_2 - \sum_i (\Phi_i^{\text{ext}})_1 \right]$$
$$+ \sum_i \int F_i^{\text{surr}} \cdot dr_i \qquad (G.15)$$

When we equate the expressions for W_{sys} given by Eqs. G.13 and G.15 and rearrange, we obtain

$$\sum_i \left(\tfrac{1}{2}m_i v_i^2\right)_2 - \sum_i \left(\tfrac{1}{2}m_i v_i^2\right)_1 + \sum_i \sum_{j>i}(\Phi_{ij})_2 - \sum_i \sum_{j>i}(\Phi_{ij})_1$$
$$+ \sum_i (\Phi_i^{\text{ext}})_2 - \sum_i (\Phi_i^{\text{ext}})_1$$
$$= \sum_i \int F_i^{\text{surr}} \cdot d\mathbf{r}_i$$

Comparison of the expression on the left side of this equation with Eq. G.11 shows the expression is the same as the change of E_{sys}. Thus, the desired expression for ΔE_{sys} is

$$\Delta E_{\text{sys}} = \sum_i \int F_i^{\text{surr}} \cdot d\mathbf{r}_i \tag{G.16}$$

The sum in Eq. G.16 includes all Particles i in the system. We defined F_i^{surr} to be the net force exerted on Particle i by the surroundings, excluding the force from any time-independent external field. Thus, ΔE_{sys} is equal to the total work done on the system by the surroundings other than work done by time-independent external fields.

In thermodynamics, we are interested in the work w_{sys} done on *macroscopic* parts of the system, rather than on individual particles. We can use the expression on the right side of Eq. G.16 for this work:

$$w_{\text{sys}} = \sum_i \int F_i^{\text{surr}} \cdot d\mathbf{r}_i$$

We must now let $d\mathbf{r}_i$ refer to the displacement of a macroscopic portion of the system (rather than a single particle) resulting from concerted motion of many particles. Energy transfer to the system by means of chaotic motions and collisions of individual particles, which is included in Eq. G.16 but not in the expression for w_{sys}, is what in thermodynamics is called *heat*. With this understanding, we can write $\Delta E_{\text{sys}} = q + w_{\text{sys}}$.

The differential form of w_{sys} is

$$dw_{\text{sys}} = \sum_i F_i^{\text{surr}} \cdot d\mathbf{r}_i \tag{G.17}$$

Note the following features brought out by the derivation of Eq. G.17:

1. The equation has been derived for a *closed* system.

2. The equation shows how we can evaluate an infinitesimal quantity of thermodynamic work dw_{sys} done on the system. We need to know the force F_i^{surr} exerted by the surroundings on each macroscopic Portion i of the system and also the displacement $d\mathbf{r}_i$ of each portion with respect to the reference frame. The scalar product $F_i^{\text{surr}} \cdot d\mathbf{r}_i$ in the equation is the same as the product of the displacement and the component of the force in the direction of the displacement.

3. We can equally well calculate thermodynamic work from the force exerted by each macroscopic portion of the *system* on the surroundings. We simply replace $\boldsymbol{F}_i^{\text{surr}}$ by $\boldsymbol{F}_i^{\text{sys}} = -\boldsymbol{F}_i^{\text{surr}}$ (from Newton's third law), where $\boldsymbol{F}_i^{\text{sys}}$ is the force exerted by Portion i of the system.

4. The work w_{sys} does not include work done internally by one part of the system on another part.

5. We do not include forces from a conservative time-independent external field in the calculation of w_{sys}.

An additional comment can be made about the transfer of energy between the system and the surroundings. We may use Eq. G.17, with appropriate redefinition of the quantities on the right side, to evaluate the work done on the *surroundings*. This work may be equal in magnitude and of opposite sign to the work w_{sys} done on the system. A necessary condition for this equality is that the interacting parts of the system and surroundings have equal displacements; that is, that there be continuity of motion at the system boundary. We expect there to be continuity of motion when a fluid contacts a moving piston or paddle.

Suppose, however, that the system moves and an interacting part of the surroundings is stationary. Then according to Eq. G.17, $\mathrm{d}w_{\text{sys}}$ is nonzero, whereas no work is done on that part of the surroundings. How can this be? One possibility, discussed by Bridgman,[6] is sliding friction at the boundary: Energy lost by the system in the form of work (negative w_{sys}) is gained by the surroundings in the form of thermal energy. Since the effect on the surroundings is the same as a flow of heat from the system, the division of energy transfer into heat and work can be ambiguous when there is sliding friction at the boundary.

Another way work can have different magnitudes for system and surroundings is a change in the potential energy shared by the system and surroundings. This shared potential energy is associated with forces acting across the boundary, other than from a time-independent external field, and is represented in Eq. G.12 by the quantity $\sum_i \sum_{k'} \Phi_{ik'}$. In the usual types of processes, $\sum_i \sum_{k'} \Phi_{ik'}$ is constant or falls off so rapidly with distance that it is negligible. Since E_{tot} is constant, during such processes with constant or negligible shared potential energy, the sum $E_{\text{sys}} + E_{\text{surr}}$ remains constant.

---

[6] Pages 47–56 (reference in Appendix I).

Appendix H

STANDARD MOLAR THERMODYNAMIC PROPERTIES

The values in this table are for a temperature of 298.15 K (25.00 °C) and a standard pressure $p° = 1$ bar. Solute standard states are based on molality. A crystalline solid is denoted by "cr." Those values of $\Delta_f H_m°$ and $S_m°$ shown with uncertainties are values recommended by the Committee on Data for Science and Technology (Cox, J. D.; Wagman, D. D.; Medvedev, V. A. *CODATA Key Values for Thermodynamics*; Hemisphere Publishing Corp.: New York, 1989; also available at http://www.codata.org).

Species	$\dfrac{\Delta_f H_m°}{\text{kJ mol}^{-1}}$	$\dfrac{S_m°}{\text{J K}^{-1}\text{mol}^{-1}}$	$\dfrac{\Delta_f G_m°}{\text{kJ mol}^{-1}}$
Inorganic substances			
Ag(cr)	0	42.55 ± 0.20	0
AgCl(cr)	−127.01 ± 0.05	96.25 ± 0.20	−109.77
C(cr, graphite)	0	5.74 ± 0.10	0
CO(g)	−110.53 ± 0.17	197.660 ± 0.004	−137.17
CO_2(g)	−393.51 ± 0.13	213.785 ± 0.010	−394.41
Ca(cr)	0	41.59 ± 0.40	0
$CaCO_3$(cr, calcite)	−1206.9	92.9	−1128.8
CaO(cr)	−634.92 ± 0.90	38.1 ± 0.4	−603.31
Cl_2(g)	0	223.081 ± 0.010	0
F_2(g)	0	202.791 ± 0.005	0
H_2(g)	0	130.680 ± 0.003	0
HCl(g)	−92.31 ± 0.10	186.902 ± 0.005	−95.30
HF(g)	−273.30 ± 0.70	173.779 ± 0.003	−275.40
HI(g)	26.50 ± 0.10	206.590 ± 0.004	1.70
H_2O(l)	−285.830 ± 0.040	69.95 ± 0.03	−237.16

(continued from previous page)

Species	$\dfrac{\Delta_f H_m^\circ}{\text{kJ mol}^{-1}}$	$\dfrac{S_m^\circ}{\text{J K}^{-1}\,\text{mol}^{-1}}$	$\dfrac{\Delta_f G_m^\circ}{\text{kJ mol}^{-1}}$
$H_2O(g)$	-241.826 ± 0.040	188.835 ± 0.010	-228.58
$H_2S(g)$	-20.6 ± 0.5	205.81 ± 0.05	-33.44
$Hg(l)$	0	75.90 ± 0.12	0
$Hg(g)$	61.38 ± 0.04	174.971 ± 0.005	31.84
$HgO(cr, red)$	-90.79 ± 0.12	70.25 ± 0.30	-58.54
$Hg_2Cl_2(cr)$	-265.37 ± 0.40	191.6 ± 0.8	-210.72
$I_2(cr)$	0	116.14 ± 0.30	0
$K(cr)$	0	64.68 ± 0.20	0
$KI(cr)$	-327.90	106.37	-323.03
$KOH(cr)$	-424.72	78.90	-378.93
$N_2(g)$	0	191.609 ± 0.004	0
$NH_3(g)$	-45.94 ± 0.35	192.77 ± 0.05	-16.41
$NO_2(g)$	33.10	240.04	51.22
$N_2O_4(g)$	9.08	304.38	97.72
$Na(cr)$	0	51.30 ± 0.20	0
$NaCl(cr)$	-411.12	72.11	-384.02
$O_2(g)$	0	205.152 ± 0.005	0
$O_3(g)$	142.67	238.92	163.14
$P(cr, white)$	0	41.09 ± 0.25	0
$S(cr, rhombic)$	0	32.054 ± 0.050	0
$SO_2(g)$	-296.81 ± 0.20	248.223 ± 0.050	-300.09
$Si(cr)$	0	18.81 ± 0.08	0
$SiF_4(g)$	-1615.0 ± 0.8	282.76 ± 0.50	-1572.8
$SiO_2(cr, \alpha\text{-quartz})$	-910.7 ± 1.0	41.46 ± 0.20	-856.3
Organic compounds			
$CH_4(g)$	-74.87	186.25	-50.77
$CH_3OH(l)$	-238.9	127.2	-166.6
$CH_3CH_2OH(l)$	-277.0	159.9	-173.8
$C_2H_2(g)$	226.73	200.93	209.21
$C_2H_4(g)$	52.47	219.32	68.43
$C_2H_6(g)$	-83.85	229.6	-32.00
$C_3H_8(g)$	-104.7	270.31	-24.3
$C_6H_6(l, benzene)$	49.04	173.26	124.54
Ionic solutes			
$Ag^+(aq)$	105.79 ± 0.08	73.45 ± 0.40	77.10
$CO_3^{2-}(aq)$	-675.23 ± 0.25	-50.0 ± 1.0	-527.90
$Ca^{2+}(aq)$	-543.0 ± 1.0	-56.2 ± 1.0	-552.8

(continued from previous page)

Species	$\dfrac{\Delta_f H_m^\circ}{\text{kJ mol}^{-1}}$	$\dfrac{S_m^\circ}{\text{J K}^{-1}\text{mol}^{-1}}$	$\dfrac{\Delta_f G_m^\circ}{\text{kJ mol}^{-1}}$
$Cl^-(aq)$	-167.08 ± 0.10	56.60 ± 0.20	-131.22
$F^-(aq)$	-335.35 ± 0.65	-13.8 ± 0.8	-281.52
$H^+(aq)$	0	0	0
$HCO_3^-(aq)$	-689.93 ± 2.0	98.4 ± 0.5	-586.90
$HS^-(aq)$	-16.3 ± 1.5	67 ± 5	12.2
$HSO_4^-(aq)$	-886.9 ± 1.0	131.7 ± 3.0	-755.4
$Hg_2^{2+}(aq)$	166.87 ± 0.50	65.74 ± 0.80	153.57
$I^-(aq)$	-56.78 ± 0.05	106.45 ± 0.30	-51.72
$K^+(aq)$	-252.14 ± 0.08	101.20 ± 0.20	-282.52
$NH_4^+(aq)$	-133.26 ± 0.25	111.17 ± 0.40	-79.40
$NO_3^-(aq)$	-206.85 ± 0.40	146.70 ± 0.40	-110.84
$Na^+(aq)$	-240.34 ± 0.06	58.45 ± 0.15	-261.90
$OH^-(aq)$	-230.015 ± 0.040	-10.90 ± 0.20	-157.24
$S^{2-}(aq)$	33.1	-14.6	86.0
$SO_4^{2-}(aq)$	-909.34 ± 0.40	18.50 ± 0.40	-744.00

Appendix I

GENERAL REFERENCES

Adkins, C. J. *Equilibrium Thermodynamics*, 3rd ed.; Cambridge University Press: Cambridge, 1983.

Bauman, R. P. *Modern Thermodynamics with Statistical Mechanics*; Macmillan: New York, 1992.

Bridgman, P. W. *The Nature of Thermodynamics*; Harvard University Press: Cambridge, MA, 1941.

Castellan, G. W. *Physical Chemistry*, 3rd ed.; Addison-Wesley: Reading, MA, 1983.

Denbigh, K. *The Principles of Chemical Equilibrium*, 4th ed.; Cambridge University Press: Cambridge, 1981.

Findlay, A.; revised by Campbell, A. N.; Smith, N. O. *The Phase Rule and Its Applications*, 9th ed.; Dover: New York, 1951.

Guggenheim, E. A. *Thermodynamics: An Advanced Treatment for Chemists and Physicists*, 8th ed.; Elsevier Science: 1985.

Hildebrand, J. H.; Scott, R. L. *The Solubility of Nonelectrolytes*, 3rd ed.; Dover: New York, 1964.

IUPAC Green Book. *Quantities, Units and Symbols in Physical Chemistry*, 2nd ed.; Blackwell: Oxford, 1993. Prepared for publication by I. Mills, T. Cvitas, K. Homann, N. Kallay, and K. Kuchitsu.

Lewis, G. N.; Randall, M.; revised by Pitzer, K. S. *Thermodynamics*, 3rd ed.; McGraw-Hill: New York, 1995.

McGlashan, M. L. *Chemical Thermodynamics*; Academic Press: London, 1979.

Pippard, A. B. *Elements of Classical Thermodynamics for Advanced Students of Physics*; Cambridge University Press: Cambridge, 1957.

Planck, M. *Treatise on Thermodynamics*, 3rd ed.; Dover: New York (translation of 7th German edition, 1922).

Prausnitz, J. M.; Lichtenthaler, R. N.; de Azevedo, E. G. *Molecular Thermodynamics of Fluid-Phase Equilibria*, 3rd ed.; Prentice Hall: Upper Saddle River, NJ, 1999.

Reid, C. E. *Chemical Thermodynamics*; McGraw-Hill: New York, 1990.

Robinson, R. A.; Stokes, R. H. *Electrolyte Solutions*, 2nd ed.; Academic Press: New York, 1959.

Wood, S. E.; Battino, R. *Thermodynamics of Chemical Systems*; Cambridge University Press: Cambridge, 1990.

Zemansky, M. W.; Dittman, R. H. *Heat and Thermodynamics*, 6th ed.; McGraw-Hill: New York, 1981.

Appendix J

ANSWERS TO SELECTED PROBLEMS

Chapter 2

2.3 $373.120\,\text{K} \pm 0.002\,\text{K}$

Chapter 3

3.2 $9.58 \times 10^3\,\text{s}$ (2 hr 40 min)

3.6 (b) $q = -w = 1.00 \times 10^5\,\text{J}$

Chapter 4

4.5 $\Delta S = 549\,\text{J K}^{-1}$ for both processes; $\int \mathrm{d}q/T_{\text{ext}} = 333\,\text{J K}^{-1}$ and 0.

4.6 (a) $q = 0$, $w = 1.50 \times 10^4\,\text{J}$, $\Delta U = 1.50 \times 10^4\,\text{J}$, $\Delta H = 2.00 \times 10^4\,\text{J}$

(c) $\Delta S = 66.7\,\text{J K}^{-1}$

4.7 $S_\text{m}^\circ \approx 151.6\,\text{J K}^{-1}\,\text{mol}^{-1}$

Chapter 5

5.6 (a) $\alpha = 8.519 \times 10^{-4}\,\text{K}^{-1}$
$\kappa_t = 4.671 \times 10^{-5}\,\text{bar}^{-1}$
$(\partial p/\partial T)_V = 18.24\,\text{bar K}^{-1}$
$(\partial U/\partial V)_T = 5437\,\text{bar}$

(b) $\Delta p \approx 1.8\,\text{bar}$

5.7 (b) $(\partial C_{p,\text{m}}/\partial p)_T$
$= -4.210 \times 10^{-8}\,\text{J K}^{-1}\,\text{Pa}^{-1}\,\text{mol}^{-1}$

5.8 (b) $8 \times 10^{-4}\,\text{K}^{-1}$

5.11 $\Delta H = 2.27 \times 10^4\,\text{J}$, $\Delta S = 39.5\,\text{J K}^{-1}$

5.12 $5.001 \times 10^3\,\text{J}$

5.13 (a) $2.56\,\text{J K}^{-1}\,\text{g}^{-1}$

5.14 (b) $f = 17.4\,\text{bar}$

5.15 (a) $\phi = 0.739$, $f = 148\,\text{bar}$

(b) $B = -7.28 \times 10^{-5}\,\text{m}^3\,\text{mol}^{-1}$

Chapter 6

6.2 (a) $S_\text{m}^\circ = 69.74\,\text{J K}^{-1}\,\text{mol}^{-1}$

(b) $\Delta_\text{vap} S_\text{m}^\circ = 118.98\,\text{J K}^{-1}\,\text{mol}^{-1}$,
$\Delta_\text{vap} H_\text{m}^\circ = 4.4035 \times 10^4\,\text{J mol}^{-1}$

6.4 $4.5 \times 10^{-3}\,\text{bar}$

6.5 $19\,\text{J mol}^{-1}$

6.6 (a) $352.82\,\text{K}$

(b) $3.4154 \times 10^4\,\text{J mol}^{-1}$

6.7 (a) $3.62 \times 10^3\,\text{Pa K}^{-1}$

(b) $3.56 \times 10^3\,\text{Pa K}^{-1}$

(c) $99.60\,°\text{C}$

6.8 (b) $\Delta_\text{vap} H_\text{m}^\circ = 4.084 \times 10^4\,\text{J mol}^{-1}$

6.9 $0.93\,\text{mol}$

Chapter 7

7.4 real gas: $p = 1.9743\,\text{bar}$
ideal gas: $p = 1.9832\,\text{bar}$

7.5 (a) $x_{\text{N}_2} = 8.79 \times 10^{-6}$
$x_{\text{O}_2} = 4.66 \times 10^{-6}$
$y_{\text{N}_2} = 0.763$
$y_{\text{O}_2} = 0.205$

(b) $x_{\text{N}_2} = 9.81 \times 10^{-6}$
$x_{\text{O}_2} = 2.64 \times 10^{-6}$
$y_{\text{N}_2} = 0.852$
$y_{\text{O}_2} = 0.116$

427

7.7 (b) $f_A = 0.03167$ bar, $f_A = 0.03040$ bar

7.8 (a) In the mixture of composition $x_A = 0.9782$, the activity coefficient is $\gamma_B \approx 11.5$.

7.9 (d) $K_{x,A} \approx 680$ kPa

7.11 Values for $m_B/m° = 20$: $\gamma_A = 1.026$, $\gamma_{m,B} = 0.526$; the limiting slopes are $d\gamma_A/d(m_B/m°) = 0$, $d\gamma_{m,B}/d(m_B/m°) = -0.09$

7.13 $p_{N_2} = 0.235$ bar,
$x_{N_2} = 0.815$,
$p_{O_2} = 0.0532$ bar,
$x_{O_2} = 0.185$,
$p = 0.288$ bar

7.14 $M_B = 187$ kg mol^{-1}
mass binding ratio $= 1.37$

Chapter 8

8.1 $\Delta_r H° = -63.94$ kJ mol^{-1}
$K = 4.41 \times 10^{-2}$

8.2 (b) $\Delta_f H_m°$: no change
$\Delta_f S_m°$: subtract 0.109 J K^{-1} mol^{-1}
$\Delta_f G_m°$: add 33 J mol^{-1}

8.3 $p(298.15\ K) = 2.6 \times 10^{-6}$ bar
$p(273.15\ K) = 2.7 \times 10^{-7}$ bar

8.4 (a) -240.34 kJ mol^{-1}, -470.36 kJ mol^{-1}, -230.02 kJ mol^{-1}

(b) -465.43 kJ mol^{-1}

(c) -39.82 kJ mol^{-1}

8.5 $\Delta H = 0.92$ kJ

8.6 (b) $\Delta_{sol} H_m^\infty = -74.75$ kJ mol^{-1}

8.7 (a) State 1:
$n_{C_6 H_{14}} = 7.822 \times 10^{-3}$ mol
$n_{H_2 O} = 0.05560$ mol
amount of O_2 consumed: 0.07431 mol
State 2:
$n_{H_2 O} = 0.11035$ mol
$n_{CO_2} = 0.04693$ mol
mass of $H_2O = 1.9880$ g

(b) $V_m(C_6 H_{14}) = 131.61$ cm^3 mol^{-1}
$V_m(H_2 O) = 18.070$ cm^3 mol^{-1}

(c) State 1: $V(C_6 H_{14}) = 1.029$ cm^3
$V(H_2 O) = 1.005$ cm^3
$V^g = 348.0$ cm^3
State 2:
$V(H_2 O) = 1.994$ cm^3
$V^g = 348.0$ cm^3

(d) State 1:
$n_{O_2} = 0.429$ mol
State 2:
$n_{O_2} = 0.355$ mol
$x_{O_2} = 0.883$
$x_{CO_2} = 0.117$

(e) State 2:
$p_2 = 27.9$ bar
$p_{O_2} = 24.6$ bar
$p_{CO_2} = 3.26$ bar

(f) $f_{H_2 O}(0.03169\ \text{bar}) = 0.03164$ bar
State 1: $f_{H_2 O} = 0.03234$ bar
State 2: $f_{H_2 O} = 0.03229$ bar

(g) State 1:
$\phi_{H_2 O} = 0.925$
$\phi_{O_2} = 0.981$
$f_{O_2} = 29.4$ bar
State 2:
$\phi_{H_2 O} = 0.896$
$\phi_{O_2} = 0.983$
$\phi_{CO_2} = 0.910$
$f_{O_2} = 24.2$ bar
$f_{CO_2} = 2.97$ bar

(h) State 1:
$n_{H_2 O}^g = 5.00 \times 10^{-4}$ mol
$n_{H_2 O}^l = 0.05510$ mol
State 2: $n_{H_2 O}^g = 5.19 \times 10^{-4}$ mol
$n_{H_2 O}^l = 0.10983$ mol

(i) State 1:
$K_{m,O_2} = 825$ bar kg mol^{-1}
$n_{O_2} = 3.57 \times 10^{-5}$ mol
State 2:
$K_{m,O_2} = 823$ bar kg mol^{-1}
$K_{m,CO_2} = 30.8$ bar kg mol^{-1}
$n_{O_2} = 5.85 \times 10^{-5}$ mol
$n_{CO_2} = 1.92 \times 10^{-4}$ mol

(j) H_2O vaporization: $\Delta U = +20.8$ J
H_2O condensation: $\Delta U = -21.6$ J

(k) O_2 dissolution: $\Delta U = -0.35$ J
O_2 desolution: $\Delta U = 0.57$ J
CO_2 desolution: $\Delta U = 3.32$ J

(l) C_6H_{14}(l) compression:
$\Delta U = -1.226$ J
solution compression: $\Delta U = -0.225$ J
solution decompression:
$\Delta U = 0.414$ J

(m) O_2 compression: $\Delta U = -81$ J
gas mixture: $dB/dT = 0.26 \times 10^{-6}$ m^3 K^{-1} mol^{-1}
gas mixture expansion: $\Delta U = 87$ J

(n) $\Delta U = 8$ J

(o) $\Delta_c U_m^\circ = -4154.4$ kJ mol^{-1}

(p) $\Delta_c H_m^\circ = -4163.1$ kJ mol^{-1}

8.8 $\Delta_f H_m^\circ = -198.8$ kJ mol^{-1}

8.9 $T_2 = 2272$ K

8.10 $p(O_2) = 2.55 \times 10^{-5}$ bar

8.11 (a) $K = 3.5 \times 10^{41}$

(b) $p_{H_2} = 2.8 \times 10^{-42}$ bar
$N_{H_2} = 6.9 \times 10^{-17}$

(c) $t = 22$ s

8.12 (c) $K = 0.15$

8.13 (a) $K = 2.5$

Chapter 9

9.1 (b) $T = 1168$ K
$\Delta_r H_m^\circ = 1.64 \times 10^5$ J mol^{-1}

9.4 $K_f = 1.860$ K kg mol^{-1}
$K_b = 0.5118$ K kg mol^{-1}

9.5 $M_B \approx 5.6 \times 10^4$ g mol^{-1}

9.6 $\Delta_{sol} H_{m,B}^\circ/$kJ mol$^{-1} = -3.06, 0, 6.35$
$\Delta_{sol} S_{m,B}^\circ/$J K^{-1} mol^{-1}
$= -121.0, -110.2, -88.4$

9.8 (a) $m_+^\alpha = m_-^\alpha = 1.20 \times 10^{-3}$ mol kg^{-1}
$m_+^\beta = 1.80 \times 10^{-3}$ mol kg^{-1}
$m_-^\beta = 0.80 \times 10^{-3}$ mol kg^{-1}
$m_P^\beta = 2.00 \times 10^{-6}$ mol kg^{-1}

9.9 (a) $p^l = 2.44$ bar

(b) $f(2.44 \text{ bar}) - f(1.00 \text{ bar})$
$= 3.4 \times 10^{-5}$ bar

9.10 (a) $x_B = 1.8 \times 10^{-7}$
$m_B = 1.0 \times 10^{-5}$ mol kg^{-1}

(b) $\Delta_{sol} H_{m,B}^\circ = -1.99 \times 10^4$ J mol^{-1}

(c) $K = 4.4 \times 10^{-7}$
$\Delta_r H_m^\circ = 9.3$ kJ mol^{-1}

9.12 0.26%

9.13 (a) $\gamma_{x,B} = 0.9826$

(b) $x_B = 4.19 \times 10^{-4}$

9.14 $K = 1.2 \times 10^{-6}$

9.15 (a) $\alpha = 0.129$
$m_+ = 1.29 \times 10^{-3}$ mol kg^{-1}

(b) $\alpha = 0.140$

9.16 $\Delta_f H_m^\circ(Cl^-, aq) = -167.15$ kJ mol^{-1}
$S_m^\circ(Cl^-, aq) = 56.46$ J K^{-1} mol^{-1}

9.17 (a) $K_s = 1.783 \times 10^{-10}$

9.18 (a) $\Delta_r H_m^\circ = -65.769$ kJ mol^{-1}

(b) $\Delta_f H_m^\circ(Ag^+, aq) = 105.84$ kJ mol^{-1}

Chapter 10

10.1 (a) $F = 4$

(b) $F = 3$

(c) $F = 2$

10.10 (a) x_B(top) $= 0.02$, x_B(bottom) $= 0.31$

(b) $n_A = 2.1$ mol, $n_B = 1.0$ mol

Chapter 11

11.2 (a) $\Delta_r G_m^\circ = -21.436$ kJ mol^{-1}
$\Delta_r S_m^\circ = -62.35$ J K^{-1} mol^{-1}
$\Delta_r H_m^\circ = -40.03$ kJ mol^{-1}

(b) $\Delta_f H_m^\circ(AgCl, s) = -127.05$ kJ mol^{-1}

(c) $S_m^\circ(AgCl, s) = 96.16$ J K^{-1} mol^{-1}
$\Delta_f S_m^\circ(AgCl, s) = -57.93$ J K^{-1} mol^{-1}
$\Delta_f G_m^\circ(AgCl, s) = -109.78$ kJ mol^{-1}

11.3 (b) $\Delta_\mathrm{f} H_\mathrm{m}^\circ(\mathrm{AgCl, s}) = -126.81\,\mathrm{kJ\,mol^{-1}}$
$\Delta_\mathrm{f} G_\mathrm{m}^\circ(\mathrm{AgCl, s}) = -109.59\,\mathrm{kJ\,mol^{-1}}$

11.4 $K_\mathrm{s} = 1.76 \times 10^{-10}$

11.5 (b) $\gamma_\pm = 0.756$

11.6 (b) $\Delta_\mathrm{f} G_\mathrm{m}^\circ = -210.72\,\mathrm{kJ\,mol^{-1}}$

(c) $K_\mathrm{s} = 1.4 \times 10^{-18}$

11.7 $E^\circ = 0.16\,\mathrm{V}$

INDEX

A boldface page number refers to a definition, "n" refers to a footnote, and "p" refers to a problem.

Absolute zero, principle of the unattainability of, 100
Acid dissociation constant, 347
Activity, **213**
  of an electrolyte solute, 223
  of a gas, 130, 215
  of an ion, 219
  of a mixture constituent, 214, 215
  of a pure liquid or solid, 215
  of a pure substance, 217
  of a solute, 215
  of a solvent, 215
  of a symmetrical electrolyte, 221
Activity coefficient, **202**
  approach to unity, 203
  of a gas, 130
  from the Gibbs–Duhem equation, 208–209
  of an ion, 219
    from the Debye–Hückel theory, 224
  in a liquid mixture, from gas fugacity, 205–208
  mean ionic:
    from the Debye–Hückel theory, 225
    of an electrolyte solute, **223**
    from the Nernst equation, 395
    from osmotic coefficients, 229
    from solubility, 330
    of a symmetrical electrolyte, **221**
    from the osmotic coefficient, 209–212
  of a pure substance, 217
  of a solute:
    in dilute solution, 204

from gas fugacity, 205–208
  of a solvent, 310n
    from gas fugacity, 205–208
    stoichiometric, **224**
Activity quotient, **287**
Additivity rule, **176**, 179, 220, 223, 244, 249, 264, 279
Adiabatic:
  boundary, **28**
  calorimeter, 111–113, 269–271
  demagnetization, **99**, 135
  flame temperature, 278–279
  process, **33**, 37, 57
Advancement, **240**
Affinity of reaction, 280n
Amount, **3**, 17
Amount of substance, **3**, 17, 402
Antoine equation, 168p
Avogadro constant, 3, 402, 403
Azeotrope, 343, **373**
  minimum-boiling, 374
  vapor-pressure curve, 374
Azeotropic behavior, 372
Azeotropy, 373

Bar, **20**
Barometric formula, 147, 231
Barotropic effect, 13
Base units, 402
Binary:
  mixture, **171**
    in equilibrium with a pure phase, 314
  solution, 171
Body, **8**
Boiling point, **153**
  curve, **154**, 371
  elevation, of a dilute solution, 319

Boltzmann's equation for entropy, 88
Bomb calorimeter, 255, 270, 272–277
Bomb calorimetry, 300p
Boundary, **27**
  adiabatic, **28**
  diathermal, **28**
Boyle temperature, **15**
Bubble-point curve, 371
Buoyant force, 56

Cailletet and Matthias, law of, **155**
Caloric theory, 40n
Calorimeter, **111**
  adiabatic, 111–113, 269–271
  bomb, 255, 270, 272–277
  Bunsen ice, 278
  combustion, 270
  constant-pressure, 113, 161
  constant-volume, 112
  continuous-flow, 116
  flame, 278
  heat-flow, 278
  isoperibol, 114, 270, 271, 278
  isothermal-jacket, 114–116, 270, 271, 278
  phase-change, 278
  reaction, 270
    constant-pressure, 270–271
Calorimetry:
  bomb, 300p
  drop, 139p
  to evaluate an equilibrium constant, 293
  reaction, 257, 269–278, 348
Carathéodory's principle of adiabatic inaccessibility, 78

431

Carnot:
  cycle, **69**, 69–71, 76
  engine, **69**, 69–71, 76, 82, 87
  heat pump, **71**
Cell:
  diagram, 388
  electrochemical, **387**
  galvanic, **387**
  potential, **389**
  reaction, **387**
  with transference, 389
  without liquid junction, 389
  without transference, 389
Celsius scale, 21
Centigrade scale, 21
Centrifugal:
  field, 230–233
  force, 231, 418
Centrifuge cell, 230–232
Charge number:
  of a cell reaction, **391**
  of an ion, 225
Charge, electric, 219, 389
Chemical amount, 3n
Chemical equation, 239
Chemical potential, 137
  of an electrolyte solute, 223–224
  as a function of electric potential, 335, 393
  as a function of elevation, 145, 146
  as a function of T and p, 162
  of an ion, 218
  of a liquid or solid, 130
  of a pure substance, **127**
  of a solvent:
    from the freezing point, 310–312
    from the molal osmotic coefficient, 310
    from osmotic pressure, 312–314
  of a species in a mixture, **180**
  standard, 213
    of a gas, 128
    of a gas constituent, 185
    of an ion, 219
    of a pure substance, **128**
    of a symmetrical electrolyte, 220
  total, 146
  in transfer equilibrium, 182
Chemical process, 239
  subscripts for, 408
Circuit:
  electrical, 53–55, 392

heater, 112, 114, 115
  ignition, 273, 275
Clapeyron equation, **165**
Clausius:
  inequality, **83**
  statement of the second law, 67
Clausius–Clapeyron equation, **166**, 309
CODATA, 348n
Coexistence curve, **151**, 162
  liquid–gas, 13
Colligative property, **315**
  to estimate solute molar mass, 317
Common-ion effect, 330, 381
Component, **358**
Components, number of, 103, 175n
Composition variables, **170**, 173
  relations at infinite dilution, 172
Compressibility factor, **15**
Compression, **32**
Compression factor, **15**
Concentration, 9, **170**
  standard, 197
Condensation curve, 372
Conditions of validity, 5n
Congruent melting, 367
Conjugate phases, 368, 380
Consolute point, 368
Constants, physical, values of, 403
Continuity of states, 13
Convergence temperature, 115
Conversion factors, 5
Coordinates:
  external, 35
  internal, 35–38
Coriolis force, 418
Coulomb's law, 415
Critical:
  curve, 377
  opalescence, 155
  point, **13**, **154**
    of partially-miscible liquids, 368
  pressure, **154**
  temperature, **154**
Cryogenics, 97–100
Cryoscopic constant, **319**
Cubic expansion coefficient, **104**, 165, 308
  of an ideal gas, 137p
  negative values of, 104n
Curie's law of magnetization, 135
Current, electric, 53, 54, 112, 387, 390, 402
Cyclic process, **33**

Dalton's law, **184**
Debye crystal theory, 93
Debye–Hückel:
  equation:
    for a mean ionic activity coefficient, 225, 227, 396
    for a single-ion activity coefficient, 225
  limiting law, **226**, 264, 266, 268, 269, 330
  theory, 224–229
    derivation, 227–229
Degree of dissociation, 347
Degrees of freedom, **148**, **356**
Deliquescence, **376**
Density, 9, 18
Dependent variables, **28**
Derivative(s), **410**
  formulas for, 410
Dew-point curve, 372
Dialysis, equilibrium, **335**
Diathermal boundary, **28**
Dieterici equation, 7p
Differential, 6, **412**
  exact, **33**, **412**
  inexact, **34**
  total, **105**, **412**
Dilution, enthalpy of, **259**
Dimensional analysis, 6
Disorder, 88
Dissipation of energy, **51**, 54, 57
Dissociation pressure of a hydrate, 375
Distribution coefficient, **334**
Donnan:
  membrane equilibrium, **335**
  potential, **335**
Duhem–Margules equation, **343**

Ebullioscopic constant, **320**
Efficiency:
  of a Carnot engine, 72–75
  of a heat engine, **72**
Efflorescence, **376**
Einstein energy relation, 35, 128n
Electric:
  charge, 219, 389
  current, 53, 54, 112, 387, 390, 402
  field, 34
  potential, 25, 53, 227, 228, 335, 337, 338, **389**, 391, 393
    inner, 335, 389
  potential difference, 53–55, 335, 389
  potential energy, 53

Index

potential function, 228
power, 116
resistance, 25, 54, 112
Electrical:
  circuit, 53–55, 392
  force, 415
  heating, 54–55, 111, 112, 116, 270
  neutrality, 180, 214, 219, 359
  resistor, 39, 54, 55
  work, 39, 53–55, 58, 60, 111–113, 116, 161, 275, 392
Electrochemical cell, **387**
Electrochemical potential, 335n
Electrode, 387, 388n
  hydrogen, 388
    standard, **397**
  potential, standard, **397**, 399
Electrolyte:
  solutions, 218–230
  symmetrical, 220–222
Electromotive force, *See* Emf
Emf, 55, **389**, 391
  standard, **393**
    evaluation of, 396
Endothermic, **270**
Energy, 415–421
  dissipation of, 50–54, 57
  electric potential, 53
  kinetic, 34, 415
  potential, 34, 416
  of the system, 34–35
  thermal, 421
    of an ideal monatomic gas, 117
  unavailable, 87
Energy dissipation, 84
Energy equivalent, **112**, 270, 271, 274, 275
Enthalpy, **59**
  change at constant pressure, 119
  of combustion, standard molar, 272, 277
  of dilution, **259**
    molar, **261**
  of formation of a solute, 262
  of formation, standard molar, 255
    of an ion, 256
    of a solute, 256
  of a gas, standard molar, **132**
  of mixing to form an ideal mixture, 246
  partial molar, 193
    in an ideal gas mixture, 186
    relative, of a solute, **263**, 266

relative, of the solvent, **262**, 264, 268
  of a solute in an ideal-dilute solution, 201
of reaction, *See* Reaction enthalpy
relative apparent, of a solute, **263**
of solution, **259**
  molar, **259**
  molar differential, 259, 260, 298, 325
  molar integral, **260**, 298
  partial molar, 259
of vaporization, **159**
Entropy, **66**, 86
  Boltzmann's equation for, 88
  change at constant pressure, 120
  change at constant volume, 119
  change during internal heat flow, 85
  an extensive property, 81
  as a measure of disorder, 88
  as a measure of unavailable energy, 87
  of mixing to form an ideal mixture, 246, 249
  molar, from calorimetry, 93
  of a nonequilibrium state, 81
  partial molar, 193
    of a solute in an ideal-dilute solution, 200
  of reaction, *See* Reaction entropy, standard molar
  residual, 96
  from a reversible process, 79–81
  standard molar, of a gas, **131**, 186
  third-law, **92**
Equation of state, 29, 30
  of a fluid, **13**
  of a gas at low pressure, 15, 188, 189, 234p, 302p
  of an ideal gas, 13
  thermodynamic, 109
  virial, 14
Equilibrium:
  dialysis, **335**
  gas–gas, 378
  liquid–gas, 338–346
  liquid–liquid, 330–338
  mechanical, 31
  phase transition, 43, 282
  reaction, 31, 346
    condition for, 280, 285
  solid–liquid, 322–330
  state, 57
  thermal, 31

transfer, 31
Equilibrium conditions, 142
  for an ideal gas mixture in a gravitational field, 231
  for a mixture in a gravitational or centrifugal field, 230
  for mixtures in two phases, 181–182
  for a pure gas in a gravitational field, 145–147
  for a pure substance in several phases, 144
  for a pure substance in two phases, 143
  for reaction, 281
  for a solution in a centrifugal field, 231
Equilibrium constant:
  on a pressure basis, **290**
  thermodynamic, **288**, 346
    of a cell reaction, 393
    for a reaction in solution, 290
    temperature dependence, 308
Equilibrium position, effect of T and p on, 296–298
Equilibrium state, **30**
Euler reciprocity relation, 413
Eutectic:
  composition, 365
  halt, 365
  point, **364**, 367, 368
  temperature, 365, 366
Exact differential, **33**, **412**
Excess molar mixing quantity, 247
Exothermic, **270**
Expansion, **32**
  free, **50**, 85
Expansion work, 38, 45–50, 58, 59, 61, 106, 112, 114, 143, 144, 182, 239, 253, 255, 275, 280, 392
  reversible, 48, 73, 106
Extensive property, **9**
Extent of reaction, **240**
External field, 28, 31, 32, 35, 37, 56, 144

Faraday constant, **391**, 403
Field:
  centrifugal, 230–233
  electric, 34
  external, 28, 31, 32, 35, 37, 56, 144
  gravitational, 17, 28, 31, 34, 35, 56, 144, 230–231

magnetic, 99, 100
Field strength, magnetic, 135
First law of thermodynamics, **36**, 47
Fluids, **11**, 12
  supercritical, 12, **155**, 159
Force, 414–421
  buoyant, 56
  centrifugal, 231, 418
  Coriolis, 418
  electrical, 415
  frictional, 43, 56
  gravitational, 17, 56, 147, 415, 418
Formation reaction, **255**
Free-energy function, 294–296, 348
Free expansion, **50**, 85, 89
Freezing point, **153**
  curve, 322, 364
    to form a solid compound, 328
    of an ideal binary mixture, 323
  depression, of a dilute solution, 317–319
    to evaluate solvent chemical potential, 310–312
    of an ideal binary mixture, 322–323
Friction:
  internal, 50, 51
  sliding, 421
Frictional force, 43, 56
Fugacity:
  effect of liquid composition, 339–342
  effect of liquid pressure, 338
  of a gas, **128**
  of a gas mixture constituent, **187**
Fugacity coefficient:
  of a gas, **130**
  of a gas mixture constituent, **188**, 190
  of a mixture constituent, 202

Galvani potential, 335, 389, 390
Galvanic cell, 55, **387**
  in an equilibrium state, 31
Gas, **12**
  ideal, 35, 184
  perfect, 35n
  solubility, 343–345
Gas constant, 403
Gibbs energy, **89**
  of formation, standard molar, **292**
  of an ion, 293
  of mixing, **245**, 284
    to form an ideal mixture, 246

molar, **245**, 252
molar, 127, 137
molar reaction, **279**
  of a cell reaction, 392
  of reaction:
    standard molar, 285
    standard molar, of a cell reaction, 393
  total differential of, for a mixture, 181
Gibbs equations, **107**
Gibbs phase rule:
  for a multicomponent system, 333n, 355–363
  for a pure substance, 148
Gibbs–Duhem equation, **177**, 179, 180, 198, 208–210, 229, 248, 268, 318, 327, 340
Gravitational:
  field, 17, 28, 31, 34, 35, 56, 144, 230–231
  field potential, 144
  force, 17, 56, 147, 415, 418
  work, 56, 58
Gravitochemical potential, 146
Green Book, *See* IUPAC Green Book

Half-cell, 388
  reaction, 388
Heat, **36**, 39–43, 420
  internal flow of, 85
  of vaporization, **159**
Heat capacity, **60**
  measurement of, by calorimetry, 111
  of reaction, molar, **258**
Heat capacity at constant pressure, **61**
  of a gas, standard molar, 132
  molar, **61**
  partial molar, 194
Heat capacity at constant volume, **60**
  of an ideal gas, 36
  of an ideal monatomic gas, 117
  molar, **60**
Heat engine, **68**, 69
Heat reservoir, **32**, 39, 68, 69, 75–77, 80, 82, 86, 87
Heater circuit, 112, 114, 115
Heating:
  at constant volume or pressure, 117–120

curve, of a calorimeter, 112, 114, 116
  electrical, 54–55, 111, 112, 116, 270
Helium, 152n
Helmholtz energy, **89**
Henry's law, **194**, 194–196
  not obeyed by electrolyte, 218
Henry's law constant, **194**
  effect of pressure on, 345
  evaluation of, 196
Henry's law constants, relations between different, 196, 234p
Hess' law, **255**, 256, 272, 293
Hydrogen electrode, 388
  standard, **397**

Ice point, **21**
Ice, high pressure forms, 153
Ideal gas, **35**, 184
  equation, 5, 13, 35
  internal pressure, 109
  mixture, **186**
    in a gravitational field, 231
    and Raoult's law, 192
Ideal mixture, **193**, 253
  mixing process, 246
  and Raoult's law, 192
Ideal solubility:
  of a gas, **344**
  of a solid, **325**
Ideal-dilute solution, **196**
  partial molar quantities in, 199–201
  solvent behavior in, 198–199, 340
Ideal-gas temperature, 21, 75
Ignition circuit, 273, 275
Independent variables, **28**, 142
  of an equilibrium state, 78
  number of, 148, 357
Indicator diagram, **48**
Inexact differential, **34**
Inner electric potential, 389
Integral(s), **410**
  formulas for, 411
Integrand, **411**
Integrating factor, 81
Intensive property, **9**
Interface surface, **10**
Internal energy, **35**
  change at constant volume, 118
  of an ideal gas, 35–36
  of mixing to form an ideal mixture, 247
  partial molar, 193

Index

Internal pressure, **108**, 108–111
  of an ideal gas, 109
Internal resistance, 55, 390
International System of Units, *See* SI
International temperature, 22
Ionic strength, **225**, 226, 228
  effect on reaction equilibrium, 347
Irreversible process, **64**, 66, 86
Isobaric process, **33**
Isochoric process, **33**
Isolated system, **28**, 30
Isoperibol calorimeter, 114, 270, 271, 278
Isopiestic:
  process, **33**
  vapor pressure technique, **212**
Isopleth, 365
Isoteniscope, 153
Isothermal:
  bomb process, **272**, 274–275, 277
  compressibility, **104**, 159, 165
    of an ideal gas, 137p
    of a liquid or solid, 126
  magnetization, 99, 135
  pressure changes, 125–126
    of a condensed phase, 126
    of an ideal gas, 125
  process, **32**
IUPAC, 1
IUPAC Green Book, 1, 195n, 202n, 323n

Joule, 40n
  coefficient, 138p
  experiment, 138p
  paddle wheel, 38, 52, 57, 65, 67
Joule–Kelvin:
  coefficient, **98**
  experiment, **97**
Joule–Thomson:
  coefficient, **98**, 122–124
  experiment, **97**

Kelvin–Planck statement of the second law, 68, 77
Kinetic energy, 415
Kirchhoff equation, **258**, 276
Konowaloff's rule, 343

Laplace equation, 167p, 233p
Law of Cailletet and Matthias, **155**
Law of rectilinear diameters, **155**
Law, scientific, 27

Le Châtelier's principle, 297, 298
Lever rule:
  for a binary phase diagram, 364, 365
  general form, 157
  for one substance in two phases, 155
  for partially-miscible liquids, 369
  for a ternary system, 380
Lewis and Randall rule, 234p
Liquid junction, 387, 389, 390, 392
  potential, 389, 390, 393
Liquids, **12**
Liquidus curve, 363, 364, 367, 369, 371–373, 377

McMillan–Mayer theory, 204n
Magnetic:
  enthalpy, 135
  field, 99, 100
  field strength, 135
Magnetization, isothermal, 99, 135
Mass, 17, 402
Mass fraction, 233p
Maxwell relations, **108**, 413
Mean ionic activity coefficient, *See* Activity coefficient, mean ionic
Mean molar:
  quantities, 188
  volume, 177
Melting point, **153**
Membrane equilibrium:
  Donnan, **335**
  osmotic, **334**
Membrane, semipermeable, 31, 182, 312
Method of intercepts, **177**, 180, 252
Miscibility gap, 254, 368, 380
Mixing process, **244**
Mixture:
  binary, **171**
  of fixed composition, 173
  ideal:
    and chemical potential, **193**
    and Raoult's law, **192**
  ideal gas, **186**
  in a gravitational field, 231
  simple, 249
Molal boiling-point elevation constant, **320**
Molal freezing-point depression constant, **319**
Molal osmotic coefficient, **209**

for evaluating a mean ionic activity coefficient, 229
evaluation, 212
Molality, **171**
standard, 197
Molar:
  differential reaction quantity, 243
  enthalpy of combustion, standard molar, 272, 277
  enthalpy of dilution, **261**
  enthalpy of formation:
    of a solute, **262**
    standard, **255**
    standard, of a solute, 256
    standard, of an ion, 256
  enthalpy of mixing to form an ideal mixture, 246
  enthalpy of solution, **259**
    differential, **259**, 260, 298, 325
    integral, **260**, 298
  enthalpy of vaporization, **160**
  enthalpy, standard, of a gas, **132**
  entropy:
    from calorimetry, 93
    standard, of a gas, **131**
  entropy of mixing to form an ideal mixture, 246
  Gibbs energy, 127, 137
  Gibbs energy of formation:
    standard, **292**
    standard, of an ion, 293
  Gibbs energy of mixing, **245**, 252
    to form an ideal mixture, 246
  heat capacities at constant volume and constant pressure, relation between, 110
  heat capacity at constant pressure, **61**
    of a gas, **132**
    of an ideal monatomic gas, 117
  heat capacity at constant volume, **60**
    of an ideal monatomic gas, 117
  heat capacity of reaction, **258**
  integral reaction quantity, 243
  internal energy of mixing to form an ideal mixture, 247
mass, 17
  from a colligative property, 317
  from sedimentation equilibrium, 233
mixing quantity, excess, **247**
quantity, **9**, 118
reaction enthalpy, **241**
  effect of temperature on, 257

standard, 308, 309
standard molar, 256
standard, of a cell reaction, 394
reaction entropy, standard, of a cell reaction, 394
reaction Gibbs energy, **279**
of a cell reaction, 392
standard, **285**
standard, of a cell reaction, 393
reaction quantity, **243**
standard, **244**
volume, 9, 18
volume of mixing:
to form an ideal mixture, 247
Mole, **3**, 17
Mole fraction, **170**
standard, 198
Mole ratio, 172

Nernst distribution law, **333**
Nernst equation, **394**
Nernst heat theorem, 92
Neutrality, electrical, 180, 214, 219, 359
Newton's law of cooling, 114
Newton's law of universal gravitation, 415
Newton's second law of motion, 44, 56, 147, 414, 418
Newton's third law of action and reaction, 45, 415, 416, 421
Nonexpansion work, 59, 91, 392
Normal:
boiling point, 154
melting point, 154

Osmosis, 312
Osmotic coefficient, *See* Molal osmotic coefficient, 321
Osmotic membrane equilibrium, **334**
Osmotic pressure, 31, 210, **313**
of a dilute solution, 321
to evaluate solvent chemical potential, 312–314
van't Hoff's equation for, **321**

Partial derivative(s), **410**
formulas for, 120–122
Partial molar:
enthalpy, 193
in an ideal gas mixture, 186
relative, of a solute, **263**, 266
relative, of the solvent, **262**, 264, 268

of a solute in an ideal-dilute solution, 201
of solution, 259
entropy, 193
in an ideal gas mixture, 186
of a solute in an ideal-dilute solution, 200
heat capacity at constant pressure, 194
internal energy, 193
quantity, **173**
of a gas mixture constituent, 189
in general, **179**
general relations, 182–183
in an ideal mixture, 193–194
in an ideal-dilute solution, 199–201
volume, 173–175, 193
in an ideal gas mixture, 186
interpretation, 174
measurement of, 177
negative value of, 175
Partial pressure, **184**
in an ideal gas mixture, 184
Partition coefficient, **334**
Pascal, **19**
Path, **32**
Path function, **33**
heat or work as a, 39
Peritectic point, 368
Perpetual motion of the second kind, 68n
Phase, **10**
boundary, **151**
coexistence, **11**
diagram:
for a binary system, 363
for a binary liquid–gas system, 369
for a binary liquid–liquid system, 368
for a binary solid–gas system, 374
for a binary solid–liquid system, 364
for a pure substance, **149**
for a system at high pressure, 376
for a ternary system, 378
rule, *See* Gibbs phase rule
separation of a liquid mixture, 251–254, 330, 368
transition, **11**
equilibrium, 12, 43, 92, 282

reversible, 12
Physical constants, values of, 403
Physical quantities, symbols for, 404–407
Physical states, 10
symbols for, 408
Plait point, 380
Plasma, 12
Plimsoll mark, 127
Potential energy, 416
electric, 53
Potential function, 416
Poynting factor, **338**
Prefixes, 3
Pressure, 19
dissociation, of a hydrate, 375
internal, **108**, 108–111
in a liquid droplet, 147
negative, 109n
partial, **184**
saturation, **153**
standard, **20**, **127**, 217, 298p, 400p
sublimation, **153**
vapor, *See* Vapor pressure
Pressure changes, isothermal, 125–126
of a condensed phase, 126
of an ideal gas, 125
Pressure factor for activity, **214**, 216–217
of an ion, 220
Pressure–volume diagram, **48**
Pressure-factor quotient, 291, 329, 347
Principle of the unattainability of absolute zero, 100
Process, **32**
adiabatic, **33**, 37, 57
chemical, 239
subscripts for, 408
compression, **32**
cyclic, **33**
expansion, **32**
impossible, **64**, 67–68, 83
irreversible, **64**, 66, 82, 86
isenthalpic, 98
isobaric, **33**
isochoric, **33**
isopiestic, **33**
isothermal, **32**
mechanical, 65, 82, 86
mixing, **244**
quasistatic, 40n

Index

reversible, 32, **40**, 42, 44, **57**, **64**, 65, 66, 86
spontaneous, **40**, 57, 58, **64**, 86
throttling, **97**
Property:
  extensive, **9**
  intensive, **9**
  molar, 118

Quantity:
  molar, **9**
  specific, **9**
Quantity calculus, 4
Quasicrystalline lattice model, 249
Quasistatic process, 40n

Raoult's law:
  conditions needed, 192
  deviations from, 341–342, 372
  for fugacity, 191, 192
    in a binary liquid mixture, 341
    in an ideal-dilute solution, 199
  for partial pressure, **191**
    in a binary system, 369
Reaction:
  cell, **387**
  in a gas phase, 289–290
  in an ideal gas mixture, 282–285
  between pure phases, 282, 289
  in solution, 290–292
Reaction enthalpy, 254–258
  molar, **241**
    effect of temperature on, 257
  standard molar, 256, 308, 309
  of a cell reaction, 394
Reaction entropy, standard molar, 349p
  of a cell reaction, 394
Reaction equation, 239
Reaction Gibbs energy, molar, **279**
  of a cell reaction, 392
Reaction quantity:
  molar, **243**
  molar differential, 243
  molar integral, 243
Reaction quotient, **287**, 394
Reciprocity relation, 100p, **108**, **413**
Rectilinear diameters, law of, **155**
Redlich–Kister series, **251**
Redlich–Kwong equation, 7p, 14
Reduction to standard states, **272**
Reference state, 216
  of an element, 255
  of an ion, 219
  of a mixture constituent, 202

of a solute, 196–197, 203
of a solvent, 202
Regular solution, 251
Relative apparent molar enthalpy of a solute, **263**
Relative partial molar enthalpy:
  of a solute, **263**, 266
  of the solvent, **262**, 264, 268
Resistance:
  electric, 25, 54, 112
  internal, 55, 390
Resistor, electrical, 39, 54, 55
Retrograde condensation, 377
Retrograde vaporization, 377
Reversible:
  adiabatic expansion of an ideal gas, 47
  adiabatic surface, **78**, 79
  expansion and compression, 44, 46
  expansion work, 48, 73, 106
  heating and cooling, 42
  isothermal expansion of an ideal gas, 47
  phase transition, 43
  process, 32, **40**, **57**, **64**, 65, 66, 86
    in a galvanic cell, 55, 392
Rubber, thermodynamics of, 141p

Salt bridge, 390
Salt effect, 330
Salting-out effect on gas solubility, 344, 351p
Saturated solution, **323**
Saturation pressure, **153**
Second law of thermodynamics:
  Clausius statement, 67
  equivalence of Clausius and Kelvin–Planck statements, 71–72
  Kelvin–Planck statement, 68, 77, 80, 82
  mathematical form, **66**
    derivation, 76–83
Sedimentation equilibrium, 233
Shear stress, **10**
SI, 1
  base units, 2, 402
  derived units, 2
  prefixes, 3
Simple mixture, 249
Solid compound, **326**, 366, 374
Solids, 10, 16
  viscoelastic, 11
Solidus curve, 363

Solubility:
  of a gas, 343–345
    ideal, **344**
  of a liquid, 331
    and Henry's law constant, 332
  of a solid, **323**
  of a solid electrolyte, 328–330
  of a solid nonelectrolyte, 323
Solubility curve, 322, 364
Solubility product, **328**
  temperature dependence, 330
Solute, **171**
  reference state, 196–197, 203
Solution, **171**
  binary, 171
  enthalpy of, **259**
  ideal-dilute, *See* Ideal-dilute solution
  regular, 251
  saturated, **323**
  solid, 190, 366
Solvent, **171**
  activity coefficient of, 310n
  behavior in an ideal-dilute solution, 198–199, 340
Species, **170**, 357
Specific quantity, **9**
Specific volume, **9**
Spontaneous process, **40**, 57, 58, **64**, 86
  conditions for, 142
    gas in a gravitational field, 145–147
    mixtures in two phases, 181–182
    pure substance in two phases, 143
    with reaction, 281
  criterion for, 91
Standard:
  boiling point, 154
  chemical potential, *See* Chemical potential, standard
  concentration, 197
  electrode potential, **397**
    from standard molar reaction Gibbs energies, 399
  emf, **393**
    evaluation, 396
  hydrogen electrode, **397**
  melting point, 154
  molality, 197
  mole fraction, 198
  potential of a cell reaction, **393**
    evaluation, 396

pressure, 20, **127**, 217, 298p, 400p
Standard molar:
    enthalpy of a gas, **132**
    enthalpy of combustion, 272, 277
    enthalpy of formation, **255**
        of an ion, 256
        of a solute, 256
    enthalpy of vaporization, 161
    entropy:
        of a gas, **131**, **186**
        of an ion, 293
    Gibbs energy of formation, **292**
    heat capacity at constant pressure of a gas, **132**
    properties, values of, 422–424
    quantities, evaluation of, 347–349
    reaction enthalpy, 256, 308, 309
        of a cell reaction, 394
    reaction entropy, 349p
        of a cell reaction, 394
    reaction Gibbs energy, **285**
        of a cell reaction, 393
    reaction quantity, **244**
Standard state:
    of a gas, 127
    of a gas mixture constituent, 184, 213
    of a liquid or solid mixture constituent, 213
    of a mixture constituent, 215
    of a pure gas, 215
    of a pure liquid or solid, 127, 215
    of a pure substance, **127**
    of a solute, 213, 215
    of a solvent, 213, 215
State:
    of aggregation, 10, 408
    equilibrium, **30**, 57
    function, **28**
        change of, 33, 255
        infinitesimal change of, 33
    physical, 10, 408
    standard, *See* Standard state
    steady, **32**, 62p
    of a system, **28**
    with nonuniform regions, 30
Statistical mechanics:
    Boyle temperature, 15
    Debye crystal theory, 93
    Debye–Hückel theory, 227
    entropy definition, 88
    ideal mixture, 204
    McMillan–Mayer theory, 204n
    mixture theory, 250

molar entropy of a gas, 96
molar heat capacity of a metal, 94
second virial coefficient, 189
translational energy of a gas, 117
virial equations, 14
Steady state, **32**, 62p
Steam engine, 70
Steam point, **21**, 26p, 154
Stoichiometric:
    activity coefficient, **224**
    coefficient, 242
    number, **242**, 245, 256, 288
Stoichiometric addition compound, *See* Solid compound
Sublimation:
    point, **153**
    pressure, **153**
    temperature, **153**
Subscripts for chemical processes, 408
Substance, **170**
Supercritical fluid, 12, **13**, **155**, 159
Superscripts, 409
Surface tension, **134**, 147
Surroundings, **27**
Symbols for physical quantities, 404–407
System, **8**, **27**
    closed, **27**
    isolated, **28**, 30, 84
    open, **27**, 136–137, 174, 175, 180
    point, **151**
    state of, **28**
Systéme International d'Unités, *See* SI

Temperature, 20
    convergence, 115
    critical, **154**
    equilibrium systems for fixed values, 22
    ideal-gas, 21, 75, 109
    international, 22
    scale, 20
    thermodynamic, 21, **75**, 109, 402
    upper consolute, 368
    upper critical solution, 368
Thermal analysis, 366
Thermal conductance, 114
Thermal energy, **40**, 421
    of an ideal monatomic gas, 117
Thermodynamic:
    equation of state, 109
    equilibrium constant, *See* Equilibrium constant, thermody-

namic
    temperature, 21, **75**, 109, 402
    work, 420
Thermometer:
    Beckmann, 25
    constant-volume gas, **22**
    liquid-in-glass, 20, 25
    optical pyrometer, 25
    quartz crystal, 25
    resistance, 25
    thermocouple, 25
Thermopile, 25, 278
Third-law entropy, **92**
Third-law method, 348
Third law of thermodynamics, 80, **92**
Throttling process, **97**
Tie line, **151**
    in a binary phase diagram, 363
    in a ternary phase diagram, 380
Total differential, **105**, **412**
    of the chemical potential, 183
    of the Gibbs energy of a mixture, 181
    of the Gibbs energy of a pure substance, 127
    of the internal energy, 136
    of the volume, 175
Triple line, 58n
Triple point, **151**
    cell, 22
    of H_2O, 21
Units, 4
    Non-SI, 2
    SI, 2
    SI derived, 2
Upper consolute temperature, 368
Upper critical solution temperature, 368

van't Hoff equation, **309**
van't Hoff's equation for osmotic pressure, **321**
Vapor, 155
Vapor pressure, **153**, 190
    curve, **154**
    of a liquid droplet, 339
    lowering, of a dilute solution, 320
Vaporization, 309
    enthalpy of, **159**
    heat of, **159**
    molar enthalpy of, **160**
Vaporus curve, 363, 369, 371–373, 377
Variables:

dependent, **28**
independent, **28**
   of a single phase, 103
natural, 107
number of independent, 29, 57
Variance, **148**, **356**
Virial equation:
   for a gas mixture, 188
   for a pure gas, 14
Virtual displacement, 143
Viscoelastic solid, 11
Volume, 17
   mean molar, 177
   of mixing to form an ideal mixture, 247
   molar, 9, 18

partial molar, 193
   in an ideal gas mixture, 186
   specific, **9**
   total differential in an open system, 175

Washburn corrections, **276**, 277, 303p
Weight fraction, 233p
Work, **36**, 37–38, 415–421
   coefficient, **38**, 57
   coordinate, **38**, 57, 78, 133, 137
   of electric polarization, 58
   electrical, 39, 53–55, 58, 60, 111–113, 116, 161, 275, 392

expansion, 38, 45–50, 58, 59, 61, 106, 112, 114, 143, 144, 182, 239, 253, 255, 275, 280, 392
   reversible, 73, 106
gravitational, 56, 58
linear mechanical, 58
of magnetization, 58, 135
nonexpansion, 59, 91, 392
reversible expansion, 48
rotational mechanical, 53, 58
stretching, 58
surface, 58
thermodynamic, 420

Zeotropic behavior, 374
Zeroth law of thermodynamics, **21**